Probability and Mathematical Statistics (C

MATTHES, KERSTAN, and MEC
MUIRHEAD • Aspects of Multiv
PARZEN • Modern Probability T
PURI and SEN • Nonparametric
PURI and SEN • Nonparametric Methods in Multivariate Analysis
RANDLES and WOLFE • Introduction to the Theory of Nonparametric Statistics
RAO • Linear Statistical Inference and Its Applications, *Second Edition*
RAO and SEDRANSK • W.G. Cochran's Impact on Statistics
ROHATGI • An Introduction to Probability Theory and Mathematical Statistics
ROHATGI • Statistical Inference
ROSS • Stochastic Processes
RUBINSTEIN • Simulation and The Monte Carlo Method
SCHEFFE • The Analysis of Variance
SEBER • Linear Regression Analysis
SEBER • Multivariate Observations
SEN • Sequential Nonparametrics: Invariance Principles and Statistical Inference
SERFLING • Approximation Theorems of Mathematical Statistics
TJUR • Probability Based on Radon Measures
WILLIAMS • Diffusions, Markov Processes, and Martingales, Volume I: Foundations
ZACKS • Theory of Statistical Inference

Applied Probability and Statistics

ABRAHAM and LEDOLTER • Statistical Methods for Forecasting
AGRESTI • Analysis of Ordinal Categorical Data
AICKIN • Linear Statistical Analysis of Discrete Data
ANDERSON, AUQUIER, HAUCK, OAKES, VANDAELE, and WEISBERG • Statistical Methods for Comparative Studies
ARTHANARI and DODGE • Mathematical Programming in Statistics
BAILEY • The Elements of Stochastic Processes with Applications to the Natural Sciences
BAILEY • Mathematics, Statistics and Systems for Health
BARNETT • Interpreting Multivariate Data
BARNETT and LEWIS • Outliers in Statistical Data, *Second Edition*
BARTHOLOMEW • Stochastic Models for Social Processes, *Third Edition*
BARTHOLOMEW and FORBES • Statistical Techniques for Manpower Planning
BECK and ARNOLD • Parameter Estimation in Engineering and Science
BELSLEY, KUH, and WELSCH • Regression Diagnostics: Identifying Influential Data and Sources of Collinearity
BHAT • Elements of Applied Stochastic Processes, *Second Edition*
BLOOMFIELD • Fourier Analysis of Time Series: An Introduction
BOX • R. A. Fisher, The Life of a Scientist
BOX and DRAPER • Evolutionary Operation: A Statistical Method for Process Improvement
BOX, HUNTER, and HUNTER • Statistics for Experimenters: An Introduction to Design, Data Analysis, and Model Building
BROWN and HOLLANDER • Statistics: A Biomedical Introduction
BUNKE and BUNKE • Statistical Inference in Linear Models, Volume I
CHAMBERS • Computational Methods for Data Analysis
CHATTERJEE and PRICE • Regression Analysis by Example
CHOW • Econometric Analysis by Control Methods
CLARKE and DISNEY • Probability and Random Processes: A First Course with Applications

(*continued on back*)

Multiple Criteria Optimization:

Theory, Computation, and Application

Multiple Criteria Optimization:

Theory, Computation, and Application

RALPH E. STEUER
College of Business Administration
University of Georgia

JOHN WILEY & SONS
New York • Chichester • Brisbane • Toronto • Singapore

Library of Congress Cataloging in Publication Data:

Steuer, Ralph E.
Multiple criteria optimization.

(Wiley series in probability and mathematical statistics. Probability and mathematical statistics, ISSN 0271-6232)
Includes bibliographies and index.
1. Mathematical optimization. 2. Programming (Mathematics) I. Title. II. Series.

QA402.5.S72 1985 519.7 85-9564
ISBN 0-471-88846-X

Printed in the United States of America

10 9 8 7 6 5 4 3 2 1

To my wife Judy, my son Evan, and
my daughters Andrea and Catherine

Preface

If there ever was an area of management science/operations research that is exciting and challenging but fraught with deceptions and pitfalls, it is multiple criteria optimization. The field is fascinating because it is clearly an art as well as a science. When studying multiple criteria optimization, we learn how not to be naive. We learn about where the "bodies are buried" in mathematical programming and how to be innovative in overcoming the variety of solution set and solution procedure difficulties that may arise.

To me, multiple criteria optimization is a breath of fresh air: We can now openly admit that a problem has multiple objectives when it possesses multiple conflicting criteria. There is no need to ignore or gloss over the fact. We can deal with multiple objectives head on, because we now have the tools to solve large-scale multiple criteria optimization problems.

Multiple criteria optimization is also interesting because of its international scope. The subject is not strictly a product of North American and Western European culture. It is also of interest to the socialist economics of Eastern Europe and Third World nations because of its application possibilities in centralized planning. The international nature of the field is apparent in the references at the conclusion of this text's chapters, as contributions to the field have come from all over the world.

Purpose of Book

This book serves both as a teaching text and as a comprehensive reference volume concerning the principles and practices of multiple criteria optimization. The pedagogical approach relies on examples and graphical illustrations; hence many graphs are presented. In addition, numerous computational examples are included to stress the computer implementability of the methods discussed.

Mathematical Background

Only a modest mathematical background is needed to reach the state of the art in multiple criteria optimization. The only prerequisites are (a) familiarity with set theory and linear algebra, (b) knowledge of the simplex method equivalent to what one would obtain in a first course in operations research, (c) a bit of calculus, and (d) an acquaintance with computers to the extent that the user can edit files and not be intimidated by mathematical programming software. Apart from calculus and computers, the book is self-contained because of the review material in Chapter 2 (Mathematical Background) and Chapter 3 (Single Objective Linear Programming).

Organization

Chapters 4 through 9 develop the theory of multiple objective linear programming (MOLP). In particular, Chapter 4 discusses the computation of all *optimal* extreme points in single objective linear programming (LP). Chapter 5, which concerns objective row parametric programming, is essentially multiple objective programming with two objectives. This chapter forms the bridge between conventional LP and MOLP.

Chapters 6, 7, and 8 discuss the subleties involved with the solution set notion of efficiency (Pareto optimality), and Chapter 9 presents the theory of linear vector-maximization for computing all efficient points. In Chapter 9, we see how generalized versions of the methods discussed in Chapter 4 can be used to compute all *efficient* extreme points.

Chapter 10 addresses goal programming. One might say that goal programming is not so much covered in Chapter 10 as it is *dissected*. Goal programming is analyzed in terms of contours, and the usefulness of the deviational variable to multiple objective programming in general is stressed. Chapter 11 discusses *filtering*, or how representative subsets of larger sets can be obtained.

Chapter 12 deals with multiple objective linear fractional programming (MOLFP), in which the objectives are fractional in the sense that they have linear numerators and linear denominators. This is a research topic, as much work remains to be done in MOLFP.

The discussion of interactive procedures in Chapters 13, 14, and 15 is the climax of the book. In Chapter 13, we discuss **STEM**, the Geoffrion-Dyer-Feinberg procedure, and the Zionts-Wallenius procedure among others. In Chapters 14 and 15, we discuss the Tchebycheff procedure. Three interactive applications are given in Chapter 16. Comments about the future, particularly in regard to the use of graphics at the computer/user interface, are made in Chapter 17. Also, Chapter 17 contains bibliographies concerning some specialized multiple criteria topics.

How to Use the Text

With my graduate course in multiple criteria optimization at the University of Georgia, we cover Chapters 6, 7, 8, and part of 9, along with Chapters 10, 11, 13, 14, and 15. We also discuss the use of computer graphics as outlined in Chapter 17. Students are asked to review material in Chapters 2 and 3 on their own in order to fill any gaps in their background. Chapters 6, 7, 8, 9, 10, and 11 provide the tools for multiple criteria analysis. Chapters 13, 14, and 15 describe their interactive application. Because the emphasis is on computer implementation, computers are used intensively in the course from day one. We use the **MINOS** linear/nonlinear programming code and the **ADBASE** vector-maximum code

(Section 9.11). Also, we use an automated package such as the one discussed in Chapter 15 for implementing the interactive weighted-sums and Tchebycheff procedures of Chapters 13 and 14. In addition, the `LAMBDA` and `FILTER` codes (Sections 11.7 and 11.9) are used for a variety of small supporting chores.

Problem Exercises

Problem exercises are found at the ends of the chapters. Those exercises whose numbers carry a "C" suffix are computer problems. Those without a "C" suffix are paper and pencil problems. A solutions manual for all problem exercises is available.

Numbering

Within a given chapter, the numbering of equations, programs, formulations, definitions, lemmas, and theorems come from the same sequence. For instance, in Chapter 9, Theorem 9.17 appears after Definition 9.13, and in Chapter 14, Lemma 14.17 falls before formulation (14.19). It is hoped that after the reader becomes accustomed to the numbering convention, it will be of some convenience, particularly in the longer chapters. Note that tables and figures are numbered separately as usual.

Notation

Set theoretic notation is used throughout the text. A listing of the most frequently used notation is given in Section 1.6. It should also be pointed out that the book utilizes two specialized items of notation. One is the *convex combination operator* γ, and the other is the *unbounded line segment operator* μ. Most readers will not be familiar with this notation; however, it is very convenient. The use of γ and μ makes the expression of convex sets in set theoretic notation particularly straightforward. The convex combination and unbounded line segment operators are described in Section 2.3.2.

Acknowledgments

I would first like to express my appreciation to all of the researchers in various disciplines from around the world who, by virtue of their contributions over the past fifteen years, have created the field of multiple criteria optimization. Without the foundations that they laid, the procedures that they pioneered, and the maturity that they have rendered to the field, a book such as this would not have been possible.

I want to thank those with whom I have conducted joint multiple criteria research. The list includes K. R. Balachandran (New York University), Eng-Ung Choo (Simon Fraser University), John P. Evans (University of North Carolina), Heinz Isermann (University of Bielefeld), Jonathan S. H. Kornbluth (Hebrew University), Kenneth D. Lawrence (AT & T Long Lines), Albert T. Schuler (U.S. Forest Service), Joe Silverman (U.S. Navy Personnel Research and Development

Center), Marc J. Wallace, Jr. (University of Kentucky), Alan W. Whisman (U.S. Navy Personnel Research and Development Center), and Eric F. Wood (Princeton University).

I want to take this opportunity to give special thanks to John P. Evans, who was my Ph.D. advisor at the University of North Carolina. He suggested multiple objective linear programming as my dissertation topic, and it was the work that we did together, much of which is covered in the first half of Chapter 9, that introduced me to the field.

I also want to give special thanks to Stanley Zionts (State University of New York at Buffalo). His colleagueship and advice over the years have been greatly appreciated. He has been a leader in the field and many of his contributions to the literature are reflected in this book, particularly in Chapters 9 and 13. In addition, special thanks go to Eng-Ung Choo for contributing the MOLFP example with a nonlinear efficient boundary in Section 12.7. Also, I want to acknowledge Erick C. Duesing (University of Scranton) who has been my very good multiple criteria friend.

Furthermore, I would like to thank Kenneth D. Lawrence, Heinz Isermann, Gary R. Reeves (University of South Carolina), and Stanley Zionts, who reviewed the manuscript of this book.

Although not explicitly covered in this book, I would like to acknowledge the work that has been conducted by Andrzej P. Wierzbicki, Manfred Gauer, and others at IIASA (International Institute for Applied Systems Analysis, Laxenburg, Austria) on the *reference point method* of multicriteria optimization. This is important work and I plan to include it in the text if this book is ever revised.

Last but not least, I want to thank the following graduate students at the University of Georgia: Lynn H. Bryant, Jack M. Dominey, Lorraine R. Gardiner, Ronald Grzybowski, Rhonda R. Hickson, Terri D. Lambert, Manwoo Lee, Frank H. Liou, Pamela E. Slaten, Wayne C. Todd, and Mark E. White. I am grateful to them for the dedicated computer and proofreading assistance that they have given me over the past two years of this project.

Ralph E. Steuer

Athens, Georgia
September 1985

Contents

1. INTRODUCTION **1**

1.1 The Multiple Objective Programming Problem 1
1.2 Multiple Criteria Examples 2
1.3 Multiple Criteria Optimization 3
1.4 Optimal versus Final Solutions 4
1.5 Relationship to Multiattribute Decision Analysis 5
1.6 Notation 5

2. MATHEMATICAL BACKGROUND **11**

2.1 Set Theory 11
2.1.1 Specifying Sets and Indicating Membership 11
2.1.2 Subsets, Supersets, and Set Equality 12
2.1.3 Disjoint Sets and Families 13
2.1.4 Set Operations 13
2.1.5 The Real Line and Intervals 13
2.1.6 Generalized Operations, Index Sets, and Cardinality 14
2.1.7 Partitions 15
2.1.8 Mappings and Functions 15
2.1.9 Ordered n-Tuples and Cartesian Products 16
2.1.10 Graph of a Function 17
2.1.11 Relations 17

2.2 Topics from Linear Algebra 18
2.2.1 Matrices 18
2.2.2 Matrix Operations 20
2.2.3 Special Matrices and Vectors 21
2.2.4 Determinants 22
2.2.5 Solving Systems of Linear Equations Using Determinants 23
2.2.6 Computing the Inverse of a Matrix 24

2.2.7 Solving Systems of Linear Equations Using Inverses 25
2.2.8 Vectors and Points in n-Space 25
2.2.9 Linear and Convex Combinations of Vectors 26
2.2.10 Linear Independence 27
2.2.11 Rank of a Matrix 28
2.2.12 Bases 29
2.2.13 Number of Solutions of a System of Linear Equations 30
2.2.14 Some Further Examples of Functions 30

2.3 Properties of Points and Sets in R^n 31
2.3.1 Open and Closed Sets 31
2.3.2 Convex Sets and Extreme Points 31
2.3.3 Images of Convex Sets 34
2.3.4 Dimensionality and Set Addition 34
2.3.5 Hyperplanes and Half-Spaces 36
2.3.6 Connected Sets and Discrete Sets 36

2.4 Cones 38
2.4.1 Generators 38
2.4.2 Dimensionality of a Cone 39
2.4.3 Extreme Rays and Polyhedral Cones 39
2.4.4 Polar Cones 40

2.5 Norms and Metrics 42
2.5.1 Family of L_p-Norms 43
2.5.2 Family of L_p-Metrics 44
2.5.3 Family of Weighted L_p-Metrics 45

3. SINGLE OBJECTIVE LINEAR PROGRAMMING 55

3.1 Gradients and Level Curves 57
3.2 A Linear Programming Example 59
3.3 Graphical Examples 60
3.4 Simplex Method 66
3.4.1 Standard Equality Constraint Format 67
3.4.2 Initial Tableau 68
3.4.3 Pivoting Procedure 70
3.4.4 Method of Rectangles 73
3.4.5 Inconsistency and Unboundedness 74
3.4.6 Degeneracy and Alternative Optima 75
3.5 Duality Theory 76
3.6 Nonpositive and Unrestricted Variables 78
3.7 MPSX Input Format 79

3.8 Anatomy of a Simplex Tableau 81
3.9 Two-Phase Method of Linear Programming 85
3.10 Dual Simplex Algorithm 87
3.11 Solving a Series of Similar LPs 89
3.11.1 When Only the Objective Function Changes 89
3.11.2 When Only the RHS Vector Changes 90
3.11.3 When a Constraint Matrix Column Changes 90
3.12 More About Linear Programming 91

4. DETERMINING ALL ALTERNATIVE OPTIMA 99

4.1 Classification of Single Objective LPs 99
4.2 Notes on Pivoting and Coding 101
4.3 Crashing 107
4.4 Master Lists 109
4.5 Phase III Numerical Example 110

5. COMMENTS ABOUT OBJECTIVE ROW PARAMETRIC PROGRAMMING 120

5.1 Conventional Parametric Programming 121
5.2 Convex Combination Parametric Programming 123
5.3 Criterion Cone Parametric Programming 127
5.4 MOLP Approach 132

6. UTILITY FUNCTIONS, NONDOMINATED CRITERION VECTORS, AND EFFICIENT POINTS 138

6.1 Utility Function Shapes 139
6.2 Monotonicity 143
6.3 Feasible Region in Criterion Space 145
6.4 Utility Function Approach 146
6.5 Dominance 147
6.6 Nondominated Criterion Vectors 148
6.7 Efficiency 149
6.8 Detecting Efficiency Using Domination Sets 150
6.9 Nonconcave Utility Functions 154
6.10 Optimality Peculiarities with Multiple Objectives 155
6.11 Efficient Extreme Points of Greatest Utility 156
6.12 Effect of Nonlinearities 158

7. POINT ESTIMATE WEIGHTED-SUMS APPROACH 165

7.1 Figure-of-Merit Interpretation 166
7.2 Mathematical Motivation 166
7.3 What Are the Weights? 168
7.4 Determining the Weights 169
7.5 Criterion Cone 170
7.5.1 When the Null Vector Condition Does Not Hold 171
7.5.2 When the Null Vector Condition Holds 173
7.6 Relative Interior of the Criterion Cone 174
7.7 Detecting Efficient Points Using Composite Gradients 175
7.8 Relationship between the Criterion Cone and Domination Set 180
7.9 Efficient Facets of the Feasible Region *S* 181
7.10 Computing Subsets of Weighting Vectors 183

8. OPTIMAL WEIGHTING VECTORS, SCALING, AND REDUCED FEASIBLE REGION METHODS 193

8.1 Optimal Weighting Vector Estimation Difficulties 193
8.1.1 Dependence on the Decision Maker's Preferences 194
8.1.2 Dependence on the Relative Lengths of the Objective Function Gradients 194
8.1.3 Dependence on the Feasible Region 196
8.2 Difficulties Caused by the Degree to Which Objectives Are Correlated 198
8.3 Additional Optimal Weighting Vector Difficulties 199
8.4 Scaling the Objective Functions 200
8.4.1 Use of 10 Raised to an Appropriate Power 200
8.4.2 Use of Range Equalization Factors 201
8.5 **e**-Constraint Reduced Feasible Region Method 202
8.5.1 Operation of the **e**-Constraint Approach 203
8.5.2 Different Types of Outcomes 204
8.5.3 Sensitivity Analysis 206
8.6 Near Optimality Analysis 206

9. VECTOR-MAXIMUM ALGORITHMS 213

9.1 Foundations 214
9.2 Vector-Maximum Theory 216
9.3 A Vector-Maximum Algorithm 220

9.4 Weak Efficiency 221
9.5 Classification of MOLPs 224
9.6 Finding an Initial Efficient Basis 225
9.6.1 Subproblem Test for Extreme Point Efficiency 225
9.6.2 Methods for Finding an Efficient Extreme Point 226
9.7 Subproblem Tests for Nonbasic Variable Efficiency 233
9.7.1 Evans-Steuer Test 233
9.7.2 Isermann's Test 233
9.7.3 Ecker's Method 235
9.7.4 Zionts-Wallenius Routine 236
9.8 Computing All Maximally Efficient Facets 242
9.9 Vector-Maximum Approach for Solving an MOLP 245
9.10 Contracting the Criterion Cone 246
9.10.1 Interval Criterion Weights Criterion Cone 247
9.10.2 Enveloping Reduced Criterion Cone 250
9.10.3 Some Specific E-Cone Contractions 252
9.11 ADBASE Program 254
9.11.1 ADBASE Input Format 255
9.11.2 Versatility of ADBASE 259
9.11.3 ADBASE Six-Field Format 260
9.11.4 An ADBASE Example 261
9.12 Computational Experience 261
9.13 Minimum Criterion Values over the Efficient Set 267
9.13.1 Payoff Tables 267
9.13.2 Properties of the Minimum Criterion Value Problem 268
9.13.3 A Simplex-Based Algorithm 269

10. GOAL PROGRAMMING **282**

10.1 Goals and Utopian Sets 283
10.2 Archimedian GP 285
10.2.1 Contours of Archimedian Objective Functions 286
10.2.2 Archimedian GP Solutions 289
10.3 OCMM Models 290
10.4 Preemptive GP 292
10.5 Lexicographic Simplex Method 295
10.6 Goal Efficiency 296
10.7 Sensitivity Issues 298
10.8 Interactive GP 298
10.9 Minimizing Maximum Deviation 299
10.10 Multiple Criterion Function GP 300
10.11 GP Software 301

11. FILTERING AND SET DISCRETIZATION **311**

11.1 Forward and Reverse Filtering 311
11.2 Weighted L_p Distance Measure 312
11.3 Range Equalization Weights 313
11.4 Mechanics of Forward Filtering 314
11.4.1 Method of First Point outside the Neighborhoods 314
11.4.2 Halving and Doubling Strategy 316
11.4.3 Initial Values of d and Δd 316
11.4.4 Method of Closest Point outside the Neighborhoods 318
11.4.5 Method of Furthest Point outside the Neighborhoods 320
11.4.6 Maximally Dispersed Subsets 320
11.4.7 Unattained Set Sizes 321
11.5 Reverse Filtering 321
11.6 `FILTER` Program 322
11.7 Interactive Forward and Reverse Filtering 324
11.8 Set Discretization 326
11.8.1 Predetermined Convex Combinations 327
11.8.2 Convex Combinations Drawn from the Uniform Distribution 328
11.8.3 Convex Combinations Drawn from the Weibull Distribution 329
11.8.4 50-50 Strategy 329
11.9 `LAMBDA` Program 330

12. MULTIPLE OBJECTIVE LINEAR FRACTIONAL PROGRAMMING **336**

12.1 Single Objective Linear Fractional Programming 337
12.1.1 Variable Transformation Method 340
12.1.2 Updated Objective Function Method 343
12.2 When the Denominator Vanishes 345
12.3 Weak and Strong Efficiency 347
12.4 An MOLFP Example and Terminology 348
12.5 Graphical Detection of Efficiency 349
12.6 Additional MOLFP Examples 351
12.7 An MOLFP Example with a Nonlinear E^w Boundary 354
12.8 MOLFP Algorithms 355

13. INTERACTIVE PROCEDURES **361**

13.1 STEM 362
13.1.1 STEM Algorithm 363
13.1.2 STEM Sample Output 365
13.1.3 STEM Comments 365
13.2 Geoffrion-Dyer-Feinberg (GDF) Procedure 367
13.2.1 A Steepest Ascent Algorithm 369
13.2.2 Frank-Wolfe Algorithm 369
13.2.3 Routine for Inducing Locally Relevant Weights 371
13.2.4 GDF Algorithm 372
13.2.5 GDF Sample Output 374
13.2.6 GDF Numerical Example 377
13.2.7 GDF Comments 379
13.3 Zionts-Wallenius (Z-W) Method 379
13.3.1 Z-W Algorithm 380
13.3.2 Z-W Numerical Example 385
13.3.3 Determining Nonbasic Variables Efficient w.r.t. $\Lambda^{(h)}$ 387
13.3.4 Z-W Comments 389
13.4 Interval Criterion Weights/Vector-Maximum Approach 389
13.4.1 Interval Criterion Weights Algorithm 390
13.4.2 Interval Criterion Weights Comments 394
13.5 Interactive Weighted-Sums/Filtering Approach 394
13.5.1 Calibration 394
13.5.2 Interactive Weighted-Sums Algorithm 396
13.5.3 Interactive Weighted-Sums Comments 399
13.6 Visual Interactive Approach of Korhonen and Laakso 399
13.6.1 Achievement Scalarizing Functions 400
13.6.2 Projecting a Line Segment onto N 401
13.6.3 Visual Interactive Algorithm 404
13.6.4 Visual Interactive Comments 405

14. INTERACTIVE WEIGHTED TCHEBYCHEFF PROCEDURE **419**

14.1 The $\mathbf{z}^{**}$ Ideal Criterion Vector 420
14.2 Selecting ε_i Values 421
14.3 Augmented Weighted Tchebycheff Metrics 422
14.3.1 Contours 423
14.3.2 Weighted and Augmented Weighted Tchebycheff Programs 424

14.3.3 Definition Points, Vertices, and Diagonals 425
14.3.4 Points of Intersection with Z 427
14.4 Selecting ρ Values 429
14.5 Diagonal Direction of a Tchebycheff Metric 430
14.6 Unsupported Nondominated Criterion Vectors 431
14.6.1 Weighted-Sums Approaches and Unsupportedness 432
14.6.2 Weighted-Sums Approaches and Supported Criterion Vectors 433
14.6.3 Subproblem Test for Unsupportedness 434
14.6.4 Domination Sets and Unsupportedness 435
14.7 Improperly Nondominated Criterion Vectors 437
14.8 Tchebycheff Theory 440
14.8.1 Theory for the Finite-Discrete Case 440
14.8.2 Theory for the Polyhedral Case 443
14.8.3 Theory for the Nonlinear and Infinite-Discrete Case 444
14.9 Tchebycheff Algorithm 446

15. TCHEBYCHEFF / WEIGHTED-SUMS IMPLEMENTATION 456

15.1 Program Formulations 456
15.2 `MPSX` Layout Matrix 459
15.3 `SAVE / REVISE / RESTORE` Sequence 459
15.4 Computing the $\mathbf{z}^*$ Vector 459
15.5 Repetitive Optimization Economies 460
15.6 Illustrative MOLP 461
15.7 Structure of Automated Package 462
15.7.1 `Mxxxx` Deck 464
15.7.2 `Pxxxx` Deck 464
15.7.3 `Lxxxxh` Deck 467
15.7.4 `Oxxxxh` File 468
15.8 Numerical Solution of the Illustrative MOLP 468
15.9 Some Final Tchebycheff Graphical Examples 472
15.10 Some Final Implementation Comments 475
15.10.1 Software Other Than `MPSX` 475
15.10.2 Criterion Value Ranges over the Efficient Set 475
15.10.3 Number of Solutions Presented at Each Iteration 476
15.10.4 Most Preferred Criterion Vector as Filtering Seed Point 476
15.10.5 Insertion of Criterion Value Lower Bounds 476
15.10.6 Selection of ε_i Values in Integer and Nonlinear Cases 477

16. APPLICATIONS **484**

16.1 Sausage Blending 484
16.1.1 A Frankfurter Blending Problem 485
16.1.2 Vector-Maximum/Filtering Solution Procedure 486
16.1.3 Frankfurter Blending Formulation 487
16.1.4 Computer Results 489
16.2 CPA Firm Audit Staff Allocation 493
16.2.1 Description of Audit Staff Problem 493
16.2.2 Audit Staff Formulation 494
16.2.3 Interval Criterion Weights/Weighted-Sums/Filtering Solution Procedure 497
16.2.4 Computer Solution 498
16.2.5 Unscaling the Final Weighting Vector 503
16.3 Managerial Compensation Planning 504
16.3.1 Current Practice 504
16.3.2 Goal Programming/Tchebycheff Solution Procedure 505
16.3.3 Managerial Compensation Formulation 506
16.3.4 Computer Results 509
16.4 Two Additional Applications 513
16.4.1 River Basin Water Quality Planning 513
16.4.2 A Markov Reservoir Release Policy Problem 515

17. FUTURE DIRECTIONS **519**

17.1 Computer/User Interface 519
17.2 Various Screen Displays 520
17.3 Trajectory Optimization 522
17.4 Multiple Objective Applications in Engineering Management 525
17.5 Other Areas of Research in Multiple Criteria Optimization 526
17.5.1 Bibliography on Bicriterion Mathematical Programming 527
17.5.2 Bibliography on Duality in Multiple Objective Programming 527
17.5.3 Bibliography on Multiple Objective Programming with Fuzzy Sets 528
17.5.4 Bibliography on Multiple Objectives in Game Theory 529
17.5.5 Bibliography on Multiple Objective Integer Programming 529

17.5.6 Bibliography on Multiple Objectives in Networks, Markov Processes, Dynamic Programming, and Location 530
17.5.7 Bibliography on Multiple Objectives in Statistics 531

INDEX **533**

Multiple Criteria Optimization:

Theory, Computation, and Application

CHAPTER 1

Introduction

A single objective constrained optimization problem is of the form

$$\begin{aligned} &\max \{f(\mathbf{x}) = z\} \\ &\text{s.t.} \qquad \mathbf{x} \in S \end{aligned}$$

where $f(\mathbf{x})$ is the objective function and S is the feasible region. The purpose of the optimization problem is to find the point in S that has the highest objective function value z. If $f(\mathbf{x})$ is linear and S is defined by linear constraints, we have a single objective *linear* programming problem. If $f(\mathbf{x})$ is linear and S is defined by linear constraints with the additional restriction that the coordinates of each point in S be integer, we have a single objective *integer* programming problem. If $f(\mathbf{x})$ or any of the constraints defining S are nonlinear, we have a single objective *nonlinear* programming problem.

1.1 THE MULTIPLE OBJECTIVE PROGRAMMING PROBLEM

Methods for solving single objective mathematical programming problems have been studied extensively for the past 40 years. However, single objective decision-making methods reflect an earlier and simpler era. The world has become more complex. As we enter the information age, we find that almost every important real-world problem involves more than one objective. More so than ever before, decision makers find it imperative to evaluate solution alternatives according to multiple criteria. Although we are well equipped with single criterion methods, we now need extensions to theory and practice so that we can address the multiple objective programming problem

$$\begin{aligned} &\max \{f_1(\mathbf{x}) = z_1\} \\ &\max \{f_2(\mathbf{x}) = z_2\} \\ &\quad \vdots \\ &\max \{f_k(\mathbf{x}) = z_k\} \\ &\text{s.t.} \qquad \mathbf{x} \in S \end{aligned}$$

whether it be linear, integer, or nonlinear. We note that a multiple objective program is just like a single objective program except that it has a stack of objective functions instead of only one.

1.2 MULTIPLE CRITERIA EXAMPLES

In single objective programming, we must settle on a single objective such as minimizing cost or maximizing profit. However, if we think long enough about any real-world application, we will almost certainly be able to identify multiple conflicting criteria. To illustrate the range of problems that may be more adequately modeled with multiple objectives and to suggest the types of measures that might be used as criteria, we have the following:

Oil Refinery Scheduling

min {cost}
min {imported crude}
min {high sulfur crude}
min {deviations from demand state}
min {flaring of gases}

Production Planning

max {total net revenue}
max {minimum net revenue in any period}
min {backorders}
min {overtime}
min {finished goods inventory}

Portfolio Selection

max {return}
min {risk}
max {dividends}
min {deviations from diversification goals}

Capital Budgeting

max {net present value}
min {capital investment requirements}
min {annual operating expenses}
max {investment in projects related to environmental protection}
max {investment in projects in a given geographical area}
max {investment in projects pertaining to a given product line}

Forest Management

max {sustained yield of timber production}
max {visitor days of dispersed recreation}
max {visitor days of hunting}

max {wildlife habitat}
max {animal-unit-months of grazing}
min {overdeviations from budget}

Determining Reservoir Release Policies

max {recreation benefits at reservoir 1}
max {flood control benefits below reservoir 1}
max {total power generation in river basin}
min {underdeviations from municipal water supply requirements in river basin}
max {recreation benefits at reservoir 2}
max {irrigation benefits below reservoir 2}

Allocating the Audit Staff in a CPA Firm

max {profit}
min {increase in audit staff}
min {decrease in audit staff}
min {excessive overtime}
min {underutilization of auditors' training}
max {time devoted to professional development}

Transportation

min {cost}
min {average shipping time to priority customers}
max {production using a given process}
min {fuel consumption}

Hot Dog, Bologna, and Salami Blending

min {cost}
min {fat}
min {deviations from a color goal}
max {protein}
min {deviations from a moisture goal}
min {deviations from a pork-to-beef goal}

1.3 MULTIPLE CRITERIA OPTIMIZATION

Excluding the trivial case in which a point exists in the feasible region S that simultaneously maximizes all k objectives, the ideal way to solve a multiple objective program would be according to the following protocol: First assess the decision maker's utility function U. Then, solve the following mathematical programming problem

$$\begin{aligned} &\max \{U(z_1, z_2, \ldots, z_k)\} \\ &\text{s.t. } f_i(\mathbf{x}) = z_i \quad 1 \leqq i \leqq k \\ &\mathbf{x} \in S \end{aligned} \tag{1.1}$$

One would expect (1.1) to be nonlinear. However, this is not the main difficulty with the approach. The difficulty is that with many problems it is not possible to obtain a mathematical representation of the decision maker's utility function U. It is about such problems that we are concerned. Despite the fact that we may never know the decision maker's utility function, the problems still have to be solved.

Therefore, without explicit knowledge of the decision maker's utility function, we have no other choice but to search somehow the "space of tradeoffs" among the objectives of

$$\begin{aligned} &\max \{f_1(\mathbf{x}) = z_1\} \\ &\max \{f_2(\mathbf{x}) = z_2\} \\ &\quad \vdots \\ &\max \{f_k(\mathbf{x}) = z_k\} \\ &\text{s.t.} \qquad \mathbf{x} \in S \end{aligned}$$

for the decision maker's *optimal* solution using only implicit information. Furthermore, the challenge is to do this in an economical and user-convenient fashion. This is not easy because the space of tradeoffs is not small. In practice, *interactive procedures* have proven to be most effective in searching tradeoff space for a final solution. These are man-machine procedures that intersperse phases of computation with phases of decision. Hence, human intervention in the solution process is one of the characteristics that distinguishes the methods of multiple criteria optimization from those of traditional (single criterion) mathematical programming.

1.4 OPTIMAL VERSUS FINAL SOLUTIONS

A few comments are in order about the properties of an optimal solution. A point in S is optimal if it maximizes the decision maker's utility function in (1.1). To be optimal, however, a point must be *efficient*. A point in S is efficient if and only if its criterion vector is *nondominated*. That is, a point is efficient if it is not possible to move feasibly from it to increase an objective without decreasing at least one of the others. *Inefficient* solutions are not candidates for optimality.

In practice, we will typically be satisfied with a "near-optimal" solution. Whether it is optimal or near-optimal, any solution that satisfactorily terminates the decision process is called a *final solution*. In the absence of a mathematical specification of the decision maker's utility function, the extensions in theory and innovations in technique that enable us to identify the decision maker's final solution of (1.1) constitute the topic of *multiple criteria optimization*.

As can be seen from the bibliographies, journal special issues, monographs, and books listed at the end of this chapter, virtually all that is known about multiple criteria optimization has been developed since the early 1970s.

1.5 RELATIONSHIP TO MULTIATTRIBUTE DECISION ANALYSIS

The topic of multiple criteria optimization is from the field of *multiple criteria decision making* (MCDM). Multiple criteria decision making is concerned with the methods and procedures by which multiple criteria can be formally incorporated into the analytical process.

Multiple criteria decision making has, however, two distinct halves. One half is *multiattribute decision analysis* and the other half is multiple criteria optimization (multiple objective mathematical programming). Multiattribute decision analysis is most often applicable to problems with a small number of alternatives in an environment of uncertainty. Multiple criteria optimization is most often applied to deterministic problems in which the number of feasible alternatives is large. Whereas multiattribute decision analysis has been most applicable in resolving difficult public policy problems (nuclear power plant siting, location of an airport, type of drug rehabilitation program), multiple criteria optimization is more useful with less controversial problems in business and government (such as indicated in Section 1.2).

Multiattribute decision analysis is most fully covered by Keeney and Raiffa (1976). The role of this book is to cover the other half, the application of optimization techniques in MCDM.

1.6 NOTATION

To familiarize the reader with the notation used in the book, we list the following:

1. LP — *Single objective linear program.*

$$\max\{\mathbf{c}^T\mathbf{x} = z \mid \mathbf{x} \in S\}$$

2. S — *Feasible region in decision space.* If S is defined by linear constraints,

$$S = \{\mathbf{x} \in R^n \mid \mathbf{A}\mathbf{x} = \mathbf{b}, \quad \mathbf{x} \geqq \mathbf{0}, \quad \mathbf{b} \in R^m\}$$

where $\mathbf{A}$ is assumed to be of full row rank.

3. MOLP — *Multiple objective linear program.*

$$\begin{aligned} &\max\{\mathbf{c}^1\mathbf{x} = z_1\} \\ &\max\{\mathbf{c}^2\mathbf{x} = z_2\} \\ &\quad\vdots \\ &\max\{\mathbf{c}^k\mathbf{x} = z_k\} \\ &\text{s.t.} \qquad \mathbf{x} \in S \end{aligned}$$

4. DM — *Decision maker.*

5. $\mathbf{C}$	$k \times n$ *criterion matrix* whose rows are the $\mathbf{c}^i$ gradients of the k objectives.
6. $\mathbf{x}^i \in R^n$	*Point in decision space* where $$\mathbf{x}^i = \begin{bmatrix} x_1^i \\ x_2^i \\ \vdots \\ x_n^i \end{bmatrix}$$ Superscripts differentiate vectors; subscripts differentiate components.
7. $\mathbf{z} \in R^k$	*Criterion vector* where $$\mathbf{z} = \begin{bmatrix} z_1 \\ z_2 \\ \vdots \\ z_k \end{bmatrix}$$
8. γ	*Convex combination operator* where the set of all convex combinations of $\mathbf{x}^1, \mathbf{x}^2, \ldots, \mathbf{x}^r$ is written $$\gamma(\mathbf{x}^1, \mathbf{x}^2, \ldots, \mathbf{x}^r) \quad \text{or} \quad \mathop{\gamma}_{i=1}^{r} (\mathbf{x}^i)$$
9. μ	*Unbounded line segment operator* where the line segment emanating from $\mathbf{x}^1$ in the direction $\mathbf{v}^a$ is written $$\mu(\mathbf{x}^1, \mathbf{v}^a)$$
10. $\Theta \subset R^n$	*Optimal set* (set of all optimal points).
Θ_x	Set of all optimal extreme points.
Θ_μ	Set of all unbounded optimal edges.
11. $E \subset R^n$	*Efficient set* (set of all efficient points).
E_x	Set of all efficient extreme points.
E_μ	Set of all unbounded efficient edges.
12. $E^w \subset R^n$	*Weakly efficient set* (set of all weakly efficient points).
E_x^w	Set of all weakly efficient extreme points.
E_μ^w	Set of all unbounded weakly efficient edges.
13. $Z \subset R^k$	*Feasible region in criterion space* (the set of all images of points in S).
14. $\mathbf{W}$	$c_j - z_j$ *reduced cost* matrix.
15. $N \subset R^k$	Set of all *nondominated criterion vectors*.
16. U	Decision maker's *utility function* where U: $R^k \to R$.
17. rel	*Relative interior* (of a set).

18. Λ Set of all *strictly positive* weighting vectors where

$$\Lambda = \left\{ \boldsymbol{\lambda} \in R^k \mid \lambda_i > 0, \quad \sum_{i=1}^{k} \lambda_i = 1 \right\}$$

19. $\bar{\Lambda}$ Set of all *nonnegative* weighting vectors where

$$\bar{\Lambda} = \left\{ \boldsymbol{\lambda} \in R^k \mid \lambda_i \geqq 0, \quad \sum_{i=1}^{k} \lambda_i = 1 \right\}$$

20. $\tilde{\Lambda}$ *Reduced* weighting vector space

$$\tilde{\Lambda} = \left\{ \boldsymbol{\lambda} \in R^k \mid \lambda_i \in \mathrm{rel}\,[\ell_i, \mu_i], \quad \sum_{i=1}^{k} \lambda_i = 1 \right\}$$

21. p Parameter of L_p-norm or L_p-metric.

22. P *Number of solutions* to be presented to the decision maker at each iteration.

23. r Weighting vector space *reduction factor.*

24. t *Intended number of iterations.*

25. $\mathbf{x}^{(h)}$ Decision space solution of the hth iteration.

26. $\mathbf{z}^{(h)}$ Criterion vector of the hth iteration where $\mathbf{z}^{(h)} = \mathbf{z}(\mathbf{x}^{(h)})$.

Readings

1. Achilles, A., K.-H. Elster, and R. Nehse (1979). "Bibliographie zur Vektoroptimierung (Theorie and Anwendungen)," *Mathematische Operationsforschung und Statistik—Series Optimization*, Vol. 10, No. 2, pp. 277–321.
2. Blair, P. D. (1979). "Multiobjective Regional Energy Planning," *Studies in Applied Regional Science*, Vol. 14, Boston: Martinus Nijhoff Publishing.
3. Carlsson, C. and Y. Kochetkov (eds.) (1983). *Theory and Practice of Multiple Criteria Decision Making*, Amsterdam: North-Holland.
4. Carlsson, C., A. Törn, and M. Zeleny (1982). *International Comparison of Multi Criteria Methods on the Basis of Three Common Case Studies*, New York: McGraw-Hill.
5. Chankong, V. and Y. Y. Haimes (1983). *Multiobjective Decision Making: Theory and Methodology*, New York: North-Holland.
6. Charnes, A., W. W. Cooper, and R. J. Niehaus (1972). *Studies in Manpower Planning*, Office of Civilian Manpower Management, Department of the Navy, Washington, D.C.
7. Cochrane, J. L. and M. Zeleny (eds.) (1973). *Multiple Criteria Decision Making*, Columbia, South Carolina: University of South Carolina Press.
8. Cohon, J. L. (1978). *Multiobjective Programming and Planning*, New York: Academic Press.

9. Cohon, J. L. and D. H. Marks (1973). "A Review and Evaluation of Multiobjective Programming Techniques," *Water Resources Research*, Vol. 11, No. 2, pp. 208–220.

10. Despontin, M., J. Moscarola, and J. Spronk (1983). "A User-Oriented Listing of Multiple Criteria Decision Methods," *Belgian Journal of Statistics, Computer Science, and Operations Research*, Vol. 23, No. 4.

11. Easton, A. (1973). *Complex Management Decisions Involving Multiple Objectives*, New York: John Wiley & Sons.

12. Fandel, G. (1972). "Optimale Entscheidung bei Mehrfacher Zielsetzung," *Lecture Notes in Economics and Mathematical Systems*, No. 76, Berlin: Springer-Verlag.

13. Fandel, G. and T. Gal (eds.) (1980). "Multiple Criteria Decision Making—Theory and Application," *Lecture Notes in Economics and Mathematical Systems*, No. 177, Berlin: Springer-Verlag.

14. French, S., R. Hartley, L. C. Thomas, and D. J. White (eds.) (1983). *Multi-Objective Decision Making*, London: Academic Press.

15. Goicoechea, A., D. R. Hansen, and L. Duckstein (1982). *Multiobjective Decision Analysis with Engineering and Business Applications*, New York: John Wiley & Sons.

16. Grauer, M., A. Lewandowski, and A. P. Wierzbicki (eds.) (1982). *Multiobjective and Stochastic Optimization*, International Institute for Applied Systems Analysis, Laxenburg, Austria.

17. Grauer, M. and A. P. Wierzbicki (eds.) (1984). "Interactive Decision Analysis," *Lecture Notes in Economics and Mathematical Systems*, No. 229, Berlin: Springer-Verlag.

18. Haimes, Y. Y. and V. Chankong (eds.) (1985). "Decision Making with Multiple Objectives," *Lecture Notes in Economics and Mathematical Systems*, No. 242, Berlin: Springer-Verlag.

19. Haimes, Y. Y., W. A. Hall, and H. T. Freedman (1975). *Multiobjective Optimization in Water Resources Systems: The Surrogate Worth Trade-Off Method*, Amsterdam: Elsevier.

20. Hansen, P. (ed.) (1983). "Essays and Surveys on Multiple Criteria Decision Making," *Lecture Notes in Economics and Mathematical Systems*, No. 209, Berlin: Springer-Verlag.

21. Hemming, T. (1978). *Multiobjective Decision Making Under Certainty*, Economic Research Institute, Stockholm School of Economics.

22. Hwang, C. L., A. S. M. Masud, S. R. Paidy, and K. Yoon (1979). "Multiple Objective Decision Making—Methods and Applications," *Lecture Notes in Economics and Mathematical Systems*, No. 164, Berlin: Springer-Verlag.

23. Ignizio, J. P. (1976). *Goal Programming and Extensions*, Lexington, Massachusetts: D. C. Heath.

24. Ignizio, J. P. (1982). *Linear Programming in Single- & Multiple-Objective Systems*, Englewood Cliffs, New Jersey: Prentice-Hall.

25. Ignizio, J. P. (ed.) (1983). *Computers and Operations Research* (Special Issue on Generalized Goal Programming), Vol. 10, No. 4.

26. Ijiri, Y. (1965). *Management Goals and Accounting for Control*, Chicago: Rand-McNally.

27. Johnsen, E. (1968). *Studies in Multiobjective Decision Models*, Economic Research Center, Lund, Sweden.

28. Keeney, R. L. and H. Raiffa (1976). *Decisions with Multiple Objectives: Preferences and Value Tradeoffs*, New York: John Wiley & Sons.

29. Lawrence, K. D. and R. E. Steuer (eds.) (1984). *IIE Transactions* (Special Report: Multiple Criteria Decision Making in Production Planning and Scheduling), Vol. 16, No. 4.

30. Lee, S. M. (1972). *Goal Programming for Decision Analysis*, Philadelphia: Auerbach.

31. Lee, S. M. and J. C. Van Horn (1983). *Academic Administration: Planning, Budgeting, and Decison Making with Multiple Objectives*, Lincoln, Nebraska: University of Nebraska Press.

32. MacCrimmon, K. R. (1973). "An Overview of Multiple Objective Decision Making." In J. L. Cochrane and M. Zeleny (eds.), *Multiple Criteria Decision Making*, Columbia, South Carolina: University of South Carolina Press, pp. 18–44.

33. Major, D. C. (1977). *Multiobjective Water Resource Planning*, Water Resources Monograph 4, American Geophysical Union, Washington, D.C.

34. Morse, J. (ed.) (1981). "Organizations: Multiple Agents with Multiple Criteria," *Lecture Notes in Economics and Mathematical Systems*, No. 190, Berlin: Springer-Verlag.

35. Nehse, R. (1982). "Bibliographie zur Vektoroptimierung—Theorie und Anwendungen (First Continuation)," *Mathematische Operationsforschung und Statistik—Series Optimization*, Vol. 13, No. 4, pp. 593–625.

36. Neumann, H.-W. (1984). *Entscheidungsunterstützung bei Mehrfachen Zielen durch Freie Algorithmenwahl im Computerdialog*, Bern, Switzerland: Verlag Peter Lang.

37. Nijkamp, P. and J. Spronk (eds.) (1981). *Multiple Criteria Analysis*, London: Gower Press.

38. Rietveld, P. (1980). *Multiple Objective Decision Methods and Regional Planning*, Studies in Regional Science and Urban Economics, Vol. 7, Amsterdam: North-Holland.

39. Roubens, M. and J. Wallenius (eds.) (1985). *European Journal of Operational Research* (Special Issue on Multiple Criteria Decision Making) (forthcoming).

40. Roy, B. (1972). "Problems and Methods with Multiple Objective Functions," *Mathematical Programming*, Vol. 1, No. 2, pp. 239–266.

41. Saaty, T. L. (1980). *The Analytic Hierarchy Process*, New York: McGraw-Hill.

42. Schniederjans, M. J. (1984). *Linear Goal Programming*, Princeton, New Jersey: Petrocelli Books.

43. Slowinski, R. and J. Weglarz (eds.) (1983). *Foundations of Control Engineering* (Special Issue Published on the Occasion of the 18th Meeting of the EURO Working Group on 'Multicriteria Decision Aid,' Poznan, October 14–15, 1983), Vol. 8, No.'s 3–4.

44. Spronk, J. (1980). *Interactive Multiple Goal Programming: Applications to Financial Planning*, Boston: Martinus Nijhoff Publishing.

45. Starr, M. K. and M. Zeleny (eds.) (1977). "Multiple Criteria Decision Making," *TIMS Studies in the Management Sciences*, Vol. 6, Amsterdam: North-Holland.

46. Tell, B. (1976). *A Comparative Study of Some Multiple-Criteria Methods*, Economic Research Institute, Stockholm School of Economics.

47. Thiriez, H. and S. Zionts (eds.) (1976). "Multiple Criteria Decision Making: Jouy-en-Josas, France," *Lecture Notes in Economics and Mathematical Systems*, No. 130, Berlin: Springer-Verlag.

48. Wallenius, J. (1975). *Interactive Multiple Criteria Decision Methods: An Investigation and an Approach*, Helsinki School of Economics.

49. White, D. J. (1982). *Optimality and Efficiency*, Chichester: John Wiley & Sons.

50. Yu, P. L. (1985). *Multiple Criteria Decision Making: Concepts, Techniques and Extensions*, New York: Plenum.

51. Zeleny, M. (ed.) (1980). *Computers and Operations Research* (Special Issue on Mathematical Programming with Multiple Objectives), Vol. 7, No.'s 1–2.

52. Zeleny, M. (1974). "Linear Multiobjective Programming," *Lecture Notes in Economics and Mathematical Systems*, No. 95, Berlin: Springer-Verlag.

53. Zeleny, M. (ed.) (1976). "Multiple Criteria Decision Making: Kyoto 1975," *Lecture Notes in Economics and Mathematical Systems*, No. 123, Berlin: Springer-Verlag.

54. Zeleny, M. (1982). *Multiple Criteria Decision Making*, New York: McGraw-Hill.

55. Zionts, S. (ed.) (1978). "Multiple Criteria Problem Solving," *Lecture Notes in Economics and Mathematical Systems*, No. 155, Berlin: Springer-Verlag.

56. Zionts, S. (1979). "MCDM—If Not a Roman Numeral, Then What?" *Interfaces*, Vol. 9, No. 4, pp. 94–101.

57. Zionts, S. and J. Spronk (eds.) (1984). *Management Science* (Special Issue on Multiple Criteria Decision Making), Vol. 30, No. 11.

CHAPTER 2

Mathematical Background

This chapter reviews basic mathematical concepts that are relied on throughout the text. The chapter characterizes the modest mathematical sophistication necessary for the study of multiple criteria optimization.

2.1 SET THEORY

A *set* is the concept of a well-defined list, collection, or class of objects. The body of conventions and notation regarding the use of sets is referred to as *set theory*. The advantage of set theory is that it facilitates clear and concise mathematical communications.

2.1.1 Specifying Sets and Indicating Membership

Sets are specified using braces { and }. There are two ways in which sets are stipulated. One way is the *roster* method and the other is the *defining-property* method. Which one is used is simply a matter of convenience.

Example 1. Sets A and B

$$A = \{a, g, h, b\}$$
$$B = \{8, 10, 12, 14, \dots\}$$
$$C = \{x \mid x = 2^n,\ n \text{ is an integer},\quad 1 \leqq n \leqq 5\}$$
$$D = \{y \mid y \text{ is a vowel}\}$$

are specified in roster form and C and D are specified in defining-property form. Set B is an example of an *infinite* set.

In defining-property form, the vertical bar | is read "such that." Also the choice of x or y (or any other symbol) as the generic element of the set is of no consequence.

Set membership is indicated by the symbol $\in$. If an element x is a member of X, we write

$$x \in X$$

This is read as "x belongs to X" or "x is in X." If x is not a member of X, we write $x \notin X$. The slash / denotes negation.

At times, a set with no elements is encountered. Such a set is called the *empty* set or *null* set. In roster form the null set is written as { }. Also, the null set is symbolically represented by $\varnothing$.

Example 2. Consider

$$F = \{x \mid -3 \leqq x \leqq 5\}$$

$$X = \{x \mid x^2 = 4, \quad 3x = 12\}$$

With regard to F, $.738 \in F$ but $-4 \notin F$. Because X is null, we can write $X = \varnothing$.

A frequent mistake when designating an empty set is to write $\{\varnothing\}$. This is incorrect because $\{\varnothing\}$ would be read as the set that contains the null set.

2.1.2 Subsets, Supersets, and Set Equality

Let F and G be sets. Then, F is a *subset* of G if and only if $x \in F$ implies $x \in G$. That is, F is a subset of G only if every element in F is also a member of G. If F is a subset of G, we would write

$$F \subset G$$

Looking at the subset relationship in reverse, we could write $G \supset F$, which would read "G is a *superset* of F." We note that the null set $\varnothing$ is a subset of every set.

Example 3. Consider the three sets

$$A = \{1, 3, 7, 8\}$$

$$B = \{3, 5, 8\}$$

$$C = \{1, 3, 7, 8, 5\}$$

Then, $A \subset C$ and $B \subset C$, but $B \not\subset A$ because $5 \notin A$. Note, for instance, that $\varnothing \subset C$.

Two sets are *equal* if and only if they contain the same elements. By definition, every set is a subset of itself. Therefore, $A = B$ if and only if $A \subset B$ and $B \subset A$.

Example 4. Sets

$$K = \{y \mid y \text{ is a borough of New York City}\}$$

$$L = \{\text{Bronx, Brooklyn, Manhattan, Queens, Staten Island}\}$$

are equal. Also $\{5, 2, 5, 7\} = \{7, 2, 5\} = \{5, 2, 7\}$ because order and duplication are unimportant.

2.1.3 Disjoint Sets and Families

If two sets A and B have no elements in common, they are *disjoint*. A *family* is a set whose elements are sets themselves.

Example 5. Of the three sets

$$A = \{7\}$$
$$B = \{10, 11\}$$
$$C = \{7, 8, 9, 10\}$$

only A and B are disjoint. The family of all subsets of $\{11, 12\}$ is

$$\Psi = \{\{11, 12\}, \{11\}, \{12\}, \varnothing\}$$

2.1.4 Set Operations

Consider the three set operations

1. Union.
2. Intersection.
3. Difference.

Let A and B be sets. Then, A *union* B (denoted $A \cup B$) is the set of all elements that belong to either A, B, or both (i.e., $A \cup B = \{x | x \in A \text{ or } x \in B\}$). A *intersect* B (denoted $A \cap B$) is the set of elements that are simultaneously in both A and B (i.e., $A \cap B = \{x | x \in A, x \in B\}$). Whereas the union includes all elements, the intersection excludes all but overlapping elements. If two sets A and B are disjoint, then $A \cap B = \varnothing$ (and if not disjoint, we would write $A \cap B \neq \varnothing$).

As for the *difference* operation, A *difference* B (written $A - B$) is the set of elements in A but not in B (i.e., $A - B = \{x | x \in A, x \notin B\}$). Note how the second set removes from the first all elements that the two share in common.

Example 6. Let

$$A = \{1, 2, 3\}$$
$$B = \{2, 5, 1\}$$
$$I = \{a, b, c, d\}$$

then

$$A \cup B = \{1, 2, 3, 5\}$$
$$A \cap B = \{1, 2\}$$
$$(I \cup \{e\}) - \{c\} = \{a, b, d, e\}$$
$$A - B = \{3\}$$

2.1.5 The Real Line and Intervals

We let R designate the set of real numbers. Since points in R can be represented as points on a line, R is sometimes referred to as the *real line*. Any continuous subset of R is an *interval*. Depending on which endpoints are or are not contained in the interval, we have the following interval characterizations:

1. *Open* (does not contain either endpoint).
2. *Closed* (contains both endpoints).
3. *Open-closed* (only contains right endpoint).
4. *Closed-open* (only contains left endpoint).

Instead of braces { and }, with intervals we use parentheses and brackets [and]. A parenthesis is used to indicate that a given endpoint is not included in its interval. A bracket is used to indicate that a given endpoint is included in its interval.

Example 7

a. $\{x \mid -2.4 < x < 5\}$ is an open interval and is written $(-2.4, 5)$.
b. $\{x \mid 0 \leqq x \leqq 1\}$ is a closed interval and is written $[0, 1]$. This interval is called the *unit* interval.
c. $\{x \mid 0 \leqq x < 33\}$ is a closed-open interval and is written $[0, 33)$.
d. $\{x \mid -\infty < x \leqq 3\}$ is an unbounded interval and is written $(-\infty, 3]$.

Unbounded intervals are designated as being open at their unbounded ends. Also, the set containing a real number can be written in interval notation. For instance, $\{a\} = [a, a]$ where $a \in R$.

2.1.6 Generalized Operations, Index Sets, and Cardinality

The union and intersection operations can be generalized to any finite number of sets. Consider

$$A_1 = \{1,2,5,7\} \qquad A_2 = \{3,4,5\} \qquad A_3 = \{1,5,7,8\}$$
$$A_4 = \{2,5,7\} \qquad A_5 = \{1,4,5\}$$

The respective union and intersection of these five sets are written

$$\bigcup_{i=1}^{5} A_i = \{1,2,3,4,5,7,8\}$$

$$\bigcap_{i=1}^{5} A_i = \{5\}$$

If we use the *index set* $I = \{1, 2, 3, 4, 5\}$, the respective union and intersection of the five sets would be written

$$\bigcup_{i \in I} A_i = \{1,2,3,4,5,7,8\}$$

$$\bigcap_{i \in I} A_i = \{5\}$$

Let J be a set. The *cardinality* of J (written $|J|$) is the number of elements in J.

Example 8. The union of A_1, A_4, and A_5 is given by

$$\bigcup_{i\in I} A_i = \{1,2,4,5,7\} \qquad I = \{1,4,5\}$$

Actually, we could use the index set idea with the summation symbol Σ. Let

$$\alpha_j = j^2 \qquad j = 1,2,3,\ldots,10$$

Then, the sum of all odd j elements could be expressed as

$$\sum_{j\in J} \alpha_j = 165 \qquad J = \{1,3,5,7,9\}$$

In the above $|I| = 3$ and $|J| = 5$.

2.1.7 Partitions

Let A be a set. Then, the family of subsets $\{A_1, A_2, A_3, \ldots, A_q\}$ is called a *partition* of A if and only if

$$\text{(i)} \qquad A = \bigcup_{i=1}^{q} A_i$$

and

$$\text{(ii)} \qquad A_i \cap A_j = \varnothing \qquad \text{for all } i \neq j$$

Condition (i) requires that the subsets *cover* A and condition (ii) requires the subsets to be mutually disjoint.

Example 9. The family of intervals $\{[-8,0],(0,17),[17,20)\}$ forms a partition of $[-8,20)$.

2.1.8 Mappings and Functions

A *mapping* f (written $f\colon A \to B$) assigns one and only one element in B to each element in A. That is, $f\colon A \to B$ *maps* elements in A into elements in B.

If $f\colon A \to B$, the set A is called the *domain* of f and B is called the *co-domain*. For each element $a \in A$ in the domain, the element $b \in B$ associated with a is called the *image* of a and is denoted $f(a)$. If f maps $a \in A$ into $b \in B$, then a is an *inverse image* of b. When A and B are sets of real numbers, the mapping $f\colon A \to B$ is a *function*.

Let $f\colon A \to B$ be a mapping. Then, the image of A is the *range*. That is, the range of f (denoted $f(A)$) is the subset of B whose elements are images of elements of A. The inverse image of the range is, of course, the domain.

A mapping $f\colon A \to B$ is called *one-to-one* if no two different elements of A have the same image. A mapping $f\colon A \to B$ is called *onto* if the range of f is the whole co-domain B.

Example 10

a. Let g associate each letter of the alphabet with its numeric position. Defining

$$A = \{a, b, c, \ldots, z\}$$
$$N = \{1, 2, 3, \ldots, 26\}$$

we write g: $A \to N$. In this case, g is a mapping but not a function.

b. Let $A = \{1, 2, 3\}$, $B = \{1, 3, 5, 7\}$, and f be defined by the formula $f(x) = 2x - 1$. Then, f: $A \to B$ is a function that is one-to-one but not onto (because $7 \in B$ but $7 \notin f(A)$).

c. Let f: $R \to [0, +\infty)$ be defined by the formula $f(x) = x^2$. Then, f is onto but not one-to-one. It is not one-to-one because, for instance, 4 has two inverse images, -2 and 2.

d. Let f: $[2, 5] \to [4, 25]$ be defined by the formula $f(x) = x^2$. Then, f is one-to-one and onto.

2.1.9 Ordered *n*-Tuples and Cartesian Products

An *ordered pair* consists of two elements (or components) of which one is designated as the first and the other is designated as the second. If the first element is $a \in A$ and the second element is $b \in B$, the ordered pair is denoted (a, b).

If we allow A and B to be two sets, the *Cartesian product* of A and B (denoted $A \times B$) is the set of all ordered pairs (a, b) such that $a \in A$ and $b \in B$. The notions of the ordered pair and Cartesian product are generalized to n elements (n components) and n sets, respectively. Thus, an *ordered n-tuple* has n ordered components

$$(x_1, x_2, \ldots, x_n)$$

and the *Cartesian product of n sets* $A_1, A_2, \ldots, A_n$ is the set of all ordered n-tuples $(x_1, x_2, \ldots, x_n)$ such that

$$x_i \in A_i \qquad i = 1, 2, \ldots, n$$

Two ordered n-tuples are equal if all corresponding elements are equal.

Consider the Cartesian product $R \times R$ (denoted R^2). R^2 represents the set of all points in two-dimensional real space (two-space). A member of R^2 is an ordered pair (of real numbers). The Cartesian product $R \times R \times R = R^3$ represents the set of all points in three-space. Hence, R^n represents the set of all points in n-space in which members of R^n are ordered n-tuples.

Example 11

a. The two ordered four-tuples $(4, 3, 5, 5)$ and $(4, 3, 4, 5)$ are not equal because their third components are not equal.

b. Let $A = \{a, b\}$, $B = \{-1, 1\}$, and $C = \{\rho, \sigma\}$. Then, $A \times B \times C = \{(a, -1, \rho), (a, -1, \sigma), (a, 1, \rho), (a, 1, \sigma), (b, -1, \rho), (b, -1, \sigma), (b, 1, \rho), (b, 1, \sigma)\}$

2.1.10 Graph of a Function

The *graph* of a function f: $A \to B$ is the set of all ordered pairs such that $a \in A$ is the first component and its image $f(a) \in B$ is the second component. That is,

$$\text{graph of } f = \{(a, b) | a \in A, \quad b = f(a)\}$$

The graph of f is a subset of Cartesian space $A \times B$. The graph of a function may or may not contain a finite number of points.

Example 12

a. Let $G = \{1, 4, 7\}$ and f: $G \to R$ be defined by $f(x) = 3x + 2$. Then, the graph of f is the set $\{(1, 5), (4, 14), (7, 23)\}$ which is a finite subset of $G \times R$.

b. Let $A = [2, 4]$ and f: $A \to R$ be defined by $f(x) = 5 - (x - 4)^2$. Since the graph of f is a subset of R^2, the graph of f can be portrayed graphically in Fig. 2.1. In this example, the graph of f only consists of the curved line segment in Fig. 2.1 because f is not defined over the intervals $(-\infty, 2)$ and $(4, +\infty)$.

c. Let $A \subset R^n$ and $B \subset R^k$. Then, the graph of the function f: $A \to B$ is a subset of R^{n+k}.

2.1.11 Relations

Let A and B be sets. Then, a "stipulation" that creates a subset of $A \times B$ is a *relation*. A relation is a notion that subsumes those of mappings and functions. While every mapping and function is a relation, not every relation is a mapping or function. In mappings and functions, every element in A must be assigned to one and only one element in B. In relations, there may be some elements in A that are not assigned to any elements in B, and there may be some elements in A that are assigned to more than one element in B. If an ordered pair (a, b) is true for the stipulation, it is included in the subset of $A \times B$. If the ordered pair is not true for the stipulation, it is not included.

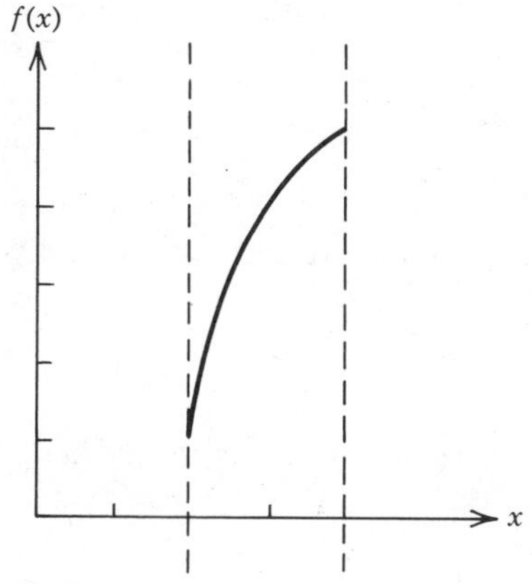

Figure 2.1. Graph of Example 12*b*.

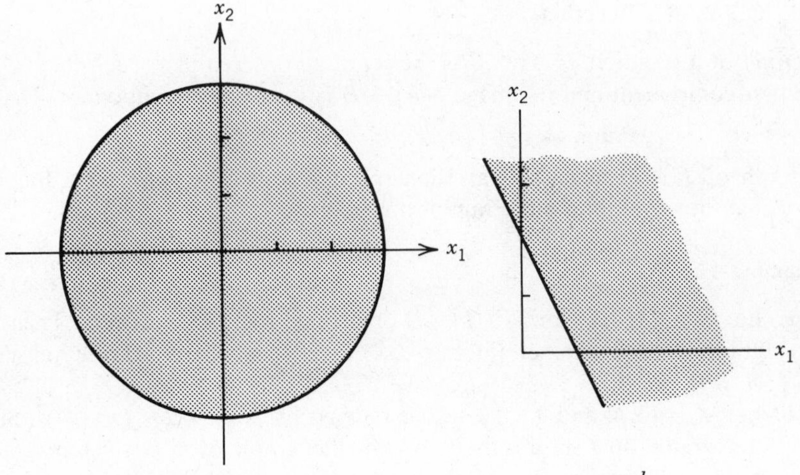

a *b*

Figure 2.2. Graphs of Example 13.

Example 13

a. Consider $R \times A$ where $A = [-4, 4]$. Then, the stipulation

$$(x_1)^2 + (x_2)^2 \leqq 9$$

where $x_1 \in R$ and $x_2 \in A$ is a relation on $R \times A$ because it creates the subset of $R \times A$ given by the shaded area in Fig. 2.2a. In Fig. 2.2*a* there are points in R that do not correspond to any points in A and there are points in R that correspond to multiple points in A. Hence, the stipulation is a relation, not a mapping or function.

b. The stipulation

$$2x_1 + x_2 \geqq 2$$

is a relation (but not a mapping or function) on R^2 because it creates the subset of R^2 given by the shaded area in Fig. 2.2*b*. Had the stipulation been $2x_1 + x_2 = 2$, the relation would have been a mapping and a function.

2.2 TOPICS FROM LINEAR ALGEBRA

In this section, we discuss topics from linear algebra that are relevant to our study of multiple objective programming.

2.2.1 Matrices

A *matrix* is a rectangular array of elements. Matrices are written with brackets [and]. Matrices are rectangular in the sense that their elements are arranged in rows and columns. This enables the *size* of a matrix to be described with two numbers, the first being the number of rows and the second being the number of columns. The *transpose* of matrix **A** (written $\mathbf{A}^T$) is the matrix obtained by writing the ith row of **A** as the ith column of $\mathbf{A}^T$.

Matrices with only a single row or column are called *vectors*. There are *column* vectors and *row* vectors. A preferred way to describe the size of a vector is to utilize the notion of *length*. For instance,

$$\mathbf{x} = \begin{bmatrix} 22 \\ -37 \\ 6 \\ -13 \end{bmatrix}$$

is a column vector of length 4 and $\mathbf{y}^T = [-2 \quad 7 \quad 10 \quad -8 \quad 9]$ a row vector of length 5. Traditionally, we label vectors with lowercase letters

$$\mathbf{b}, \mathbf{d}, \mathbf{x}, \mathbf{z}, \boldsymbol{\lambda}, \boldsymbol{\alpha}, \ldots$$

and matrices (that are not vectors) with uppercase letters

$$\mathbf{A}, \mathbf{C}, \mathbf{D}, \mathbf{N}, \mathbf{T}, \ldots$$

Also, it is customary for a vector (unless carrying a transpose T superscript) to be considered a column vector unless specified to the contrary. To signify symbolically elements in matrices and vectors, we use subscripts.

Example 14. For **A**, an $m \times n$ matrix; **b**, a column vector of length m; and $\mathbf{d}^T$, a row vector of length n, we write

$$\mathbf{A} = \begin{bmatrix} a_{11} & a_{12} & \cdots & a_{1n} \\ a_{21} & & & \\ \vdots & & & \\ a_{m1} & & & a_{mn} \end{bmatrix}$$

$$\mathbf{b} = \begin{bmatrix} b_1 \\ b_2 \\ \vdots \\ b_m \end{bmatrix} \qquad \mathbf{d}^T = [d_1 \quad d_2 \quad \ldots \quad d_n]$$

Concerning *equality* among matrices, two matrices **A** and **B** are equal if and only if

1. **A** and **B** are of the same size.
2. Their corresponding elements are equal (i.e., $a_{ij} = b_{ij}$ for each pair of double subscripts i, j).

In the use of inequalities with vectors, some authors utilize the following convention that distinguishes between *single-* and *double-*bar inequalities.

1. $\mathbf{x} < \mathbf{y}$ denotes $x_i < y_i$ for all i.
2. $\mathbf{x} \leq \mathbf{y}$ denotes $x_i \leqq y_i$ for all i with $x_i < y_i$ for at least one i.
3. $\mathbf{x} \leqq \mathbf{y}$ denotes $x_i \leqq y_i$ for all i.

However, in this book, we will not use the single-bar inequality with vectors.

Example 15. Let

$$\mathbf{x} = \begin{bmatrix} 2 \\ 4 \\ -2 \\ 0 \end{bmatrix} \qquad \mathbf{y} = \begin{bmatrix} 2 \\ 4 \\ -1 \\ 0 \end{bmatrix}$$

To preserve the distinction that $\mathbf{x} \le \mathbf{y}$ (single bar), in this book we will instead write $\mathbf{x} \leqq \mathbf{y}$, $\mathbf{x} \neq \mathbf{y}$.

2.2.2 Matrix Operations

In this section, we discuss matrix addition, matrix subtraction, scalar multiplication, and matrix multiplication.

Matrix Addition and Subtraction. Matrix addition (or subtraction) is performed by adding (or subtracting) corresponding elements in matrices of the same size.

Scalar Multiplication. In scalar multiplication, we multiply a matrix by a scalar, which means each element inside **A** is simply multiplied by the scalar.

Matrix Multiplication. Matrix multiplication is only defined when the number of columns of the first matrix **A** (where **A** is $m \times r$) equals the number of rows of the second matrix **B** (where **B** is $r \times n$). Elements c_{ij} in the *product matrix* $\mathbf{AB} = \mathbf{C}$ (where **C** is $m \times n$) are given by

$$c_{ij} = \sum_{k=1}^{r} a_{ik} b_{kj}$$

Thus, c_{ij} is computed by summing the products arrived at by multiplying the elements in the ith row of **A** by their corresponding elements in the jth column of **B**.

Example 16. Premultiplying matrix **B** by matrix **A** (or postmultiplying **A** by **B**), we have

$$\mathbf{AB} = \begin{bmatrix} 1 & 2 \\ 3 & 4 \\ 5 & 6 \end{bmatrix} \begin{bmatrix} 2 & 2 & 1 & -1 \\ 0 & 0 & 3 & -3 \end{bmatrix} = \begin{bmatrix} 2 & 2 & 7 & -7 \\ 6 & 6 & 15 & -15 \\ 10 & 10 & 23 & -23 \end{bmatrix}$$

Because of dimensionality problems, it is not possible to postmultiply **B** by **A**.

Using horizontal and vertical lines, we can partition a matrix into component *submatrices* as in

$$\left[\begin{array}{ccc|cc} 2 & 2 & 1 & 4 & 5 \\ 0 & 0 & 5 & 0 & 2 \\ \hline 1 & 0 & 5 & 7 & 7 \\ 3 & 2 & 1 & 4 & 2 \\ \hline 6 & 3 & 1 & 1 & 1 \end{array}\right]$$

When partitioned in this way, the component submatrices are called *blocks*. Matrix addition, subtraction, and multiplication can be carried out in terms of blocks as long as the operand matrices have been partitioned in a compatible fashion.

Example 17. With the number of columns in **A** and **B** equaling the number of rows in **E** and **F**, respectively, we have

$$\left[\begin{array}{c|c} \mathbf{A} & \mathbf{B} \\ \hline \mathbf{C} & \mathbf{D} \end{array}\right]\left[\begin{array}{c} \mathbf{E} \\ \hline \mathbf{F} \end{array}\right] = \left[\begin{array}{c} \mathbf{AE} + \mathbf{BF} \\ \hline \mathbf{CE} + \mathbf{DF} \end{array}\right]$$

A *vector product* (scalar product or dot product) is the scalar quantity (element in the 1×1 matrix) that results whenever we multiply a row vector first by a column vector second, according to the rules of matrix multiplication.

Example 18. We have the following two illustrations of the vector product operation.

$$\mathbf{a}^T\mathbf{b} = [5 \quad 2 \quad -1 \quad 0]\begin{bmatrix} 6 \\ -1 \\ 4 \\ 7 \end{bmatrix} = 24$$

$$\mathbf{g}^T\mathbf{x} = [2 \quad 4 \quad -3]\begin{bmatrix} x_1 \\ x_2 \\ x_3 \end{bmatrix} = 2x_1 + 4x_2 - 3x_3$$

With regard to the transpose operation, we note the following four properties:

1. $(\mathbf{A} + \mathbf{B})^T = \mathbf{A}^T + \mathbf{B}^T$
2. $(\mathbf{A}^T)^T = \mathbf{A}$
3. $(k\mathbf{A})^T = k\mathbf{A}^T$ where k is a scalar
4. $(\mathbf{AB})^T = \mathbf{B}^T\mathbf{A}^T$

2.2.3 Special Matrices and Vectors

A *null* matrix is a matrix, all of whose elements are zero. A *square* matrix has the same number of rows as columns. A square matrix of size $n \times n$ is said to be of *order n*. A square matrix whose only nonzero elements appear on its *main diagonal* (line from its upper left-hand corner to its lower right-hand corner) is called a *diagonal* matrix. A diagonal matrix whose main diagonal consists only of ones is called an *identity* matrix. A *triangular* matrix is any square matrix whose nonzero elements are (a) on or above, or (b) on or below the main diagonal. An *upper triangular* matrix is a matrix whose nonzero elements are all on or above the main diagonal.

The identity matrix (denoted **I**) functions as the matrix counterpart of the number 1 in regular arithmetic. Whenever a matrix is pre- or postmultiplied by an identity matrix of appropriate order, the matrix is returned unaffected by the multiplication. That is, if **A** is $m \times n$,

$$\mathbf{I}_m\mathbf{A} = \mathbf{A} \quad \text{and} \quad \mathbf{A}\mathbf{I}_n = \mathbf{A}$$

The *null* vector (denoted **0**) is a vector of zeros. The *sum* vector (denoted **e**) is a vector of ones. When used in the vector product $\mathbf{e}^T\mathbf{v}$, the sum vector sums the elements in **v**, as in

$$\mathbf{e}^T\mathbf{v} = [1 \quad 1 \quad 1 \quad 1 \quad 1]\begin{bmatrix} 1 \\ 9 \\ 2 \\ -3 \\ 4 \end{bmatrix} = 13$$

2.2.4 Determinants

The *determinant* is a number associated with a square matrix. Matrices that are not square do not have determinants. The determinant of matrix **A** is written with straight vertical lines as $|\mathbf{A}|$ or as

$$\begin{vmatrix} a_{11} & a_{12} & \cdots & a_{1n} \\ a_{21} & & & \\ \vdots & & & \\ a_{n1} & & & a_{nn} \end{vmatrix}$$

The determinant of a 1×1 matrix is defined to be the number inside the 1×1 matrix. The determinant of a 2×2 matrix is given by the product of its main diagonal elements less the product of its *secondary diagonal* (upward at a 45° angle) elements. The determinant of a 3×3 matrix is obtained by recopying the first two columns of the array on the right and then subtracting the sum of the three secondary diagonal products from the sum of the three main diagonal products. To calculate the determinant of 4×4 and larger matrices, other methods are used that are not discussed here.

Example 19. The determinant of **A** is given by

$$|\mathbf{A}| = \begin{vmatrix} 2 & 2 \\ -1 & -4 \end{vmatrix} = [2(-4)] - [(-1)2] = -6$$

The determinant of **B** is given by

$$\begin{aligned} |\mathbf{B}| &= \begin{vmatrix} 2 & 3 & -4 \\ 0 & -4 & 2 \\ 1 & -1 & 5 \end{vmatrix} \begin{matrix} 2 & 3 \\ 0 & -4 \\ 1 & -1 \end{matrix} \\ &= [2(-4)5 + 3(2)1] - [1(-4)(-4) + (-1)2(2)] \\ &= -46 \end{aligned}$$

The determinants of square matrices of any size have the following properties.

1. If any two rows (or any two columns) are interchanged, the new matrix has the same determinant as the old matrix except that its sign is reversed.
2. $|\mathbf{A}| = |\mathbf{A}^T|$.
3. If there exists any row (or any column) all of whose elements are zeros, the determinant is 0.
4. If the elements in any two rows (or any two columns) are the same, the determinant is 0.
5. If the elements of any row (or any column) are multiplied by a scalar, then the determinant of the new matrix is equal to the determinant of the old matrix multiplied by the scalar.
6. The determinant of a triangular matrix is computed by taking the product of its main diagonal elements.
7. Any scalar multiple of any row (or column) can be added to any other row (or column) without changing the value of the determinant.

2.2.5 Solving Systems of Linear Equations Using Determinants

Consider a square system $\mathbf{Ax} = \mathbf{b}$ in which there are as many equations as unknowns. If $|\mathbf{A}| = 0$, the system $\mathbf{Ax} = \mathbf{b}$ either has no solutions or has an infinite number of solutions. If $|\mathbf{A}| \neq 0$, the system $\mathbf{Ax} = \mathbf{b}$ has a unique solution and we can solve for the unique solution using *Cramer's Rule*, as in Example 20.

Example 20. Consider the $\mathbf{Ax} = \mathbf{b}$ system

$$\begin{aligned} 2x_1 + 3x_2 + 4x_3 &= 8 \\ x_1 \qquad - 6x_3 &= -7 \\ -x_1 + x_2 + 4x_3 &= 7 \end{aligned}$$

For the numerator determinant for x_i, we replace the ith column of $\mathbf{A}$ with the right-hand side (RHS) vector $\mathbf{b}$.

RHS Vector **b** (column 1)

$$x_1 = \frac{\begin{vmatrix} 8 & 3 & 4 \\ -7 & 0 & -6 \\ 7 & 1 & 4 \end{vmatrix}}{|\mathbf{A}|} = \frac{-22}{22} = -1$$

RHS Vector **b** (column 2)

$$x_2 = \frac{\begin{vmatrix} 2 & 8 & 4 \\ 1 & -7 & -6 \\ -1 & 7 & 4 \end{vmatrix}}{|\mathbf{A}|} = \frac{44}{22} = 2$$

Thus, the unique solution of the system is $(-1, 2, 1)$.

2.2.6 Computing the Inverse of a Matrix

Let $\mathbf{A}$ and $\mathbf{B}$ be square matrices of order n. Then, $\mathbf{B}$ is the *inverse* of $\mathbf{A}$ (denoted $\mathbf{A}^{-1}$) and $\mathbf{A}$ is the inverse of $\mathbf{B}$ (denoted $\mathbf{B}^{-1}$) if and only if $\mathbf{AB} = \mathbf{I}_n$ and $\mathbf{BA} = \mathbf{I}_n$. If a matrix has an inverse, the inverse is unique. Only matrices with nonzero determinants have inverses. Although there are numerous methods for obtaining the inverse of a matrix, we will only discuss the Gauss-Jordan method that uses the following three *elementary row operations*.

1. Interchange rows.
2. Multiply a row by a nonzero scalar.
3. Add k times the jth row to the ith row (where k is a nonzero scalar).

Let $\mathbf{A}$ be an $m \times m$ matrix. To use the Gauss-Jordan method, adjoin an $m \times m$ identity matrix on the right to form the $m \times 2m$ structure

$$[\mathbf{A} \mid \mathbf{I}_m]$$

Then, we perform elementary row operations on the $m \times 2m$ matrix until we have $\mathbf{I}_m$ on the left. The $m \times m$ matrix on the right is then $\mathbf{A}^{-1}$.

Example 21. Let $\mathbf{A}$ be the 3×3 matrix on the left.

$$\left[\begin{array}{rrr|rrr} 1 & 2 & 3 & 1 & 0 & 0 \\ -1 & 0 & 4 & 0 & 1 & 0 \\ 0 & 2 & 2 & 0 & 0 & 1 \end{array}\right]$$

Let R_i be the ith row of the 3×6 matrix. Add R_1 to R_2 to obtain

$$\left[\begin{array}{rrr|rrr} 1 & 2 & 3 & 1 & 0 & 0 \\ 0 & 2 & 7 & 1 & 1 & 0 \\ 0 & 2 & 2 & 0 & 0 & 1 \end{array}\right]$$

Multiply R_2 by $1/2$ to obtain

$$\left[\begin{array}{rrr|rrr} 1 & 2 & 3 & 1 & 0 & 0 \\ 0 & 1 & \frac{7}{2} & \frac{1}{2} & \frac{1}{2} & 0 \\ 0 & 2 & 2 & 0 & 0 & 1 \end{array}\right]$$

Add $-2R_2$ to R_3, add $-2R_2$ to R_1, and then multiply R_3 by $-1/5$ to obtain

$$\left[\begin{array}{rrr|rrr} 1 & 0 & -4 & 0 & -1 & 0 \\ 0 & 1 & \frac{7}{2} & \frac{1}{2} & \frac{1}{2} & 0 \\ 0 & 0 & 1 & \frac{1}{5} & \frac{1}{5} & -\frac{1}{5} \end{array}\right]$$

Add $-7/2R_3$ to R_2 and add $4R_3$ to R_1 to obtain $\mathbf{A}^{-1}$ on the right

$$\left[\begin{array}{rrr|rrr} 1 & 0 & 0 & \frac{4}{5} & -\frac{1}{5} & -\frac{4}{5} \\ 0 & 1 & 0 & -\frac{1}{5} & -\frac{1}{5} & \frac{7}{10} \\ 0 & 0 & 1 & \frac{1}{5} & \frac{1}{5} & -\frac{1}{5} \end{array}\right]$$

Because

$$\mathbf{A}^{-1}\mathbf{A} = \mathbf{A}\mathbf{A}^{-1} = \mathbf{I}$$

the inverse of matrix **A** can be verified by multiplying it by **A**. In the case of Example 21,

$$\mathbf{A}\mathbf{A}^{-1} = \begin{bmatrix} 1 & 2 & 3 \\ -1 & 0 & 4 \\ 0 & 2 & 2 \end{bmatrix} \begin{bmatrix} \frac{4}{5} & -\frac{1}{5} & -\frac{4}{5} \\ -\frac{1}{5} & -\frac{1}{5} & \frac{7}{10} \\ \frac{1}{5} & \frac{1}{5} & -\frac{1}{5} \end{bmatrix} = \begin{bmatrix} 1 & 0 & 0 \\ 0 & 1 & 0 \\ 0 & 0 & 1 \end{bmatrix}$$

Inverses have the following two additional properties:

1. Let $\mathbf{A}^{-1}$ exist. Then, $(\mathbf{A}^{-1})^{-1} = \mathbf{A}$.
2. Let **A** and **B** be square matrices of the same size such that $|\mathbf{A}| \neq 0$ and $|\mathbf{B}| \neq 0$. Then, $(\mathbf{AB})^{-1} = \mathbf{B}^{-1}\mathbf{A}^{-1}$.

2.2.7 Solving Systems of Linear Equations Using Inverses

Starting with $\mathbf{Ax} = \mathbf{b}$, we have

$$\begin{aligned} \mathbf{A}^{-1}\mathbf{Ax} &= \mathbf{A}^{-1}\mathbf{b} \quad (\text{premultiply both sides by } \mathbf{A}^{-1}) \\ \mathbf{Ix} &= \mathbf{A}^{-1}\mathbf{b} \quad (\text{recognize that } \mathbf{A}^{-1}\mathbf{A} = \mathbf{I}) \\ \mathbf{x} &= \mathbf{A}^{-1}\mathbf{b} \quad (\text{recognize that } \mathbf{Ix} = \mathbf{x}) \end{aligned}$$

Thus, we can obtain the solution to a system of linear equations by simply premultiplying the RHS vector **b** by $\mathbf{A}^{-1}$.

Example 22. Consider the following system of linear equations

$$\begin{aligned} x_1 + 2x_2 + 3x_3 &= 1 \\ -x_1 \qquad + 4x_3 &= -3 \\ 2x_2 + 2x_3 &= -2 \end{aligned}$$

Since we know $\mathbf{A}^{-1}$ from Example 21,

$$\mathbf{x} = \mathbf{A}^{-1}\mathbf{b} = \begin{bmatrix} \frac{4}{5} & -\frac{1}{5} & -\frac{4}{5} \\ -\frac{1}{5} & -\frac{1}{5} & \frac{7}{10} \\ \frac{1}{5} & \frac{1}{5} & -\frac{1}{5} \end{bmatrix} \begin{bmatrix} 1 \\ -3 \\ -2 \end{bmatrix} = \begin{bmatrix} 3 \\ -1 \\ 0 \end{bmatrix}$$

2.2.8 Vectors and Points in *n*-Space

We can consider ordered *n*-tuples $(x_1, x_2, \ldots, x_n)$ and vectors $\mathbf{x} = [x_1\ x_2\ \ldots\ x_n]$ of length n to be equivalent. Thus, we can represent vectors as either points in R^n (hence, we can write $\mathbf{x} \in R^n$) or as arrows from the origin to $(x_1, x_2, \ldots, x_n) \in R^n$. When considered an arrow, a vector has *direction* and *magnitude*. The direction

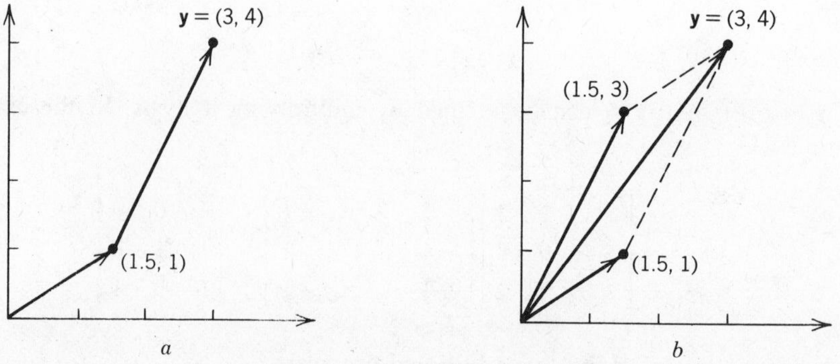

Figure 2.3. Graphs of Example 23.

of a vector is established by the arrow's orientation away from the origin. The magnitude of the vector is a measure of the distance between $\mathbf{0} \in R^n$ and $\mathbf{x} \in R^n$. When multiplying a vector by a scalar, the magnitude of the vector is multiplied by the absolute value of the scalar. If the scalar is negative, the direction reverses.

When graphically portraying the addition and subtraction of vectors, we can either use the "tail-to-head" or "completion of the parallelogram" methods. In the tail-to-head method, one of the vectors is drawn emanating from the origin. Then, the others are simply drawn tail-to-head (preserving each vector's length and direction). In the completion of the parallelogram method, the vectors are drawn, two at a time, emanating from the origin. Then, by "completing the parallelogram" the sum of the two vectors is obtained.

Example 23. In Fig. 2.3a, we graphically determine

$$\mathbf{y} = .5\begin{bmatrix} 3 \\ 2 \end{bmatrix} + 1.5\begin{bmatrix} 1 \\ 2 \end{bmatrix}$$

according to the tail-to-head method. In Fig. 2.3b, $\mathbf{y}$ is determined according to the completion-of-the-parallelogram method.

2.2.9 Linear and Convex Combinations of Vectors

Consider the vectors $\mathbf{a}^1, \mathbf{a}^2, \ldots, \mathbf{a}^n \in R^m$ and the scalars $\alpha_i \in R$ for $1 \leqq i \leqq n$. Then, the weighted-sum

$$\alpha_1\mathbf{a}^1 + \alpha_2\mathbf{a}^2 + \cdots + \alpha_n\mathbf{a}^n$$

is a *linear combination* of the vectors $\mathbf{a}^1, \mathbf{a}^2, \ldots, \mathbf{a}^n$. Depending on the weights α_i, different adjectives are applied to different linear combinations. For example, a *strictly positive* linear combination of vectors is formed when $\alpha_i > 0$ for all i; a *nonnegative* linear combination of vectors is formed when $\alpha_i \geqq 0$ for all i, and so

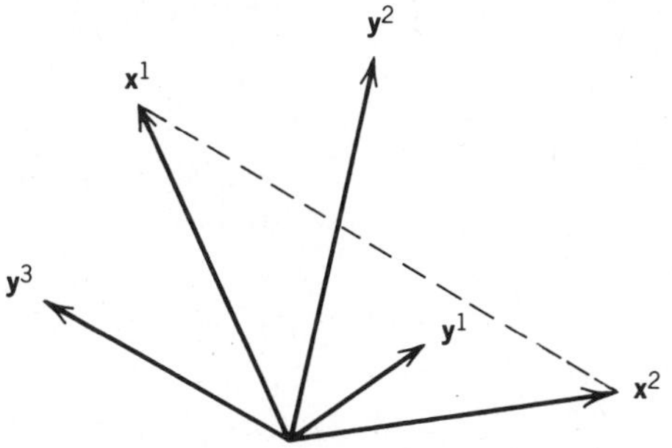

Figure 2.4. Graph of Example 24.

forth. Linear combinations of vectors

$$\sum_{i=1}^{n} \lambda_i \mathbf{a}^i \quad \text{where} \quad \lambda_i \geqq 0, \quad \sum_{i=1}^{n} \lambda_i = 1$$

whose weights are nonnegative and sum to one are *convex combinations*. If each λ_i in the above were greater than zero, we would have a *strictly convex* combination.

Example 24. In Fig. 2.4, the set of all convex combinations of $\mathbf{x}^1$ and $\mathbf{x}^2$ is the dashed line. The linear combination producing $\mathbf{y}^1$ has positive weights that sum to less than one. The linear combination producing $\mathbf{y}^2$ has positive weights that sum to more than one. If any of the weights are less than zero, the resulting linear combination lies outside the "cone" defined by $\mathbf{x}^1$ and $\mathbf{x}^2$ as shown, for instance, by $\mathbf{y}^3$.

2.2.10 Linear Independence

A set of vectors $\{\mathbf{a}^1, \mathbf{a}^2, \ldots, \mathbf{a}^n\}$, where $\mathbf{a}^i \in R^m$ for all i, is *linearly independent* if and only if the only set of scalars $\{\alpha_1, \alpha_2, \ldots, \alpha_n\}$ for which

$$\alpha_1 \mathbf{a}^1 + \alpha_2 \mathbf{a}^2 + \cdots + \alpha_n \mathbf{a}^n = \mathbf{0} \in R^m$$

is the set in which $\alpha_1 = \alpha_2 = \cdots = \alpha_n = 0$. Otherwise, the set of vectors $\{\mathbf{a}^1, \mathbf{a}^2, \ldots, \mathbf{a}^n\}$ is *linearly dependent*.

Example 25. Consider $\mathbf{x}^1 = (1, 0, 0)$, $\mathbf{x}^2 = (0, -1, 0)$, and $\mathbf{x}^3 = (0, 9, -1)$. The set of vectors $\{\mathbf{x}^1, \mathbf{x}^2, \mathbf{x}^3\}$ is linearly independent because the only set of weights for which

$$\alpha_1 \mathbf{x}^1 + \alpha_2 \mathbf{x}^2 + \alpha_3 \mathbf{x}^3 = \mathbf{0} \in R^3$$

is the set in which $\alpha_1 = \alpha_2 = \alpha_3 = 0$. Consider $\mathbf{y}^1 = (1, 2, 3)$ and $\mathbf{y}^2 = (0, 0, 0)$.

The set of vectors $\{\mathbf{y}^1, \mathbf{y}^2\}$ is linearly dependent because

$$\alpha_1\mathbf{x}^1 + \alpha_2\mathbf{x}^2 = \mathbf{0} \in R^3$$

whenever $\alpha_1 = 0$ and $\alpha_2 \neq 0$. Thus, any set of vectors containing the null vector is linearly dependent.

When a set of vectors $\{\mathbf{a}^1, \mathbf{a}^2, \ldots, \mathbf{a}^n\}$ is linearly dependent, at least one vector exists in the set that can be expressed as a linear combination of the others. Such a vector is linearly dependent *on* the set $\{\mathbf{a}^1, \mathbf{a}^2, \ldots, \mathbf{a}^n\}$. A vector $\mathbf{b} \in R^m$ that cannot be written as a linear combination of the vectors in $\{\mathbf{a}^1, \mathbf{a}^2, \ldots, \mathbf{a}^n\}$ is linearly independent *of* the set of vectors $\{\mathbf{a}^1, \mathbf{a}^2, \ldots, \mathbf{a}^n\}$. It is clear that any subset of a set of linearly independent vectors is linearly independent, and any superset of a set of linearly dependent vectors is linearly dependent. An $m \times m$ matrix $\mathbf{A}$ whose columns comprise a linearly independent set of vectors is *invertible* (has an inverse because $|\mathbf{A}| \neq 0$).

Whenever a vector $\mathbf{b}$ is written as a linear combination of a set of linearly independent vectors $\{\mathbf{a}^1, \mathbf{a}^2, \ldots, \mathbf{a}^n\}$, the weights α_i in

$$\mathbf{b} = \sum_{i=1}^{n} \alpha_i \mathbf{a}^i$$

are unique. This is proven by supposing that there is another different set of weights $\{\beta_1, \beta_2, \ldots, \beta_n\}$ such that

$$\mathbf{b} = \sum_{i=1}^{n} \beta_i \mathbf{a}^i$$

By equating and rearranging we obtain

$$\sum_{i=1}^{n} (\alpha_i - \beta_i)\mathbf{a}^i = \mathbf{0}$$

Since the $\mathbf{a}^i$ are linearly independent, $\alpha_i = \beta_i$ for all i, which is a contradiction. Thus, there is only one way to write a given vector $\mathbf{b}$ as a linear combination of a particular set of linearly independent vectors $\{\mathbf{a}^1, \mathbf{a}^2, \ldots, \mathbf{a}^n\}$.

2.2.11 Rank of a Matrix

The *rank* of a matrix $\mathbf{A}$ (denoted $r(\mathbf{A})$) is given by the maximum number of linearly independent rows in $\mathbf{A}$. In a matrix, the maximum number of linearly independent rows equals the maximum number of linearly independent columns. Thus, the minimum dimension of a matrix places an upper bound on the rank of the matrix. Therefore, the largest number of vectors in R^m that can be linearly independent of one another is m. Consequently, any set of $m + 1$ or more vectors in R^m is linearly dependent.

One method for determining the rank of a matrix is to use elementary row operations to transform the matrix in question into *echelon form* as below.

$$\begin{bmatrix} 0 & 2 & 1 & 5 & 2 & -1 \\ 0 & 0 & 0 & 3 & 0 & -3 \\ 0 & 0 & 0 & 0 & -4 & 0 \\ 0 & 0 & 0 & 0 & 0 & 2 \\ 0 & 0 & 0 & 0 & 0 & 0 \end{bmatrix}$$

A matrix is in echelon form when the number of leading zeros in each row increases, row by row, until either the bottom of the matrix has been reached or only rows of zeros remain. Once in echelon form, the rank of the original matrix is given by the number of nonzero rows. Another way to find the rank of a matrix $\mathbf{A}$ is to determine the order of the largest square submatrix in $\mathbf{A}$ whose determinant is nonzero.

Example 26. Consider the set of vectors $\{\mathbf{b}^1, \mathbf{b}^2, \mathbf{b}^3\}$ that comprise the columns of

$$\mathbf{B} = \begin{bmatrix} 1 & 2 & 3 \\ 0 & -1 & 4 \\ -3 & -3 & -21 \end{bmatrix}$$

The rank of $\mathbf{B} = 2$ because $|\mathbf{B}| = 0$, but there is a 2×2 submatrix of $\mathbf{B}$ (e.g., the submatrix whose elements are b_{21}, b_{23}, b_{31}, and b_{33}) whose determinant is nonzero.

2.2.12 Bases

A *basis* in R^m is any set $\{\mathbf{a}^1, \mathbf{a}^2, \ldots, \mathbf{a}^m\}$ of m linearly independent m-vectors.

Example 27. Consider $\mathbf{a}^1 = (1, 0, 0)$, $\mathbf{a}^2 = (0, 0, 1)$, and $\mathbf{a}^3 = (0, 1, 0)$. Then, the set $\{\mathbf{a}^1, \mathbf{a}^2, \mathbf{a}^3\}$ is a basis in R^3. Consider $\mathbf{b}^1 = (1, 4, -1, 0)$ and $\mathbf{b}^2 = (1, 0, 0, 0)$. Although $\{\mathbf{b}^1, \mathbf{b}^2\}$ is a linearly independent set of vectors, the set $\{\mathbf{b}^1, \mathbf{b}^2\}$ is not a basis in R^4 because there are not enough $\mathbf{b}^i$.

The interesting property of a basis $\{\mathbf{a}^1, \mathbf{a}^2, \ldots, \mathbf{a}^m\}$ in R^m is that the basis vectors $\mathbf{a}^1, \mathbf{a}^2, \ldots, \mathbf{a}^m$ *span* all of m-space. That is, any vector $\mathbf{b} \in R^m$ can be expressed as a linear combination of the basis vectors $\mathbf{a}^1, \mathbf{a}^2, \ldots, \mathbf{a}^m$ as in

$$\alpha_1 \mathbf{a}^1 + \alpha_2 \mathbf{a}^2 + \cdots + \alpha_m \mathbf{a}^m = \mathbf{b}$$

How would the linear combination weights α_i be found? Invert the matrix whose columns are the basic vectors $\mathbf{a}^1, \mathbf{a}^2, \ldots, \mathbf{a}^m$. Then, by premultiplying the vector $\mathbf{b} \in R^m$ by the inverse, the α_i weights are determined.

2.2.13 Number of Solutions of a System of Linear Equations

A system of linear equations $\mathbf{Ax} = \mathbf{b}$ in which $\mathbf{A}$ is $m \times n$ is either

1. *Inconsistent* (does not have a solution).
2. *Consistent* with a *unique* solution.
3. *Consistent* with an *infinite* number of solutions.

Which of the mutually exclusive cases pertain is determined by applying the notion of rank to the coefficient matrix $\mathbf{A}$ and to the augmented matrix $(\mathbf{A}, \mathbf{b})$.

1. If $r(\mathbf{A}) < r(\mathbf{A}, \mathbf{b})$, the system is *inconsistent* because $\mathbf{b}$ is linearly independent of the columns of $\mathbf{A}$. Thus, no solution vector $\mathbf{x} \in R^n$ exists because there is no linear combination of the columns of $\mathbf{A}$ that span to $\mathbf{b}$.
2. If $r(\mathbf{A}) = r(\mathbf{A}, \mathbf{b}) = n$, the system is *consistent* with a *unique* solution. The system is consistent because $r(\mathbf{A}) = r(\mathbf{A}, \mathbf{b})$. That is, the columns of $\mathbf{A}$ are able to span to $\mathbf{b}$. The solution is unique because there is only one way to write a given vector $\mathbf{b}$ as a linear combination of a particular set of linearly independent vectors $\{\mathbf{a}^1, \mathbf{a}^2, \ldots, \mathbf{a}^n\}$.
3. If $r(\mathbf{A}) = r(\mathbf{A}, \mathbf{b}) < n$, the system is, of course, *consistent* [because $r(\mathbf{A}) = r(\mathbf{A}, \mathbf{b})$], but possesses an *infinite* number of solutions. There are an infinite number of solutions because of the presence of $[n - r(\mathbf{A})]$ linearly dependent columns in $\mathbf{A}$. The variables x_i associated with such linearly dependent columns are sometimes known as *free* variables.

2.2.14 Some Further Examples of Functions

From set theory we know that a mapping $f\colon A \to B$ is a function if $A \subset R^n$ and $B \subset R^m$ because R^n and R^m are sets of real numbers. Examples of functions in which n or m (or both) are greater than one are now provided.

Example 28

a. Let $\mathbf{x} \in S$, $S \subset R^n$, and $\mathbf{z} \in R^k$. Then, the mapping $f\colon S \to R^k$ defined by
$$\mathbf{Cx} = \mathbf{z}$$
(where $\mathbf{C}$ is $k \times n$) is a function. The graph of f is a subset of R^{n+k}.

b. Let $S \subset \{\mathbf{x} \in R^n \mid x_i \geqq 0 \text{ for all } i\}$ and let $Z = \{\mathbf{z} \in R^k \mid \mathbf{z} = \mathbf{Cx}, \mathbf{x} \in S\}$. Then, the mapping $U\colon Z \to R$ defined by
$$U(\mathbf{z}) = z_1 z_2 \cdots z_k$$
is a function. The graph of U is a subset of R^{k+1}.

c. Let $S \subset R^n$ and $q < n$. Then, the function $\pi\colon S \to R^q$ that defines the image of $(x_1, x_2, \ldots, x_q, \ldots, x_n) \in S$ to be $(x_1, x_2, \ldots, x_q) \in R^q$ is called a *q-space projection*.

2.3 PROPERTIES OF POINTS AND SETS IN R^n

This section reviews some concepts pertaining to points and sets in R^n beyond those discussed in Section 2.1.

2.3.1 Open and Closed Sets

With radius $\delta > 0$, an n-dimensional *open hypersphere* centered at $\bar{\mathbf{x}} \in R^n$ is the set

$$H_s = \left\{ \mathbf{x} \in R^n | (x_1 - \bar{x}_1)^2 + (x_2 - \bar{x}_2)^2 + \cdots + (x_n - \bar{x}_n)^2 < \delta^2 \right\}$$

In R an open hypersphere is an open interval. In R^2 an open hypersphere is a disk that does not contain its periphery. In R^3 an open hypersphere is a sphere that does not contain its surface, and so forth.

A point $\bar{\mathbf{x}} \in S \subset R^n$ is an *interior* point of S if and only if $\bar{\mathbf{x}}$ belongs to an n-dimensional open hypersphere H_s centered at $\bar{\mathbf{x}}$ such that $H_s \subset S$. A set is *open* if all of its members are interior points. If S is an open set, then its *complement* (i.e., the set consisting of all points in R^n that are not in S) is a *closed* set, and vice versa. A closed set contains all of its boundary points. A *boundary* point of $S \subset R^n$ is a point $\bar{\mathbf{x}} \in R^n$ such that every n-dimensional open hypersphere centered at $\bar{\mathbf{x}}$ contains points in S and points not in S. The *closure* of S is the union of S and its set of boundary points.

The null set $\varnothing$ and R^n are both open *and* closed by the following argument. Since all points in R^n are interior points, R^n is open and thus $\varnothing$ is closed. But since R^n contains all of its boundary points, R^n is closed and thus $\varnothing$ is open.

A set $S \subset R^n$ is *bounded* if and only if there exists an n-dimensional hypersphere that contains S. If a set is not bounded, it is *unbounded*.

Example 29. The set of Fig. 2.5*a* is closed and bounded but contains no interior points. The set of Fig. 2.5*b* is bounded but is neither open nor closed. The set of Fig. 2.5*c* is closed and unbounded. The set of Fig. 2.5*d* is open and unbounded.

2.3.2 Convex Sets and Extreme Points

A set $S \subset R^n$ is *convex* if and only if for any $\mathbf{x}^1, \mathbf{x}^2 \in S$ the point $(\lambda\mathbf{x}^1 + (1 - \lambda)\mathbf{x}^2) \in S$ for all $\lambda \in [0, 1]$. Otherwise, the set is *nonconvex*. That is, a set is convex whenever all points on the line connecting any two points in the set are also in the set. Consequently, a convex set cannot have any "dents," "depressions," or "holes." The null set $\varnothing$ is defined to be convex.

If we allow $S \subset R^n$, a point $\bar{\mathbf{x}} \in S$ is an *extreme point* (sometimes called a *vertex*) of S if and only if points $\mathbf{x}^1, \mathbf{x}^2 \in S$, $\mathbf{x}^1 \neq \mathbf{x}^2$ do not exist such that $\bar{\mathbf{x}} = \lambda\mathbf{x}^1 + (1 - \lambda)\mathbf{x}^2$ for some $\lambda \in (0, 1)$. That is, an extreme point cannot be expressed as a strictly convex combination of any two different points in the set in question. Because of its definition, an extreme point cannot be an interior point. Therefore, all extreme points are boundary points.

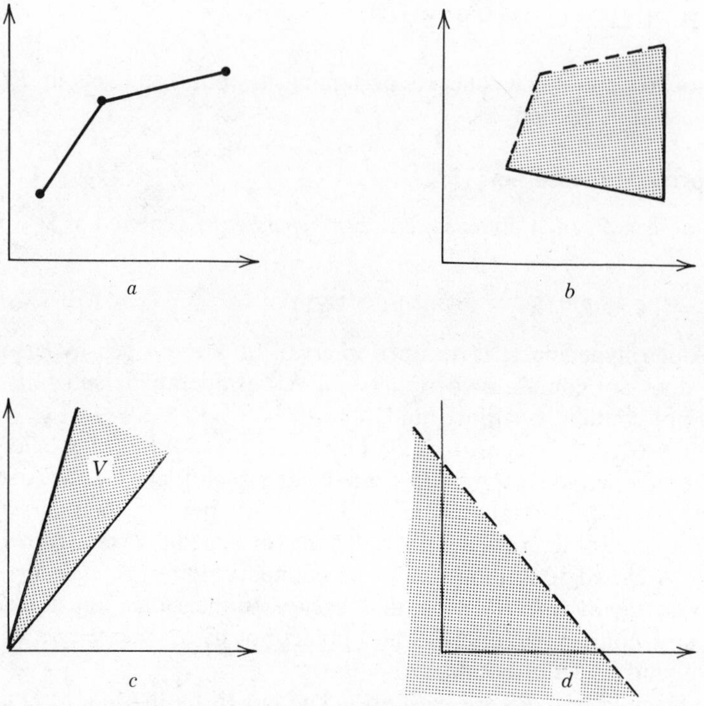

Figure 2.5. Graphs of Example 29.

Example 30. The set of Fig. 2.6*a* is convex with five extreme points as indicated. The set of Fig. 2.6*b* is convex with an infinite number of extreme points (all points along its boundary). The sets of Figs. 2.6*c* and *d* are nonconvex. We do not talk about extreme points of nonconvex sets.

The *convex combination* operator γ is used in the sense that the set of all convex combinations of $\mathbf{x}^1, \mathbf{x}^2, \ldots, \mathbf{x}^q$ is written

$$\gamma(\mathbf{x}^1, \mathbf{x}^2, \ldots, \mathbf{x}^q) \quad \text{or} \quad \mathop{\gamma}_{i=1}^{q} (\mathbf{x}^i)$$

The *unbounded line segment* operator μ is used in the sense that the unbounded line segment emanating from $\mathbf{x} \in R^n$ in the direction $\mathbf{v} \in R^n$ is written

$$\mu(\mathbf{x}, \mathbf{v})$$

We note that a convex set is the set of all convex combinations of its extreme points and points along its unbounded (one-dimensional) *edges*.

Example 31. In Figs. 2.7*a* and *b*, the nonconvex set X and the unbounded convex set Y are respectively given by

$$X = \gamma(\mathbf{x}^1, \mathbf{x}^2) \cup \mathop{\gamma}_{i=2}^{4} (\mathbf{x}^i)$$

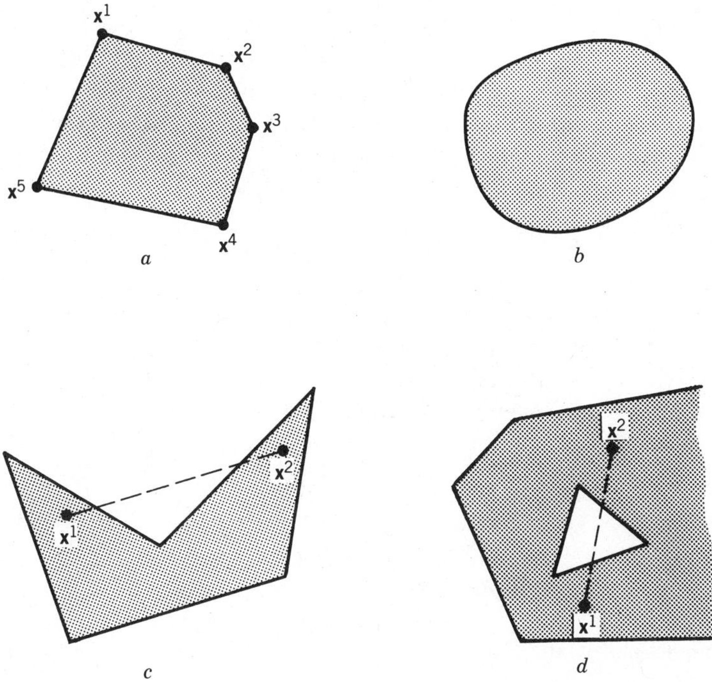

Figure 2.6. Graphs of Example 30.

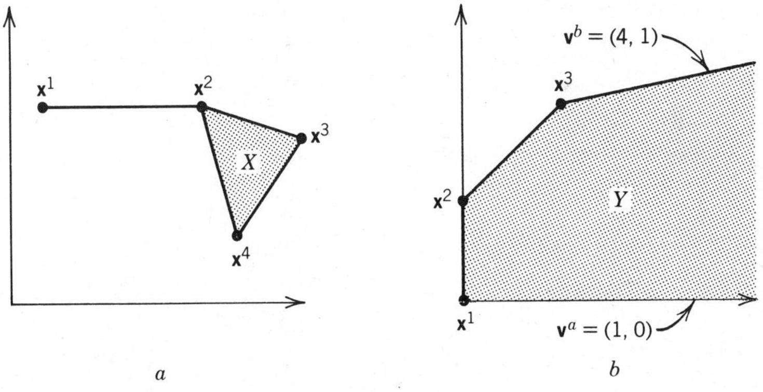

Figure 2.7. Graphs of Example 31.

and

$$Y = \gamma(\mathbf{x}^2, \mathbf{y}, \mathbf{z})$$

where $\mathbf{y} \in \mu(\mathbf{x}^1, \mathbf{v}^a)$ and $\mathbf{z} \in \mu(\mathbf{x}^3, \mathbf{v}^b)$.

With convex sets, extreme points that are connected by an edge are *adjacent*. The *path of adjacent extreme points* from $\mathbf{x}^1$ to $\mathbf{x}^3$ in Fig. 2.7*b* is $\{\mathbf{x}^1, \mathbf{x}^2, \mathbf{x}^3\}$. The *path of edges* from $\mathbf{x}^1$ to $\mathbf{x}^3$ is $\gamma(\mathbf{x}^1, \mathbf{x}^2)$ and $\gamma(\mathbf{x}^2, \mathbf{x}^3)$.

2.3.3 Images of Convex Sets

If we let $M \subset R^n$, the *convex hull* of M is the intersection of all convex sets containing M. The convex hull of M is the smallest convex set containing M. A *polyhedral* set is the convex hull of a finite set of points and a finite number of unbounded line segments. A polyhedral set is a *polyhedron* if it is the convex hull of a finite set of points. Whereas a polyhedral set need not be bounded, a polyhedron is bounded.

Let $S \subset R^n$ and $f\colon S \to R^k$. Then, if S is convex and f is linear, the range $f(S)$, the image of S under f, is convex. If S is polyhedral, the image of S under f is given by the set of all convex combinations of the images of the extreme points and points along the unbounded edges of S.

Example 32. Let $f\colon S \to R^2$ be defined by $\mathbf{Cx} = \mathbf{z}$ where S is as graphed in Fig. 2.8*a* and

$$\mathbf{C} = \begin{bmatrix} 1 & 2 \\ 1 & -1 \end{bmatrix}$$

Then, with $\mathbf{Cx}^i = \mathbf{z}^i$, the image of S under f (denoted Z) is given by $\gamma(\mathbf{z}^1, \mathbf{z}^2, \mathbf{z}^3, \mathbf{z}^4)$ as graphed in Fig. 2.8*b*.

2.3.4 Dimensionality and Set Addition

Let $\bar{\mathbf{x}} \in S \subset R^n$. Then, the maximal number of linearly independent vectors in

$$\{(\mathbf{x} - \bar{\mathbf{x}}) \in R^n \,|\, \mathbf{x} \in S\}$$

specifies the *dimensionality* of S.

Let X and Y be sets in R^n. Then, the *set addition* of X and Y (denoted $X \oplus Y$) is given by

$$X \oplus Y = \{\mathbf{z} \in R^n \,|\, \mathbf{z} = \mathbf{x} + \mathbf{y}, \quad \mathbf{x} \in X, \quad \mathbf{y} \in Y\}$$

That is, every point in X is added to every point in Y.

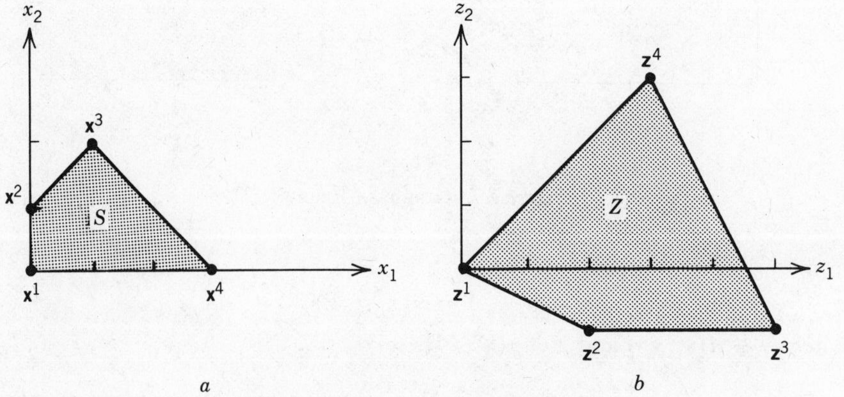

Figure 2.8. Graphs of Example 32.

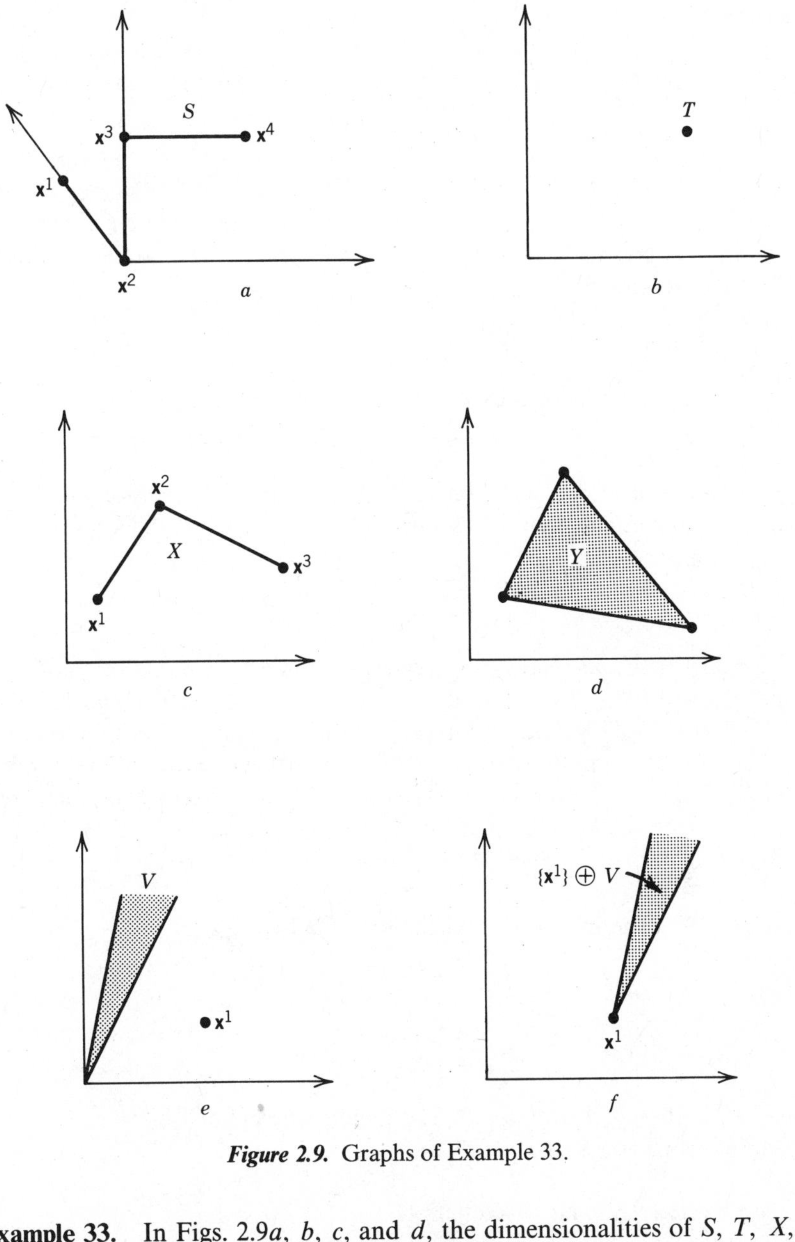

Figure 2.9. Graphs of Example 33.

Example 33. In Figs. 2.9a, b, c, and d, the dimensionalities of S, T, X, and Y are 3, 0, 2, and 2, respectively. For instance, the individual line segment $\gamma(\mathbf{x}^1, \mathbf{x}^2)$ in Fig. 2.9c is of dimensionality 1. In Figs. 2.9e and f, the set addition of $\{\mathbf{x}^1\}$ and V effects a *translation* of V to $\mathbf{x}^1$.

Let $S \subset R^n$ be closed and convex with q extreme points $\mathbf{x}^i$ and r different unbounded edge directions $\mathbf{v}^j$. Then, $\bar{\mathbf{x}} \in S$ if and only if $\bar{\mathbf{x}}$ can be written as the sum of two terms in which the first term is some convex combination of the q extreme points and the second term is some nonnegative linear combination of the r unbounded edge directions. Thus, S can be written as the following set addition

$$\underset{i=1}{\overset{q}{\gamma}}\,(\mathbf{x}^i) \oplus \left\{\mathbf{x} \in R^n \,|\, \mathbf{x} = \sum_{j=1}^{r} \alpha_j \mathbf{v}^j; \quad \alpha_j \geqq 0 \quad \text{for all } j\right\}$$

In this way, a convex set can be *characterized* by its extreme points and unbounded edge directions.

2.3.5 Hyperplanes and Half-Spaces

The *hyperplane* in R^n defined by $\mathbf{c} \neq \mathbf{0} \in R^n$ and $\bar{z} \in R$ fixed is the set

$$\{\mathbf{x} \in R^n \,|\, \mathbf{c}^T\mathbf{x} = \bar{z}\}$$

A hyperplane is a closed and convex $n - 1$ dimensional subset of R^n. With regard to the hyperplane defined by $\mathbf{c} \neq \mathbf{0} \in R^n$ and $\bar{z} \in R$, the sets

$$\{\mathbf{x} \in R^n \,|\, \mathbf{c}^T\mathbf{x} \leqq \bar{z}\}$$

$$\{\mathbf{x} \in R^n \,|\, \mathbf{c}^T\mathbf{x} < \bar{z}\}$$

are *closed* and *open half-spaces*, respectively. Note that the intersection of the two half-spaces $\{\mathbf{x} \in R^n \,|\, \mathbf{c}^T\mathbf{x} \leqq \bar{z}\}$ and $\{\mathbf{x} \in R^n \,|\, \mathbf{c}^T\mathbf{x} \geqq \bar{z}\}$ defines the hyperplane $\{\mathbf{x} \in R^n \,|\, \mathbf{c}^T\mathbf{x} = \bar{z}\}$. Sets that are formed by the intersection of a finite number of closed half-spaces are polyhedral (with "flat sides") and have a finite number of extreme points and unbounded edges. Let $S \subset R^n$ and H be a hyperplane in R^n. Then, H is a *supporting* hyperplane of S (i.e., H *supports* S) at $\bar{\mathbf{x}} \in S$ if $\bar{\mathbf{x}} \in H$ and all of S is in one of the closed half-spaces defined by H.

Example 34. The intersection of the three half-spaces

$$\{\mathbf{x} \in R^2 \,|\, 2x_1 + x_2 \geqq 12\}$$

$$\{\mathbf{x} \in R^2 \,|\, x_2 \leqq 6\}$$

$$\{\mathbf{x} \in R^2 \,|\, x_2 \geqq 0\}$$

in Fig. 2.10 is the polyhedral set S. The hyperplane defined by $-x_1 + 3x_2 = 15$ (dashed line) supports S at $\mathbf{x}^2 = (3, 6)$.

2.3.6 Connected Sets and Discrete Sets

Let $S \subset R^n$. Then, S is *disconnected* if and only if there exist two open sets D and H such that

1. $D \cap S$ and $H \cap S$ are nonempty disjoint sets.
2. $S = (D \cap S) \cup (H \cap S)$.

Otherwise S is *connected*. The null set $\varnothing$ is considered a connected set.

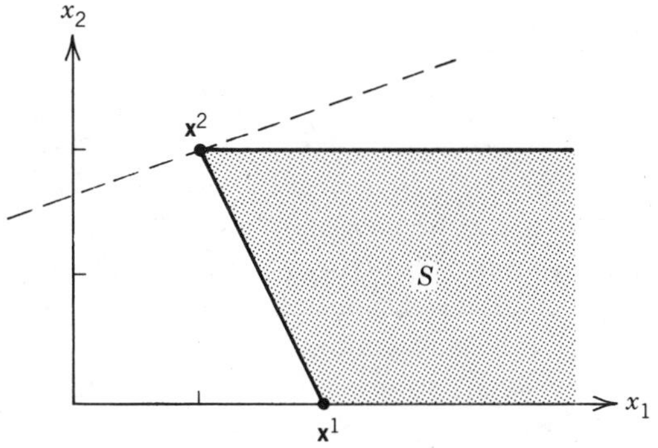

Figure 2.10. Graph of Example 34.

Example 35. Set S in Fig. 2.11 is not connected because D and H are open sets such that conditions 1 and 2 above hold.

As an illustration of connectedness, let $A = \{\mathbf{x} \in R^n | \mathbf{c}^T\mathbf{x} \leqq \bar{z}\}$ and $B = \{\mathbf{x} \in R^n | \mathbf{c}^T\mathbf{x} > \bar{z}\}$. Although $A \cap B = \varnothing$, they are connected because it is not possible to enclose A in an open set M and B in an open set N, such that M and N are disjoint.

A connected set is sometimes said to be a *continuous* set. A *discrete* set is a set for which there exists a one-to-one mapping into the set of all positive integers. If a discrete set contains an infinite number of elements, it is said to be *countably infinite*. If a discrete set contains more than one point, it is disconnected. The null set $\varnothing$ is considered a discrete set.

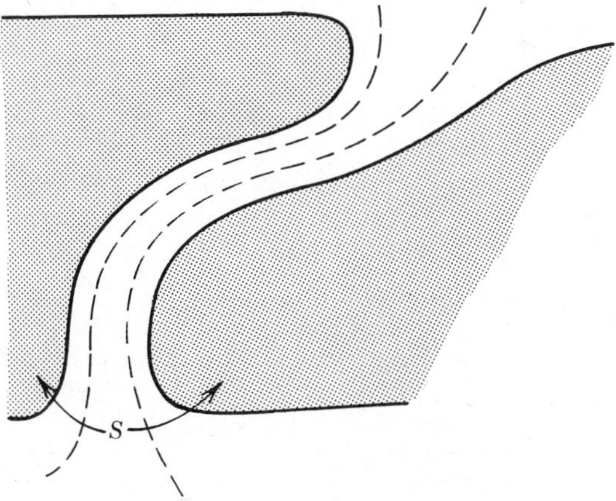

Figure 2.11. Graph of Example 35.

2.4 CONES

Let $\mathbf{v} \in V \subset R^n$, $V \neq \varnothing$. Then, V is a *cone* if and only if $\alpha\mathbf{v} \in V$ for all scalars $\alpha \geqq 0$. The origin $\mathbf{0} \in R^n$ is contained in every cone. All cones are connected sets. Moreover, other than for the *singleton* set containing the origin, all cones are unbounded sets. However, a cone need not be convex.

A half-ray emanating from the origin is a cone. As other examples, the closed half-space $\{\mathbf{x} \in R^n | \mathbf{c}^T\mathbf{x} \leqq 0\}$ is a convex cone but the open half-space $\{\mathbf{x} \in R^n | \mathbf{c}^T\mathbf{x} < 0\}$ is not a cone because it does not contain the origin. The *nonnegative orthant* $\{\mathbf{x} \in R^n | \mathbf{x} \geqq \mathbf{0}\}$ is a convex cone. R^n is another example of a convex cone.

2.4.1 Generators

Consider a set of n-vectors $\{\mathbf{v}^1, \mathbf{v}^2, \ldots, \mathbf{v}^k\}$ and the set V where

$$V = \left\{ \mathbf{v} \in R^n | \mathbf{v} = \sum_{i=1}^{k} \alpha_i \mathbf{v}^i, \quad \alpha_i \geqq 0 \right\}$$

V consists of all nonnegative linear combinations of the $\mathbf{v}^i$ and is the convex cone *generated* by the set $\{\mathbf{v}^1, \mathbf{v}^2, \ldots, \mathbf{v}^k\}$. The $\mathbf{v}^i$ are said to be *generators* of V. The only cone for which the set of generators is unique is $\{\mathbf{0} \in R^n\}$.

Let $\{\mathbf{v}^1, \mathbf{v}^2, \ldots, \mathbf{v}^k\}$ be a set of generators for the convex cone V and let $\mathbf{v}^r \in \{\mathbf{v}^1, \mathbf{v}^2, \ldots, \mathbf{v}^k\}$. Then, $\mathbf{v}^r$ is a *nonessential* generator if without $\mathbf{v}^r$ the other $\mathbf{v}^i$ are still able to generate V. A nonessential generator is one that can be expressed as a nonnegative linear combination of other generators and an *essential* generator is one that cannot. The smallest number of essential generators that generate a cone is called the *minimal number of generators* for that cone.

Example 36. In Fig. 2.12a, $\mathbf{v}^2$ is a nonessential generator of V because V is just as ably generated by $\{\mathbf{v}^1, \mathbf{v}^3\}$ as it is by $\{\mathbf{v}^1, \mathbf{v}^2, \mathbf{v}^3\}$. The minimal number of generators is two. In Fig. 2.12b, either $\mathbf{v}^2$ or $\mathbf{v}^3$ is nonessential but not both at the

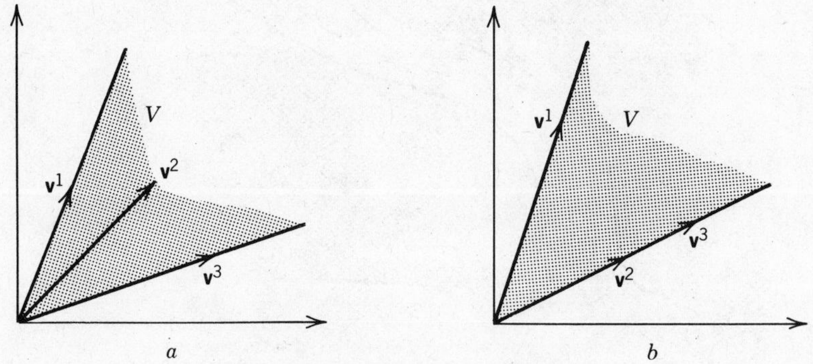

Figure 2.12. Graphs of Example 36.

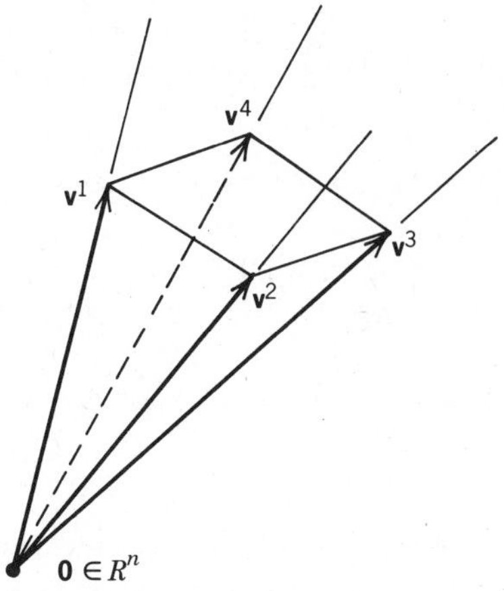

Figure 2.13. Graph of Example 37.

same time. For this cone it suffices for the sets $\{\mathbf{v}^1, \mathbf{v}^2\}$ or $\{\mathbf{v}^1, \mathbf{v}^3\}$ to be minimal sets of generators.

2.4.2 Dimensionality of a Cone

The *dimensionality* of a cone $V \subset R^n$ is given by the number of linearly independent vectors in V. For example, the dimensionality of the singleton cone $\{\mathbf{0} \in R^n\}$ is 0, and the dimensionality of the convex cone generated by a set of k linearly independent vectors is k. A method for determining the dimensionality of a cone is to compute the rank of the matrix whose rows (or columns) are generators of the cone.

A cone that has an extreme point (vertex) is said to be a *pointed* cone. Cones such as the closed half-space $\{\mathbf{x} \in R^n | \mathbf{c}^T\mathbf{x} \leqq 0\}$, a hyperplane, and R^n are *nonpointed* cones because they do not possess vertices.

Example 37. In Fig. 2.13, we see that just because every member in a set of generators (e.g., $\{\mathbf{v}^1, \mathbf{v}^2, \mathbf{v}^3, \mathbf{v}^4\}$) is essential, this does not mean that the set of generators is linearly independent. Although the cone is of dimensionality three, the minimal number of generators is four.

2.4.3 Extreme Rays and Polyhedral Cones

Let $V \in R^n$ be a closed convex cone and $\bar{\mathbf{v}} \in V$. Then, the set

$$\{\mathbf{y} \in R^n | \mathbf{y} = \alpha\bar{\mathbf{v}}, \quad \alpha \geqq 0\}$$

is an *extreme ray* of V if and only if $\bar{\mathbf{v}} \neq \mathbf{0}$ and there *do not* exist two other vectors $\mathbf{v}^1, \mathbf{v}^2 \in V$, $\mathbf{v}^1 \neq \beta\mathbf{v}^2$ for all scalars $\beta \neq 0$, such that $\bar{\mathbf{v}} = \alpha_1\mathbf{v}^1 + \alpha_2\mathbf{v}^2$ for some scalars $\alpha_1, \alpha_2 > 0$. A one-dimensional edge of a polyhedral cone is an extreme ray of the cone. We have the following relationships among the minimal number of generators, the number of extreme rays, and the dimensionality of polyhedral cones.

1. The minimal number of generators of a cone cannot be less than the dimensionality of the cone.
2. For a pointed convex cone of dimensionality one or greater, the minimal number of generators equals the number of extreme rays.
3. For a nonpointed convex cone $V \subset R^n$, $V \neq \varnothing$, the minimal number of generators is one greater than the dimensionality of the cone.

Example 38

a. For a cone of dimensionality zero (i.e., the singleton cone $\{\mathbf{0} \in R^n\}$, there is one extreme point, no extreme rays, and the minimal number of generators is one.

b. A hyperplane passing through the origin in R^n is a cone of dimensionality $n - 1$. The hyperplane cone is polyhedral because it can be formed by the intersection of two closed half-spaces. The minimal number of generators for such a cone is n, yet the cone does not possess an extreme point or any extreme rays.

c. Consider the cone R^n. Of dimensionality n, it does not have an extreme point or any extreme rays. The following $n + 1$ vectors

$$\mathbf{v}^1 = \begin{bmatrix} 1 \\ 0 \\ 0 \\ \vdots \\ 0 \end{bmatrix} \quad \mathbf{v}^2 = \begin{bmatrix} 0 \\ 1 \\ 0 \\ \vdots \\ 0 \end{bmatrix} \quad \ldots \mathbf{v}^n = \begin{bmatrix} 0 \\ 0 \\ 0 \\ \vdots \\ 1 \end{bmatrix} \quad \mathbf{v}^{n+1} = \begin{bmatrix} -1 \\ -1 \\ -1 \\ \vdots \\ -1 \end{bmatrix}$$

suffice as a minimal set of generators because some nonnegative linear combination of the $\mathbf{v}^i$ can reach any point in R^n.

2.4.4 Polar Cones

Let $V \subset R^n$ be a cone. Then, the *nonnegative polar* of V (denoted $V^{\geqq}$) is the convex cone

$$V^{\geqq} = \{\mathbf{y} \in R^n \mid \mathbf{y}^T\mathbf{v} \geqq 0 \quad \text{for all} \quad \mathbf{v} \in V\}$$

That is, all vectors in $V^{\geqq}$ make a less than or equal to 90° angle with each of the vectors in V. $V^{\geqq}$ is convex regardless of whether or not V is convex.

Example 39. In Figs. 2.14*a*, *b*, and *c*, the cones of reference V are generated by the $\mathbf{v}^i$ and the nonnegative polars $V^{\geqq}$ are generated by the $\mathbf{y}^i$. In Fig. 2.14*d*

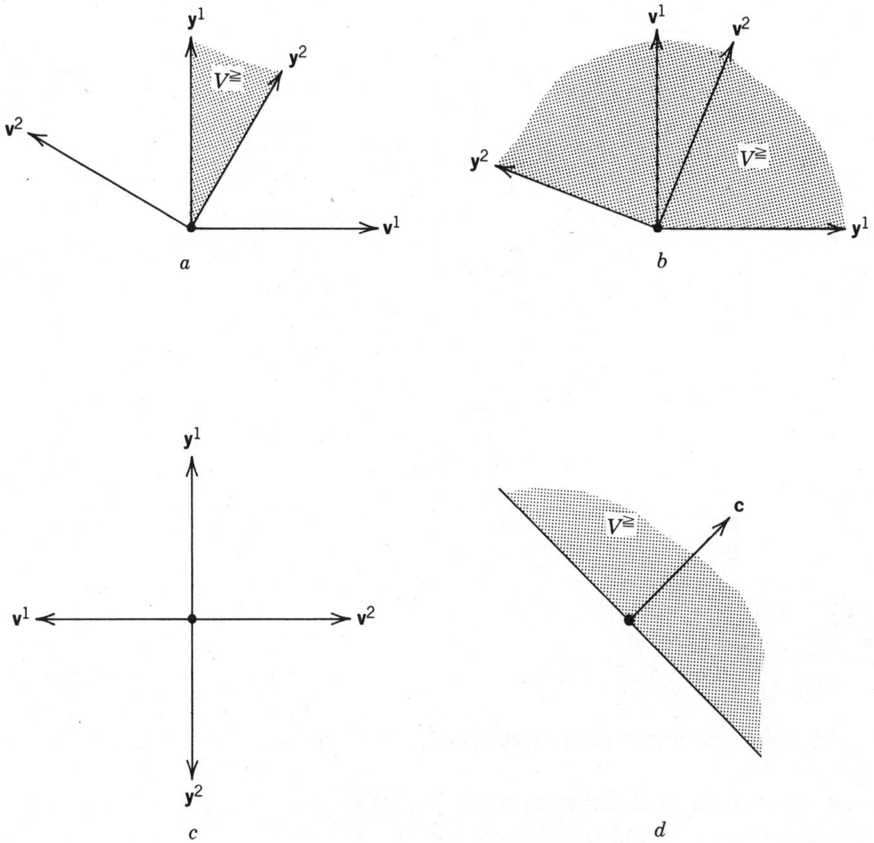

Figure 2.14. Graphs of Example 39.

the nonnegative polar of the cone generated by $\mathbf{c} \in R^n$ is the closed half-space $\{\mathbf{x} \in R^n | \mathbf{c}^T\mathbf{x} \geqq 0\}$.

Consider the convex cone $V \subset R^n$ generated by $\{\mathbf{v}^1, \mathbf{v}^2, \ldots, \mathbf{v}^k\}$. V satisfies the *null vector condition* if there exists an $\boldsymbol{\alpha} \in R^k$, $\alpha_i > 0$ for all i such that

$$\sum_{i=1}^{k} \alpha_i \mathbf{v}^i = \mathbf{0} \in R^n$$

Other than for the singleton cone $\{\mathbf{0} \in R^n\}$, a cone that satisfies the null vector condition is nonpointed. The nonnegative polar of a cone V for which the null vector condition holds is the subspace of R^n orthogonal to V as illustrated in Fig. 2.14*c*. For example, the nonnegative polar of $\{\mathbf{0} \in R^n\}$ is R^n. If V is a convex cone, then the nonnegative polar of $V^{\geqq}$ is V again.

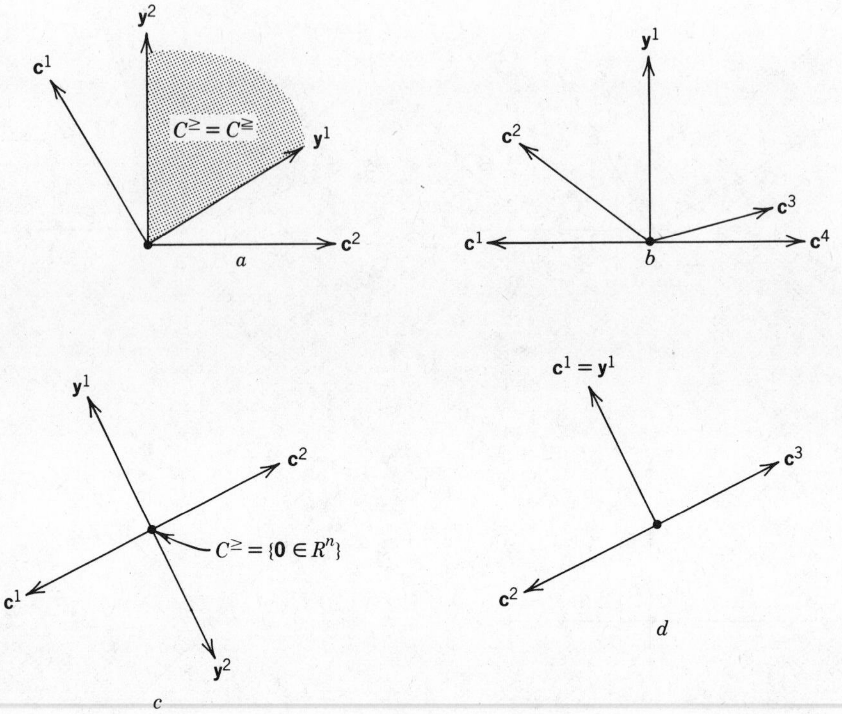

Figure 2.15. Graphs of Example 40.

Let $V \subset R^n$ be a cone generated by $\{\mathbf{v}^1, \mathbf{v}^2, \ldots, \mathbf{v}^k\}$. Then, the *semipositive polar* of V (denoted $V^{\geq}$) is the convex cone

$$V^{\geq} = \{\mathbf{y} \in R^n \,|\, \mathbf{y}^T\mathbf{v}^i \geqq 0 \quad \text{for all } i \text{ and}$$
$$\mathbf{y}^T\mathbf{v}^i > 0 \quad \text{for at least one } i\} \cup \{\mathbf{0} \in R^n\}$$

Note that for a vector $\mathbf{y}$ to be in $V^{\geq}$ it must have a positive vector product with at least one of the $\mathbf{v}^i$. The reason the origin $\{\mathbf{0} \in R^n\}$ is included is because $V^{\geq}$ would not be a cone without it.

Example 40. In Fig. 2.15, the cones of reference C are generated by the $\mathbf{c}^i$, and the nonnegative polars $C^{\geqq}$ are the cones generated by the $\mathbf{y}^i$. In Fig. 2.15*a*, $C^{\geq} = C^{\geqq}$ as shown. In Fig. 2.15*b*, $C^{\geqq}$ is the half-ray generated by $\mathbf{y}^1$ and $C^{\geq} = C^{\geqq}$. In Fig. 2.15*c*, $C^{\geqq}$ is the line generated by $\mathbf{y}^1$ and $\mathbf{y}^2$ and $C^{\geq}$ is the origin. In Fig. 2.15*d*, $C^{\geqq}$ is the half-ray generated by $\mathbf{y}^1$ and $C^{\geq} = C^{\geqq}$.

2.5 NORMS AND METRICS

Norms measure the lengths of vectors and *metrics* measure the distance between points in R^n.

2.5.1 Family of L_p-Norms

A *norm* on R^n is a function that assigns to each $\mathbf{v} \in R^n$ a scalar $\|\mathbf{v}\| \in R$ if and only if the function satisfies for all $\mathbf{w} \in R^n$ and $k \in R$ the following axioms:

1. $\|\mathbf{v}\| \geqq 0$.
2. $\|\mathbf{v}\| = 0$ if and only if $\mathbf{v} = \mathbf{0} \in R^n$.
3. $\|\mathbf{v} + \mathbf{w}\| \leqq \|\mathbf{v}\| + \|\mathbf{w}\|$.
4. $\|k\mathbf{v}\| = k\|\mathbf{v}\|$.

Consider the family of L_p*-norms*. The L_p-norm of $\mathbf{v} \in R^n$ is given by

$$\|\mathbf{v}\|_p = \left[\sum_{i=1}^{n} |v_i|^p\right]^{1/p} \qquad p \in \{1, 2, 3, \ldots\} \cup \{\infty\}$$

The L_p-norms of $\mathbf{v} \in R^n$ when p equals 1, 2 , and ∞ are

$$\|\mathbf{v}\|_1 = \sum_{i=1}^{n} |v_i|$$

$$\|\mathbf{v}\|_2 = \sqrt{\sum_{i=1}^{n} |v_i|^2}$$

$$\|\mathbf{v}\|_\infty = \max_{i=1,2,\ldots,n} \{|v_i|\}$$

The expression for the L_∞-norm results because the component of largest magnitude totally dominates when raised to the $p = \infty$ power.

Example 41. Consider the vector $\mathbf{v} = (4, -5, 1)$. The L_1-, L_2-, and L_∞-norms of $\mathbf{v}$ are

$$\begin{aligned}
\|\mathbf{v}\|_1 &= |4| + |-5| + |1| \\
&= 10 \\
\|\mathbf{v}\|_2 &= \sqrt{|4|^2 + |-5|^2 + |1|^2} \\
&= 6.48 \\
\|\mathbf{v}\|_\infty &= \max\{|4|, |-5|, |1|\} \\
&= 5
\end{aligned}$$

We note that the L_1- and L_∞-norms are easy to compute.

Dividing each component of a vector by the norm of the vector *normalizes* a vector. The resultant length of the normalized vector is one.

Example 42. Normalizing the vector $\mathbf{v} = (2, -3, -7)$ according to the L_∞-norm, we obtain

$$\frac{\mathbf{v}}{\|\mathbf{v}\|_\infty} = \frac{(2, -3, -7)}{7} = \left(\tfrac{2}{7}, -\tfrac{3}{7}, -1\right)$$

and the L_∞-norm of the normalized vector is one.

2.5.2 Family of L_p-Metrics

A *metric* on R^n is a distance function that assigns to each pair of vectors $\mathbf{x}, \mathbf{y} \in R^n$ a scalar $\|\mathbf{x} - \mathbf{y}\| \in R$ if and only if the function satisfies for all $\mathbf{z} \in R^n$ the following four axioms:

1. $\|\mathbf{x} - \mathbf{y}\| \geqq 0$ and $\|\mathbf{x} - \mathbf{x}\| = 0$.
2. $\|\mathbf{x} - \mathbf{y}\| = \|\mathbf{y} - \mathbf{x}\|$.
3. $\|\mathbf{x} - \mathbf{y}\| \leqq \|\mathbf{x} - \mathbf{z}\| + \|\mathbf{z} - \mathbf{y}\|$.
4. If $\mathbf{x} \neq \mathbf{y}$, then $\|\mathbf{x} - \mathbf{y}\| > 0$.

Consider the family of L_p*-metrics*. For an L_p-metric, the distance between two points, $\mathbf{x}, \mathbf{y} \in R^n$, is given by

$$\|\mathbf{x} - \mathbf{y}\|_p = \left[\sum_{i=1}^{n} |x_i - y_i|^p \right]^{1/p} \qquad p \in \{1, 2, 3, \ldots\} \cup \{\infty\}$$

Example 43. Consider the two vectors $\mathbf{x} = (2, 0, -1)$ and $\mathbf{y} = (-7, 3, -4)$. The distances between $\mathbf{x}$ and $\mathbf{y}$ according to the L_1-, L_2-, and L_∞-metrics are

$$\begin{aligned}
\|\mathbf{x} - \mathbf{y}\|_1 &= |2 + 7| + |0 - 3| + |-1 + 4| \\
&= 15 \\
\|\mathbf{x} - \mathbf{y}\|_2 &= \sqrt{|2 + 7|^2 + |0 - 3|^2 + |-1 + 4|^2} \\
&= 9.949 \\
\|\mathbf{x} - \mathbf{y}\|_\infty &= \max\{|2 + 7|, |0 - 3|, |-1 + 4|\} \\
&= 9
\end{aligned}$$

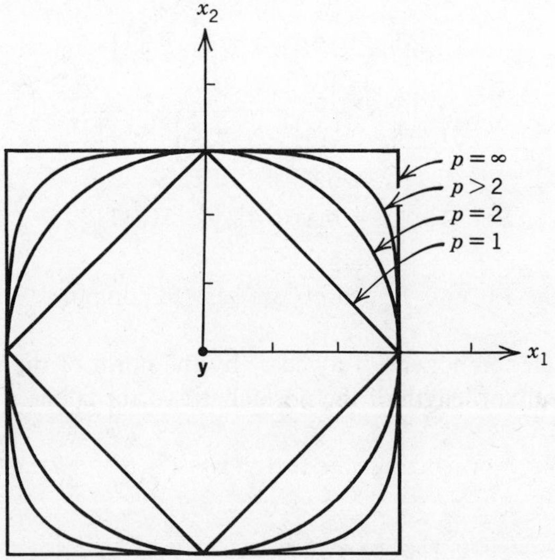

Figure 2.16. Contours of the L_p family of metrics.

Graphing, for instance, the loci of points of distance three away from $\mathbf{y} \in R^2$ according to the family of L_p-metrics, we have the representative *contours* of Fig. 2.16. We note that the L_∞-metric is often called the Tchebycheff metric.

2.5.3 Family of Weighted L_p-Metrics

Let $\mathbf{x}, \mathbf{y} \in R^n$. For a *weighted* L_p-metric, the distance between $\mathbf{x}$ and $\mathbf{y}$ is given by

$$\|\mathbf{x} - \mathbf{y}\|_p^{\boldsymbol{\lambda}} = \left[\sum_{i=1}^{n} (\lambda_i |x_i - y_i|)^p\right]^{1/p} \qquad p \in \{1, 2, 3, \ldots\} \cup \{\infty\}$$

where $\boldsymbol{\lambda} \in R^n$ is a nonnegative vector of weights.

Example 44. Let $\mathbf{x} = (1, 4, -3)$ and $\mathbf{y} = (6, 3, -4)$. Then, with $\boldsymbol{\lambda} = (.2, .3, .5)$,

$$\|\mathbf{x} - \mathbf{y}\|_1^{\boldsymbol{\lambda}} = 1.8$$

$$\|\mathbf{x} - \mathbf{y}\|_2^{\boldsymbol{\lambda}} = 1.158$$

$$\|\mathbf{x} - \mathbf{y}\|_\infty^{\boldsymbol{\lambda}} = 1.0$$

Observe that the (unweighted) family of L_p-metrics is the weighted family of L_p-metrics with $\boldsymbol{\lambda} = (1, 1, \ldots, 1)$. Graphing, for instance, the loci of points of distance two away from $\mathbf{y} \in R^2$ according to the family of weighted L_p-metrics with $\boldsymbol{\lambda} = (\frac{1}{4}, \frac{3}{4})$, we have the representative contours of Fig. 2.17.

With contours of weighted L_p-metrics, the dimensions with the smallest weights are the most elongated and the dimensions with the largest weights are the most compressed.

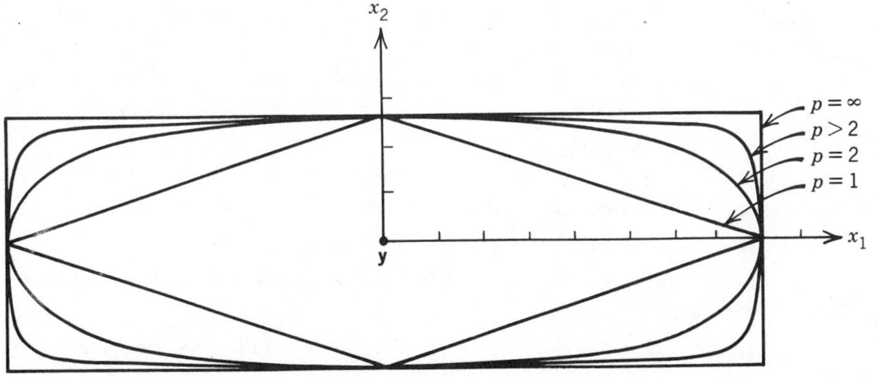

Figure 2.17. Contours of the L_p family of weighted metrics.

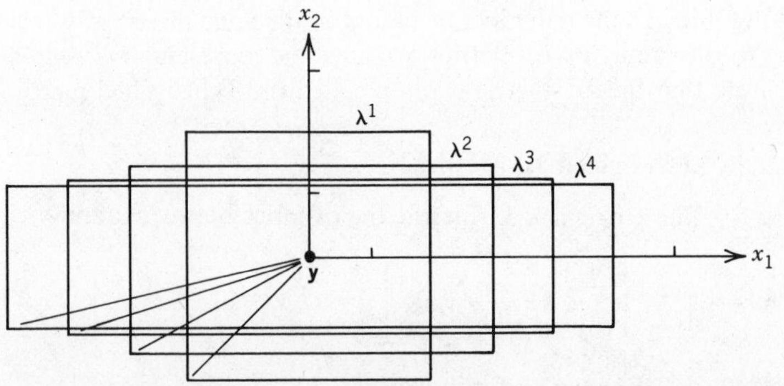

Figure 2.18. Weighted L_∞ (Tchebycheff) contours of Example 45.

Example 45. The contours of distance one away from $\mathbf{y} \in R^2$ according to the weighted L_∞ (Tchebycheff) metric with

$$\boldsymbol{\lambda}^1 = (\tfrac{1}{2}, \tfrac{1}{2})$$

$$\boldsymbol{\lambda}^2 = (\tfrac{1}{3}, \tfrac{2}{3})$$

$$\boldsymbol{\lambda}^3 = (\tfrac{1}{4}, \tfrac{3}{4})$$

$$\boldsymbol{\lambda}^4 = (\tfrac{1}{5}, \tfrac{4}{5})$$

are given in Fig. 2.18. In Fig. 2.18, notice how the angle of the line from $\mathbf{y}$ to the lower left-hand corners of the contours varies with $\boldsymbol{\lambda}$.

Readings

1. Campbell, H. G. (1977). *An Introduction to Matrices, Vectors and Linear Programming*, Englewood Cliffs, New Jersey: Prentice-Hall.
2. Childress, R. L. (1974). *Sets, Matrices and Linear Programming*, Englewood Cliffs, New Jersey: Prentice-Hall.
3. Goldman, A. J. and A. W. Tucker (1956). "Polyhedral Convex Cones." In H. W. Kuhn and A. W. Tucker (eds.), *Linear Inequalities and Related Systems*, Annals of Mathematical Studies, No. 38, pp. 19–40.
4. Grawoig, D. E., B. Fielitz, J. Robinson, D. Tabor (1976). *Mathematics: A Foundation for Decisions*, Reading, Massachusetts: Addison-Wesley.
5. Hadley, G. (1962). *Linear Algebra*, Reading, Massachusetts: Addison-Wesley.
6. Lipschutz, S. (1964). *Set Theory and Related Topics*, Schaum's Outline Series, New York: McGraw-Hill.
7. Lipschutz, S. (1965). *General Topology*, Schaum's Outline Series, New York: McGraw-Hill.
8. Lipschutz, S. (1968). *Linear Algebra*, Schaum's Outline Series, New York: McGraw-Hill.

9. Rockafellar, R. T. (1970). *Convex Analysis*, Princeton, New Jersey: Princeton University Press.

10. Schneider, D. M., M. Steeg, and F. H. Young (1982). *Linear Algebra*, New York: Macmillan.

PROBLEM EXERCISES

2-1 In both roster and defining-property form, specify:

(a) The set of all nonnegative integers less than 10.

(b) The set of all even numbers from 4 to 16 inclusive, except 8.

(c) The set of all states of the United States beginning with the letter "A."

2-2 Specify all subset relationships among:

(a) $A = \{2,5,7\}$	**(b)** $A = \{6,7,8,9\}$
$B = \{2,7\}$	$B = \{6,10\}$
$C = \{2,5,7,9\}$	$C = \varnothing$
$D = \{2,7,9\}$	$D = \{10\}$

2-3 Of the four sets:

$$A = \{9,14\}$$

$$B = \{8,10\}$$

$$C = \{10,11,12\}$$

$$D = \varnothing$$

which pairs of sets are disjoint?

2-4 Specify the family of subsets of $\{a,b,c,d\}$ that have two or fewer members.

2-5 Let

$$A_1 = \{4,5,6,7\}$$

$$A_2 = \{4,6\}$$

$$A_3 = \{3,4,7,8\}$$

$$A_4 = \{8,9\}$$

$$A_5 = \{4,6,7,8\}$$

Specify

(a) $A_5 \cap \varnothing$

(b) $(A_1 \cup (A_2 - A_3)) \cap A_5$

(c) $\bigcap_{i=1}^{5} A_i$

(d) $\bigcup_{i \in I} A_i$ where $I = \{2,4,5\}$

2-6 Let $I = \{1,7,9,12,86,142\}$ and $J = (I - \{10,12\}) \cup \{6,3\}$. What is:

(a) $|I|$

(b) $|J|$

2-7 Do the following sets:

(a) $[-1,1)$, $\{1,2\}$, $[1,2)$, $(2,8]$

(b) $[-1,3)$, $(3,5]$, $(5,6)$, $[6,8]$

(c) $[-1,8)$, $\{8\}$

partition the interval $[-1,8]$? If not, why not?

2-8 Let $A = [0,3]$. Consider the function $f\colon A \to R$ defined by $f(x) = x^2 - x$.

(a) What is the domain of f?

(b) What is the co-domain of f?

(c) What is the image of $2 \in A$?

(d) What is the inverse image of 6?

(e) What is the range of f?

(f) Specify the graph of f in defining-property form.

(g) Is f one-to-one? If not, why not?

(h) Is f onto? If not, why not?

2-9 Consider the function $f\colon R \to R$ defined by $f(x) = x^2$.

(a) What is the image of -2?

(b) What is the inverse image of 0?

(c) What are the inverse images of 16?

2-10 Let $A = \{2,3,4\}$ and $B = \{4,5,6,7,8\}$. Consider the function $f\colon A^2 \to B$ defined by $f(x_1, x_2) = x_1 + x_2$.

(a) Specify the graph of f in roster form.

(b) How many dimensions are needed to plot the graph of f?

(c) Is f one-to-one? If not, why not?

(d) Is f onto? If not, why not?

(e) What is the image of $(2,4)$?

(f) What is the inverse image of 8?

(g) What are the inverse images of 6?

2-11 **(a)** Let:

$$A_1 = \{1,4,6\}$$
$$A_2 = \{2,4,8\}$$
$$A_3 = \{2,6,8\}$$

What is $(A_1 \cup A_2) \cap (A_1 - A_3)$?

(b) Are the two sets $(-8,0)$ and $(0,4)$ a partition of $(-8,4)$? If not, why not?

(c) Let $U\colon R^2 \to R$ be the function defined by

$$U(z_1, z_2) = z_1 z_2$$

Is U one-to-one? Onto? If not, why not?

2-12 Which of the following terms

relation

mapping

function

apply to:

(a) The relationship on $A \times B$ described by

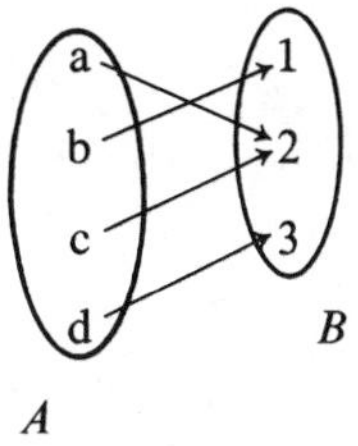

(b) The relationship on R^2 given by $(x_1)^2 + (x_2)^2 = 16$.

2-13 Compute:

(a) $\begin{bmatrix} 3 & 1 \\ 2 & -4 \end{bmatrix} + \begin{bmatrix} 5 & -2 \\ -6 & -4 \end{bmatrix}$ **(b)** $\begin{bmatrix} 4 & -3 \\ -2 & -4 \end{bmatrix} - \begin{bmatrix} -2 & -4 \\ 7 & 3 \end{bmatrix}$

2-14 Multiply:

(a) $\begin{bmatrix} 0 & -6 \\ 7 & 4 \end{bmatrix}\begin{bmatrix} 1 & 0 \\ 7 & -7 \end{bmatrix}\begin{bmatrix} 1 & 0 & 6 \\ 1 & -1 & 1 \end{bmatrix}$

(b) $[.1 \quad .6 \quad .3]\begin{bmatrix} 2 & 3 & 1 & 2 \\ 4 & 0 & 0 & -1 \\ -1 & -1 & 1 & -2 \end{bmatrix}$

(c) $\begin{bmatrix} 2 & 0 & 4 \\ 7 & 0 & -1 \\ 3 & -2 & 4 \end{bmatrix}\begin{bmatrix} x_1 \\ x_2 \\ x_3 \end{bmatrix}$

(d) $\begin{bmatrix} 4 \\ 7 \\ 8 \end{bmatrix}[5 \quad 9 \quad -7]$

2-15 Calculate the vector products:

(a) $\mathbf{a}^T\mathbf{b}$ where $\mathbf{a} = (2, -3, 4, 5)$ and $\mathbf{b} = (2, -3, 0, -1)$.

(b) $\mathbf{e}^T\mathbf{v}$ where $\mathbf{v} = (13, 9, 17)$.

2-16 Calculate:

(a) $\boldsymbol{\lambda}^T\mathbf{C}\mathbf{x}$ where

$$\boldsymbol{\lambda} = \begin{bmatrix} .2 \\ .5 \\ .1 \\ .2 \end{bmatrix} \qquad \mathbf{C} = \begin{bmatrix} 2 & 5 & 7 & 3 & 1 \\ 0 & 0 & -1 & 2 & 4 \\ -1 & 0 & -1 & 0 & 1 \\ 2 & 1 & 4 & 0 & 0 \end{bmatrix} \qquad \mathbf{x} = \begin{bmatrix} x_1 \\ x_2 \\ x_3 \\ x_4 \\ x_5 \end{bmatrix}$$

(b) $\mathbf{D} = \dfrac{\mathbf{A}^T - \mathbf{BC}}{\mathbf{a}^T\mathbf{b}}$ where $\mathbf{a} = (2, 8, 1)$ and $\mathbf{b} = (-2, -7, -11)$,

$$\mathbf{A} = \begin{bmatrix} 6 & 0 \\ 0 & -6 \end{bmatrix} \quad \mathbf{B} = \begin{bmatrix} 4 & 0 & -2 \\ 6 & 8 & 0 \end{bmatrix} \quad \mathbf{C} = \begin{bmatrix} 3 & 3 \\ 6 & 6 \\ 9 & 9 \end{bmatrix}$$

2-17 Let **P** and **Q** be $n \times n$, **A** be $m \times m$, and $m < n$. Compute **PQ** where

$$\mathbf{P} = \left[\begin{array}{c|c} \mathbf{A} & \mathbf{B} \\ \hline \mathbf{C} & \mathbf{D} \end{array}\right] \quad \mathbf{Q} = \left[\begin{array}{c|c} \mathbf{A}^{-1} & \mathbf{F} \\ \hline \mathbf{G} & \mathbf{H} \end{array}\right]$$

2-18 Calculate the determinants of:

(a) $[-6]$ **(b)** $\begin{bmatrix} 16 & 2 \\ -8 & 0 \end{bmatrix}$

(c) $\begin{bmatrix} -1 & 4 & 3 \\ 4 & 1 & -7 \\ 2 & -8 & -6 \end{bmatrix}$ **(d)** $\begin{bmatrix} 2 & -2 & -7 \\ 0 & \frac{1}{4} & 4 \\ 0 & 0 & -\frac{1}{3} \end{bmatrix}$

2-19 Using Cramer's Rule, solve the $\mathbf{Ax} = \mathbf{b}$ system of linear equations

$$\mathbf{A} = \begin{bmatrix} 4 & 2 & 0 \\ 1 & 1 & 1 \\ -2 & 0 & 4 \end{bmatrix} \quad \mathbf{b} = \begin{bmatrix} 2 \\ 4 \\ -14 \end{bmatrix}$$

2-20 Compute the inverses of:

(a) $\mathbf{A} = \begin{bmatrix} 6 & 4 \\ 5 & 7 \end{bmatrix}$ **(b)** $\mathbf{B} = \begin{bmatrix} -1 & 2 & 6 \\ -4 & 2 & 1 \\ 7 & 0 & 0 \end{bmatrix}$

(c) $\mathbf{C} = \begin{bmatrix} 4 & 2 & 0 \\ 1 & 1 & 1 \\ -2 & 0 & 2 \end{bmatrix}$

Verify your inverses by multiplying them by the original matrices.

2-21 Using inverses, solve the following systems of linear equations where $\mathbf{Ax} = \mathbf{b}$.

(a) $\mathbf{A} = \begin{bmatrix} 1 & 2 & 2 \\ -1 & -4 & 0 \\ 2 & 7 & 4 \end{bmatrix}$ $\mathbf{b} = \begin{bmatrix} -16 \\ 26 \\ -35 \end{bmatrix}$

(b) $\mathbf{A} = \begin{bmatrix} -5 & 4 \\ 3 & 4 \end{bmatrix}$ $\mathbf{b} = \begin{bmatrix} -33 \\ 7 \end{bmatrix}$

2-22 Let $\mathbf{x}^1 = (1, 3)$, $\mathbf{x}^2 = (5, 1)$, and $\mathbf{x}^3 = (-1, 6)$. Using the tail-to-head method, portray (a) and (b). Using the completion of the parallelogram method, portray (c), (d), and (e).

(a) $\mathbf{x}^1 + \mathbf{x}^2 + \mathbf{x}^3$

(b) $\mathbf{x}^1 - 4\mathbf{x}^2 + 2\mathbf{x}^3$

(c) $.1\mathbf{x}^1 + .9\mathbf{x}^2$

(d) $.4\mathbf{x}^1 + .6\mathbf{x}^2$

(e) $.8\mathbf{x}^1 + .2\mathbf{x}^2$

2-23 Determine whether or not the following sets of vectors are linearly independent. Show your analysis.

(a) $(2, -4, 2), (2, 2, -1), (7, -4, 1)$

(b) $(1, 1, 1, 1), (2, 4, 6, 8), (-1, -3, 2, 1)$

(c) $(0, 1, 0), (1, 0, 0), (0, 0, 1)$

(d) $(1, 2), (-2, -4)$

(e) $(6, 3), (-1, -3), (4, -7)$

(f) $(1, 0, 0, 2)$

(g) $(1, 7, 3), (0, 0, 0), (6, 1, 2)$

2-24 Determine whether or not the following form bases for their respective spaces.

(a) $(1, 1, 1), (1, -1, 5)$

(b) $(6, 0, 5), (4, 0, 9), (7, 0, 4)$

(c) $(1, 1), (2, -3), (4, 5)$

(d) $(3, 2, 1), (1, -2, 0), (1, 2, 1)$

(e) (-31)

(f) $(1, 3), (16, 2)$

2-25 Determine the weights that cause the following basis vectors to span to the point $(6, -7, 2)$.

(a) $(1, 0, 0), (0, 1, 0), (0, 0, 1)$

(b) $(3, 1, 0), (0, 3, 2), (0, 0, 3)$

(c) $(2, 0, 1), (3, 2, -3), (-1, -3, 5)$

2-26 Determine the maximal number of linearly independent rows in

(a) $$\begin{bmatrix} 2 & 4 & 4 & -16 \\ -1 & 2 & 6 & -16 \\ 3 & -1 & -8 & -30 \end{bmatrix}$$

(b) $$\begin{bmatrix} 4 & 4 & 2 & 0 \\ -1 & 1 & -1 & 0 \\ 3 & 17 & -2 & 0 \end{bmatrix}$$

2-27 By determining the largest square submatrix whose determinant is nonzero, determine the rank of

(a) $$\mathbf{A} = \begin{bmatrix} 6 & 8 & 14 \\ 6 & 8 & 5 \\ 12 & 16 & 2 \end{bmatrix}$$

(b) $$\mathbf{B} = \begin{bmatrix} 1 & 2 & 0 & -1 \\ 2 & 6 & -3 & -3 \\ 3 & 10 & -6 & -5 \end{bmatrix}$$

2-28 Can any of the rows in

$$\mathbf{C} = \begin{bmatrix} 1 & 3 & 1 \\ 1 & 3 & -1 \\ -1 & 3 & 1 \\ -1 & 3 & -1 \end{bmatrix}$$

be expressed as convex combinations of the other rows in **C**?

2-29 Using the concept of rank, determine whether the following systems of linear equations are inconsistent, consistent-unique, or consistent-infinite.

(a)
$$\begin{aligned} -x_1 + 3x_2 - 4x_3 &= -12 \\ -3x_1 - x_2 - 2x_3 &= -2 \\ -3x_1 - 11x_2 + 8x_3 &= -14 \end{aligned}$$

(b)
$$\begin{aligned} -2x_1 + 4x_2 &= 4 \\ 3x_1 + 4x_3 &= 5 \\ -x_1 + 2x_2 + x_3 &= 2 \end{aligned}$$

(c)
$$\begin{aligned} 3x_1 - 7x_2 - 2x_3 &= -33 \\ -4x_1 + 3x_2 + 3x_3 &= 46 \\ 3x_1 - x_2 + x_3 + x_4 &= -15 \end{aligned}$$

2-30 Which of the points

(a) $(2, -3, 4, 2)$ **(b)** $(4, 4, 1, 0)$
(c) $(0, -1, -1, 0)$ **(d)** $(7, 2, 4, 1)$

are in the open hypersphere H_s of radius 5 centered at $(4, -1, 2, 0)$?

2-31 Specify which of the following terms

open	polyhedral
closed	connected
bounded	disconnected
unbounded	discrete
convex	continuous
nonconvex	countably infinite

apply to:

(a) $\{\mathbf{x} \in R^n \mid x_i \geqq 0\}$.

(b) R^n.

(c) $\{\mathbf{x} \in R^n \mid x_i \in [0, 10],\ x_i \text{ integer}\}$.

(d) $\{1, \frac{1}{2}, \frac{1}{4}, \frac{1}{8}, \ldots\}$.

2-32 Let S denote the intersection of the closed half-spaces

$$\{\mathbf{x} \in R^2 \mid -x_1 + x_2 \leqq 0\}$$
$$\{\mathbf{x} \in R^2 \mid x_1 \leqq 4\}$$
$$\{\mathbf{x} \in R^2 \mid x_2 \geqq 0\}$$
$$\{\mathbf{x} \in R^2 \mid x_1 + x_2 \geqq 2\}$$

(a) Graph S.

(b) How many extreme points has S?

(c) Using γ, specify S.

(d) Specify the boundary of S.

2-33 Let $\mathbf{x}^1 = (1,1)$, $\mathbf{x}^2 = (5,0)$, $\mathbf{x}^3 = (4,2)$, and $\mathbf{x}^4 = (4,4)$.

(a) Graph the convex hull of $\mathbf{x}^1, \mathbf{x}^2, \mathbf{x}^3, \mathbf{x}^4$.

(b) What is the dimensionality of $\gamma(\mathbf{x}^1,\mathbf{x}^2,\mathbf{x}^3)$?

(c) What is the dimensionality of $\gamma(\mathbf{x}^1,\mathbf{x}^2) \cup \gamma(\mathbf{x}^2,\mathbf{x}^3)$?

(d) What is the dimensionality of $\{\mathbf{x}^1\}$?

2-34 Express the sets of Fig. 2.19 in terms of γ and μ.

2-35 Graph:

(a) $\gamma(\mathbf{z}^1,\mathbf{z}^2,\mathbf{z}^3) \oplus \{\mathbf{z} \in R^2 \mid \mathbf{z} \leqq \mathbf{0}\}$ where $\mathbf{z}^1 = (2,4)$, $\mathbf{z}^2 = (3,3)$, and $\mathbf{z}^3 = (4,1)$.

(b) $\{\mathbf{x}^1\} \oplus V$ where $\mathbf{x}^1 = (3,1)$ and V is the cone generated by $\mathbf{v}^1 = (-1,4)$ and $\mathbf{v}^2 = (2,3)$.

2-36 Graph the image of S under $f\colon S \to R^2$ where $S = \{\mathbf{x} \geqq \mathbf{0} \mid x_2 \leqq 2;\ x_1 + x_2 \leqq 4\}$ and f is defined by $\mathbf{C}\mathbf{x} = \mathbf{z}$ where:

$$\mathbf{C} = \begin{bmatrix} -1 & 2 \\ 1 & 2 \end{bmatrix}$$

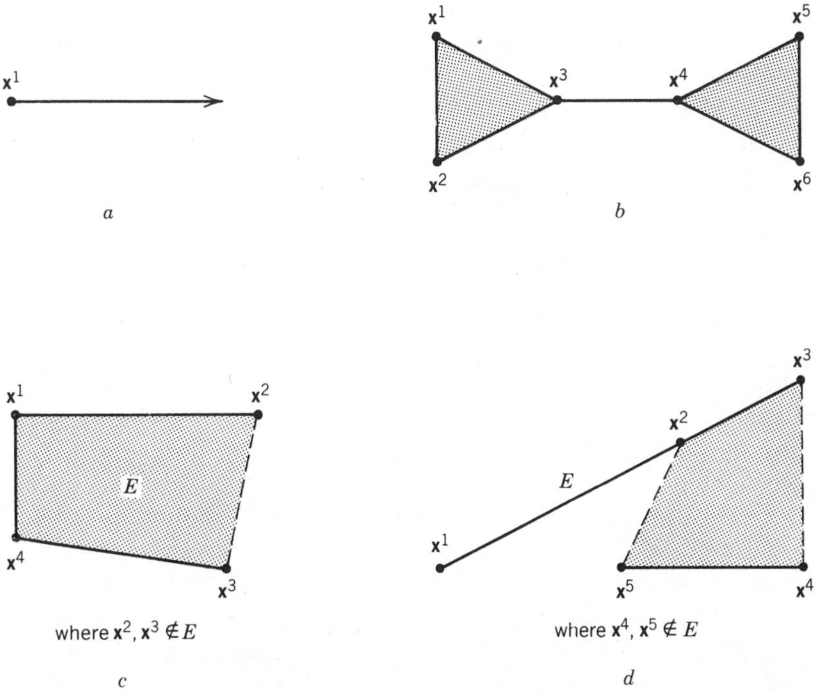

Figure 2.19. Graphs of Problem 2-34.

2-37 Determine the inequalities that define the convex hull of:

$$\mathbf{x}^1 = (4,4,0)$$
$$\mathbf{x}^2 = (0,10,0)$$
$$\mathbf{x}^3 = (3,3,3)$$
$$\mathbf{x}^4 = (0,4,4)$$
$$\mathbf{x}^5 = (0,0,0)$$

2-38 Determine the inequalities that define the cone generated by:

$$\mathbf{v}^a = (1,4,0)$$
$$\mathbf{v}^b = (6,6,0)$$
$$\mathbf{v}^c = (3,4,2)$$
$$\mathbf{v}^d = (1,4,2)$$

2-39 Let $Z = \{(0,0),(0,4),(1,3),(3,2),(4,2)\}$.

(a) Which points of Z possess supporting hyperplanes?

(b) Which do not?

2-40 Graph the nonnegative polars of the cones generated by:

(a) $(4,1)$ and $(-4,1)$ **(b)** $(2,3)$

2-41 Normalize each of the following vectors according to the L_1-, L_2-, and L_∞-norms.

(a) $(3,5,2)$ **(b)** $(8,-10,-3,-1)$

(c) $(-1,0,0,1)$

2-42 Calculate the distances, according to the L_1-, L_2-, and L_∞-metrics, between $\mathbf{x}^1$ and $\mathbf{x}^2$ where:

(a) $\mathbf{x}^1 = (2,3,0,-4,-1)$ and $\mathbf{x}^2 = (3,-2,2,-6,1)$

(b) $\mathbf{x}^1 = (2,1,0,2)$ and $\mathbf{x}^2 = (1,0,1,2)$

(c) $\mathbf{x}^1 = (4,4,4)$ and $\mathbf{x}^2 = (0,0,4)$

2-43 Consider:

$$\mathbf{x}^1 = (1,2,0,1)$$
$$\mathbf{x}^2 = (4,1,3,5)$$
$$\mathbf{x}^3 = (2,2,2,2)$$
$$\mathbf{x}^4 = (-1,2,0,-1)$$

along with $\boldsymbol{\lambda} = (.1,.2,.4,.3)$. Using the weighted L_2-metric, rank $\mathbf{x}^2$, $\mathbf{x}^3$, and $\mathbf{x}^4$ according to their distances away from $\mathbf{x}^1$.

2-44 Centered at $\bar{\mathbf{x}} = (8,6)$, draw the $\|\mathbf{x} - \bar{\mathbf{x}}\|_1^{\boldsymbol{\lambda}} = 4$ contour for:

(a) $\boldsymbol{\lambda} = (\frac{1}{3}, \frac{2}{3})$ **(b)** $\boldsymbol{\lambda} = (\frac{3}{4}, \frac{1}{4})$

2-45 Centered at $\bar{\mathbf{x}} = (0,0,0)$, draw the $\|\mathbf{x} - \bar{\mathbf{x}}\|_\infty^{\boldsymbol{\lambda}} = 2$ contour for $\boldsymbol{\lambda} = (\frac{1}{4}, \frac{1}{8}, \frac{5}{8})$.

2-46 Let $Z = \{\mathbf{z} \in R^2 \mid z_1, z_2 \geqq 0;\ z_1 + z_2 \leqq 3\}$ and let $\mathbf{z}^{**} = (4,4)$. What are the coordinates of the point in Z at which the lowest numbered $\|\mathbf{z} - \mathbf{z}^{**}\|_\infty^{\boldsymbol{\lambda}}$ contour intersects Z for:

(a) $\boldsymbol{\lambda} = (\frac{1}{5}, \frac{4}{5})$ **(b)** $\boldsymbol{\lambda} = (\frac{2}{3}, \frac{1}{3})$

CHAPTER 3

Single Objective Linear Programming

The following is a single objective linear program (LP):

$$
\begin{array}{rll}
\max \{c_1 x_1 + c_2 x_2 + \cdots + c_n x_n = z\} & & \text{Objective} \\
\text{subject to: } a_{11} x_1 + a_{12} x_2 + \cdots + a_{1n} x_n \leqq b_1 & & \\
a_{21} x_1 + a_{22} x_2 + \cdots + a_{2n} x_n \leqq b_2 & & \text{Main constraints} \\
\vdots \qquad\qquad\qquad \vdots & & \\
a_{m1} x_1 + a_{m2} x_2 + \cdots + a_{mn} x_n \leqq b_m & & \\
x_1, x_2, \ldots, x_n \geqq 0 & & \text{Nonnegativity restrictions}
\end{array}
$$

As seen, an LP has three parts: an *objective*, *main constraints*, and *nonnegativity restrictions*. In addition, we call

$c_1x_1 + c_2x_2 + \cdots + c_nx_n$	Objective function
the c_j	Objective function (or criterion) coefficients
z	Objective function (or criterion) value
the x_i	*Structural* (or original or *decision*) variables
the b_j	Right-hand side (RHS) values
the a_{ij}	Constraint coefficients

Since the above formulation only contains structural variables, it is said to be in *structural variable format*.

Frequently, the nonnegativity restrictions are referred to as *nonnegativity constraints* or *nonnegativity requirements*. They require that none of the decision variables be negative. A point $(x_1, x_2, \ldots, x_n)$ that simultaneously satisfies *all* of the constraints (main and nonnegativity) is called a *feasible point*. The set of all feasible points is called the *feasible region* and is denoted with the uppercase letter S.

The purpose of an LP is to determine the feasible point (or points) that maximize the objective function. That is, the endeavor is to find the point (or points) in S that generate (when "plugged into" the objective function) the largest criterion value z.

Within the three-part framework of an objective, main constraints, and nonnegativity restrictions, an LP can be formulated in different ways. It can be formulated with a "min" objective function and any mixture of $\leqq$, $=$, or $\geqq$ constraints. With reference to the criterion value z, the *optimal* (i.e., *maximizing* or *minimizing*) value of the objective function is denoted z^*. Any feasible point that generates z^* is called an *optimal point*.

With regard to the feasible region S, sometimes it contains no points, sometimes it contains only one, but usually it is a multidimensional subset of R^n containing an infinite number of points.

Because S is formed by the intersection of a finite number of hyperplanes (equality constraints) and closed half-spaces ($\leq$ and $\geq$ inequality constraints), S is polyhedral. Thus, S is closed and convex with a finite number of extreme points. Both the graphical and simplex methods described in this chapter for solving LPs exploit the polyhedral properties of S in the sense that optimal solutions can be found without having to examine all points in S.

In an LP, the term "linear" is used because all relationships in the formulation are linear. "Program" dates back to the field of theoretical economics in the 1930s and 40s. At that time, it was the term used by economists to refer to problems concerned with the optimal allocation of limited resources.

Linear programming is an outgrowth of this early work in economics. However, the field of linear programming did not develop as such until Dantzig formulated the general linear programming problem and the *simplex method* of solution in 1947. Soon thereafter, other important theoretical contributions were made by Gale, Kuhn, and Tucker. Orchard-Hays was one of the first to extend the computer feasibility of the simplex method. Since the publication of the two-volume set by Charnes and Cooper (1961), the popularity of linear programming and its applicability in both the private and public sectors have been remarkable. With regard to textbooks, the most significant contributions in the field of linear programming have been made by Dantzig (1963), Hadley (1962), and Gass (whose first edition was published in 1958).

Although our review of linear programming will be, for the most part, conventional, our approach will involve some nonstandard aspects. Linear programming, as expressed in Chapters 3, 4, and 5, is not presented as an end in itself. It sets the stage for *multiple objective* linear programming. There are a number of differences that reflect this change in role. For example, we will make frequent use of the convex combination operator γ and the unbounded line segment operator μ (see Section 2.3.2). But most significantly, we will pay more than the usual small amount of attention to the phenomenon of "alternative optima." In fact, Chapter 4 is devoted to methods by which we can compute all optimal points when alternative optima exist. Chapter 5 discusses some parametric linear programming topics that form the transition from LP to MOLP.

We will motivate linear programming by means of the sequence of graphical examples in Sections 3.2 and 3.3. Then, with a geometric understanding of the various situations that can occur in LP, the simplex method is introduced starting in Section 3.4.

3.1 GRADIENTS AND LEVEL CURVES

In linear programming, we always arrange terms with unknowns on the left, leaving only constants on the right. In this way, we can refer to the left-hand side (LHS) and right-hand side (RHS) of a linear expression.

In our graphical illustrations, we will frequently employ the convention of "gradients and level curves." A *level curve* (or *contour*) of a linear expression is the locus of points that satisfies a given RHS value. In R^2, level curves are (one-dimensional) straight lines. In R^3, level curves are (two-dimensional) planes. In R^n, level curves are $n - 1$ dimensional hyperplanes. A useful feature of level curves is that all level curves of a given linear expression are parallel to one another.

The gradient of a linear expression is obtained by forming the vector of LHS coefficients of the unknowns. For instance, the gradient of $-\frac{1}{2}x_1 + 4x_2 = z$ is $(-\frac{1}{2}, 4)$ and the gradient of $-3x_1 - 5x_3 = z$ in R^3 is $(-3, 0, -5)$. All gradients are *normal* (orthogonal) to their level curves. The significance of the gradient is that when it is drawn as an arrow, it points in the direction of greatest increase of the RHS value. Thus, by knowing the direction of the gradient, we know which parallel shifts will lead to higher numbered level curves and which will lead to lower numbered level curves.

Example 1. Consider the number 2 level curve of $-x_1 + 2x_2 = z$. Attached to an arbitrary point on the level curve, let us draw the *gradient arrow* (or more simply, the *gradient*) as in Fig. 3.1. The gradient $(-1, 2)$ is drawn by moving away from a point on the level curve in the proportion of -1 unit in the x_1 direction for every 2 units in the x_2 direction. As shown, the gradient is normal to

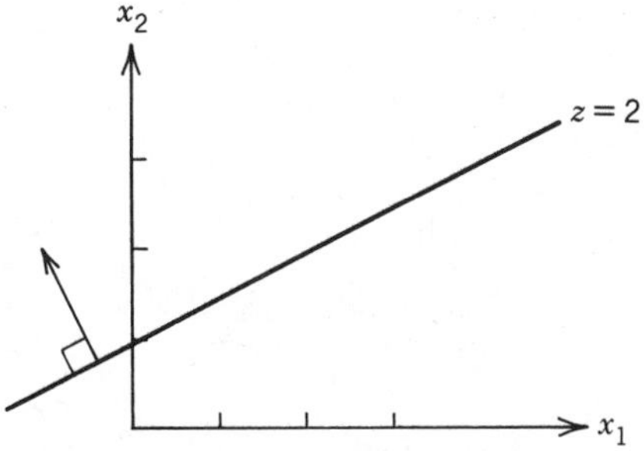

Figure 3.1. Graph of Example 1.

the level curve. All points in the open half-space containing the gradient arrow (attached to the level curve) lie on higher numbered level curves. All points in the open half-space not containing the gradient arrow lie on lower numbered level curves.

We will need to identify regions that are formed by the intersection of a number of closed half-spaces. A systematic method for doing this is to use gradients and level curves.

Example 2. Consider the set of points

$$S = \{\mathbf{x} \in R^2 | 2x_1 + x_2 \geqq 3; \quad x_1 + x_2 \leqq 4; \quad x_1 \geqq 0; \quad x_2 \geqq 0\}$$

formed by the intersection of the closed half-spaces defined by the four specified inequalities. Respectively attaching the gradients (2, 1) and (1, 1) to the number 3 and number 4 *bounding* level curves of the first two inequalities, we find that S is that portion of the nonnegative orthant illustrated by the shaded region in Fig. 3.2. That is, within the nonnegative orthant, S is formed by the intersection of the closed half-space containing the gradient of the first inequality and the closed half-space *not* containing the gradient of the second inequality. In terms of the convex combination operator γ,

$$S = \mathop{\gamma}_{i=1}^{4} (\mathbf{x}^i)$$

In this example, S has four extreme points

$$\mathbf{x}^1, \mathbf{x}^2, \mathbf{x}^3, \quad \text{and} \quad \mathbf{x}^4$$

and the set of boundary points of S is given by

$$\gamma(\mathbf{x}^1, \mathbf{x}^2) \cup \gamma(\mathbf{x}^2, \mathbf{x}^3) \cup \gamma(\mathbf{x}^3, \mathbf{x}^4) \cup \gamma(\mathbf{x}^4, \mathbf{x}^1)$$

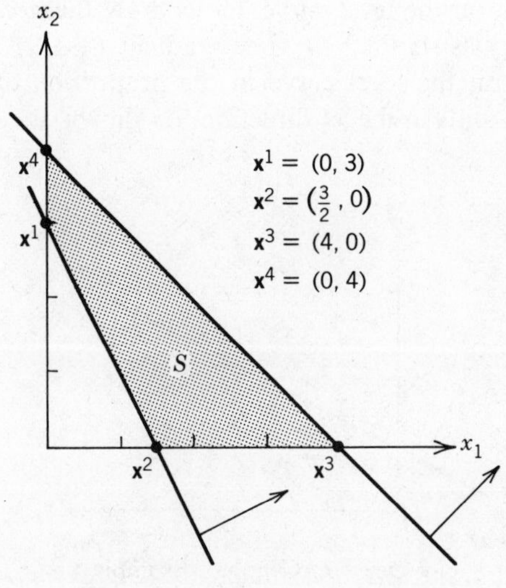

Figure 3.2. Graph of Example 2.

3.2 A LINEAR PROGRAMMING EXAMPLE

As an example of a problem that can be solved using linear programming, consider the situation of Marco Polo Caravans, Inc.

> Marco Polo Caravans, Inc., uses dromedaries (one hump) and camels (two humps) to carry dried figs from Baghdad to Mecca. A camel can carry 1,000 lbs. and a dromedary can carry 500 lbs. While on the trip, a camel consumes 3 bales of hay and 100 gallons of water. A dromedary consumes 4 bales of hay and 80 gallons of water. Marco Polo's supply facilities, located in various oases along the way, have available only 1600 gallons of water and 60 bales of hay.
>
> The camels and dromedaries are rented from a herdsman near Baghdad, the rent being 11 pazuzas per camel and 5 pazuzas per dromedary. If Marco Polo Caravans has a load of 10,000 lbs. of figs that must be transported from Baghdad to Mecca, how many camels and dromedaries should be used to minimize the rent paid to the herdsman? [Adapted from C. R. Carr and C. W. Howe, *Quantitative Decision Procedures in Management and Economics*, New York: McGraw-Hill, 1964, p. 182]

Let x_1 be the number of dromedaries and x_2 be the number of camels. Then, the Marco Polo problem is formulated as follows:

$$\begin{array}{llll} \min\{ & 5x_1 + 11x_2 = z\} & & \text{Rent} \\ \text{s.t.} & 500x_1 + 1000x_2 \geqq & 10000 & \text{Caravan capacity} \\ & 4x_1 + 3x_2 \leqq & 60 & \text{Hay} \\ & 80x_1 + 100x_2 \leqq & 1600 & \text{Water} \\ & x_1, x_2 \geqq & 0 & \end{array}$$

where s.t. is an abbreviation for "subject to." Pursuing a graphical solution, we identify the feasible region S in Fig. 3.3 by graphing the bounding level curves of the main constraints and attaching gradients. Being closed, convex, and polyhedral, the feasible region has four extreme points $\mathbf{x}^1$, $\mathbf{x}^2$, $\mathbf{x}^3$, and $\mathbf{x}^4$.

Because it is a minimization LP, the endeavor is to find the point (or points) in S that minimize $5x_1 + 11x_2 = z$. This is accomplished by finding the lowest numbered level curve of the objective function that intersects the feasible region.

To establish the *slope* of level curves of the objective function, let us draw the 0 level curve of the objective function (that goes through the origin), portraying it with a dashed line. Let us now attach the gradient of the objective function (5, 11) to the origin. Being a minimization problem, this establishes the *direction of optimization* as the opposite of that of the objective function gradient. Hence, we see that extreme point $\mathbf{x}^2 = (12, 4)$ minimizes the LP as indicated by the $z^* = 104$ level curve of the objective function. Extreme point $\mathbf{x}^2$ is uniquely optimal because all other points in S lie on higher numbered level curves. Thus, to achieve the minimum rental of 104 pazuzas, a caravan of 12 dromedaries and 4 camels would have to be formed to transport the 10,000 lbs. of figs.

It is no coincidence that the optimal solution occurred at an extreme point. Because S is polyhedral (i.e., has a finite number of extreme points) and the level curves of the objective function are linear, we arrive at Observation 3.1.

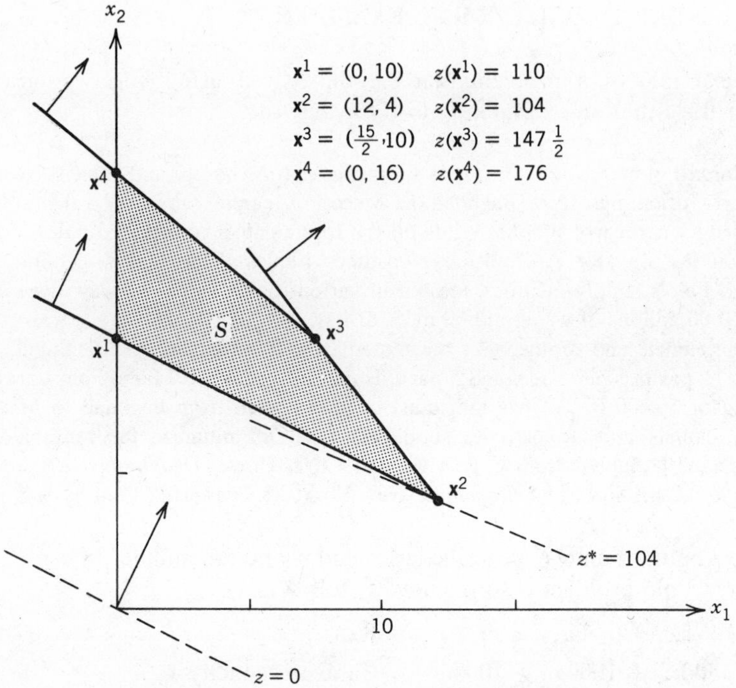

Figure 3.3. Marco Polo Caravans example.

Observation 3.1. If an LP has an optimal solution, at least one of the extreme points of S is optimal.

The significance of Observation 3.1 is that if an optimal solution exists, one can be found by searching among the extreme points. Let us now define the following notation Θ, Θ_x, and Θ_μ where

1. Θ is the set of all *optimal* points.
2. Θ_x is the set of all *optimal extreme* points.
3. Θ_μ is the set of all *unbounded optimal edges* of S.

In the Marco Polo Caravans example, we thus have $\Theta = \{\mathbf{x}^2\}$, $\Theta_x = \{\mathbf{x}^2\}$, and $\Theta_\mu = \varnothing$.

3.3 GRAPHICAL EXAMPLES

In addition to the usual situation in which a unique, bounded criterion value solution is obtained over a nonempty, bounded feasible region (as in the Marco Polo example), various other situations can occur in single objective linear programming as illustrated in Examples 3 through 8.

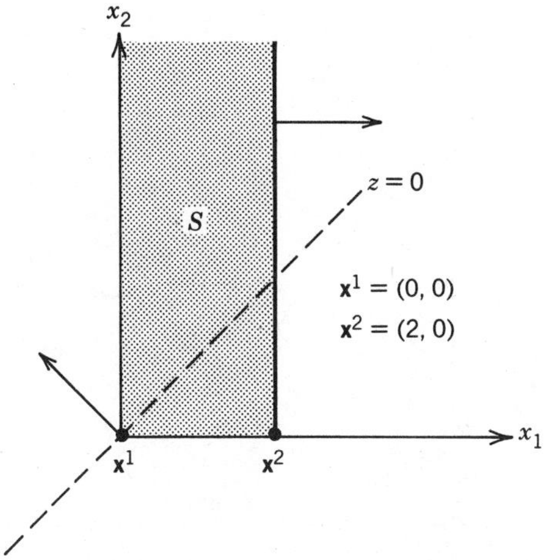

Figure 3.4. Graph of Example 3.

Example 3. Consider the LP

$$
\begin{aligned}
&\max \{ -x_1 + x_2 = z \} \\
&\text{s.t.} \quad x_1 \leqq 2 \\
&\qquad x_1, x_2 \geqq 0
\end{aligned}
$$

that is graphed in Fig. 3.4. This problem does not have an optimal solution. Consequently, Observation 3.1 does not pertain. Because of the unboundedness of S and the direction of optimization, no matter what point in S we choose, there will always be other points in S that lie on higher numbered level curves of the objective function. Such a problem is said to have an *unbounded optimal* z-value. In summary, $\Theta = \Theta_x = \Theta_\mu = \varnothing$ and z^* is *positive-unbounded.*

Although an unbounded optimal z-value implies that S is unbounded, Example 4 shows that an unbounded S does not necessarily mean that z^* is unbounded. Also, Example 4 illustrates that it is possible for an LP to have more than one optimal solution.

Example 4. Consider the LP

$$
\begin{aligned}
&\min \{ \; 2x_1 - 2x_2 = z \} \\
&\text{s.t.} \quad -x_1 + x_2 \leqq 2 \\
&\qquad -2x_1 + x_2 \leqq 1 \\
&\qquad x_1, x_2 \geqq 0
\end{aligned}
$$

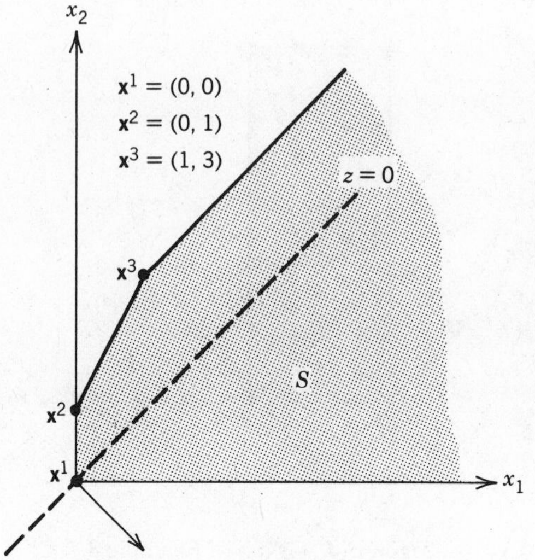

Figure 3.5. Graph of Example 4.

that is graphed in Fig. 3.5. Since the gradients of the objective function and the first constraint are nonzero scalar multiples of one another, their level curves are parallel. Being a minimization LP, all points on the unbounded edge $\mu(\mathbf{x}^3, \mathbf{v})$ where $\mathbf{v} = (1, 1)$ are optimal. Although S is unbounded, in contrast to Example 3, $\Theta = \mu(\mathbf{x}^3, \mathbf{v})$, $\Theta_x = \{\mathbf{x}^3\}$, $\Theta_\mu = \{\mu(\mathbf{x}^3, \mathbf{v})\}$ and $z^* = -4$.

Whenever an LP has more than one optimal point, it has an infinite number of optimal points. However, in contrast to Example 4, Example 5 shows that Θ need not be unbounded.

Example 5. Consider the LP

$$\begin{aligned} \max\{ \quad x_1 + x_2 &= z\} \\ \text{s.t.} \quad x_1 + x_2 &\leqq 5 \\ -x_1 + 3x_2 &\leqq 9 \\ 3x_1 - x_2 &\leqq 9 \\ x_1, x_2 &\geqq 0 \end{aligned}$$

that is graphed in Fig. 3.6. In this problem, the $\gamma(\mathbf{x}^1, \mathbf{x}^2)$ edge is optimal. Although the optimal set contains an infinite number of points, it is not unbounded. Thus, we have $\Theta = \gamma(\mathbf{x}^1, \mathbf{x}^2)$, $\Theta_x = \{\mathbf{x}^1, \mathbf{x}^2\}$, $\Theta_\mu = \varnothing$, and $z^* = 5$.

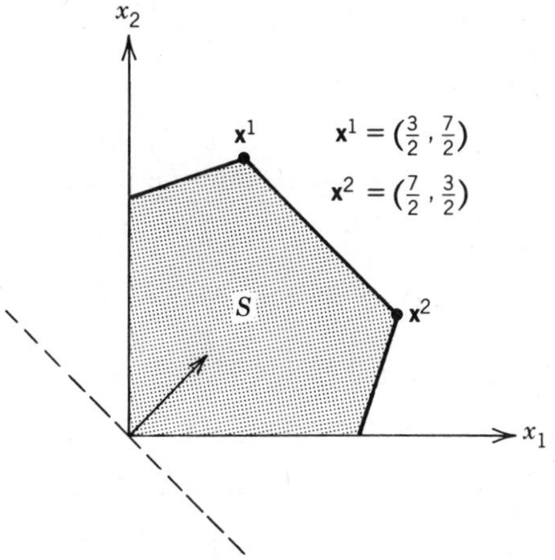

Figure 3.6. Graph of Example 5.

As illustrated in Example 6, the only time interior points of S are optimal in an LP is when we have a *null* objective function (that is, when all objective function coefficients are zero).

Example 6. Consider the LP

$$\max \{0x_1 + 0x_2 = z\}$$
$$\text{s.t.} \qquad \mathbf{x} \in S$$

of Fig. 3.7.

With a null objective function, it is not possible to draw a level curve and gradient of the objective function in the usual way. In this problem all points,

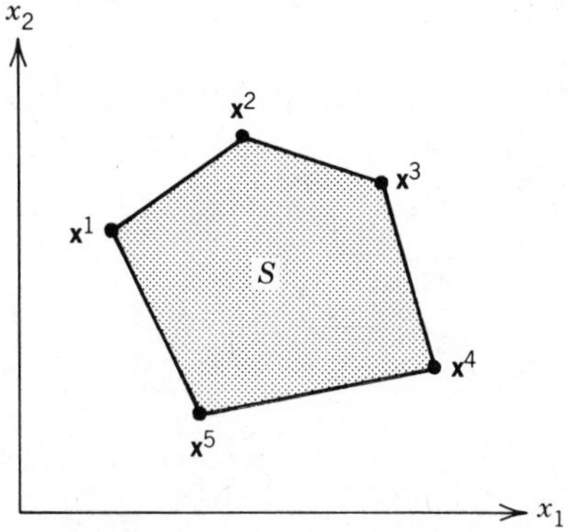

Figure 3.7. Graph of Example 6.

when plugged into the objective function, yield a z-value of zero. Thus, all points in S are in a massive tie for optimality. In this problem, we have $\Theta = S$, $\Theta_x = \{\mathbf{x}^1, \mathbf{x}^2, \mathbf{x}^3, \mathbf{x}^4, \mathbf{x}^5\}$, $\Theta_\mu = \varnothing$, and $z^* = 0$.

Sometimes an LP may be defined such that the feasible region does not exist. Such LPs are said to be *inconsistent*.

Example 7. Consider the LP

$$\begin{aligned} \max\{\ \text{whatever} &= z\} \\ \text{s.t.} \quad x_1 + \ x_2 &\leqq 2 \\ 2x_1 + 2x_2 &\geqq 8 \\ x_1, x_2 &\geqq 0 \end{aligned}$$

that is graphed in Fig. 3.8.

Since no points exist that satisfy the constraints, it does not matter what the objective function is because S is empty. Since this LP is inconsistent, $\Theta = \Theta_x = \Theta_\mu = \varnothing$ and z^* does not exist.

Examples 3 and 7 show us that there are two ways for Θ to be empty. One is when we have an unbounded optimal z-value problem, and the other is when the LP is inconsistent.

Examples 8 and 9 involve *redundant* constraints and *degenerate* points. A constraint is *redundant* when the feasible region is the same with or without it. Although some of the nonnegativity constraints may be redundant, we are only concerned about main constraints that are redundant. A point $\mathbf{x} \in S$ is *degenerate* if the number of constraints *active* at $\mathbf{x}$ is greater than the dimensionality of the space of the decision variables. A constraint is said to be active (or *binding*) at

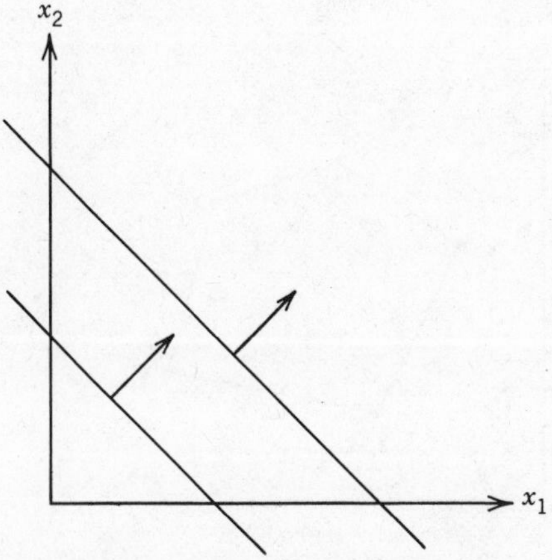

Figure 3.8. Graph of Example 7.

$\mathbf{x} \in S$ if the constraint is satisfied with equality at $\mathbf{x}$. We must remember to include nonnegativity constraints (if binding at the point in question) when counting active constraints to determine degeneracy.

Example 8. Consider the LP

$$\begin{aligned} \max\{ \quad x_1 \qquad\quad &= z\} \\ \text{s.t.} \quad 2x_1 + 4x_2 &= 8 \\ -2x_1 + x_2 &\leqq 2 \\ x_2 &\leqq 3 \\ x_1, x_2 &\geqq 0 \end{aligned}$$

that is graphed in Fig. 3.9.

In this problem, S is the line segment $\gamma(\mathbf{x}^1, \mathbf{x}^2)$. With all main constraints in the problem, $\mathbf{x}^1$ is degenerate (because the first main constraint, the second main constraint, and the x_1 nonnegativity restriction are active at $\mathbf{x}^1$). In this problem, $\Theta = \{\mathbf{x}^2\}$ and $z^* = 4$. Note that the feasible region is the same with or without the last two main constraints. Thus, constraints two and three are redundant and can be deleted from the problem if so desired.

As another example of degeneracy, let us consider one of the four-sided pyramids of Egypt. The apex of the pyramid is degenerate because we are in three-space and all four sides of the pyramid are active there. The four base extreme points are not degenerate because each is defined by only three surfaces, the bottom and two sides. Although the apex is degenerate, it is not the result of main constraint redundancy. Let us consider a five-sided pyramid. The apex of this pyramid possesses two *degrees* of degeneracy because two more constraints are active at the apex than the dimensionality of the space. A six-sided pyramid has three degrees of degeneracy at its apex, and so on.

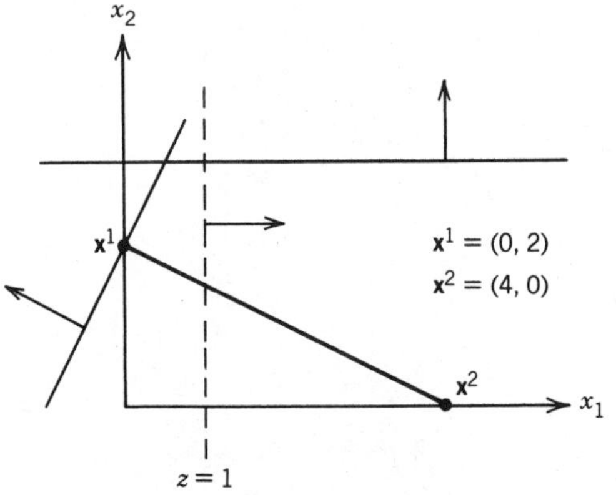

Figure 3.9. Graph of Example 8.

3.4 SIMPLEX METHOD

The *simplex method* is a computational procedure for solving linear programming problems. The simplex method iterates by moving from one extreme point to another in search of an optimal extreme point. When we perform an iteration of the simplex method, it is said that we *pivot*.

The simplex method begins its search for an optimal extreme point from the origin. However, the origin may not be feasible. We define two phases of activity. In Phase I, we move from the origin to an *initial* extreme point of the feasible region. Phase I is signified by the dashed line from the origin to $\mathbf{x}^1$ in Fig. 3.10. In Phase II, we pivot from the initial extreme point to an optimal extreme point (if one exists). Phase II is signified by the arrows along the *edge-connected* path $\mathbf{x}^1 \rightarrow \mathbf{x}^2 \rightarrow \mathbf{x}^3 \rightarrow \mathbf{x}^4 \rightarrow \mathbf{x}^5$. If Phase I cannot find an initial extreme point, it is because the feasible region S is empty. If Phase II cannot find an optimal extreme point, it is because the problem has an unbounded optimal z-value solution.

From extreme point $\mathbf{x}^1$ in Fig. 3.10, Phase II works as follows. Each of the edges $\gamma(\mathbf{x}^1, \mathbf{x}^2)$ and $\gamma(\mathbf{x}^1, \mathbf{x}^6)$ emanating from $\mathbf{x}^1$ is evaluated in terms of its *z-value improvement* per unit movement along the edge away from $\mathbf{x}^1$. Selecting $\gamma(\mathbf{x}^1, \mathbf{x}^2)$ because it is the edge with the largest positive z-value improvement index, we pivot to extreme point $\mathbf{x}^2$. We now evaluate each of the edges $\gamma(\mathbf{x}^2, \mathbf{x}^1)$ and $\gamma(\mathbf{x}^2, \mathbf{x}^3)$ emanating from $\mathbf{x}^2$. Since $\gamma(\mathbf{x}^2, \mathbf{x}^3)$ is the only one with a positive z-value improvement index, we pivot along $\gamma(\mathbf{x}^2, \mathbf{x}^3)$ to $\mathbf{x}^3$, and so forth. We continue iterating (pivoting) in this fashion until we arrive at $\mathbf{x}^5$. Since there are no edges emanating from $\mathbf{x}^5$ that have positive z-value improvement indices, $\mathbf{x}^5$ is optimal. We know, because of the convexity of S, that when an extreme point has no emanating edges with positive z-value improvement indices, it is optimal. If an LP has an unbounded optimal z-value, this is recognized in Phase II by the existence of an unbounded edge with a positive z-value improvement index.

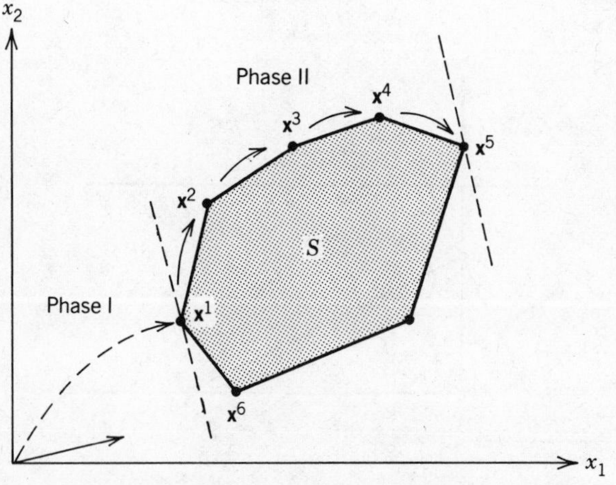

Figure 3.10. Phases I and II of the simplex method.

3.4.1 Standard Equality Constraint Format

To prepare an LP for solution by the simplex method, the LP is placed into *standard equality constraint format*. A formulation is in standard equality constraint format when it has

1. A "max" objective function.
2. All main constraints in equality form.
3. All RHS values nonnegative.
4. An $m \times m$ *initial identity matrix* embedded in the last m columns of the *constraint matrix* (where the constraint matrix is the matrix of LHS coefficients of the main constraints).

A "min" objective function is converted to a "max" objective function by multiplying its LHS by -1 and remembering that the original z-value is the negative of the new z-value. Each $\leqq$ main constraint is transformed into an equality constraint by *adding* an s-variable (sometimes called a *slack* variable) to its LHS. Each $\geqq$ main constraint is transformed into an equality constraint by *subtracting* an s-variable from its LHS. In this way, the LHSs of the inequality constraints are brought into equality with their RHSs. Each s-variable entered into the problem must be nonnegative. A negative RHS value is made positive by multiplying the constraint in question by -1.

We now discuss the initial identity matrix. We adopt the convention that structural variable columns *not* be used in forming the initial identity matrix. Since an s-variable column with a -1 coefficient is unusable as an identity matrix column, how do we obtain an $m \times m$ initial identity matrix at the end of the constraint matrix if there are not enough s-variables with $+1$ coefficients? We introduce *artificial* variables (a-variables). That is, we add an a-variable to each main constraint not possessing an s-variable with a $+1$ coefficient. Artificial variables must always be nonnegative. Also, each artificial variable inserted into the problem appears in the (maximization) objective function with a $-M$ coefficient. M is a *sufficiently large* positive number. We try to use as few artificial variables as possible.

Example 9. Consider the structural variable formulation

$$\min\{-x_1 \qquad = z\}$$

$$\begin{aligned} \text{s.t.} \quad 2x_1 - x_2 &\leqq 1 \\ - x_2 &\geqq -4 \\ -x_1 - x_2 &= -5 \\ -x_1 + x_2 &\geqq 2 \\ x_1, x_2 &\geqq 0 \end{aligned}$$

In standard equality constraint format, we have

$$\begin{aligned}
\max\{\ & x_1 & & & & - Ma_6 - Ma_7 = z\} \\
\text{s.t.}\quad & 2x_1 - x_2 \quad + s_4 & & & & = 1 \\
& \qquad x_2 & & + s_5 & & = 4 \\
& x_1 + x_2 & & & + a_6 & = 5 \\
& -x_1 + x_2 - s_3 & & & & + a_7 = 2 \\
& & & & & \text{all vars} \geqq 0
\end{aligned}$$

In this problem, the s-variable in the fourth constraint is entered first so as not to be in the way of the 4×4 initial identity matrix formed by the s_4, s_5, a_6, and a_7 columns.

Artificial variables are required when the origin is not a member of the feasible region (or is not recognized as a member of S because of degeneracy). The purpose of the $-M$ objective function coefficient of each artificial variable is to penalize very severely the objective function value whenever any artificial variables are positive. The use of $-M$ coefficients with artificial variables is sometimes called the "big M" method. The $-M$'s of the big M method cause the simplex method to give first priority to the task of driving all artificial variables to zero. Once all artificial variables are zero, we are feasible and have arrived at an extreme point of S. From this extreme point, we commence with Phase II. When we cannot drive all artificial variables to zero, it is because the LP is inconsistent.

3.4.2 Initial Tableau

The calculations of the simplex method are performed in *tableaus*. When we pivot from the *initial* tableau, we generate a *second* tableau. When we pivot from the *second* tableau, we generate a *third* tableau, and so forth. Associated with each tableau is an extreme point. Thus, by pivoting, we move through a path of *adjacent* extreme points in search of an optimal extreme point. A simplex tableau is organized as in Fig. 3.11.

In matrix notation, an LP in standard equality constraint format is written as

$$\begin{aligned}
\max\{\ & \mathbf{cx} = z\} \\
\text{s.t.}\quad & \mathbf{Ax} = \mathbf{b} \\
& \mathbf{x} \geqq \mathbf{0}
\end{aligned}$$

where $\mathbf{c}$ is the *row* vector of objective function coefficients, $\mathbf{A}$ is $m \times n$, an identity matrix constitutes the last m columns of $\mathbf{A}$, and $\mathbf{b} \geqq \mathbf{0}$. To load the initial simplex tableau, we place $\mathbf{A}$ in the $\mathbf{Y}$-matrix area and $\mathbf{b}$ in the $\bar{\mathbf{b}}$-column. The

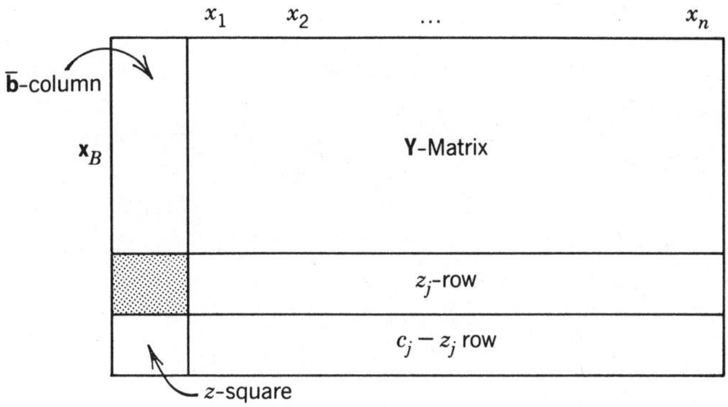

Figure 3.11. Organization of a simplex tableau.

vector of *basic* variables $\mathbf{x}_B \in R^m$ consists of those variables whose columns make up the initial identity matrix. All variables not in $\mathbf{x}_B$ are *nonbasic* variables. Let $\mathbf{c}_B \in R^m$ be the row vector of objective function coefficients pertaining to the variables in $\mathbf{x}_B$, and let $\mathbf{y}_j$ be the jth column of $\mathbf{Y}$. Then, we have

1. $\mathbf{c}_B\bar{\mathbf{b}}$ as the z-square value.
2. $\mathbf{c}_B\mathbf{y}_j$ as the jth element in the z_j-row.
3. $c_j - \mathbf{c}_B\mathbf{y}_j$ as the jth element in the $c_j - z_j$ row.

Example 10. Consider the LP

$$\begin{aligned} \max\{\ 4x_1 + 7x_2 &= z\} \\ \text{s.t.} \quad 2x_1 + 3x_2 &\leqq 12 \\ -x_1 + x_2 &\geqq 2 \\ x_1, x_2 &\geqq 0 \end{aligned}$$

With $\mathbf{c}_B = (0, -M)$, the initial simplex tableau is given in Table 3.1.

TABLE 3.1
Initial Simplex Tableau of Example 10

		x_1	x_2	s_3	s_4	a_5	
s_4	12	2	3	0	1	0	
a_5	2	-1	1	-1	0	1	
		M	$-M$	M	0	$-M$	z_j-row
z	$-2M$	$4-M$	$7+M$	$-M$	0	0	$c_j - z_j$ row

3.4.3 Pivoting Procedure

In the tableau of a nondegenerate extreme point (degeneracy is discussed later), each nonbasic variable corresponds to an emanating edge. The $c_j - z_j$ elements in the columns of the nonbasic variables are the z-value improvement indices for the emanating edges. Therefore, from each extreme point, we will want to pursue the emanating edge with the largest positive $c_j - z_j$ value to the extreme point at its other end. This is accomplished by making basic the nonbasic variable in question, and making nonbasic one of the currently basic variables. The *pivoting procedure* by which this is done is as follows:

1. The *entering* nonbasic variable is specified by the column of the tableau that has the largest positive $c_j - z_j$ element in it. If no positive elements exist, go to Step 6.
2. To select the *remove* variable, consider only *positive* elements in the entering column of the **Y**-matrix. For each such positive y_{ij}-element, form a ratio with the y_{ij} element and the corresponding value in the $\bar{\mathbf{b}}$-column, with the y_{ij}-element as the denominator. The smallest ratio determines the variable to be removed from the *basis* (where the set of basic variables refers to the basis of a tableau).
3. Let the row of the remove variable be termed the *remove row*. Let the y_{ij}-element specified by the intersection of the remove row and entering column be termed the *pivot element*. Now, start a new tableau.
4. Using row operations involving the remove row, compute the $\bar{\mathbf{b}}$-column, the **Y**-matrix, and $c_j - z_j$ row.
5. Compute the z-square value and the elements in the z_j-row from their definitions. Then, go to Step 1.
6. The present tableau is an *optimal* tableau. Stop.

To illustrate the pivoting procedure, consider the LP

$$\begin{aligned} \max\{\ x_1 \qquad &= z\} \\ \text{s.t.} \qquad x_1 + x_2 &= 3 \\ 2x_1 - x_2 &\leqq 1 \\ 2x_2 &\geqq 1 \\ x_1, x_2 &\geqq 0 \end{aligned}$$

whose graph is in Fig. 3.12. With $\mathbf{c}_B = (-M, 0, -M)$, the initial simplex tableau is given in Table 3.2. In Table 3.2, for instance,

$$\begin{aligned} z\text{-square} = \mathbf{c}_B\bar{\mathbf{b}} \qquad &= -M(3) + 0(1) - M(1) = -4M \\ z_1 = \mathbf{c}_B\mathbf{y}_1 \qquad &= -M(1) + 0(2) - M(0) = -M \\ c_1 - z_1 = c_1 - \mathbf{c}_B\mathbf{y}_1 = 1 - (-M) \qquad &= 1 + M \end{aligned}$$

In Step 1 of the pivoting procedure, x_2 is determined to be the entering nonbasic variable because $3M$ is the largest positive $c_j - z_j$ element. By forming

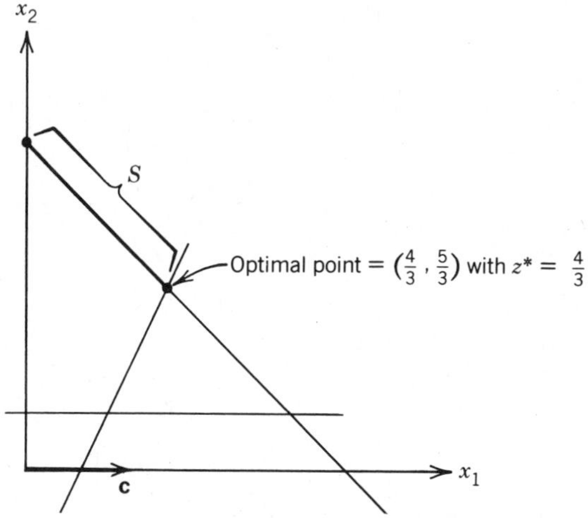

Figure 3.12. Graph of illustrative LP.

the ratios $\frac{3}{1}$ and $\frac{1}{2}$ in accordance with Step 2, a_6 is determined to be the variable to be removed from the basis. In Step 3, we put a square around the pivot element and start a new (second) tableau. In Step 4, we perform row operations on the rows of the most recent tableau to form the rows of the new tableau. All row operations involve the remove row. The row operations are carried out so that in the new tableau, the column that was the entering column in the most recent tableau has

1. An identity matrix column in the **Y**-matrix with the one where the pivot element was.
2. A zero as its $c_j - z_j$ element.

Let R_{a_4} be the a_4 row of the most recent tableau and R'_{a_4} (with a prime) the a_4 row of the new tableau. Thus, to form the $\bar{\mathbf{b}}$-column, **Y**-matrix, and the $c_j - z_j$

TABLE 3.2
First Simplex Tableau

		x_1	x_2	s_3	a_4	s_5	a_6	
a_4	3	1	1	0	1	0	0	
s_5	1	2	-1	0	0	1	0	
a_6	1	0	$\boxed{2}$	-1	0	0	1	
		$-M$	$-3M$	M	$-M$	0	$-M$	z_j
z	$-4M$	$1+M$	$3M$	$-M$	0	0	0	$c_j - z_j$

TABLE 3.3
Second Simplex Tableau

		x_1	x_2	s_3	a_4	s_5	a_6	
a_4	$\frac{5}{2}$	1	0	$\frac{1}{2}$	1	0	$-\frac{1}{2}$	
s_5	$\frac{3}{2}$	$\boxed{2}$	0	$-\frac{1}{2}$	0	1	$\frac{1}{2}$	
x_2	$\frac{1}{2}$	0	1	$-\frac{1}{2}$	0	0	$\frac{1}{2}$	
		$-M$	0	$-M/2$	$-M$	0	$M/2$	z_j
z	$-5M/2$	$1+M$	0	$M/2$	0	0	$-3M/2$	$c_j - z_j$

row, we perform the following row operations:

$$R'_{a_4} = R_{a_4} - \tfrac{1}{2}R_{a_6}$$

$$R'_{s_5} = R_{s_5} + \tfrac{1}{2}R_{a_6}$$

$$R'_{x_2} = \tfrac{1}{2}R_{a_6}$$

$$[c_j - z_j \text{ row}]' = [c_j - z_j \text{ row}] - \frac{3M}{2}R_{a_6}$$

Calculating the z-square and z_j-row elements from their definitions as in Step 5, we have Table 3.3.

In the second tableau, x_1 is the entering variable, s_5 is the remove variable, and the pivot element is as indicated. Performing the row operations

$$R'_{a_4} = R_{a_4} - \tfrac{1}{2}R_{s_5}$$

$$R'_{x_1} = \tfrac{1}{2}R_{s_5}$$

$$R'_{x_2} = R_{x_2}$$

$$[c_j - z_j \text{ row}]' = [c_j - z_j \text{ row}] - \frac{1+M}{2}R_{s_5}$$

and computing the z-square and z_j-row elements from their definitions, we have the third tableau in Table 3.4. Continuing the pivoting procedure with y_{13} as the pivot element and

$$R'_{s_3} = \tfrac{4}{3}R_{a_4}$$

$$R'_{x_1} = R_{x_1} + \tfrac{1}{3}R_{a_4}$$

$$R'_{x_2} = R_{x_2} + \tfrac{2}{3}R_{a_4}$$

$$[c_j - z_j \text{ row}]' = [c_j - z_j \text{ row}] - (\tfrac{1}{3} + M)R_{a_4}$$

we have the fourth tableau in Table 3.5. In the fourth tableau, we not only

TABLE 3.4
Third Simplex Tableau

		x_1	x_2	s_3	a_4	s_5	a_6	
a_4	$\frac{7}{4}$	0	0	$\boxed{\frac{3}{4}}$	1	$-\frac{1}{2}$	$-\frac{3}{4}$	
x_1	$\frac{3}{4}$	1	0	$-\frac{1}{4}$	0	$\frac{1}{2}$	$\frac{1}{4}$	
x_2	$\frac{1}{2}$	0	1	$-\frac{1}{2}$	0	0	$\frac{1}{2}$	
		1	0	$-\frac{1}{4} - 3M/4$	$-M$	$\frac{1}{2} + M/2$	$\frac{1}{4} + 3M/4$	z_j
z	$\frac{3}{4} - 7M/4$	0	0	$\frac{1}{4} + 3M/4$	0	$-\frac{1}{2} - M/2$	$-\frac{1}{4} - 7M/4$	$c_j - z_j$

TABLE 3.5
Fourth Simplex Tableau

		x_1	x_2	s_3	a_4	s_5	a_6	
s_3	$\frac{7}{3}$	0	0	1	$\frac{4}{3}$	$-\frac{2}{3}$	-1	
x_1	$\frac{4}{3}$	1	0	0	$\frac{1}{3}$	$\frac{1}{3}$	0	
x_2	$\frac{5}{3}$	0	1	0	$\frac{2}{3}$	$-\frac{1}{3}$	0	
		1	0	0	$\frac{1}{3}$	$\frac{1}{3}$	0	z_j
z	$\frac{4}{3}$	0	0	0	$-\frac{1}{3} - M$	$-\frac{1}{3}$	$-M$	$c_j - z_j$

achieve feasibility at $(x_1, x_2) = (4/3, 5/3)$, but this point turns out to be optimal as well. Thus, we are done with the problem. Normally, however, we expect about $2m$ Phase II pivots (where m is the number of main constraints) before the problem is solved.

3.4.4 Method of Rectangles

When forming a new tableau, the column that was the entering column in the most recent tableau is easily formed (an identity matrix column with a zero in the $c_j - z_j$ row). Also, the row that was the remove row in the most recent tableau is easily calculated (just divide by the pivot element). As an alternative to using row operations to calculate the other $\bar{\mathbf{b}}$-column, $\mathbf{Y}$-matrix, and $c_j - z_j$ elements, we can use the *Method of Rectangles*.

In the Method of Rectangles, we imagine rectangles in the most recent tableau anchored at the pivot element as in Fig. 3.13. The elements at the corners of the rectangles are used to calculate the elements in the next tableau at the location of the corner diagonally opposite the pivot element.

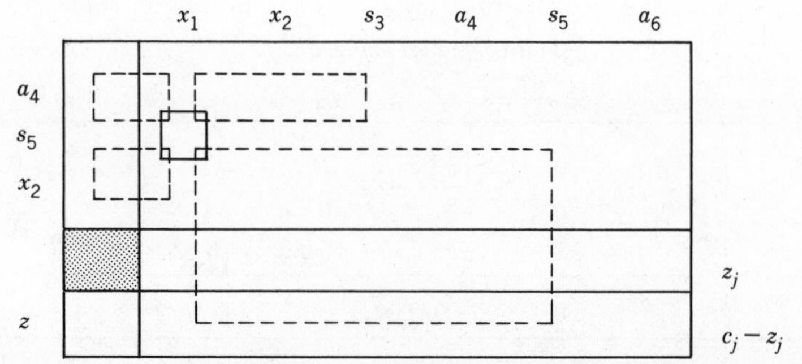

Figure 3.13. Method of rectangles.

Let us call the element at the corner diagonally opposite the pivot element in the most recent tableau the *old element*, and the element at the same location, but in the new tableau, the *new element*. Then, the formula for calculating the new element is

$$\text{New element} = \text{Old element} - \frac{\text{Product of other two diagonal elements}}{\text{Pivot element}}$$

Let us superimpose the rectangles of Fig. 3.13 onto the second tableau in Table 3.3 to calculate the new elements in the third tableau. Then, we have

$$\bar{b}_1 = \tfrac{5}{2} - \frac{(\tfrac{3}{2})(1)}{2} = \tfrac{7}{4}$$

$$y_{13} = \tfrac{1}{2} - \frac{(1)(-\tfrac{1}{2})}{2} = \tfrac{3}{4}$$

$$c_5 - z_5 = 0 - \frac{(1+M)(1)}{2} = -\tfrac{1}{2} - M/2$$

$$\bar{b}_3 = \tfrac{1}{2} - \frac{(0)(\tfrac{3}{2})}{2} = \tfrac{1}{2}$$

and so forth. Please note that the Method of Rectangles does not work on the entering column, remove row, z-square, and z_j-row elements.

3.4.5 Inconsistency and Unboundedness

In the simplex method, *inconsistency* (empty feasible region) is recognized when an artificial variable exists at a positive value in an *optimal* tableau. Figure 3.14 portrays how inconsistency might appear.

If, in the simplex method, we have a nonpositive **Y**-matrix column, S is unbounded. If we have a nonpositive **Y**-matrix column accompanied by a positive $c_j - z_j$ element, then our LP has an unbounded optimal z-value. Notice that we cannot form a pivot element in such a column. Sooner or later, we will encounter such a condition in an unbounded optimal z-value problem. Figure 3.15 portrays the occurrence of such a situation.

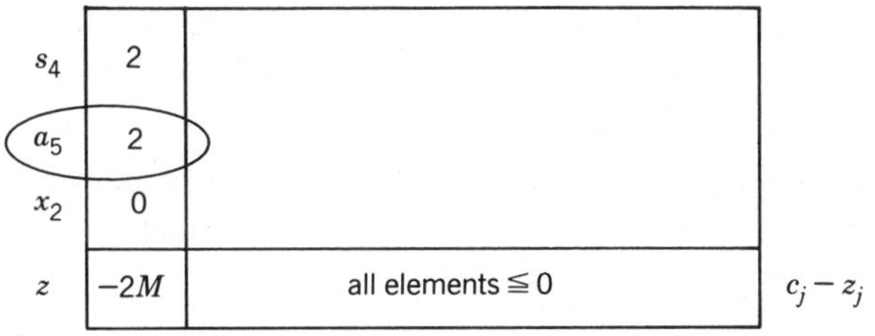

Figure 3.14. Simplex example of inconsistency.

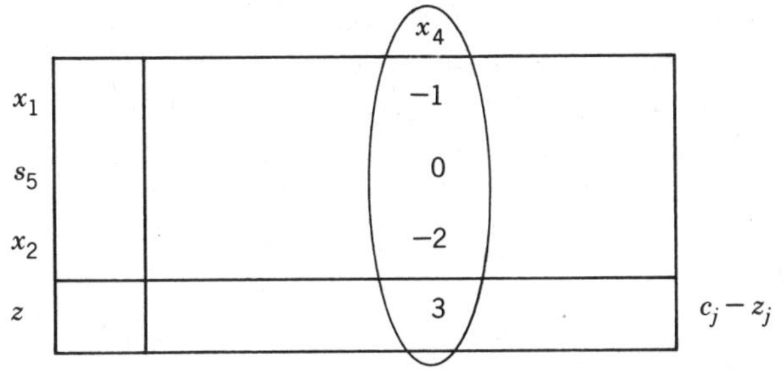

Figure 3.15. Simplex example of an unbounded optimal z-value problem.

3.4.6 Degeneracy and Alternative Optima

An extreme point is *degenerate* if one or more of its basic variable values is zero. In the simplex method, an extreme point is degenerate if and only if in its tableau one or more $\bar{\mathbf{b}}$-column elements are zero. The extreme point of the tableau of Fig. 3.16 is degenerate. Whereas there is only one tableau that pertains to a given *nondegenerate* extreme point, several different tableaus might pertain to a given degenerate extreme point.

The term *alternative optima* means there is more than one optimal solution. From a nondegenerate optimal extreme point, the issue is definitive (Case 1). From a degenerate optimal extreme point, the issue is *not* definitive (Case 2).

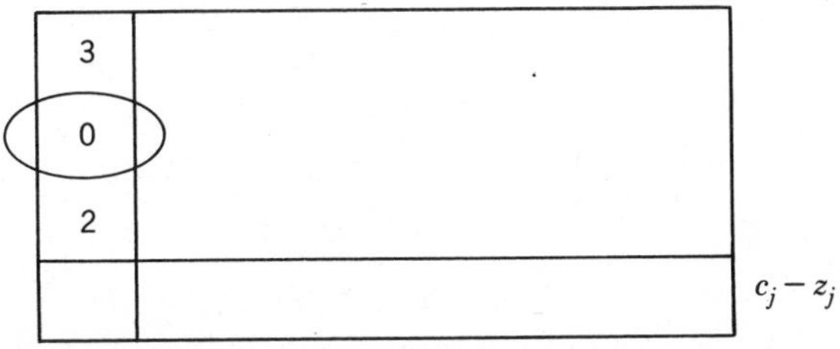

Figure 3.16. Tableau of a degenerate extreme point.

Case 1: If the extreme point of an optimal tableau is *nondegenerate*, alternative optima *exist* if and only if there exists a nonbasic $c_j - z_j$ element whose value is zero.

Case 2: If the extreme point of an optimal tableau is *degenerate*, alternative optima *may exist* if there exists a nonbasic $c_j - z_j$ element whose value is zero.

To pivot to another *optimal* extreme point in Case 1, we use as the entering variable a nonbasic variable whose $c_j - z_j$ element is zero. If a pivot element cannot be found in this column, it is because the emanating edge pertaining to the nonbasic variable is unbounded. To pivot to another *optimal basis* in Case 2, we use a nonbasic variable whose $c_j - z_j$ element is zero as the entering variable. Whether this leads to another extreme point or simply another optimal tableau pertaining to the same extreme point depends on the problem. Suffice it to say here that any optimal extreme point can be reached from any other optimal extreme point by only pivoting among optimal bases. More is said about alternative optima in Chapter 4.

3.5 DUALITY THEORY

Every LP has an associated LP that is, in a sense, its opposite. If one LP is called the *primal*, its "opposite" is called the *dual*. Since we toggle between the primal and the dual (the dual of the dual is the primal), it does not matter whether we call a given problem the dual or the primal. To form the dual of an LP, we use the following *duality lists*:

Primal (or Dual)	***Dual (or Primal)***
max	min
objective function coefficients	RHS
RHS	objective function coefficients
jth column of constraint coefficients	jth row of constraint coefficients
ith row of constraint coefficients	ith column of constraint coefficients
jth variable $\geqq 0$	jth constraint $\geqq$
jth variable unrestricted	jth constraint $=$
jth variable $\leqq 0$	jth constraint $\leqq$
ith constraint $\leqq$	ith variable $\geqq 0$
ith constraint $=$	ith variable unrestricted
ith constraint $\geqq$	ith variable $\leqq 0$

To obtain the dual of a *maximization* LP, we convert from left to right. To obtain the dual of a *minimization* LP, we convert from right to left. From the duality lists we have, in summary, the following: (a) if we are maximizing in the primal, we are minimizing in the dual; (b) the objective function coefficients of the primal become the RHS values of the dual; (c) the ith row of constraint coefficients in the primal becomes the ith column of constraint coefficients in the dual; (d) if the jth variable is unrestricted in the primal, the jth constraint in the dual is an equality, and so forth. Noting that there is a *dual variable* for each primal main constraint, we arrive at Example 11.

Example 11. Let the following LP be the primal (with the dual variable for each main constraint being written on the left):

$$\begin{array}{l|rl}
 & \max\{ & 3x_1 + 2x_2 = z\} \\
w_1 & \text{s.t.} & 4x_1 + 5x_2 \leqq 25 \\
w_2 & & -x_1 + x_2 \leqq 2 \\
w_3 & & x_1 \qquad \leqq 4 \\
 & & x_1, x_2 \geqq 0
\end{array}$$

Converting from the left duality list to the right duality list, the dual is

$$\begin{aligned}
&\min\{25w_1 + 2w_2 + 4w_3 = v\} \\
\text{s.t.}\quad & 4w_1 - w_2 + w_3 \geqq 3 \\
& 5w_1 + w_2 \qquad \geqq 2 \\
& w_1, w_2, w_3 \geqq 0
\end{aligned}$$

Some important relationships between a primal (P) and its dual (D) are as follows.

1. The following four situations are mutually exclusive and collectively exhaustive:
   a. (P) inconsistent, (D) inconsistent.
   b. (P) inconsistent, (D) unbounded optimal z-value.
   c. (P) unbounded optimal z-value, (D) inconsistent.
   d. (P) bounded optimal z-value, (D) bounded optimal z-value.
2. If $\mathbf{x}$ is a feasible solution vector in a maximization (P) and $\mathbf{w}$ is a feasible solution vector in its (D), the objective function value of (D) at $\mathbf{w}$ is *greater than or equal* to the objective function value of (P) at $\mathbf{x}$.
3. If $\mathbf{x}^*$ is optimal in (P) and $\mathbf{w}^*$ is optimal in (D), the objective function value of (P) at $\mathbf{x}^*$ is *equal* to the objective function value of (D) at $\mathbf{w}^*$. (That is, $z^* = \mathbf{cx}^* = \mathbf{b}^T\mathbf{w}^* = v^*$.)
4. An optimal solution $\mathbf{w}^*$ of (D) is found in the last m elements of the z_j-row of an optimal primal tableau, and an optimal solution $\mathbf{x}^*$ of (P) is found in the last n elements of the z_j-row of an optimal dual tableau.

Example 12. Consider the LP of Section 3.4.3

$$
\begin{array}{r|l}
 & \max\{\ x_1 \qquad\quad = z\} \\
w_1 & \text{s.t.} \quad x_1 + x_2 = 3 \\
w_2 & \qquad 2x_1 - x_2 \leqq 1 \\
w_3 & \qquad\quad 2x_2 \geqq 1 \\
 & \qquad x_1, x_2 \geqq 0
\end{array}
$$

Its dual is

$$
\begin{aligned}
\min\{3w_1 + w_2 + w_3 &= v\} \\
\text{s.t.} \quad w_1 + 2w_2 \qquad &\geqq 1 \\
w_1 - w_2 + 2w_3 &\geqq 0 \\
w_1 &\ \text{unrestricted} \\
w_2 &\geqq 0 \\
w_3 &\leqq 0
\end{aligned}
$$

From the optimal primal tableau of Table 3.5, $\mathbf{w}^* = (\frac{1}{3}, \frac{1}{3}, 0)$ and $z^* = v^* = \frac{4}{3}$.

3.6 NONPOSITIVE AND UNRESTRICTED VARIABLES

Since the simplex method presumes that all variables are nonnegative, how do we solve an LP that has nonpositive or unrestricted variables? For each occurrence of a nonpositive variable x_ℓ, we perform the substitution $-x_\ell^-$ where $x_\ell^- \geqq 0$. For each occurrence of an unrestricted variable x_ℓ, we perform the substitution $x_\ell^+ - x_\ell^-$ where $x_\ell^+, x_\ell^- \geqq 0$.

Example 13. Consider the LP

$$
\begin{aligned}
\max\{\ -x_1 + 2x_2 - x_3 &= z\} \\
\text{s.t.} \quad 5x_1 + 4x_2 - 6x_3 &\leqq 20 \\
-4x_1 + 2x_2 - x_3 &\leqq 16 \\
x_1 - x_2 + 6x_3 &\leqq 18 \\
x_2 \qquad &\geqq 1 \\
x_2 \qquad &\leqq 4 \\
x_3 &\leqq 0 \\
x_1 &\ \text{unrestricted}
\end{aligned}
$$

Using a *tabular* presentation to present the reformulation of the problem with all nonnegative variables, we have Table 3.6.

In Table 3.6, $x_1 = x_1^+ - x_1^-$ and $x_3 = -x_3^-$. Since the optimal solution of the reformulated problem is $x_1^+ = 0.0$, $x_1^- = 1.51724$, $x_2 = 4.0$, and $x_3^- = 1.93103$,

TABLE 3.6
Tabular Presentation of the Reformulated LP of Example 13

	x_1^+	x_1^-	x_2	x_3^-		RHS
Obj	-1	1	2	1	max	
	5	-5	4	6	$\leqq$	20
	-4	4	2	1	$\leqq$	16
s.t.	1	-1	-1	-6	$\leqq$	18
			1		$\geqq$	1
			1		$\leqq$	4

all vars $\geqq 0$

the optimal solution to the original LP is $x_1 = -1.51724$, $x_2 = 4.0$, and $x_3 = -1.93103$.

3.7 MPSX INPUT FORMAT

With regard to the computer solution of an LP, most of the widely used commercial LP packages require problems to be specified in **MPSX** *input format*. In **MPSX** input format, the problem itself, each row, and each column are given alphanumeric names of up to eight characters. In addition, each row is assigned one of the following *row-type indicators*:

- **N** If row is nonconstrained.
- **G** If row is $\geqq$
- **L** If row is $\leqq$
- **E** If row is $=$

Also, with *bounded variable cards*, **MPSX** allows for the convenient specification of variables that have lower or upper bounds, are unrestricted in sign, and are to be held fixed at a certain value. This is accomplished by using the following *variable indicators*:

- **LO** Variable has a lower bound.
- **UP** Variable has an upper bound.
- **FR** Variable is *free* (unrestricted).
- **FX** Variable is to be fixed in value.

With regard to the **UP** variable indicator, **MPSX** format has the convention that if a variable is specified with an upper bound of zero, the variable is assumed to be nonpositive. The different parts of an LP input deck in **MPSX** format are illustrated in Fig. 3.17.

In the LHS Coefficient and RHS sections, it is good style to place only one coefficient on a card. This makes proofing and subsequent modifications to the deck easier. Note that the cards for each column of data must be grouped together in the LHS coefficients section. Note that if **FR** is used in a bounded

First Card

NAME 1–4	Name of MPSX Deck 15–22

Second Card

ROWS 1–4

Third Card

N 2	Objective Name 5–12

Next m Cards

G, L or E 2	Row Name 5–12

Next Card

COLUMNS 1–7

LHS Coefficient Cards

Column Name 5–12	Row Name 15–22	Coefficient Value 25–36	Row Name 40–47	Coefficient Value 50–61

Next Card

RHS 1–3

RHS Value Cards

RHS Name 5–12	Row Name 15–22	RHS Value 25–36	Row Name 40–47	RHS Value 50–61

Next Card (Optional)

BOUNDS 1–6

Bounded Variable Cards (Optional)

LO, UP, FR or FX 2–3	Bounds Vector Name 5–12	Variable Name 15–22	Upper or Lower Bound 25–36

Last Card

ENDATA 1–6

Figure 3.17. Different parts of an MPSX input deck.

```
NAME          ORIGINAL
ROWS
 N  OBJ
 L  R1
 L  R2
 L  R3
COLUMNS
    X1        OBJ          -1.0
    X1        R1            5.0
    X1        R2           -4.0
    X1        R3            1.0
    X2        OBJ           2.0
    X2        R1            4.0
    X2        R2            2.0
    X2        R3           -1.0
    X3        OBJ          -1.0
    X3        R1           -6.0
    X3        R2           -1.0
    X3        R3            6.0
RHS
    RVECT     R1           20.0
    RVECT     R2           16.0
    RVECT     R3           18.0
BOUNDS
 FR BNDVECT   X1
 LO BNDVECT   X2            1.0
 UP BNDVECT   X2            4.0
 UP BNDVECT   X3            0.0
ENDATA
```

Figure 3.18. **MPSX** deck for original LP with a bounds section.

variable card, nothing is entered in columns 25–36. If **FX** is used, the value at which the variable is to be fixed is entered in columns 25–36. More about **MPSX** input format is described in [IBM (1976, 1979)].

Example 14. The **MPSX** input deck for the original LP of Example 13 is given in Fig. 3.18. Without a **BOUNDS** section, an **MPSX** input deck for the reformulated LP of Example 13 is given in Fig. 3.19. In Fig. 3.19, note that it is possible to include special symbols in the alphanumeric name specifications.

3.8 ANATOMY OF A SIMPLEX TABLEAU

Consider the LP

$$\begin{aligned} \max\{ \ \mathbf{cx} &= z \} \\ \text{s.t.} \quad \mathbf{Ax} &= \mathbf{b} \\ \mathbf{x} &\geqq \mathbf{0} \end{aligned}$$

where $\mathbf{c}$ is the row vector of objective function coefficients and $\mathbf{A}$ is of full row

```
NAME          REFORM
ROWS
 N  OBJ.TIVE
 L  R1
 L  R2
 L  R3
 G  R4
 L  R5
COLUMNS
    X(1)+     OBJ.TIVE     -1.0
    X(1)+     R1            5.0
    X(1)+     R2           -4.0
    X(1)+     R3            1.0
    X(1)-     OBJ.TIVE      2.0
    X(1)-     R1           -5.0
    X(1)-     R2            4.0
    X(1)-     R3           -1.0
    X(2)      OBJ.TIVE      2.0
    X(2)      R1            4.0
    X(2)      R2            2.0
    X(2)      R3           -1.0
    X(2)      R4            1.0
    X(2)      R5            1.0
    X(3)-     OBJ.TIVE      1.0
    X(3)-     R1            6.0
    X(3)-     R2            1.0
    X(3)-     R3           -6.0
RHS
    RVECT     R1           20.0
    RVECT     R2           16.0
    RVECT     R3           18.0
    RVECT     R4            1.0
    RVECT     R5            4.0
ENDATA
```

Figure 3.19. MPSX deck for reformulated LP without a bounds section.

rank. Let $\mathbf{x} \in R^n$ be partitioned such that $\mathbf{x} = (\mathbf{x}_N, \mathbf{x}_B)$ where $\mathbf{x}_N$ is the subvector of nonbasic variables and $\mathbf{x}_B$ is the subvector of basic variables. Thus, let $\mathbf{A}$ be partitioned such that $\mathbf{A} = [\mathbf{N}; \mathbf{B}]$, where $\mathbf{N}$ consists of the nonbasic columns of $\mathbf{A}$ and $\mathbf{B}$ consists of the basic columns. Similarly, let $\mathbf{c} = (\mathbf{c}_N, \mathbf{c}_B)$. Since $\mathbf{B}$ is a basis in R^m, $\mathbf{B}^{-1}$ exists. Hence, we have

$$\mathbf{b} = \mathbf{A}\mathbf{x}$$

$$\mathbf{b} = \mathbf{N}\mathbf{x}_N + \mathbf{B}\mathbf{x}_B$$

$$\mathbf{B}^{-1}\mathbf{b} = \mathbf{B}^{-1}\mathbf{N}\mathbf{x}_N + \mathbf{B}^{-1}\mathbf{B}\mathbf{x}_B$$

$$\mathbf{B}^{-1}\mathbf{b} = \mathbf{B}^{-1}\mathbf{N}\mathbf{x}_N + \mathbf{I}\mathbf{x}_B$$

$$\mathbf{B}^{-1}\mathbf{b} = \mathbf{x}_B \qquad \text{(because } \mathbf{x}_N = \mathbf{0}\text{)}$$

TABLE 3.7
Partitioned Simplex Tableau Specifying Quantities in Terms of Their Definitions

		$\mathbf{x}_N$	$\mathbf{x}_B$	
$\mathbf{x}_B$	$\mathbf{B}^{-1}\mathbf{b}$	$\mathbf{B}^{-1}\mathbf{N}$	$\mathbf{I}$	
		$\mathbf{c}_B\mathbf{B}^{-1}\mathbf{N}$	$\mathbf{c}_B$	z_j-row
z	$\mathbf{c}_B\mathbf{B}^{-1}\mathbf{b}$	$\mathbf{c}_N - \mathbf{c}_B\mathbf{B}^{-1}\mathbf{N}$	$\mathbf{0}$	$c_j - z_j$ row

Similarly,

$$z = \mathbf{c}_B\mathbf{B}^{-1}\mathbf{b} \qquad \left(\text{substituting } \mathbf{x}_B = \mathbf{B}^{-1}\mathbf{b}\right)$$

Let $\mathbf{y}_j$ be the jth column of $\mathbf{Y} = [\mathbf{B}^{-1}\mathbf{N}; \mathbf{I}]$. Then, with $z_j = \mathbf{c}_B\mathbf{y}_j$ and $c_j - z_j = c_j - \mathbf{c}_B\mathbf{y}_j$,

$$\begin{aligned} z_j\text{-row} &= \mathbf{c}_B\mathbf{Y} \\ &= \mathbf{c}_B[\mathbf{B}^{-1}\mathbf{N}; \mathbf{I}] \\ &= \left(\mathbf{c}_B\mathbf{B}^{-1}\mathbf{N}, \mathbf{c}_B\mathbf{I}\right) \end{aligned}$$

and

$$\begin{aligned} c_j - z_j \text{ row} &= c_j\text{-row} - z_j\text{-row} \\ &= \left(\mathbf{c}_N, \mathbf{c}_B\right) - \left(\mathbf{c}_B\mathbf{B}^{-1}\mathbf{N}, \mathbf{c}_B\mathbf{I}\right) \\ &= \left(\mathbf{c}_N - \mathbf{c}_B\mathbf{B}^{-1}\mathbf{N}, \mathbf{c}_B - \mathbf{c}_B\mathbf{I}\right) \\ &= \left(\mathbf{c}_N - \mathbf{c}_B\mathbf{B}^{-1}\mathbf{N}, \mathbf{0}\right) \end{aligned}$$

Thus, in terms of the definitions of the quantities involved, we have the partitioned simplex tableau of Table 3.7.

Note that the objective function coefficients only affect the z-square, z_j-row, and $c_j - z_j$ row. Let us also note that in the partitioned simplex tableau, the columns of $\mathbf{B}^{-1}$ are found in the columns of the current $\mathbf{Y}$-matrix under the variables that were basic in the initial tableau. This is due to the fact that $\mathbf{Y} = \mathbf{B}^{-1}\mathbf{A}$ and that the columns of $\mathbf{A}$ pertaining to the basis of the initial tableau formed an identity matrix.

Let $J_B = \{i \mid x_i \text{ is a basic variable}\}$. Then, by knowing $\mathbf{A}$, $\mathbf{b}$, $\mathbf{c}$, J_B, and the basis inverse $\mathbf{B}^{-1}$, any element of the simplex tableau can be calculated *on demand* from its definitions.

TABLE 3.8
Tableau of Example 15

		x_1	x_2	s_3	s_4	
s_3	9	0	$\frac{19}{5}$	1	$-\frac{3}{5}$	
x_1	2	1	$\frac{2}{5}$	0	$\frac{1}{5}$	
		5	2	0	1	z_j-row
z	10	0	1	0	-1	$c_j - z_j$ row

Example 15. Consider the LP

$$\begin{aligned} \max \{5x_1 + 3x_2 & = z\} \\ \text{s.t.} \quad 3x_1 + 5x_2 + s_3 & \leqq 15 \\ 5x_1 + 2x_2 \quad + s_4 & \leqq 10 \\ x_1, x_2, s_3, s_4 & \geqq 0 \end{aligned}$$

for which

$$\mathbf{c} = (5, 3, 0, 0)$$

$$\mathbf{A} = [\mathbf{a}_1, \mathbf{a}_2, \mathbf{a}_3, \mathbf{a}_4] = \begin{bmatrix} 3 & 5 & 1 & 0 \\ 5 & 2 & 0 & 1 \end{bmatrix}$$

$$\mathbf{b} = \begin{bmatrix} 15 \\ 10 \end{bmatrix}$$

In the tableau of Table 3.8, we verify that

$$J_B = \{3, 1\}$$

$$\mathbf{B}^{-1} = [\mathbf{a}_3, \mathbf{a}_1]^{-1} = \begin{bmatrix} 1 & 3 \\ 0 & 5 \end{bmatrix}^{-1} = \begin{bmatrix} 1 & -\frac{3}{5} \\ 0 & \frac{1}{5} \end{bmatrix} = [\mathbf{y}_3, \mathbf{y}_4]$$

$$z = \mathbf{c}_B \mathbf{B}^{-1} \mathbf{b} = [0 \quad 5] \begin{bmatrix} 1 & -\frac{3}{5} \\ 0 & \frac{1}{5} \end{bmatrix} \begin{bmatrix} 15 \\ 10 \end{bmatrix} = 10$$

$$\mathbf{y}_1 = \mathbf{B}^{-1} \mathbf{a}_1 = \begin{bmatrix} 1 & -\frac{3}{5} \\ 0 & \frac{1}{5} \end{bmatrix} \begin{bmatrix} 3 \\ 5 \end{bmatrix} = \begin{bmatrix} 0 \\ 1 \end{bmatrix}$$

$$\mathbf{y}_2 = \mathbf{B}^{-1} \mathbf{a}_2 = \begin{bmatrix} 1 & -\frac{3}{5} \\ 0 & \frac{1}{5} \end{bmatrix} \begin{bmatrix} 5 \\ 2 \end{bmatrix} = \begin{bmatrix} \frac{19}{5} \\ \frac{2}{5} \end{bmatrix}$$

Also, in nonbasic/basic (x_2, s_4, s_3, x_1) order

$$(\mathbf{c}_B \mathbf{B}^{-1} \mathbf{N}, \mathbf{c}_B \mathbf{I}) = (2, 1, 0, 5)$$

$$(\mathbf{c}_N - \mathbf{c}_B \mathbf{B}^{-1} \mathbf{N}, \mathbf{0}) = (1, -1, 0, 0)$$

which after rearrangement into (x_1, x_2, s_3, s_4) order yields

$$z_j\text{-row} = (5, 2, 0, 1)$$

and

$$c_j - z_j \text{ row} = (0, 1, 0, -1)$$

respectively.

3.9 TWO-PHASE METHOD OF LINEAR PROGRAMMING

In Section 3.4, big M's were used to drive all artificial variables to zero and hence achieve a feasible extreme point. Then, with all big M terms in the objective function suppressed, we moved from the initial extreme point in search of an optimal extreme point. Driving all artificial variables to zero was called *Phase I*, and moving from the initial extreme point in search of an optimal solution was called *Phase II*.

In this section, we formalize the treatment of Phases I and II into two distinct optimization stages. If there are no artificial variables, there is no Phase I and we start immediately with Phase II. However, for the rest of this section, let us assume that there are artificial variables. In Phase I, we minimize the sum of all artificial variables (called the *Phase I artificial* objective function). If a Phase I artificial objective function value of zero cannot be achieved, the problem is inconsistent and we stop. If the problem is consistent, we will arrive at a Phase I optimal tableau characterizing an initial feasible extreme point. At this extreme point, we convert to the *real* objective function.* This is done in the tableau by recalculating the z-square; z_j-row (if the z_j-row is being carried); and $c_j - z_j$ row from their definitions of $\mathbf{c}_B\mathbf{B}^{-1}\mathbf{b}$, $(\mathbf{c}_B\mathbf{B}^{-1}\mathbf{N}, \mathbf{c}_B)$, and $(\mathbf{c}_N - \mathbf{c}_B\mathbf{B}^{-1}\mathbf{N}, \mathbf{0})$, respectively. All other quantities in the tableau are unaffected by the change of objective function. Then, with the real objective function in place, we proceed with Phase II under the proviso that no artificial variables ever reenter the basis.

Example 16. Consider the LP

$$\begin{aligned} \max\{-x_1 + 3x_2 &= z\} \\ \text{s.t.} \quad -x_1 + x_2 &\geqq 2 \\ x_1 + 2x_2 &= 10 \\ x_1, x_2 &\geqq 0 \end{aligned}$$

With $-a_4 - a_5$ as the Phase I artificial objective function, we have the Phase I tableaus of Tables 3.9, 3.10, and 3.11. Then, with the real objective function

*See Hadley (1962, p. 152) about what to do if one or more artificial variables are in the basis of an optimal Phase I tableau.

TABLE 3.9
Initial Phase I Tableau of Example 16

		x_1	x_2	s_3	a_4	a_5	
a_4	2	-1	[1]	-1	1	0	
a_5	10	1	2	0	0	1	
	-12	0	3	-1	0	0	$c_j - z_j$

TABLE 3.10
Second Phase I Tableau

		x_1	x_2	s_3	a_4	a_5	
x_2	2	-1	1	-1	1	0	
a_5	6	[3]	0	2	-2	1	
	-6	3	0	2	-3	0	$c_j - z_j$

TABLE 3.11
Final Phase I Tableau at $(x_1, x_2) = (2, 4)$

		x_1	x_2	s_3	a_4	a_5	
x_2	4	0	1	$-\frac{1}{3}$	$\frac{1}{3}$	$\frac{1}{3}$	
x_1	2	1	0	$\frac{2}{3}$	$-\frac{2}{3}$	$\frac{1}{3}$	
	0	0	0	0	-1	-1	$c_j - z_j$

$-x_1 + 3x_2$, we have the Phase II tableaus of Tables 3.12 and 3.13. The conversion from Table 3.11 to Table 3.12 at the initial feasible extreme point $(x_1, x_2) = (2, 4)$ is accomplished by recalculating the z-square and $c_j - z_j$ row from their definitions as follows:

$$z = \mathbf{c}_B \mathbf{B}^{-1} \mathbf{b} = 10$$

$$c_j - z_j \text{ row} = (\mathbf{c}_N - \mathbf{c}_B \mathbf{B}^{-1} \mathbf{N}, \mathbf{0}) = (\tfrac{5}{3}, -\tfrac{5}{3}, -\tfrac{2}{3}, 0, 0)$$

Note that it is necessary to rearrange the above $c_j - z_j$ elements to match the nonbasic/basic partition at $(x_1, x_2) = (2, 4)$ before placing them in the tableau of Table 3.12.

TABLE 3.12
Starting Phase II Tableau at $(x_1, x_2) = (2,4)$

		x_1	x_2	s_3	a_4	a_5	
x_2	4	0	1	$-\frac{1}{3}$	$\frac{1}{3}$	$\frac{1}{3}$	
x_1	2	1	0	$\boxed{\frac{2}{3}}$	$-\frac{2}{3}$	$\frac{1}{3}$	
z	10	0	0	$\frac{5}{3}$	—	—	$c_j - z_j$

TABLE 3.13
Optimal Tableau of Example 16

		x_1	x_2	s_3	a_4	a_5	
x_2	5	$\frac{1}{2}$	1	0	0	$\frac{1}{2}$	
s_3	3	$\frac{3}{2}$	0	1	-1	$\frac{1}{2}$	
z	15	$-\frac{5}{2}$	0	0	—	—	$c_j - z_j$

3.10 DUAL SIMPLEX ALGORITHM

In the *primal* simplex method of Section 3.4, we start with all basic variable values nonnegative, keeping them that way until all $c_j - z_j$ elements are nonpositive. In the *dual simplex method*, we start with all $c_j - z_j$ elements nonpositive, keeping them that way until all basic variable values are nonnegative. The dual simplex algorithm is as follows.

Step 1: We must be able to set up a simplex tableau with all $c_j - z_j$ elements $\leqq 0$. (The initial basis matrix **B** need not be an identity matrix).

Step 2: The variable to be *removed* from the basis is specified by the negative value in the $\bar{\mathbf{b}}$-column of greatest magnitude. If no negative values exist, go to Step 5.

Step 3: To determine the *entering* variable, consider all negative elements in the remove row of the **Y**-matrix. With the negative remove row elements, form ratios with the corresponding $c_j - z_j$ elements, with the $c_j - z_j$ elements on top. The ratio of smallest algebraic value signifies the variable to enter the basis.

Step 4: Pivot as usual. Go to Step 2.

Step 5: The present tableau is an optimal tableau. Stop.

TABLE 3.14
First Tableau of Example 17

		x_1	x_2	s_3	s_4	
s_3	-1	-1	-1	1	0	
s_4	-2	-2	$\boxed{-3}$	0	1	
		-3	-1	0	0	$c_j - z_j$

Example 17. Consider the LP

$$
\begin{aligned}
\max \{ -3x_1 - x_2 & \qquad\qquad = z \} \\
\text{s.t.} \quad x_1 + x_2 - s_3 & \qquad = 1 \\
2x_1 + 3x_2 \qquad - s_4 & = 2 \\
x_1, x_2, s_3, s_4 & \geqq 0
\end{aligned}
$$

Starting with s_3 and s_4 as basic variables

$$\mathbf{B} = \begin{bmatrix} -1 & 0 \\ 0 & -1 \end{bmatrix} \quad \text{and} \quad \mathbf{B}^{-1} = \begin{bmatrix} -1 & 0 \\ 0 & -1 \end{bmatrix}$$

Thus, we have the starting tableau of Table 3.14.

Note that the z-square is not meaningful until we have arrived at an optimal tableau. Removing s_4 from the basis because it has the negative value of greatest magnitude, x_2 is the entering variable because associated with it is the ratio of smallest value. Forming the second tableau, we obtain Table 3.15.

With s_3 as the remove variable and s_4 as the entering variable, we have the optimal tableau of Table 3.16.

Inconsistency (an empty feasible region) is detected if, associated with a negative basic variable, all **Y**-matrix elements in that row are nonnegative. In other words, we would not be able to form any ratios if we were to remove that

TABLE 3.15
Second Tableau of Example 17

		x_1	x_2	s_3	s_4	
s_3	$-\frac{1}{3}$	$-\frac{1}{3}$	0	1	$\boxed{-\frac{1}{3}}$	
x_2	$\frac{2}{3}$	$\frac{2}{3}$	1	0	$-\frac{1}{3}$	
		$-\frac{7}{3}$	0	0	$-\frac{1}{3}$	$c_j - z_j$

TABLE 3.16
Optimal Tableau of Example 17

		x_1	x_2	s_3	s_4	
s_4	1	1	0	−3	1	
x_2	1	1	1	−1	0	
z	−1	−2	0	−1	0	$c_j - z_j$

variable from the basis. An unbounded optimal z-value problem does not generally appear in the dual simplex method because of the starting requirement that all $c_j - z_j$ elements be nonpositive.

Although the dual simplex method was originally designed to avoid the use of artificial variables, it is not usually used to solve LPs from scratch because of the difficulty in finding a starting basis such that all $c_j - z_j$ elements are $\leqq 0$. However, the dual simplex method has special applications, as discussed in Section 3.11.2.

3.11 SOLVING A SERIES OF SIMILAR LPs

It is often necessary to solve a series of LPs in which there are only small differences, one problem to another. In such a series of LPs, it is likely that the optimal solution of one problem will not be many pivots away from the optimal solution of another. To solve such a series of LPs, it is advisable to *cold start* (from the origin) one of the problems and then *warm start* each subsequent problem from the optimal basis of the LP solved immediately before it. What must be done between problems to accomplish our warm starting purposes when the objective function changes, the RHS vector changes, or an **A**-matrix column changes is discussed in Sections 3.11.1 through 3.11.3.

3.11.1 When Only the Objective Function Changes

Recall from Section 3.8 that the only parts of a simplex tableau subject to change when the objective function changes are the z-square, z_j-row, and $c_j - z_j$ row. Now, from a cold start, let us assume that we have just pivoted to an optimal tableau of one of the problems. To warm start the next LP, we only need to recompute the z-square, z_j-row, and $c_j - z_j$ row with the new objective function. If the new $c_j - z_j$ row is nonpositive, we have an optimal tableau of the new problem. If the $c_j - z_j$ row contains positive elements, we pivot in the usual way until all $c_j - z_j$ elements are nonpositive, at which point we have an optimal tableau for the new problem. Notice that we change objective functions in the same way we changed from the artificial objective function to the real objective function in the two-phase method of linear programming (see Section 3.9). We continue replacing objective functions between problems in this way until all LPs in the series are solved.

3.11.2 When Only the RHS Vector Changes

Recall that the only parts of a simplex tableau subject to change when the RHS vector $\mathbf{b} \in R^m$ changes are the $\bar{\mathbf{b}}$-column and the z-square. Now, from a cold start, let us assume that we have just pivoted to an optimal tableau of one of the problems. To warm start the next LP, we only need to recompute the $\bar{\mathbf{b}}$-column and z-square with the new RHS vector. If the new $\bar{\mathbf{b}}$-column is nonnegative, we have an optimal tableau of the new problem. If the $\bar{\mathbf{b}}$-column contains negative elements, we apply the dual simplex algorithm (see Section 3.10) until the $\bar{\mathbf{b}}$-column is nonnegative, at which point we have an optimal tableau of the new problem. We continue using the dual simplex algorithm between problems in this way until all LPs in the series are solved.

3.11.3 When a Constraint Matrix Column Changes

Let us assume that we have just obtained an optimal tableau. Then, $\mathbf{a}_\ell$, the ℓth column of the **A**-matrix, changes. There are several cases.

1. *ℓth column is not in the optimal basis*. Only the ℓth column in the tableau changes. Recompute the new $\mathbf{y}_\ell$. This may cause the new $c_\ell - z_\ell$ element to be positive, in which case, we pivot as usual until optimality is restored.

2. *ℓth column is in the optimal basis*. Let $\bar{\mathbf{B}}$ be the same as the optimal basis matrix **B**, except that the old $\mathbf{a}_\ell$ is replaced by the new $\mathbf{a}_\ell$.

2.1. *$\bar{\mathbf{B}}$ is a basis*. Compute $\bar{\mathbf{B}}^{-1}$. Let $\mathbf{x}_{\bar{B}} = \bar{\mathbf{B}}^{-1}\mathbf{b}$, the new **Y**-matrix $= [\bar{\mathbf{B}}^{-1}\mathbf{N}; \mathbf{0}]$, and the new $c_j - z_j$ row $= (\mathbf{c}_N - \mathbf{c}_B\bar{\mathbf{B}}^{-1}\mathbf{b}, \mathbf{0})$.

2.1.1. If $\mathbf{x}_{\bar{B}} \geqq \mathbf{0}$ and the new $c_j - z_j$ row $\leqq \mathbf{0}$, then no pivots are required. Install the new **Y**-matrix and the new $c_j - z_j$ row, and replace the $\bar{\mathbf{b}}$-column with $\mathbf{x}_{\bar{B}}$. Now, we have an optimal tableau of the modified problem.

2.1.2. If $\mathbf{x}_{\bar{B}} \geqq \mathbf{0}$ and the new $c_j - z_j$ row $\nleqq \mathbf{0}$, install the new **Y**-matrix and the new $c_j - z_j$ row, and replace the $\bar{\mathbf{b}}$-column with $\mathbf{x}_{\bar{B}}$. Then, pivot as usual to restore optimality.

2.1.3. If $\mathbf{x}_{\bar{B}} \ngeqq \mathbf{0}$ and the new $c_j - z_j$ row $\leqq \mathbf{0}$, install the new **Y**-matrix and the new $c_j - z_j$ row, and replace the $\bar{\mathbf{b}}$-column with $\mathbf{x}_{\bar{B}}$. Then, apply the dual simplex algorithm to restore optimality.

2.1.4. If $\mathbf{x}_{\bar{B}} \ngeqq \mathbf{0}$ and the new $c_j - z_j$ row $\nleqq \mathbf{0}$, add an extra variable x_{n+1} to the tableau that has the new $\mathbf{a}_\ell$ as its **A**-matrix column. Compute $\mathbf{y}_{n+1}$. Then, treat x_ℓ as an artificial variable and use the two-phase method of linear programming to restore optimality.

2.2. *$\bar{\mathbf{B}}$ is not a basis*. In this case, it may be best to resolve the problem from scratch.

3.12 MORE ABOUT LINEAR PROGRAMMING

In Sections 3.4 through 3.11, we have reviewed the simplex method of single objective linear programming. However, there is much more to the mechanics of linear programming. To obtain more information about pivoting, duality, bounded variables, the revised simplex method, and other topics in LP, students should consult such texts as Hadley (1962), Dantzig (1963), Zionts (1974), and Gass (1985).

Readings

1. Bazaraa, M. S. and J. J. Jarvis (1977). *Linear Programming and Network Flows*, New York: John Wiley & Sons.
2. Charnes, A. and W. W. Cooper (1961). *Management Models and Industrial Applications of Linear Programming*, Vols. I and II, New York: John Wiley & Sons.
3. Dantzig, G. B. (1963). *Linear Programming and Extensions*, Princeton, New Jersey: Princeton University Press.
4. Dantzig, G. B. (1982). "Reminiscences About the Origins of Linear Programming," *Operations Research Letters*, Vol. 1, No. 2, pp. 43–48.
5. Gale, D. (1960). *The Theory of Linear Economic Models*, New York: McGraw-Hill.
6. Gass, S. I. (1970). *An Illustrated Guide to Linear Programming*, New York: McGraw-Hill.
7. Gass, S. I. (1985). *Linear Programming: Methods and Applications*, New York: McGraw-Hill.
8. Hadley, G. (1962). *Linear Programming*, Reading, Massachusetts: Addison-Wesley.
9. Hillier, F. S. and G. J. Lieberman (1980). *Introduction to Operations Research*, San Francisco: Holden-Day.
10. IBM Document No. SH19–1127-0 (1976). "IBM Mathematical Programming System Extended (MPSX/370): Basic Reference Manual," IBM Corporation, Data Processing Division, White Plains, New York.
11. IBM Document No. GH19–1091-1 (1979). "IBM Mathematical Programming System Extended/370: Primer," IBM Corporation, Data Processing Division, White Plains, New York.
12. Orchard-Hays, W. (1968). *Advanced Linear Programming Computing Techniques*, New York: McGraw-Hill.
13. Randolph, P. H. and H. D. Meeks (1978). *Applied Linear Optimization*, Columbus, Ohio: Grid Publishing.
14. Schrage, L. (1981). *Linear Programming Models with LINDO*, Palo Alto, California: Scientific Press.
15. Wagner, H. M. (1975). *Principles of Operations Research With Applications to Managerial Decisions*, Englewood Cliffs, New Jersey: Prentice-Hall.
16. Zionts, S. (1974). *Linear and Integer Programming*, Englewood Cliffs, New Jersey: Prentice-Hall.

PROBLEM EXERCISES

3-1 For each of the following sets of main and nonnegativity constraints:

(a)
$$\begin{aligned} x_1 \qquad &\geqq 1 \\ x_2 &\geqq 1 \\ x_1 + x_2 &\leqq 4 \\ x_1, x_2 &\geqq 0 \end{aligned}$$

(b)
$$\begin{aligned} x_1 + x_2 &\geqq 1 \\ x_2 &\leqq 5 \\ x_1 \qquad &\leqq 6 \\ 7x_1 + 9x_2 &\leqq 63 \\ x_1, x_2 &\geqq 0 \end{aligned}$$

(c)
$$\begin{aligned} x_1 + 3x_2 &\geqq 3 \\ x_1 + x_2 &\geqq 2 \\ x_1, x_2 &\geqq 0 \end{aligned}$$

(i) Attach a gradient to the bounding level curve of each main constraint.

(ii) Graphically indicate the feasible region S by shading.

(iii) Express S in terms of the convex combination and unbounded line segment operators γ and μ.

(iv) Specify the boundary points of S.

(v) Specify the coordinates of each extreme point of S.

3-2 Indicating the feasible region by shading, specify Θ, Θ_x, Θ_μ, and z^* for the LPs:

(a)
$$\begin{aligned} \max \{-2x_1 - 3x_2 &= z\} \\ \text{s.t.} \quad x_1 + x_2 &\leqq 4 \\ 6x_1 + 2x_2 &\geqq 8 \\ x_1 + 5x_2 &\geqq 4 \\ x_1 \qquad &\leqq 3 \\ x_2 &\leqq 3 \\ x_1, x_2 &\geqq 0 \end{aligned}$$

(b)
$$\begin{aligned} \max \{\ 4x_1 + x_2 &= z\} \\ \text{s.t.} \quad -x_1 + x_2 &\leqq 2 \\ -2x_1 + x_2 &\geqq 4 \\ x_1, x_2 &\geqq 0 \end{aligned}$$

(c)
$$\begin{aligned} \max \{-4x_1 + 5x_2 &= z\} \\ \text{s.t.} \quad -x_1 + x_2 &\geqq 0 \\ x_1 \qquad &\leqq 0 \\ x_1, x_2 &\geqq 0 \end{aligned}$$

(d)
$$\begin{aligned} \min \{-2x_1 - 2x_2 &= z\} \\ \text{s.t.} \quad x_1 + x_2 &\leqq 5 \\ x_2 &\leqq 4 \\ -x_1 + x_2 &\leqq 2 \\ x_1, x_2 &\geqq 0 \end{aligned}$$

(e)
$$\begin{aligned} \min \{-x_1 + x_2 &= z\} \\ \text{s.t.} \quad 2x_1 + 3x_2 &\leqq 7 \\ 3x_1 + 3x_2 &\geqq 6 \\ x_2 &\geqq \tfrac{1}{2} \\ x_1 \qquad &\geqq 1 \\ x_1, x_2 &\geqq 0 \end{aligned}$$

3-3 Graphing the following LPs,

(i) Indicate the feasible region by shading.

(ii) Specify Θ, Θ_x, Θ_μ, and z^*.

(iii) Identify any redundant main constraints.

(iv) Identify any degenerate extreme points.

(a) $\max \{x_1 = z\}$

s.t. $x_1 + 4x_2 \leqq 4$

$\frac{1}{2}x_1 + \frac{1}{2}x_2 = 1$

$-x_1 + x_2 \geqq -2$

$x_1, x_2 \geqq 0$

(b) $\max \{-x_1 + x_2 = z\}$

s.t. $-2x_1 + 2x_2 \leqq 0$

$2x_1 + x_2 \leqq 9$

$x_1 \leqq 5$

$x_1, x_2 \geqq 0$

(c) $\max \{-6x_1 - 4x_2 = z\}$

s.t. $-x_1 \leqq -4$

$x_1 + x_2 \leqq 4$

$x_1 - x_2 = 4$

$x_1, x_2 \geqq 0$

(d) $\max \{-x_1 + 2x_2 = z\}$

s.t. $-x_1 + 2x_2 \leqq 4$

$x_1, x_2 \geqq 0$

(e) $\min \{-x_1 + x_2 = z\}$

s.t. $2x_1 + 3x_2 \leqq 7$

$-4x_1 - 6x_2 \leqq -14$

$x_2 \geqq \frac{1}{2}$

$x_1 + x_2 \geqq 2$

$x_1, x_2 \geqq 0$

3-4 Transform the following LPs into standard equality constraint format and load the initial simplex tableau:

(a) $\max \{7x_1 - 3x_2 = z\}$

s.t. $3x_1 + 2x_2 \leqq -3$

$-4x_1 = 2$

$x_1, x_2 \geqq 0$

(b) $\min \{x_1 - x_3 = z\}$

s.t. $6x_1 - 5x_2 = -8$

$4x_1 - 7x_2 + x_3 \leqq -9$

$-12x_2 + 7x_3 \leqq 10$

$3x_1 + 4x_3 \geqq -8$

$x_1, x_2, x_3 \geqq 0$

3-5 Solve each of the following LPs by the simplex method and specify:

(i) The coordinates of an optimal extreme point.

(ii) The optimal z-value.

(a) $\max \{-x_1 + x_2 = z\}$

s.t. $x_1 \leqq 2$

$x_2 \geqq 1$

$x_1, x_2 \geqq 0$

(b) $\max \{-x_1 + 2x_2 = z\}$

s.t. $x_1 + x_2 \leqq 2$

$2x_1 + 2x_2 \geqq 8$

$x_1, x_2 \geqq 0$

(c) $\max \{-2x_1 - x_2 = z\}$

s.t. $x_2 \leqq 3$

$-x_1 + x_2 \geqq 0$

$x_1 + x_2 \geqq 4$

$x_1, x_2 \geqq 0$

(d) $\min \{-2x_1 - x_2 = z\}$

s.t. $3x_1 + 5x_2 \leqq 15$

$6x_1 + 2x_2 \leqq 24$

$x_1, x_2 \geqq 0$

3-6 Solve the following LPs:

(i) Graphically.

(ii) By the simplex method and associate with each tableau its extreme point of S.

(a)
$$\begin{aligned} \min \{34 + 2x_1 - x_2 = z\} \\ \text{s.t.} \quad x_1 + x_2 \leqq 4 \\ x_2 = 3 \\ x_1, x_2 \geqq 0 \end{aligned}$$

(b)
$$\begin{aligned} \min \{-5x_1 - 3x_2 = z\} \\ \text{s.t.} \quad 900x_1 + 1500x_2 \leqq 4500 \\ 40x_1 + 16x_2 \leqq 80 \\ x_1, x_2 \geqq 0 \end{aligned}$$

(c)
$$\begin{aligned} \max \{x_2 = z\} \\ \text{s.t.} \quad -x_1 + x_2 \leqq 2 \\ x_1 \geqq 1 \\ x_1, x_2 \geqq 0 \end{aligned}$$

3-7 Write the duals of:

(a)
$$\begin{aligned} \max \{2x_1 - x_2 = z\} \\ \text{s.t.} \quad -4x_1 + x_2 = 3 \\ x_1 \leqq 2 \\ x_1 + x_2 \leqq 3 \\ x_1, x_2 \geqq 0 \end{aligned}$$

(b)
$$\begin{aligned} \min \{4x_1 - 6x_2 + 3x_3 = z\} \\ \text{s.t.} \quad 3x_1 + 2x_2 + 3x_3 \geqq 1 \\ -2x_1 - 5x_2 + x_3 = -1 \\ x_1, x_3 \geqq 0 \\ x_2 \text{ unrestricted} \end{aligned}$$

3-8 **(a)** Using the simplex method, solve:

$$\begin{aligned} \max \{x_1 + 2x_2 = z\} \\ \text{s.t.} \quad -x_1 + x_2 \leqq 2 \\ 3x_1 + x_2 \leqq 6 \\ x_1, x_2 \geqq 0 \end{aligned}$$

(b) Reading from the optimal tableau, specify the optimal solution to the dual.

(c) Specify the optimal dual objective function value.

3-9 **(a)** Formulate the following LP with nonnegative variables:

$$\begin{aligned} \max \{x_1 - 4x_2 = z\} \\ \text{s.t.} \quad x_1 - x_2 \leqq 2 \\ -4x_1 + x_2 \leqq 16 \\ x_1 \text{ unrestricted} \\ x_2 \leqq 0 \end{aligned}$$

(b) Solve using the simplex method.

3-10C **(a)** Form MPSX input decks for the following:

(i)
$$\begin{aligned} \max \{3x_1 + 4x_2 = z\} \\ \text{s.t.} \quad -x_1 + x_2 \leqq 4 \\ x_1 + 4x_2 \leqq 21 \\ x_1 \leqq 5 \\ x_2 \geqq 2 \\ x_1, x_2 \geqq 0 \end{aligned}$$

(ii)
$$\begin{aligned} \max \{2x_1 + 3x_3 = z\} \\ \text{s.t.} \quad 4x_1 + 6x_2 - 7x_3 \leqq 10 \\ -2x_1 + 3x_2 = 14 \\ x_1, x_2, x_3 \geqq 0 \end{aligned}$$

(iii)
$$\begin{aligned} \max \ & \{9x_1 + x_2 = z\} \\ \text{s.t.} \ & 2x_1 + x_2 \leqq 12 \\ & x_1 - x_2 \geqq 2 \\ & -x_1 + 4x_2 = 6 \\ & x_2 \geqq 7 \\ & x_1, x_2 \geqq 0 \end{aligned}$$

(iv)
$$\begin{aligned} \min \ & \{12x_1 + 2x_2 + 6x_3 = z\} \\ \text{s.t.} \ & 2x_1 + x_2 - x_3 \geqq 7 \\ & x_1 - x_2 + 4x_3 \geqq 2 \\ & x_1 \leqq 0 \\ & x_2 \geqq 0 \\ & x_3 \text{ unrestricted} \end{aligned}$$

(b) Solve the above problems on the computer.

3-11C (a) Form an MPSX input deck for the following LP:

	x_1	x_2	z_1	z_2	α		
Obj					1	min	
s.t.			0.3		1	$\geqq$	1.8
				0.7	1	$\geqq$	0.7
	−2	3	−1			=	0
	1			−1		=	0
	1	1				$\leqq$	8
		1				$\leqq$	7

$x_1, x_2 \geqq 0 \quad z_1, z_2, \alpha$ unrestricted

(b) Solve the above LP for the optimal solution values of x_1, x_2, z_1, z_2, and α.

3-12C Solve:

$$\begin{aligned} \min \ & \{-3x_1 + x_2 + x_3 = z\} \\ \text{s.t.} \ & 2x_1 + x_2 \leqq 3 \\ & x_2 - 3x_3 = 2 \\ & 3x_1 + 4x_3 \geqq 10 \\ & x_1, x_3 \geqq 0 \\ & x_2 \text{ unrestricted} \end{aligned}$$

on the computer and specify the result. Now, formulate its dual and solve it on the computer and specify the result.

3-13C (a) Graph the LP:

$$\begin{aligned} \max \ & \{x_3 = z\} \\ \text{s.t.} \ & x_1 \geqq 2 \\ & -x_1 + x_2 \leqq 0 \\ & x_1 \leqq 9 \\ & x_1 + 5x_2 \leqq 24 \\ & x_3 \leqq 4 \\ & x_1, x_2, x_3 \geqq 0 \end{aligned}$$

(b) Specify the optimal set.

(c) Specify all optimal extreme points.

(d) Using an LP code, solve the problem of part (a).

(e) Are alternative optima indicated on the printout? If yes, describe how.

3-14C Let S be defined by:

$$\begin{aligned} x_1 + x_2 \quad\quad &\leqq 3 \\ x_2 + x_3 &\leqq 3 \\ x_1, x_2, x_3 &\geqq 0 \end{aligned}$$

Use an LP code to find the point in S on the line $\{\mathbf{x} \in R^3 \,|\, \mathbf{x} = \mathbf{x}^1 + \alpha(\mathbf{x}^2 - \mathbf{x}^1), \alpha \in R\}$ that is closest to $\mathbf{x}^2$ where $\mathbf{x}^1 = (1,0,0)$ and $\mathbf{x}^2 = (3,3,3)$.

3-15 Consider the LP:

$$\begin{aligned} \max \{5x_1 + 3x_2 &= z\} \\ \text{s.t.} \quad x_1 + x_2 &\leqq 5 \\ x_1 \quad\quad &\leqq 4 \\ x_2 &\geqq 3 \\ x_1, x_2 &\geqq 0 \end{aligned}$$

(a) After converting to standard equality constraint format, specify **c**, **A**, and **b**.

(b) At the extreme point characterized by

$$J_B = \{2,1,5\} \quad \text{and} \quad \mathbf{B}^{-1} = \begin{bmatrix} 1 & -1 & 0 \\ 0 & 1 & 0 \\ -1 & 1 & 1 \end{bmatrix}$$

compute $\mathbf{x}_B$, z, z_1, and $c_4 - z_4$ from their definitions.

3-16 Consider the LP:

$$\begin{aligned} \max \{3x_1 + 2x_2 - 2x_3 &= z\} \\ \text{s.t.} \quad x_1 + 6x_2 + 2x_3 &\leqq 24 \\ 2x_1 - 5x_2 + x_3 &\leqq 14 \\ x_1, x_2, x_3 &\geqq 0 \end{aligned}$$

Construct the simplex tableau at the extreme point characterized by

$$J_B = \{2,1\} \quad \text{and} \quad \mathbf{B}^{-1} = \begin{bmatrix} \frac{2}{17} & -\frac{1}{17} \\ \frac{5}{17} & \frac{6}{17} \end{bmatrix}$$

from the definitions of the $\mathbf{x}_B$, z, **Y**-matrix, z_j-row, and $c_j - z_j$ row.

3-17 Using the simplex method, solve the following problems using the two-phase method:

(a)
$$\begin{aligned} \max\{\ 4x_1 + 3x_2 &= z\} \\ \text{s.t.} \quad x_1 + x_2 &\geqq 1 \\ 3x_1 + 2x_2 &\leqq 8 \\ x_1, x_2 &\geqq 0 \end{aligned}$$

(b)
$$\begin{aligned} \max\{\ 2x_1 + 3x_2 \qquad &= z\} \\ \text{s.t.} \quad -x_1 + x_2 - 2x_3 &= 1 \\ x_1 + 2x_2 - 3x_3 &= 2 \\ x_1, x_2, x_3 &\geqq 0 \end{aligned}$$

(c)
$$\begin{aligned} \max\{\ x_1 + x_2 &= z\} \\ \text{s.t.} \quad x_1 + x_2 &\geqq 3 \\ -x_1 + 2x_2 &\geqq 0 \\ -x_1 + 2x_2 &\leqq 2 \\ x_1, x_2 &\geqq 0 \end{aligned}$$

(d)
$$\begin{aligned} \max\{\ 4x_1 - x_2 + x_3 &= z\} \\ \text{s.t.} \quad 6x_1 + x_2 - x_3 &= 9 \\ -18x_1 - 3x_2 + 3x_3 &= 2 \\ x_1, x_2, x_3 &\geqq 0 \end{aligned}$$

3-18 Using the dual simplex algorithm, solve:

(a)
$$\begin{aligned} \max\{-5x_1 - 3x_2 &= z\} \\ \text{s.t.} \quad -x_1 + x_2 &\leqq 1 \\ x_2 &\geqq 2 \\ x_1, x_2 &\geqq 0 \end{aligned}$$

(b)
$$\begin{aligned} \max\{-x_1 - x_2 &= z\} \\ \text{s.t.} \quad x_1 + x_2 &\geqq 2 \\ 2x_1 + x_2 &\leqq 1 \\ x_1, x_2 &\geqq 0 \end{aligned}$$

3-19 **(a)** Graph the LP:

$$\begin{aligned} \max\{-x_1 + 4x_2 &= z\} \\ \text{s.t.} \quad x_2 &\leqq 3 \\ x_1 + x_2 &\leqq 4 \\ 4x_1 + x_2 &\leqq 12 \\ x_1, x_2 &\geqq 0 \end{aligned}$$

(b) Using the simplex method, solve the above problem.

(c) Using a warm start from the optimal tableau of (b), solve the problem of (a) but with $4x_1 + 5x_2$ as the objective function.

(d) Using a warm start from the optimal tableau of (c), solve the problem of (a) with $5x_1 + x_2$ as the objective function.

3-20 Using the simplex method, solve:

$$\begin{aligned} \max\{\ x_1 + x_2 &= z\} \\ \text{s.t.} \quad x_1 + 2x_2 &\leqq 6 \\ 2x_1 + x_2 &\leqq 6 \\ x_1, x_2 &\geqq 0 \end{aligned}$$

Now change the RHS vector to

$$\mathbf{b} = \begin{bmatrix} 5 \\ 13 \end{bmatrix}$$

From the optimal tableau of the problem just solved, use the dual simplex algorithm to solve the problem with the new RHS.

3-21 **(a)** Using the simplex method, solve:

$$\begin{aligned} \max \{ -4x_1 + 7x_2 &= z\} \\ \text{s.t.} \quad -4x_1 + 8x_2 &\leqq 32 \\ x_1 \qquad &\leqq 4 \\ x_1, x_2 &\geqq 0 \end{aligned}$$

(b) Using a warm start from the optimal tableau of (a), solve the above problem but with the coefficient of x_2 in the first main constraint being 6 instead of 8.

CHAPTER 4

Determining All Alternative Optima

This chapter provides a Phase III, to be appended to the linear programming Phases I and II of the previous chapter. Phase III is applicable whenever the optimal basis terminating Phase II is nonunique. Phase III consists of pivoting among all optimal bases to compute all optimal extreme points and all unbounded optimal edges. Since the optimal set Θ is the set of all convex combinations of the optimal extreme points and points on unbounded optimal edges, Phase III enables us to *characterize* Θ with the computation of Θ_x and Θ_μ. With a Phase III, a linear programming procedure is equipped with an "all alternative optima" capability. Not only is Phase III important to linear programming, but other (similar) Phase III's will be developed in Chapters 5 and 9 to explore and characterize portions of the surface of the feasible region for more generalized purposes.

4.1 CLASSIFICATION OF SINGLE OBJECTIVE LPs

Consider the following mutually exclusive and collectively exhaustive classification of single objective LPs:

1. $S = \varnothing$.
2. $S \neq \varnothing$, $\Theta = \varnothing$.
3. $S \neq \varnothing$, Θ contains only one point.
4. $S \neq \varnothing$, Θ is bounded but contains an infinite number of points.
5. $S \neq \varnothing$, Θ is unbounded.

Classification 1 is inconsistency. Classification 2 is that the LP has an unbounded z-value solution. Classifications 3, 4, and 5 are variations of the case in which the LP has an optimal solution. Examples 7, 3, 8, 6, and 4 of Chapter 3 respectively illustrate the five classifications.

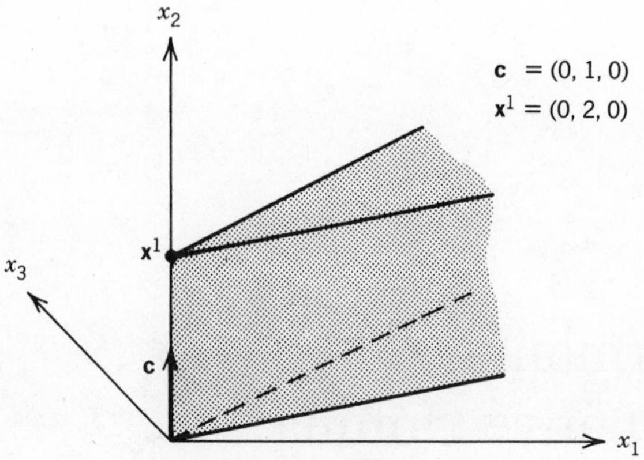

Figure 4.1. Graph of Example 1.

In conventional linear programming, pivoting stops as soon as an optimal basis is obtained. Because of this, conventional linear programming may not be able to categorize LPs that have nonunique optimal bases with regard to classifications 3, 4, or 5.

Example 1. Consider the LP

$$
\begin{array}{lrrrl}
\max\{ & & x_2 & & = z\} \\
\text{s.t.} & -x_1 & & + \; x_3 & \leqq 0 \\
 & x_1 & & - \; 2x_3 & \leqq 0 \\
 & & x_2 & & \leqq 2 \\
 & & & x_1, x_2, x_3 & \geqq 0
\end{array}
$$

that is graphed in Fig. 4.1. The feasible region is the "wedge" as drawn, and the optimal set is the top of the wedge for which

$$\Theta = \gamma(\mathbf{w}, \mathbf{y}) \quad \text{where} \quad \mathbf{w} \in \mu(\mathbf{x}^1, \mathbf{v}^a), \quad \mathbf{y} \in \mu(\mathbf{x}^1, \mathbf{v}^b),$$

$$\mathbf{v}^a = (1, 0, 1), \quad \text{and} \quad \mathbf{v}^b = (2, 0, 1)$$

$$\Theta_x = \{\mathbf{x}^1\}$$

$$\Theta_\mu = \{\mu(\mathbf{x}^1, \mathbf{v}^a), \mu(\mathbf{x}^1, \mathbf{v}^b)\}$$

At $\mathbf{x}^1 = (0, 2, 0)$, we have the Phase II optimal tableau of Table 4.1. Note the zero $c_j - z_j$ elements in the x_1 and x_3 columns. This suggests the possibility of alternative optima. However, the existence of other optimal points cannot be detected from this tableau because of the degeneracy at $\mathbf{x}^1$. Thus, we are unable to determine whether this is a class 3, 4, or 5 LP without further pivoting.

Table 4.1
Phase II Optimal Tableau of Example 1

		x_1	x_2	x_3	s_4	s_5	s_6	
s_4	0	−1	0	1	1	0	0	
s_5	0	1	0	−2	0	1	0	
x_2	2	0	1	0	0	0	1	
z	2	0	0	0	0	0	−1	$c_j - z_j$

Because commercial LP codes (as of this writing) do not possess Phase III capabilities, they can neither guarantee (a) the characterization of the optimal set Θ in terms of Θ_x and Θ_μ, nor (b) the categorization of LPs in terms of classes 3, 4, or 5 when alternative optima are present.

4.2 NOTES ON PIVOTING AND CODING

This section lists ten notes concerning technical points used in Phase III procedures. The notes deal with pivoting among optimal bases and the coding of bases, extreme points, and unbounded edge directions.

Note 1. We can pivot on a negative element if the corresponding basic variable is degenerate. This is because the $\bar{\mathbf{b}}$-column remains the same. This is shown by the Method of Rectangles (Section 3.4.4) in Fig. 4.2 in which −2 is the pivot element. Since one of the "other diagonal elements" is zero (the degenerate basic variable), none of the elements in the $\bar{\mathbf{b}}$-column change during the pivoting process.

Note 2. Elements in the $c_j - z_j$ row do not change when pivoting among optimal tableaus (i.e., when pivoting into the basis of an optimal tableau a nonbasic variable whose $c_j - z_j$ element is zero). This is shown by the Method of

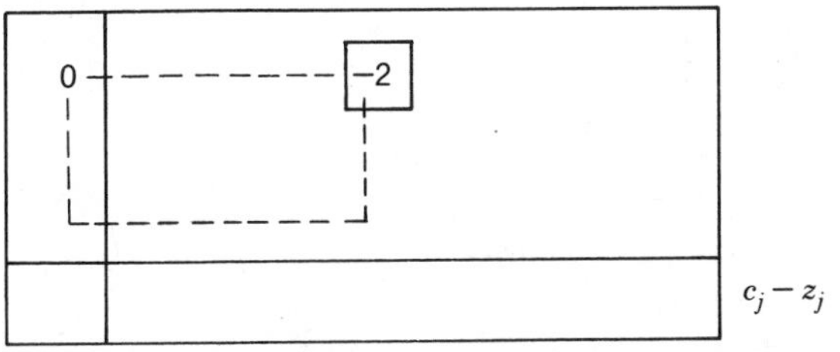

Figure 4.2. Illustrative tableau of Note 1.

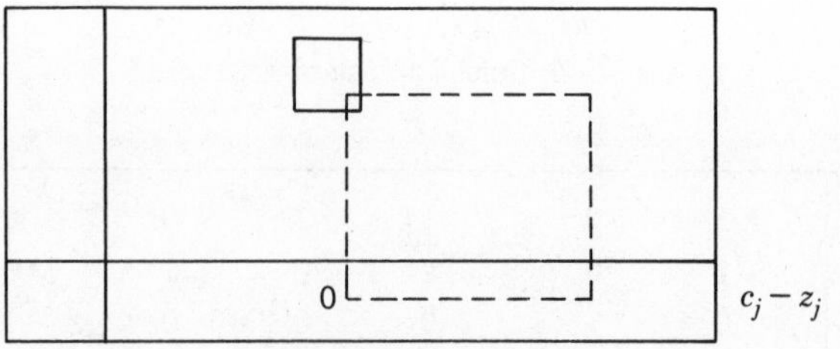

Figure 4.3. Illustrative tableau of Note 2.

TABLE 4.2
Tableau of Note 3

		x_1	x_2	x_3	s_4	s_5	s_6	
s_4	0	[−1]	0	[1]	1	0	0	
s_5	0	[1]	0	[−2]	0	1	0	
x_2	2	0	1	0	0	0	1	
z	2	0	0	0	0	0	−1	$c_j - z_j$
		↑		↑				

Rectangles in Fig. 4.3 in which the pivot element is in the row indicated by the arrow. Since the nonbasic $c_j - z_j$ "other diagonal element" is zero, none of the elements in the $c_j - z_j$ row change during the pivoting process.

Note 3. Assume an optimal tableau. Then, any pivot that results in another optimal tableau is an *optimal pivot*. Recall the optimal tableau of Example 1 that is repeated in Table 4.2. There are four optimal pivots from this tableau. Thus, the basis of this tableau has four *adjacent* optimal bases.

Note 4. To determine Θ_x and Θ_μ, we must pivot to all optimal bases. Pivoting to all optimal bases can be likened to exploring all branches of a tree. Consider the *tree of optimal bases* in Fig. 4.4, in which the bases are numbered in the order in which they are encountered. With ①, the first (i.e., Phase II) optimal basis,

1. Two optimal bases are adjacent to ①.
2. Three new optimal bases are adjacent to ②.
3. One new optimal basis is adjacent to ③.
4. No new optimal bases are adjacent to ④ or ⑤.
5. Two new optimal bases are adjacent to ⑥.
6. No new optimal bases are adjacent to ⑦.
7. No new optimal bases are adjacent to ⑧ and ⑨.

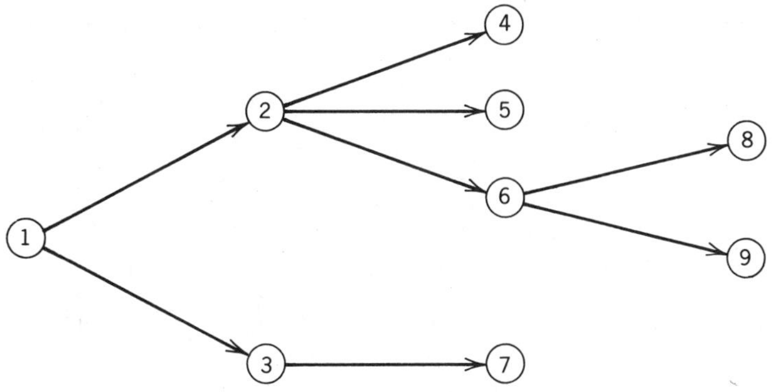

Figure 4.4. Tree of optimal bases.

After we examine the last of the bases in the tree and find no new optimal bases, all optimal bases have been found.

Note 5. To efficiently explore all branches of trees as in Fig. 4.4, we must develop a *Phase III bookkeeping system* that "remembers where we have been and where we have yet to go." Since it is not practical to store more than one tableau at a time, our Phase III bookkeeping system will be predicated on *coded bases*.

Note 6. We utilize the subscripts of the basic variables to code a basis. Consider the tableau whose basic variables are s_8, x_3, s_{10}, and x_5. Then,

$$\begin{aligned}\text{Code of basis } \{8, 3, 10, 5\} &= 2^8 + 2^3 + 2^{10} + 2^5 \\ &= 1320\end{aligned}$$

As long as the code is less than or equal to $2^{31} - 1$ (i.e., the largest subscript is less than or equal to 31), the code can be stored in a 4-byte integer storage location as

1320

We could even store 32 subscripts in a storage word if we were willing to utilize the sign bit.

Note 7. To work with round numbers, let us only store 30 subscripts in a storage word. Let h be the largest subscript. Then, the number of 4-byte integer storage words required to store a coded basis is the largest integer less than

$$\frac{h}{30} + 1$$

To illustrate, the code of basis $\{1, 11, 5, 20, 54, 36, 55\}$ requires two 4-byte words

$$\begin{aligned} \text{1st word} &= 2^1 + 2^5 + 2^{11} + 2^{20} \\ &= 1050698 \\ \text{2nd word} &= 2^{(36-30)} + 2^{(54-30)} + 2^{(55-30)} \\ &= 2^6 + 2^{24} + 2^{25} \\ &= 50331712 \end{aligned}$$

and is stored as

1050698	50331712

Note 8. A basis is *decoded* by successively dividing by 2. With six structural variables and no artificial variables, basis 1320 is decoded as follows:

		Remainder	***Basic Variable***
2	1320		
2	660		
2	330	0	
2	165	0	
2	82	1	x_3
2	41	0	
2	20	1	x_5
2	10	0	
2	5	0	
2	2	1	s_8
2	1	0	
	0	1	s_{10}

The remainder of the first division is ignored because the subscripting of the variables uses *1-origin*, rather than *0-origin*, indexing. Thus, the code of a basis is even since the term 2^0 never occurs in the construction of the code. Had 1320 been the code of the second storage word, the subscripts of the basic variables would be 33, 35, 38, and 40.

Note 9. Using the subscripts of the nonzero basic variables, an extreme point is coded. Let $\mathbf{B}$ be the basis of the current tableau, θ_j the pivot ratio pertaining to the x_j entering nonbasic variable, and $\mathbf{y}_j$ the jth column of the $\mathbf{Y}$-matrix of the current tableau. The coordinates of the *adjacent* extreme point pertaining to the given pivot element in the column of the x_j nonbasic variable are given by

$$\begin{bmatrix} \mathbf{x}_B \\ 0 \\ \vdots \\ \\ 0 \end{bmatrix} + \theta_j \begin{bmatrix} -\mathbf{y}_j \\ 0 \\ \vdots \\ 0 \\ 1 \\ 0 \\ \vdots \\ 0 \end{bmatrix} \begin{matrix} \\ \\ \\ \\ \leftarrow \text{pertains to } x_j \text{ nonbasic variable} \\ \\ \\ \\ \end{matrix}$$

In the above, the coordinates of the adjacent extreme point associated with the currently basic components are given by $\mathbf{x}_B - \theta_j \mathbf{y}_j$. The coordinates associated with all currently nonbasic components are zero except for the jth nonbasic component that is θ_j. The direction of the emanating edge pertaining to the x_j nonbasic variable is

$$\begin{bmatrix} -\mathbf{y}_j \\ 0 \\ \vdots \\ 0 \\ 1 \\ 0 \\ \vdots \\ 0 \end{bmatrix}$$

Note 10. As with extreme points, unbounded edge directions are coded using only nonzero components.

Example 2. Let Table 4.3 be the tableau at an extreme point designated $\mathbf{x}^1$. For the nonbasic x_2 column with 6 as the pivot element (and $\theta_2 = \frac{1}{2}$ as the pivot

TABLE 4.3
Tableau of Example 2 at x^1

		x_1	x_2	x_3	$\cdots$	s_9
x_1	3		[6]	-1		
s_6	2		[4]	-3		
x_5	0		[-1]	0		
s_8	7		3	-2		
			↑	↑		

ratio), the coordinates of the adjacent extreme point are

$$\begin{bmatrix} 3 \\ 0 \\ 0 \\ 0 \\ 0 \\ 2 \\ 0 \\ 7 \\ 0 \end{bmatrix} + \left(\tfrac{1}{2}\right) \begin{bmatrix} -6 \\ 1 \\ 0 \\ 0 \\ 1 \\ -4 \\ 0 \\ -3 \\ 0 \end{bmatrix} = \begin{bmatrix} 0 \\ \frac{1}{2} \\ 0 \\ 0 \\ \frac{1}{2} \\ 0 \\ 0 \\ \frac{11}{2} \\ 0 \end{bmatrix}$$

For the nonbasic x_2 column with 4 as the pivot element (and $\theta_2 = \frac{1}{2}$), we compute the same adjacent extreme point. For the nonbasic x_2 column with -1 as the pivot element (and $\theta_2 = 0$), the "adjacent" extreme point is the current extreme point. The cause of an "adjacent" extreme point being the current extreme point is degeneracy. Since a pivot element cannot be found in the x_3 column, the edge

$$\mu(\mathbf{x}^1, \mathbf{v})$$

emanating from the current extreme point where

$$\mathbf{v} = \begin{bmatrix} 1 \\ 0 \\ 1 \\ 0 \\ 0 \\ 3 \\ 0 \\ 2 \\ 0 \end{bmatrix}$$

is unbounded. Thus, we have

$$\text{Code of basis } \{1,6,5,8\} = 2^1 + 2^6 + 2^5 + 2^8 = 354$$

$$\text{Code of extreme point } \mathbf{x}^1 = 2^1 + 2^6 + 2^8 = 322$$

$$\text{Code of unbounded edge direction } \mathbf{v} = 2^1 + 2^3 + 2^6 + 2^8 = 330$$

4.3 CRASHING

Crashing is the process of moving, in the minimal number of pivots, from one basis to another without regard to intermediate feasibility.

Suppose we reach the end of a branch in a tree of optimal bases (tableaus). To explore other unexamined branches, we may have to move from our current basis to other bases that may not be adjacent. For example, after reaching the end of the branch whose endpoint is ⑦ in Fig. 4.2, we have to move to ⑧, which may very well be a nonadjacent basis.

Suppose we wish to crash from our current basis $\mathbf{B}_1$ to basis $\mathbf{B}_2$. For the two bases, let the sets of basic variable indices be given by I_{B_1} and I_{B_2}, respectively. Let us define *unneeded variables* as variables that have to leave the current basis and *missing variables* as variables that have to come into the current basis. Relevant to $\mathbf{B}_1$ and $\mathbf{B}_2$, we define the *unneeded index set* $I_U = \{i \in I_{B_1} | i \notin I_{B_2}\}$ and the *missing index set* $I_M = \{i \in I_{B_2} | i \notin I_{B_1}\}$. The number of unneeded variables equals the number of missing variables, and denote this number by ℓ.

The routine for crashing is iterative. The first iteration begins by selecting the column of a missing variable. Intersecting the missing variable column in a row of an unneeded variable, we select a nonzero pivot element. We then pivot. Relevant to the resulting (intermediate) basis, we perform the second iteration. In the second iteration, we note that there is one fewer missing variables and one fewer unneeded variables. By pivoting in this fashion $\ell - 1$ additional times, the number of unneeded and missing variables is reduced to zero and we have arrived at the desired basis.

A flowchart of the crashing routine is given in Fig. 4.5. In Block 7, we reset I_{B_1} to $I_{B_1} - \{r\} \cup \{e\}$ to reflect the newly resulting (intermediate) basis. When crashing, it does not matter if there are negative elements in the $\bar{\mathbf{b}}$-column of an intermediate tableau. The intermediate bases are only a means to an end and are not of significance. With the help of Lemma 4.1, in Block 5 we are assured by Theorem 4.2 that we will always be able to find a nonzero pivot element.

Lemma 4.1. Let $\mathbf{B} = \{\mathbf{a}_{j_1}, \ldots, \mathbf{a}_{j_m}\}$ be a basis in R^m; I_B the set of basic variable indices; $\mathbf{a}_e \in R^m$ such that $\mathbf{a}_e \notin B$; and $J = \{i \in I_B | y_{ie} = 0\}$. Then, $\mathbf{a}_e$ is linearly dependent on

$$\{\mathbf{a}_i\}_{i \in (I_B - J)}$$

Proof. Since $\mathbf{B}$ is a basis

$$\sum_{i \in I_B} \mathbf{a}_i y_{ie} = \mathbf{a}_e$$

Since $y_{ie} = 0$ for all $i \in J$,

$$\sum_{i \in (I_B - J)} \mathbf{a}_i y_{ie} = \mathbf{a}_e$$

Thus, $\mathbf{a}_e$ is linearly dependent on $\{\mathbf{a}_i\}_{i \in (I_B - J)}$. ◀

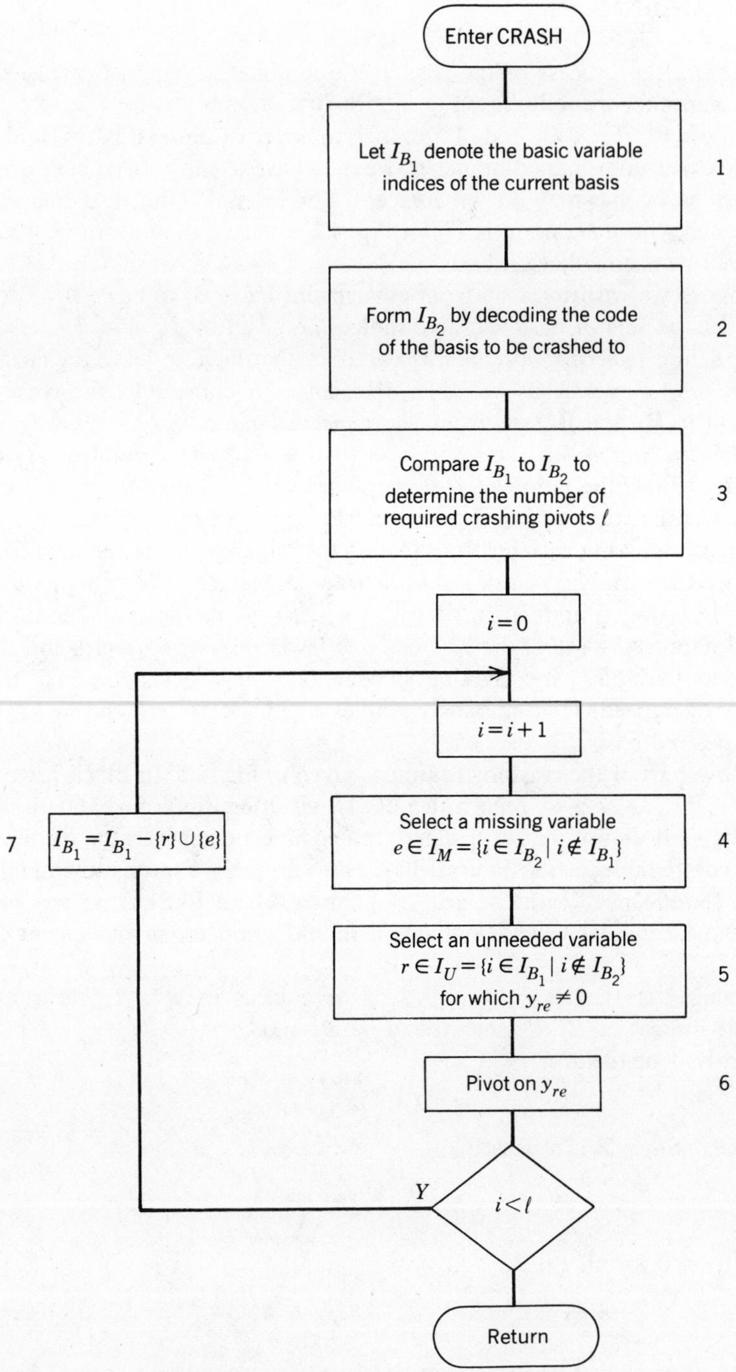

Figure 4.5. Flowchart of crashing routine.

Theorem 4.2. Let $\mathbf{B}_1$ and $\mathbf{B}_2$ be two bases and let I_{B_1}, I_{B_2}, I_U, and I_M be as defined earlier. Let $e \in I_M$ be a missing variable index. Then, there exists an $i \in I_U$ such that $y_{ie} \neq 0$.

Proof. Suppose $y_{ie} = 0$ for all $i \in I_U$. By Lemma 4.1, this implies that $\mathbf{a}_e$ is linearly dependent on

$$\{\mathbf{a}_i\}_{i \in (I_{B_1} - I_U)}$$

Since

$$\{\mathbf{a}_e\} \cup \{\mathbf{a}_i\}_{i \in (I_{B_1} - I_U)}$$

is a subset of the set of vectors constituting basis $\mathbf{B}_2$, $\mathbf{a}_e$ is linearly independent of

$$\{\mathbf{a}_i\}_{i \in (I_{B_1} - I_U)}$$

Since this is a contradiction, there exists an $i \in I_U$ such that $y_{ie} \neq 0$. ◀

4.4 MASTER LISTS

In the bookkeeping system of the Phase III procedure for enumerating the contents of Θ_x and Θ_μ, it is necessary to construct the four *master lists*:

- L_B Contains all coded optimal bases.
- L_x Contains the coded optimal extreme points of the corresponding entries in L_B.
- L_e Contains the coded optimal extreme points from which unbounded optimal edges emanate.
- L_μ Contains the coded directions of the unbounded optimal edges emanating from the corresponding entries in L_e.

Corresponding elements in L_B and L_x are associated as *pairs*, as are corresponding elements in L_e and L_μ.

The heart of the Phase III bookkeeping system is the L_B master list. Each detected optimal basis is stored in L_B. By crashing from entry to entry in L_B and "processing" each optimal basis for new adjacent optimal bases, we have a means of remembering where we have been and where we have yet to go. Where we "have been" is denoted by the bases in L_B to which we have crashed. Where we "have yet to go" is denoted by bases in L_B to which we have not yet crashed.

At first, L_B grows faster than we can process the entries in the list. Sooner or later, however, the crashing process will catch up with the end of the list. As soon as the last basis in the list is processed and no new adjacent optimal bases are detected, no further Phase III pivoting is required and we are done.

4.5 PHASE III NUMERICAL EXAMPLE

The Phase III procedure is illustrated by means of the following LP:

$$\begin{aligned} \max\{\ 0x_1 + 0x_2 + 0x_3 &= z\} \\ \text{s.t.}\quad -x_1 \qquad + x_3 &\leqq 0 \\ -x_1 \qquad + 2x_3 &\leqq 2 \\ x_1, x_2, x_3 &\geqq 0 \end{aligned}$$

that is graphed in Fig. 4.6. In this LP, $\Theta = S$. Thus, $\Theta_x = \{\mathbf{x}^1, \mathbf{x}^2\}$ and Θ_μ contains the two unbounded edges emanating from $\mathbf{x}^1$ and the two unbounded edges emanating from $\mathbf{x}^2$. Since all tableaus are optimal, the initial tableau is the Phase II optimal tableau. Thus, Phase III commences from the initial tableau and is described step by step as follows.

1. Table 4.4 is the initial Phase III optimal tableau.
2. For the initial Phase III tableau, we store 48 in L_B and 32 in L_x, and print out the coordinates of extreme point 32.
3. When we examine the x_1 nonbasic column, the optimal basis (x_1, s_5) is detected and 34 is stored in L_B. Noting that

$$\begin{bmatrix} 1 \\ 0 \\ 0 \\ 1 \\ 1 \end{bmatrix}$$

is an unbounded optimal edge direction, we store (32, 50) as the first (L_e, L_μ) pair and print out the coordinates of extreme point 32 and unbounded edge direction 50.

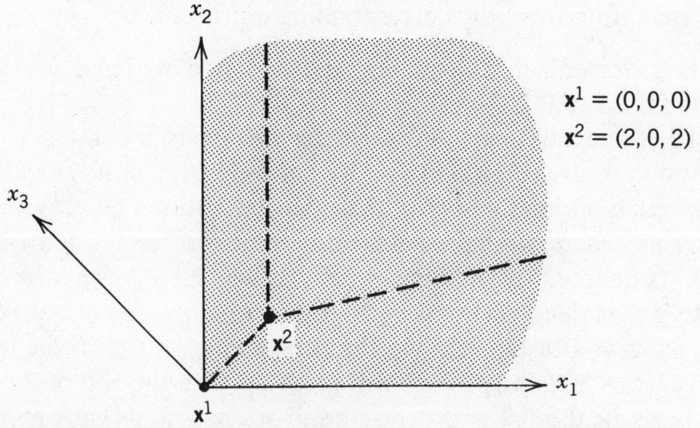

Figure 4.6. Graph of Phase III numerical example.

TABLE 4.4
Initial Phase III Optimal Tableau

		x_1	x_2	x_3	s_4	s_5	
s_4	0	$\boxed{-1}$	0	$\boxed{1}$	1	0	
s_5	2	-1	0	2	0	1	
z	0	0	0	0	0	0	$c_j - z_j$
		↑ $\begin{pmatrix} x_1 \\ s_5 \end{pmatrix}_{34}$	↑	↑ $\begin{pmatrix} x_3 \\ s_5 \end{pmatrix}_{40}$			

4. Examining the x_2 nonbasic column, we note that

$$\begin{bmatrix} 0 \\ 1 \\ 0 \\ 0 \\ 0 \end{bmatrix}$$

is an unbounded optimal edge direction. We store (32, 4) as the second (L_e, L_μ) pair and print out the coordinates of extreme point 32 and unbounded edge direction 4.

5. When we examine the x_3 nonbasic column, the optimal basis (x_3, s_5) is detected and 40 is stored in L_B.

6. We crash to coded basis 34 and obtain the second Phase III optimal tableau of Table 4.5.

7. Corresponding to the 34 entry in L_B, we store 32 in L_x. However, we do not print out the coordinates of extreme point 32 because there is an

TABLE 4.5
Second Phase III Optimal Tableau

		x_1	x_2	x_3	s_4	s_5	
x_1	0	1	0	$\boxed{-1}$	$\boxed{-1}$	0	
s_5	2	0	0	$\boxed{1}$	-1	1	
z	0	0	0	0	0	0	$c_j - z_j$
			↑	↑ ${}_{40}\begin{pmatrix} x_3 \\ s_5 \end{pmatrix}\begin{pmatrix} x_1 \\ x_3 \end{pmatrix}_{10}$	↑ $\begin{pmatrix} s_4 \\ s_5 \end{pmatrix}_{48}$		

TABLE 4.6
Third Phase III Optimal Tableau

		x_1	x_2	x_3	s_4	s_5	
x_3	0	$\boxed{-1}$	0	1	$\boxed{1}$	0	
s_5	2	$\boxed{1}$	0	0	-2	1	
z	0	0	0	0	0	0	$c_j - z_j$
		↑ $_{34}\begin{pmatrix} x_1 \\ s_5 \end{pmatrix}\begin{pmatrix} x_3 \\ x_1 \end{pmatrix}_{10}$	↑		↑ $\begin{pmatrix} s_4 \\ s_5 \end{pmatrix}_{48}$		

earlier 32 in L_x signifying that the extreme point has already been printed out.

8. Examining the x_2 nonbasic column, we detect the unbounded optimal edge direction 4 emanating from extreme point 32. It is not recorded because the (L_e, L_μ) pair of (32, 4) is already in the lists.
9. Examining the x_3 nonbasic column, we detect optimal bases 40 and 10. Although 40 had been detected earlier, 10 is new and is entered in L_B.
10. Examining the x_4 nonbasic column, we detect optimal basis 48 that is not recorded because 48 is already in L_B. We also detect the unbounded optimal edge (32, 50) that is not recorded because the (L_e, L_μ) pair is already in the lists.
11. We crash to coded basis 40 and obtain the third Phase III optimal tableau of Table 4.6.
12. Corresponding to the 40 entry in L_B, we store 32 in L_x. However, we do not print out the coordinates of extreme point 32 because there are earlier 32's in L_x signifying that the extreme point has already been printed out.
13. Examining the x_1 nonbasic column, we detect optimal bases 34 and 10 that are not recorded because they are already in L_B.
14. Examining the x_2 nonbasic column, we detect the (32, 4) unbounded optimal edge that is not recorded because the (L_e, L_μ) pair is already in the lists.
15. Examining the x_4 nonbasic column, we detect optimal basis 48 that is not recorded because it is already in L_B.
16. We crash to coded basis 10 and obtain the fourth Phase III optimal tableau of Table 4.7.
17. Corresponding to the 10 entry in L_B, we store 10 in L_x and print out the coordinates of extreme point 10.
18. Examining the x_2 nonbasic column, we detect the unbounded optimal edge (10, 4). Being new to the lists, it is stored as the third (L_e, L_μ) pair,

TABLE 4.7
Fourth Phase III Optimal Tableau

		x_1	x_2	x_3	s_4	s_5	
x_3	2	0	0	1	-1	[1]	
x_1	2	1	0	0	-2	[1]	
z	0	0	0	0	0	0	$c_j - z_j$
			↑		↑	↑	

$${}_{34}\begin{pmatrix} s_5 \\ x_1 \end{pmatrix}\begin{pmatrix} x_3 \\ s_5 \end{pmatrix}_{40}$$

and the coordinates of extreme point 10 and unbounded edge direction 4 are printed out.

19. Examining the x_4 nonbasic column, we note that

$$\begin{bmatrix} 2 \\ 0 \\ 1 \\ 1 \\ 0 \end{bmatrix}$$

is an unbounded optimal edge direction. We store (10, 26) as the fourth (L_e, L_μ) pair and print out the coordinates of extreme point 10 and unbounded edge direction 26.

20. Examining the x_5 nonbasic column, we find nothing new. Since we have crashed to the end of L_B and no new optimal bases have been detected, Phase III terminates.

The completed master lists appear in Table 4.8. It is just a coincidence that there are as many (L_B, L_x) pairs as there are (L_e, L_μ) pairs. The results of Phase III are

$$\Theta_x = \{\mathbf{x}^1, \mathbf{x}^2\}$$

$$\Theta_\mu = \{\mu(\mathbf{x}^1, \mathbf{v}^a), \mu(\mathbf{x}^1, \mathbf{v}^b), \mu(\mathbf{x}^2, \mathbf{v}^c), \mu(\mathbf{x}^2, \mathbf{v}^d)\}$$

$$\text{where } \mathbf{v}^a = (1, 0, 0)$$

$$\mathbf{v}^b = (0, 1, 0)$$

$$\mathbf{v}^c = (2, 0, 1)$$

$$\mathbf{v}^d = (0, 1, 0)$$

The reason each new extreme point and each new unbounded edge are printed out when first detected is to save the core storage that would be required to accumulate all of the extreme points and unbounded edges for printing out all at

TABLE 4.8
Completed Master Lists of Phase III Numerical Example

L_B	L_x	L_e	L_μ
48	32	32	50
34	32	32	4
40	32	10	4
10	10	10	26

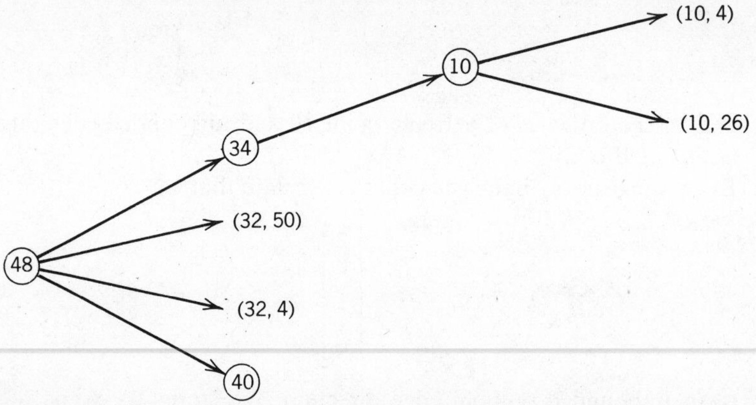

Figure 4.7. Tree of optimal bases and unbounded optimal edges for the Phase III numerical example.

once at the end of the run. The tree of optimal bases and unbounded optimal edges is presented in Fig. 4.7.

Since all convex combinations of the optimal extreme points and points on the unbounded optimal edges constitute Θ, Phase III enables the characterization of the optimal set by the specification of Θ_x and Θ_μ.

Readings

1. Gass, S. I. (1985). *Linear Programming: Methods and Applications*, New York: McGraw-Hill.
2. Hadley, G. (1962). *Linear Programming*, Reading, Massachusetts: Addison-Wesley.
3. Orchard-Hays, W. (1968). *Advanced Linear-Programming Computing Techniques*, New York: McGraw-Hill.

4. van de Panne, C. (1971). *Linear Programming and Related Topics*, Amsterdam: North-Holland.

PROBLEM EXERCISES

4-1 Formulate the LPs of Fig. 4.8 that have:

(a) $\gamma(\mathbf{x}^1, \mathbf{x}^2, \mathbf{x}^3)$ **(b)** $\mu(\mathbf{x}^4, \mathbf{v})$ **(c)** S

as their optimal sets, respectively.

4-2 Graphically determine Θ, Θ_x, and Θ_μ for:

(a)
$$\begin{aligned} \max\{ \quad & x_2 = z\} \\ \text{s.t.} \quad & x_1 \leqq 3 \\ & x_2 = 2 \\ & x_3 \leqq 3 \\ & x_1 - x_3 \leqq 1 \\ & x_1, x_2, x_3 \geqq 0 \end{aligned}$$

(b)
$$\begin{aligned} \max\{ \quad & x_3 = z\} \\ \text{s.t.} \quad & -x_1 + x_2 \leqq 2 \\ & -x_1 + x_2 \leqq 1 \\ & x_1, x_2, x_3 \geqq 0 \end{aligned}$$

(c)
$$\begin{aligned} \max\{ -3x_1 + 3x_2 & = z\} \\ \text{s.t.} \quad x_2 & \leqq 3 \\ x_3 & \leqq 2 \\ -x_1 + x_2 & \leqq 2 \\ x_1, x_2, x_3 & \geqq 0 \end{aligned}$$

4-3 Determine the adjacent tableaus by pivoting on the:

(a) -3 element in the x_1 column

(b) -1 element in the x_3 column

(c) 3 element in the x_2 column

(d) -2 element in the x_2 column

in

		x_1	x_2	x_3	s_4	s_5	s_6	
s_4	0	1	0	-1	1	0	0	
s_5	6	2	3	4	0	1	0	
s_6	0	-3	-2	3	0	0	1	
z		-6	0	-4	0	0	0	$c_j - z_j$

4-4 Code the bases whose variable indices are:

(a) $\{3, 4\}$ **(b)** $\{7, 1, 4\}$ **(c)** $\{10, 2, 3, 6, 8\}$

4-5 For an LP that has 126 variables, specify the contents of the coded storage words as they would appear in the L_B master list for basis $\{6, 14, 93, 57, 42, 7\}$.

$\mathbf{x}^1 = (0, 3, 0)$
$\mathbf{x}^2 = (2, 3, 0)$
$\mathbf{x}^3 = (0, 3, 4)$

a

$\mathbf{x}^4 = (1, 3, 0)$
$\mathbf{x}^5 = (3, 4, 0)$
$\mathbf{v} \; = (0, 0, 1)$

b

$\mathbf{x}^6 = (1, 1, 0)$
$\mathbf{x}^7 = (3, 1, 0)$
$\mathbf{x}^8 = (3, 1, 2)$
$\mathbf{x}^9 = (1, 1, 2)$
$\mathbf{x}^{10} = (2, 4, 1)$

c

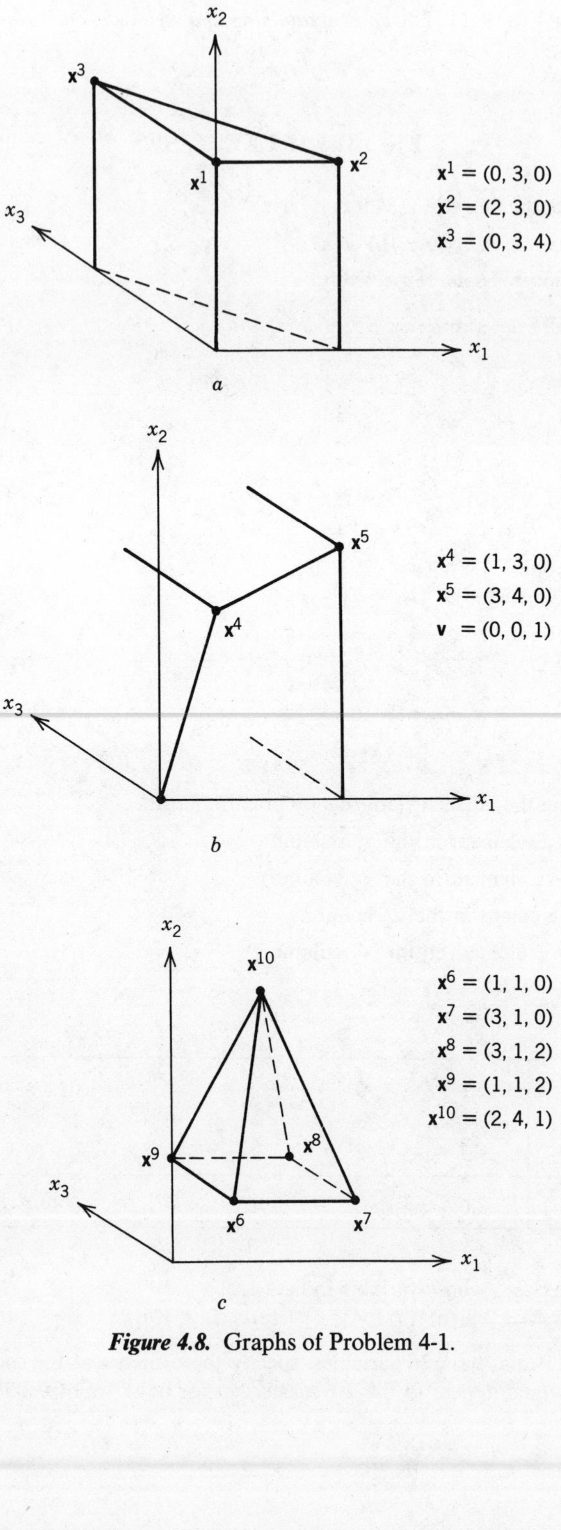

Figure 4.8. Graphs of Problem 4-1.

4-6 Determine the variables comprising the coded bases:

(a) 1st word

1st word
5420

(b)

1st word	2nd word
0	5420

(c)

1st word	2nd word
436906	70

(d)

1st word	2nd word	3rd word
17301648	0	65568

4-7 Determine which variables are basic and which are nonbasic when the basis codes are:

(a) 68 and the number of simplex variables is 6.

(b) 626 and the number of simplex variables is 12.

4-8 From coded basis 240 whose tableau is:

		x_1	x_2	x_3	s_4	s_5	s_6	s_7
s_4	8	0	0	2	1	0	0	0
s_5	4	1	0	0	0	1	0	0
s_6	0	3	0	−3	0	0	1	0
s_7	4	0	1	0	0	0	0	1

crash to coded basis 78.

4-9 With regard to the tableau of coded basis 240 in Problem 4-8, code:

(a) The extreme point.

(b) The adjacent extreme point along the edge associated with the x_2 nonbasic variable.

(c) The direction of the emanating edge associated with x_2.

(d) The adjacent extreme point along the edge associated with the x_3 nonbasic variable.

(e) The direction of the emanating edge associated with x_3.

4-10 Using the methodology of crashing down the list of coded optimal bases, determine all optimal extreme points and all unbounded optimal edges of:

(a)
$$\begin{aligned} \max \{2x_1 + 2x_2 &= z\} \\ \text{s.t.} \quad x_1 + x_2 &\leqq 4 \\ x_1 + 2x_3 &\leqq 8 \\ x_1, x_2, x_3 &\geqq 0 \end{aligned}$$

(b)
$$\begin{aligned} \max \{-3x_2 &= z\} \\ \text{s.t.} \quad -x_1 + 2x_2 &\leqq 6 \\ -x_1 + 2x_2 + x_3 &\leqq 6 \\ 2x_1 + 3x_2 + 2x_3 &\geqq 3 \\ x_1, x_2, x_3 &\geqq 0 \end{aligned}$$

(c) $\max\{-x_1 + x_2 = z\}$

$$\begin{aligned} \text{s.t.} \quad -x_1 + x_2 \quad &\leqq 2 \\ -x_1 \quad + 2x_3 &\leqq 6 \\ x_1, x_2, x_3 &\geqq 0 \end{aligned}$$

4-11C Consider the LP:

	x_1	x_2	x_3		
Obj		1		max	
s.t.		1		≦	4
	−1	1	−1	≦	2
	1		−1	≦	0
	−1		1	≦	4
	1	1	−1	≦	4

(a) Solve using a commercial LP code.

(b) By graphing the LP, determine Θ, Θ_x, and Θ_μ.

4-12C How many extreme points are optimal in the following LP:

	x_1	x_2	x_3	x_4	x_5		
Obj	0	0	0	0	0	max	
s.t.	1			3		≦	10
				2		≦	22
					1	≦	16
	−1	3				≦	18
			2			≦	21
			1		5	≦	14

4-13C Solve for all extreme points of the feasible region defined by:

	x_1	x_2	x_3	x_4	x_5		
s.t.	5	5		1		≦	17
		−1	2		5	≦	14
				2	1	≦	13
		2		2		≦	19

4-14C How many bases and extreme points are there to the feasible region defined by:

	x_1	x_2	x_3	x_4		
s.t.		2	1	2	≦	9
		5	6	2	≦	9
	1	−4	6		≦	5

4-15C How many bases, extreme points, and unbounded edges are optimal in the following LP:

	x_1	x_2	x_3	x_4	x_5	x_6	x_7		
Obj	0	0	0	0	0	0	0	max	
		4		3		-1		$\leq$	8
s.t.	1			5			2	$\leq$	4
	5	4		2		-1		$\leq$	4
			4		2		3	$\leq$	7

CHAPTER 5

Comments About Objective Row Parametric Programming

This chapter comments on the parameterization of the objective row of a single objective LP. Rather than provide a broad review of parametric programming, the purpose is to focus on the connection between the row parameterization of an LP objective function and multiple objective linear programming. We will see how objective row parametric linear programming problems can be formulated as bicriterion linear programs. In this way, parametric linear programming forms a bridge between the current practice of single objective linear programming and multiple objective linear programming.

Consider the LP

$$\max \{\mathbf{c}^1\mathbf{x} = z\}$$

$$\text{s.t.} \qquad \mathbf{x} \in S$$

where $\mathbf{c}^1$ (without a transpose superscript) is the *row* vector of objective function coefficients. To analyze solution changes as a result of objective function changes, let

1. $\mathbf{c}^2 \in R^n$ be the *change vector*.
2. $\mathbf{c}^+ \in R^n$ be the *parameterized* objective function gradient.

Four methods are discussed for parameterizing the objective function gradient. The first method is conventional objective row parametric programming. The second and third methods treat variations of the conventional approach culminating in the multiple objective interpretation of the fourth method.

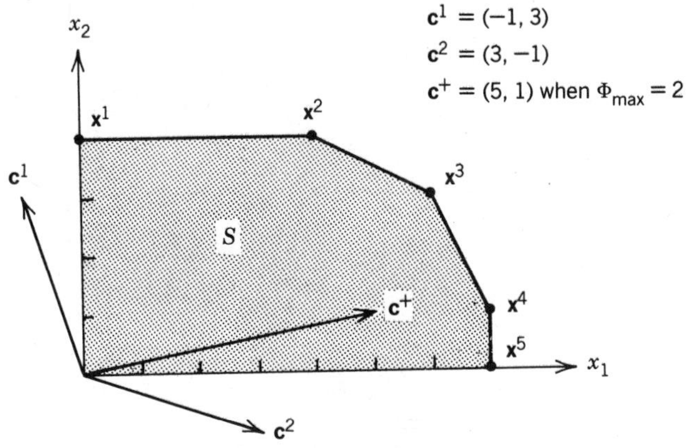

Figure 5.1. Graph of Example 1.

5.1 CONVENTIONAL PARAMETRIC PROGRAMMING

In this method, we define the parameterized objective function gradient

$$\mathbf{c}^+ = \mathbf{c}^1 + \Phi\mathbf{c}^2$$
$$\text{where } \Phi \in [0, \Phi_{max}]$$

for some fixed positive scalar value Φ_{max}. When using this method, the purpose is to determine a *path* of extreme points (and edges) that is *parametrically optimal* as Φ goes from $0 \to \Phi_{max}$. A point $\bar{\mathbf{x}} \in S$ is parametrically optimal if it maximizes

$$\max\{\mathbf{c}^+\mathbf{x} \mid \mathbf{x} \in S\}$$

for some $\Phi \in [0, \Phi_{max}]$.

Example 1. In Fig. 5.1, we use conventional objective row parametric programming with $\Phi_{max} = 2$. Thus, the parametrically optimal path traced out as Φ goes from $0 \to \Phi_{max}$ is

$$\gamma(\mathbf{x}^1, \mathbf{x}^2) \cup \gamma(\mathbf{x}^2, \mathbf{x}^3) \cup \gamma(\mathbf{x}^3, \mathbf{x}^4)$$

If, however, Φ_{max} had equalled $\frac{1}{2}$, the parametrically optimal path would have been

$$\gamma(\mathbf{x}^1, \mathbf{x}^2)$$

and if Φ_{max} had equalled infinity (theoretically), the parametrically optimal path would have been

$$\bigcup_{i=1}^{4} \gamma(\mathbf{x}^i, \mathbf{x}^{i+1})$$

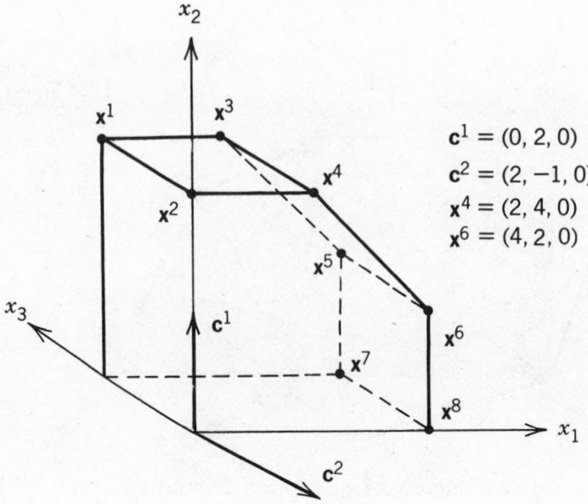

Figure 5.2. Graph of Example 2.

Let the *parametrically optimal set* be the set of all parametrically optimal points. Example 2 shows that the parametrically optimal path generated by conventional objective row parametric programming may be less than the parametrically optimal set.

Example 2. In Fig. 5.2, with $\Phi_{\max} = 2$, the parametrically optimal set is

$$\mathop{\gamma}_{i=1}^{4}(\mathbf{x}^i) \cup \mathop{\gamma}_{i=3}^{6}(\mathbf{x}^i) \cup \mathop{\gamma}_{i=5}^{8}(\mathbf{x}^i)$$

However, the parametrically optimal path traced out by conventional objective row parametric programming as Φ goes from $0 \to 2$ is

$$\gamma(\mathbf{x}^2, \mathbf{x}^4) \cup \gamma(\mathbf{x}^4, \mathbf{x}^6)$$

The pivoting stops at $\mathbf{x}^6$ because $\mathbf{c}^+ = (4, 0, 0)$ when $\Phi_{\max} = 2$ and traditional LP systems such as MPSX [5] do not continue onward to compute alternative optima. Consequently, of the eight parametrically optimal extreme points, only three are detected (i.e., $\mathbf{x}^1$, $\mathbf{x}^3$, $\mathbf{x}^5$, $\mathbf{x}^7$, and $\mathbf{x}^8$ are missed).

It is sometimes of interest to determine *critical values* of Φ, at which new bases (extreme points) become parametrically optimal. In Example 2, we have two critical values of Φ. They are

$$\Phi_1 = 0 \quad \left\{ \begin{array}{l} \text{This is the critical value between} \\ \mathbf{x}^2 \text{ and } \mathbf{x}^4 \text{ because } \mathbf{c}^+ = \mathbf{c}^1 + \Phi_1 \mathbf{c}^2 \\ \text{is normal to edge } \gamma(\mathbf{x}^2, \mathbf{x}^4). \end{array} \right.$$

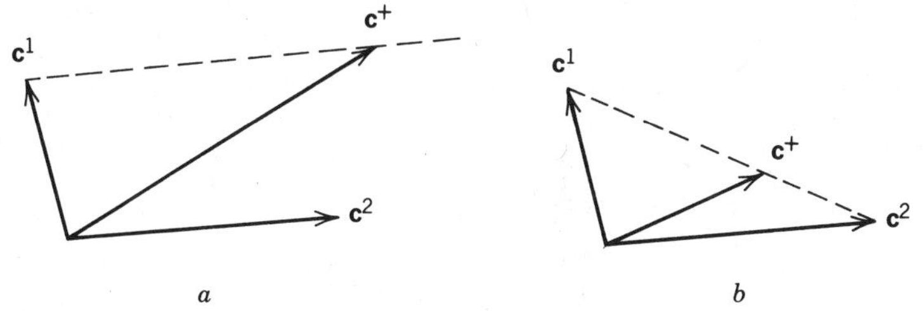

Figure 5.3. Lines traced by headpoint of $\mathbf{c}^+$.

and

$$\Phi_2 = \tfrac{2}{3} \quad \begin{cases} \text{This is the critical value between} \\ \mathbf{x}^4 \text{ and } \mathbf{x}^6 \text{ because } \mathbf{c}^+ = \mathbf{c}^1 + \Phi_2\mathbf{c}^2 \\ \text{is normal to edge } \gamma(\mathbf{x}^4, \mathbf{x}^6). \end{cases}$$

In Fig. 5.3*a*, we note that the headpoint of $\mathbf{c}^+$ traces the dashed line parallel to $\mathbf{c}^2$ as Φ increases. Hence, no matter how large a finite number we make $\Phi_{\max}$, $\mathbf{c}^+$ will never point exactly in the direction of the change vector $\mathbf{c}^2$.

5.2 CONVEX COMBINATION PARAMETRIC PROGRAMMING

Let $\boldsymbol{\lambda} = (\lambda_1, \lambda_2)$ such that

$$\boldsymbol{\lambda} \in \overline{\Lambda} = \left\{ \boldsymbol{\lambda} \in R^2 | \lambda_i \geqq 0, \quad \sum_{i=1}^{2} \lambda_i = 1 \right\}$$

Then, in this version of conventional objective row parametric programming, the parameterized objective function gradient is written as a convex combination of $\mathbf{c}^1$ and $\mathbf{c}^2$

$$\mathbf{c}^+ = \lambda_1\mathbf{c}^1 + \lambda_2\mathbf{c}^2$$
$$\text{where } \boldsymbol{\lambda} \in \overline{\Lambda}$$

In this method, the headpoint of $\mathbf{c}^+$ traces the dashed line in Fig. 5.3*b* as λ_1 goes from $1 \to 0$ (i.e., as λ_2 goes from $0 \to 1$). We also note that in this method, $\mathbf{c}^+$ can point in exactly the same direction of $\mathbf{c}^2$. We now expand the two phase method of LP to the following three phases:

Phase I: Find an initial extreme point of S in the usual way.

Phase II: Solve $\max\{\mathbf{c}^1\mathbf{x} | \mathbf{x} \in S\}$ to obtain an initial parametrically optimal basis (extreme point).

Phase III: Changing the objective gradient to $\mathbf{c}^+ = \lambda_1\mathbf{c}^1 + \lambda_2\mathbf{c}^2$, compute all other parametrically optimal bases (extreme points) by varying λ_2 from $0 \to 1$.

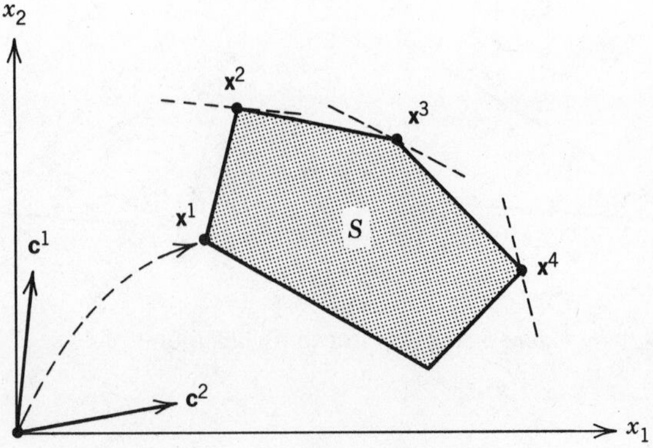

Figure 5.4. Graph of Example 3.

Example 3. If we start at the origin in Fig. 5.4, Phase I locates $\mathbf{x}^1$. Then, Phase II moves us from $\mathbf{x}^1$ to $\mathbf{x}^2$. As we increase λ_2 from $0 \to 1$ in Phase III, we encounter $\mathbf{x}^3$ and $\mathbf{x}^4$. In this way, we have found the set of all parametrically optimal extreme points $\{\mathbf{x}^2, \mathbf{x}^3, \mathbf{x}^4\}$.

In Phase III, because we have changed the objective function to

$$\mathbf{c}^+ = \lambda_1 \mathbf{c}^1 + \lambda_2 \mathbf{c}^2$$

the reduced cost row for the parameterized objective function is given by the convex combination of the reduced cost rows for $\mathbf{c}^1$ and $\mathbf{c}^2$

$$c_j^+ - z_j^+ \text{ reduced cost row} = \sum_{i=1}^{2} \lambda_i \left[c_j^i - z_j^i \text{ reduced cost row} \right]$$

where $(\lambda_1, \lambda_2) \in \bar{\Lambda}$. At a Phase III parametrically optimal basis for a given value of λ_2 (denoted $\bar{\lambda}_2$), we have two cases:

Case 1: All nonbasic $c_j^2 - z_j^2$ elements are less than or equal to zero. In this case, no convex combination with $\lambda_2 \geqq \bar{\lambda}_2$ of the two reduced cost rows can create a positive element in the $c_j^+ - z_j^+$ reduced cost row. Therefore, we terminate Phase III at the current parametrically optimal extreme point.

Case 2: There exists a nonbasic $c_j^2 - z_j^2$ element that is positive. In this case, there exists a convex combination with $\lambda_2 \geqq \bar{\lambda}_2$ of the two reduced cost rows that produces a positive element in the $c_j^+ - z_j^+$ reduced cost row. Then, we pivot into the basis the nonbasic variable associated with the "first" $c_j^+ - z_j^+$ reduced cost element to go positive as we increase λ_2 above $\bar{\lambda}_2$.

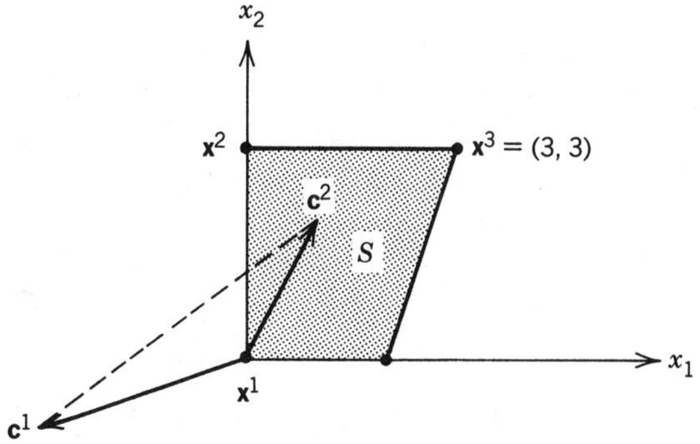

Figure 5.5. Graph of Example 4.

To compute the *critical* values of λ_2, at which new bases become parametrically optimal, we use the *rule*

$$\text{Next larger critical } \lambda_2 = \min_{j \in J} \frac{-\left(c_j^1 - z_j^1\right)}{\left(c_j^2 - z_j^2\right) - \left(c_j^1 - z_j^1\right)}$$

where $J = \{ j | (c_j^2 - z_j^2) > 0 \text{ and } (c_j^1 - z_j^1) \leqq 0\}$. Not only does this rule tell us the next larger critical λ_2 value, but the "minimizing" j specifies the next nonbasic variable to pivot into the basis to continue the parameterization of λ_2.

Example 4. With $\mathbf{c}^1 = (-3, -1)$ and $\mathbf{c}^2 = (1, 2)$, consider the parametric LP subject to

$$x_2 \leqq 3$$
$$3x_1 - x_2 \leqq 6$$
$$x_1, x_2 \geqq 0$$

that is graphed in Fig. 5.5. If we use the procedure of this section, Phase I is unnecessary and Phase II results in the tableau of Table 5.1. To begin Phase III, we add a reduced cost row for $\mathbf{c}^2$ to the ending Phase II tableau. This results in

TABLE 5.1
Phase II Tableau of Example 4

		x_1	x_2	s_3	s_4	
s_3	3	0	1	1	0	
s_4	6	3	-1	0	1	
		-3	-1	0	0	$c_j^1 - z_j^1$

TABLE 5.2
Initial Phase III Tableau of Example 4

		x_1	x_2	s_3	s_4	
s_3	3	0	$\boxed{1}$	1	0	
s_4	6	3	−1	0	1	
		−3	−1	0	0	$c_j^1 - z_j^1$
		1	2	0	0	$c_j^2 - z_j^2$

TABLE 5.3
Second Phase III Tableau of Example 4

		x_1	x_2	s_3	s_4	
x_2	3	0	1	1	0	
s_4	9	$\boxed{3}$	0	1	1	
		−3	0	1	0	$c_j^1 - z_j^1$
		1	0	−2	0	$c_j^2 - z_j^2$

Table 5.2. Using the rule for computing critical λ_2 values, we have $J = \{1,2\}$ that results in

$$\text{1st critical } \lambda_2 = \min\left\{\frac{3}{1+3}, \frac{1}{2+1}\right\} = \tfrac{1}{3}$$

Notice that for $\boldsymbol{\lambda} = (\frac{2}{3}, \frac{1}{3})$, $\mathbf{c}^+$ points in the direction normal to edge $\gamma(\mathbf{x}^1, \mathbf{x}^2)$ and the $c_j^+ - z_j^+$ row equals $(-\frac{5}{3}, 0, 0, 0)$. Pivoting x_2 into the basis, we obtain Table 5.3. With $J = \{1\}$,

$$\text{2nd critical } \lambda_2 = \min\left\{\frac{3}{1+3}\right\} = \tfrac{3}{4}$$

Notice that for $\boldsymbol{\lambda} = (\frac{1}{4}, \frac{3}{4})$, $\mathbf{c}^+$ points in the direction normal to edge $\gamma(\mathbf{x}^2, \mathbf{x}^3)$ and the $c_j^+ - z_j^+$ row equals $(0, 0, -\frac{5}{4}, 0)$. Pivoting x_1 into the basis, we obtain Table 5.4.

Since $J = \varnothing$, we are done. Thus, we have the following subsets of $\bar{\Lambda}$ pertaining to the different parametrically optimal extreme points and edges:

$$\bar{\Lambda}_{x^1} = \{\boldsymbol{\lambda} \in \bar{\Lambda} \mid \lambda_1 \in [\tfrac{2}{3}, 1], \lambda_2 \in [0, \tfrac{1}{3}]\}$$
$$\bar{\Lambda}_{\gamma(x^1, x^2)} = \{\boldsymbol{\lambda} \in \bar{\Lambda} \mid \lambda_1 = \tfrac{2}{3}, \lambda_2 = \tfrac{1}{3}\}$$
$$\bar{\Lambda}_{x^2} = \{\boldsymbol{\lambda} \in \bar{\Lambda} \mid \lambda_1 \in [\tfrac{1}{4}, \tfrac{2}{3}], \lambda_2 \in [\tfrac{1}{3}, \tfrac{3}{4}]\}$$
$$\bar{\Lambda}_{\gamma(x^2, x^3)} = \{\boldsymbol{\lambda} \in \bar{\Lambda} \mid \lambda_1 = \tfrac{1}{4}, \lambda_2 = \tfrac{3}{4}\}$$
$$\bar{\Lambda}_{x^3} = \{\boldsymbol{\lambda} \in \bar{\Lambda} \mid \lambda_1 \in [0, \tfrac{1}{4}], \lambda_2 \in [\tfrac{3}{4}, 1]\}$$

TABLE 5.4
Final Phase III Tableau of Example 4

		x_1	x_2	s_3	s_4	
x_2	3	0	1	1	0	
x_1	3	1	0	$\frac{1}{3}$	$\frac{1}{3}$	
		0	0	2	1	$c_j^1 - z_j^1$
		0	0	$-\frac{7}{3}$	$-\frac{1}{3}$	$c_j^2 - z_j^2$

Because

$$\frac{1}{\lambda_1}\left[\lambda_1\mathbf{c}^1 + \lambda_2\mathbf{c}^2\right] = \mathbf{c}^1 + \Phi\mathbf{c}^2$$

the relationship between Φ of Section 5.1 and λ_2 of this section is

$$\Phi = \frac{\lambda_2}{1 - \lambda_2}$$

Therefore, the critical Φ values pertaining to the critical λ_2 values of $\frac{1}{3}$ and $\frac{3}{4}$ of Example 4 are $\frac{1}{2}$ and 3, respectively.

5.3 CRITERION CONE PARAMETRIC PROGRAMMING

In this variation, we view objective row parametric programming as the "rotation" of $\mathbf{c}^+$ from one extreme ray ($\mathbf{c}^1$) of a two-dimensional *criterion cone* to the other ($\mathbf{c}^2$), where the criterion cone is the cone generated by $\{\mathbf{c}^1, \mathbf{c}^2\}$. Then, the path of parametrically optimal extreme points and edges is computed in the same way as in the three-phase convex combination method of the previous section.

Example 5. The parametrically optimal set in Fig. 5.6 is $\gamma(\mathbf{x}^1, \mathbf{x}^2)$. The criterion cone is the cone generated by $\mathbf{c}^1$ and $\mathbf{c}^2$. With $\mathbf{a}^j$ normal to the $\gamma(\mathbf{x}^1, \mathbf{x}^2)$ edge of S,

a. The subset cone generated by $\{\mathbf{c}^1, \mathbf{a}^j\}$ *pertains* to $\mathbf{x}^1$.
b. The cone generated by $\mathbf{a}^j$ pertains to all points in $\gamma(\mathbf{x}^1, \mathbf{x}^2)$ and *uniquely* pertains to points such as $\mathbf{x}^3$ in the relative interior of $\gamma(\mathbf{x}^1, \mathbf{x}^2)$.
c. The subset cone generated by $\{\mathbf{a}^j, \mathbf{c}^2\}$ pertains to $\mathbf{x}^2$.

A cone "pertains" to a point $\bar{\mathbf{x}} \in S$ if $\bar{\mathbf{x}}$ optimizes

$$\max\{\mathbf{c}^+\mathbf{x} \mid \mathbf{x} \in S\}$$

for all $\mathbf{c}^+$ that are positive scalar multiples of convex combinations of generators of the cone.

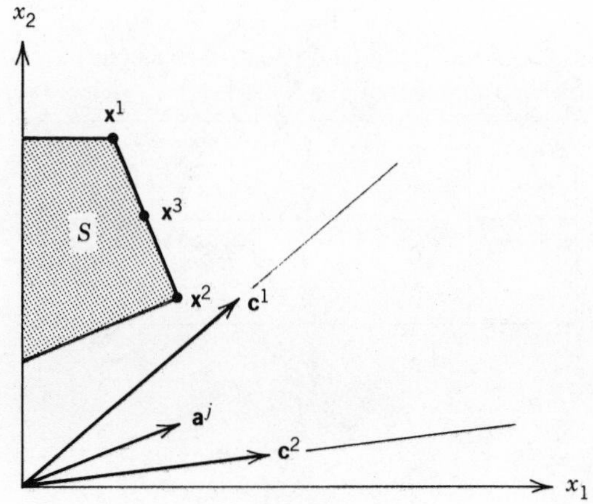

Figure 5.6. Graph of Example 5.

Example 6. Consider the "minimum cost feedmix problem"

$$\begin{aligned} \min\{\ c_1x_1 &+ c_2x_2 = z\} \\ \text{s.t.} \quad 3x_1 &+ x_2 \geqq 3 \\ x_1 &+ 3x_2 \geqq 3 \\ x_1&, x_2 \geqq 0 \end{aligned}$$

where, because of supply and price uncertainties of the commodities, $c_1 \in [0, 2]$ and $c_2 \in [1, 2]$. Assume that our purpose is to determine all points that might be optimal for some combination of the c_i from the intervals. Converting the objective function to maximization form, we have the cross-hatched *rectangle* of possible objective function coefficients in Fig. 5.7. Also shown is the criterion cone whose extreme rays (generators) are $\mathbf{c}^1 = (-2, -1)$ and $\mathbf{c}^2 = (0, -2)$. In this problem, the parametrically optimal set is

$$\gamma(\mathbf{x}^1, \mathbf{x}^2) \cup \mu(\mathbf{x}^2, \mathbf{v})$$

where $\mathbf{v} = (1, 0)$. The cross-hatched area in the subset cone generated by $\{\mathbf{c}^1, -\mathbf{a}^2\}$ is the portion of the rectangle of objective function coefficients that pertains to $\mathbf{x}^1$. Likewise, the cross-hatched portion in the subset cone generated by $\{-\mathbf{a}^2, \mathbf{c}^2\}$ pertains to $\mathbf{x}^2$. The portions of the rectangle on the rays generated by $-\mathbf{a}^2$ and $\mathbf{c}^2$ pertain to $\gamma(\mathbf{x}^1, \mathbf{x}^2)$ and $\mu(\mathbf{x}^2, \mathbf{v})$, respectively.

Let us now consider Example 7. Example 7 can be solved by the method of this section because its criterion cone is of two dimensions.

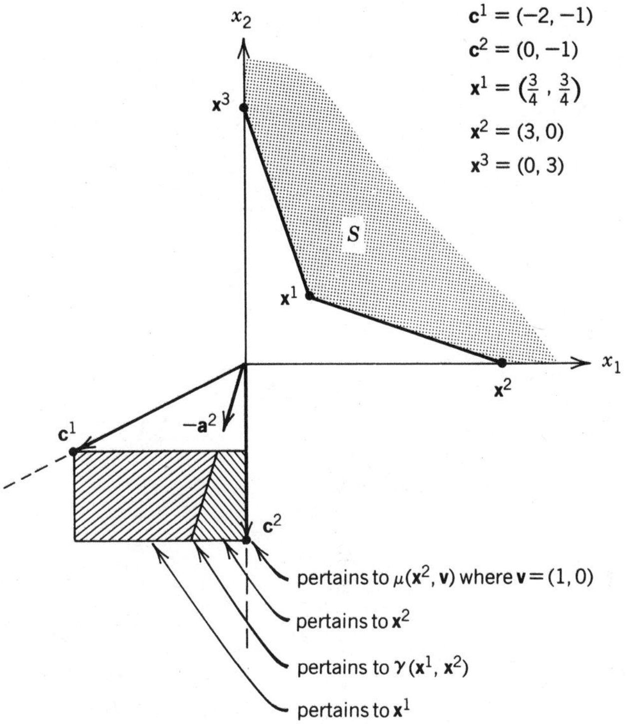

Figure 5.7. Graph of Example 6.

Example 7. Let $\mathbf{c}^1 = (-1, 3)$, $\mathbf{c}^2 = (3, 3)$, $\mathbf{c}^3 = (1, 2)$ and let S be defined by

$$x_2 \leqq 4$$

$$x_1 + 2x_2 \leqq 10$$

$$2x_1 + x_2 \leqq 10$$

$$x_1, x_2 \geqq 0$$

as in Fig. 5.8. We note that the criterion cone is two-dimensional and generated by $\{\mathbf{c}^1, \mathbf{c}^2\}$. With regard to

$$\mathbf{c}^+ = \lambda_1 \mathbf{c}^1 + \lambda_2 \mathbf{c}^2 + \lambda_3 \mathbf{c}^3$$

where $(\lambda_1, \lambda_2, \lambda_3) \in \bar{\Lambda} = \{\boldsymbol{\lambda} \in R^3 \mid \lambda_i \geqq 0, \Sigma_{i=1}^3 \lambda_i = 1\}$, let us compute the subsets of $\bar{\Lambda}$ pertaining to the different parametrically optimal extreme points and edges. Adding reduced cost rows for $\mathbf{c}^2$ and $\mathbf{c}^3$ to the tableau of $\mathbf{x}^1$ (the initial parametrically optimal extreme point), we obtain the first Phase III tableau of Table 5.5.

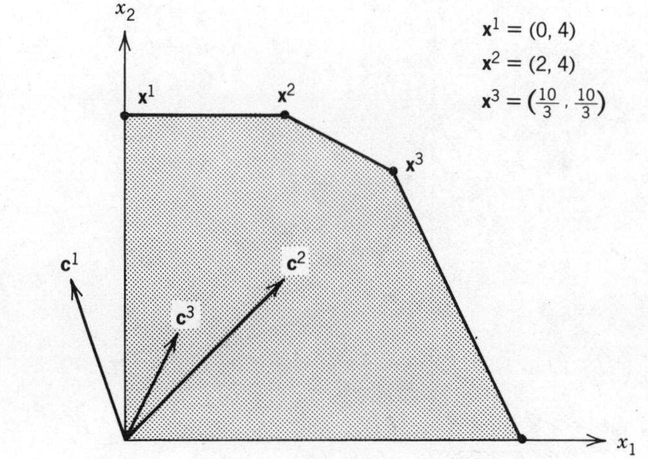

Figure 5.8. Graph of Example 7.

TABLE 5.5
First Phase III Tableau of Example 7

		x_1	x_2	s_3	s_4	s_5	
x_2	4	0	1	1	0	0	
s_4	2	[1]	0	−2	1	0	
s_5	6	2	0	−1	0	1	
		−1	0	−3	0	0	$c_j^1 - z_j^1$
		3	0	−3	0	0	$c_j^2 - z_j^2$
		1	0	−2	0	0	$c_j^3 - z_j^3$

In this problem,

$$c_j^+ - z_j^+ \text{ reduced cost row} = \sum_{i=1}^{3} \lambda_i \left[c_j^i - z_j^i \text{ reduced cost row} \right]$$

Since the nonbasic components of the $c_j^+ - z_j^+$ row must be nonpositive for those $\boldsymbol{\lambda}$-vectors in the subset $\overline{\Lambda}_{x^1} \subset \overline{\Lambda}$ associated with $\mathbf{x}^1$, we work with the nonbasic columns of the *reduced cost matrix*. Thus, to define $\overline{\Lambda}_{x^1}$, we form

$$\begin{array}{rcl} -\lambda_1 + 3\lambda_2 + \lambda_3 \leqq 0 & & 2\lambda_1 - 2\lambda_2 \geqq 1 \\ -3\lambda_1 - 3\lambda_2 - 2\lambda_3 \leqq 0 & \Leftrightarrow & -\lambda_1 - \lambda_2 \leqq 2 \\ \boldsymbol{\lambda} \in \overline{\Lambda} & & \lambda_1 + \lambda_2 \leqq 1 \\ & & \lambda_1, \lambda_2 \geqq 0 \end{array}$$

where the right-hand system is formed from the left-hand system by using the

Table 5.6
Second Phase III Tableau of Example 7

		x_1	x_2	s_3	s_4	s_5	
x_2	4	0	1	1	0	0	
x_1	2	1	0	-2	1	0	
s_5	2	0	0	$\boxed{3}$	-2	1	
		0	0	-5	1	0	$c_j^1 - z_j^1$
		0	0	3	-3	0	$c_j^2 - z_j^2$
		0	0	0	-1	0	$c_j^3 - z_j^3$

substitution $\lambda_3 = 1 - \lambda_1 - \lambda_2$. Using the *rule* of Section 5.2, we bring x_1 into the basis and pivot to $\mathbf{x}^2 = (2, 4)$. Thus, we have Table 5.6.

Working with the nonbasic columns of the reduced cost matrix, we form the systems

$$\begin{array}{rl} -5\lambda_1 + 3\lambda_2 & \leqq 0 \\ \lambda_1 - 3\lambda_2 - \lambda_3 & \leqq 0 \\ \boldsymbol{\lambda} \in \overline{\Lambda} & \end{array} \quad \Leftrightarrow \quad \begin{array}{rl} -5\lambda_1 + 3\lambda_2 & \leqq 0 \\ 2\lambda_1 - 2\lambda_2 & \leqq 1 \\ \lambda_1 + \lambda_2 & \leqq 1 \\ \lambda_1, \lambda_2 & \geqq 0 \end{array}$$

to define $\overline{\Lambda}_{x^2}$. Next, pivoting to $\mathbf{x}^3 = (\frac{10}{3}, \frac{10}{3})$, we obtain Table 5.7.

Working with the nonbasic reduced cost columns, we form the systems

$$\begin{array}{rl} -\frac{7}{3}\lambda_1 - \lambda_2 - \lambda_3 & \leqq 0 \\ \frac{5}{3}\lambda_1 - \lambda_2 & \leqq 0 \\ \boldsymbol{\lambda} \in \overline{\Lambda} & \end{array} \quad \Leftrightarrow \quad \begin{array}{rl} -\frac{4}{3}\lambda_1 & \leqq 1 \\ \frac{5}{3}\lambda_1 - \lambda_2 & \leqq 0 \\ \lambda_1 + \lambda_2 & \leqq 1 \\ \lambda_1, \lambda_2 & \geqq 0 \end{array}$$

to define $\overline{\Lambda}_{x^3}$. Graphing the right-hand systems (that are expressed in terms of λ_1

Table 5.7
Third Phase III Tableau of Example 7

		x_1	x_2	s_3	s_4	s_5	
x_2	$\frac{10}{3}$	0	1	0	$\frac{2}{3}$	$-\frac{1}{3}$	
x_1	$\frac{10}{3}$	1	0	0	$-\frac{1}{3}$	$\frac{2}{3}$	
s_3	$\frac{2}{3}$	0	0	1	$-\frac{2}{3}$	$\frac{1}{3}$	
		0	0	0	$-\frac{7}{3}$	$\frac{5}{3}$	$c_j^1 - z_j^1$
		0	0	0	-1	-1	$c_j^2 - z_j^2$
		0	0	0	-1	0	$c_j^3 - z_j^3$

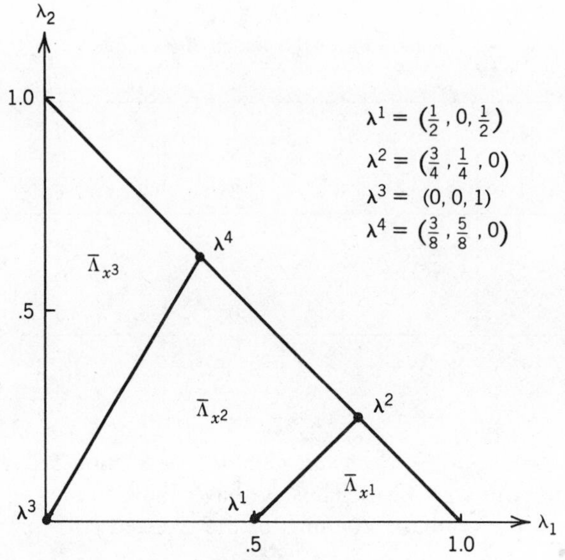

Figure 5.9. Parametric diagram of Example 7.

and λ_2), we portray the subsets of $\bar{\Lambda}$ corresponding to the three parametrically optimal extreme points in the *parametric diagram* of Fig. 5.9. In the parametric diagram, the subsets of $\bar{\Lambda}$ pertaining to edges $\gamma(\mathbf{x}^1, \mathbf{x}^2)$ and $\gamma(\mathbf{x}^2, \mathbf{x}^3)$ are given by $\gamma(\lambda^1, \lambda^2)$ and $\gamma(\lambda^3, \lambda^4)$, respectively. More will be said about parametric diagrams in Chapter 7.

5.4 MOLP APPROACH

Of the three objective row parametric programming approaches of Sections 5.1, 5.2, and 5.3, the (two-dimensional) criterion cone method is the most illuminating with regard to our interest in multiple objectives. However, the difficulty with the three approaches is that they may not necessarily "sweep out" all parametrically optimal points. They merely construct a (nonlooping) path of extreme points and edges from an extreme point that maximizes one of the generators of the criterion cone to an extreme point that maximizes the other generator. As we saw in Section 5.1, this is not sufficient for characterizing the whole parametrically optimal set in Example 2.

In this section, we show how an MOLP (multiple objective linear programming) solution procedure can be applied to a parametric LP since objective row parametric programming is essentially multiple objective linear programming with two objectives. If we use such a solution procedure, all parametrically optimal points can be obtained.

Let $V \subset R^n$ be a closed, convex cone. Then, $\bar{\mathbf{x}} \in S$ is *weakly efficient* (with respect to V) if and only if, for some $\mathbf{c}^+$, $\bar{\mathbf{x}}$ solves

$$\max \{\mathbf{c}^+\mathbf{x} \mid \mathbf{x} \in S\} \tag{5.1}$$

where $\mathbf{c}^+$ *points into* (e.g., is a convex combination of the generators of) V. In Chapter 9, a procedure that computes all points that solve (5.1) where $\mathbf{c}^+$ points into a convex cone is called a *vector-maximum* algorithm. In the case where V is a two-dimensional closed, convex cone, the set of all weakly efficient points is the same as the set of all parametrically optimal points. Therefore, we can use a vector-maximum algorithm that solves for all weakly efficient points of an MOLP to solve for all parametrically optimal points of an objective row parametric LP.

Thus, to solve an objective row parametric LP, the problem must be set up as the following MOLP:

$$\begin{aligned} &\max \{\mathbf{c}^1\mathbf{x} = z_1\} \\ &\max \{\mathbf{c}^2\mathbf{x} = z_2\} \\ &\text{s.t.} \qquad \mathbf{x} \in S \end{aligned}$$

where $\mathbf{c}^1$ is one generator of the two-dimensional criterion cone and $\mathbf{c}^2$ is the other. This means that in Example 2, with $\mathbf{c}^1 = (0, 2, 0)$ and $\mathbf{c}^2 = (2, -1, 0)$, all eight, not just three, of the parametrically optimal extreme points would be enumerated with this approach.

We can now see the transition from LP to MOLP via objective row parametric LP. In LP, we have zero-dimensional (null objective function) and one-dimensional criterion cones. In objective row parametric LP, we have two-dimensional criterion cones. In MOLP, we have higher dimensional criterion cones.

Readings

1. Gal, T. (1979). *Postoptimal Analysis, Parametric Programming and Related Topics*, New York: McGraw-Hill.
2. Gal, T. (1980). "A Historiogramme of Parametric Programming," *Journal of the Operational Research Society*, Vol. 31, No. 5, pp. 449–451.
3. Gass, S. I. (1985). *Linear Programming: Methods and Applications*, New York: McGraw-Hill.
4. Hadley, G. (1962). *Linear Programming*, Reading, Massachusetts: Addison-Wesley.
5. IBM Document No. GH19–1091-1 (1979). "IBM Mathematical Programming System Extended/370: Primer," IBM Corporation, Data Processing Division, White Plains, New York.
6. McKeown, P. G. and R. A. Minch (1983). "Multiplicative Interval Variation of Objective Function Coefficients in Linear Programming," *Management Science*, Vol. 28, No. 12, pp. 1462–1470.

7. Steuer, R. E. (1981). "Algorithms for Linear Programming Problems with Interval Objective Function Coefficients," *Mathematics of Operations Research*, Vol. 6, No. 3, pp. 333–348.

8. Steuer, R. E., K. D. Lawrence, and P. G. McKeown (1986). "Using Vector-Maximum Technology to Solve the Minimum Cost Feedmix Problem," College of Business Administration, University of Georgia, Athens, Georgia.

9. Wendell, R. E. (1984). "Using Bounds on the Data in Linear Programming: The Tolerance Approach to Sensitivity Analysis," *Mathematical Programming*, Vol. 29, No. 3, pp. 304–322.

10. Wendell, R. E. (1985). "The Tolerance Approach to Sensitivity Analysis in Linear Programming," *Management Science*, Vol. 31, No. 5, pp. 564–578.

11. Yu, P. L. and M. Zeleny (1976). "Linear Multiparametric Programming by Multicriteria Simplex Method," *Management Science*, Vol. 23, No. 2, pp. 159–170.

PROBLEM EXERCISES

5-1 Let:

$$\mathbf{c}^1 = (0,1)$$
$$\mathbf{c}^2 = (\tfrac{1}{3}, -\tfrac{1}{3})$$

(a) Graph $\mathbf{c}^+ = \mathbf{c}^1 + \Phi\mathbf{c}^2$ for $\Phi = 1, 2, 3, \ldots, 8$.

(b) Graph $\mathbf{c}^+ = \lambda_1\mathbf{c}^1 + (1 - \lambda_1)\mathbf{c}^2$ for $\lambda_1 = \frac{7}{8}, \frac{3}{4}, \frac{5}{8}, \ldots, \frac{1}{8}$.

(c) With regard to the convex cone generated by $\{\mathbf{c}^1, \mathbf{c}^2\}$, are the $\mathbf{c}^+$ directions more radially dispersed in (a) or (b)?

5-2 What convex combination of:

(a) $\mathbf{c}^1 = (-1, 3)$ **(b)** $\mathbf{c}^1 = (0, 1)$

$\mathbf{c}^2 = (2, 1)$ $\mathbf{c}^2 = (8, 8)$

points in the same direction as $\mathbf{c}^+ = (1, 5)$?

5-3 What convex combination of:

$$\mathbf{c}^1 = (3,1,1)$$
$$\mathbf{c}^2 = (-2,2,0)$$
$$\mathbf{c}^3 = (0,0,1)$$

points in the same direction as $\mathbf{c}^+ = (-1, 6, 4)$?

5-4 What convex combination of:

$$\mathbf{c}^1 = (-1,2,0)$$
$$\mathbf{c}^2 = (2,6,2)$$
$$\mathbf{c}^3 = (-1,1,2)$$

points in the same direction as $\mathbf{c}^+ = (0, 4, 3)$?

5-5 Consider the LP:

$$\min \{c_1x_1 + c_2x_2 = z\}$$
$$\text{s.t.} \quad -6x_1 + x_2 \leqq 0$$
$$x_1 + x_2 \geqq 3$$
$$x_1 + 2x_2 \geqq 4$$
$$5x_1 + 2x_2 \geqq 10$$
$$x_1, x_2 \geqq 0$$
$$\text{where } c_1 \in [1, 5]$$
$$c_2 \in [1, 4]$$

(a) Graph the rectangle of permissible objective function gradients.

(b) Determine the parametrically optimal set.

(c) Subdivide the rectangle of permissible objective function gradients according to the portions that pertain to each parametrically optimal extreme point.

(d) By "plugging into" the subdivided rectangle of permissible objective function gradients, determine the portions of the parametrically optimal set associated with:

(i) $c_1 = 3.75 \quad c_2 = 1.50$

(ii) $c_1 = 3.62 \quad c_2 = 3.97$

(iii) $c_1 = 2.68 \quad c_2 = 2.68$

(iv) $c_1 = 1.00 \quad c_2 = 2.76$

(v) $c_1 = 4.95 \quad c_2 = 1.25$

5-6 With $\mathbf{c}^1 = (-2, 1)$ and $\mathbf{c}^2 = (2, 1)$, consider the maximization LP whose feasible region is defined by:

$$x_1 + x_2 \geqq 4$$
$$x_1 + 3x_2 \leqq 6$$
$$x_1, x_2 \geqq 0$$

Using graphical means, determine the:

(a) subsets of $\overline{\Lambda}$ **(b)** ranges of Φ

that pertain to the different parametrically optimal extreme points and edges.

5-7 Allowing the feasible region to be defined by

$$x_1 + 2x_2 \leqq 10$$
$$x_1 \leqq 5$$
$$x_1, x_2 \geqq 0$$

consider the parametric program in which:

$$\mathbf{c}^1 = (0, 1)$$
$$\mathbf{c}^2 = (2, -1)$$
$$\mathbf{c}^3 = (4, 3)$$

With λ_1 and λ_2 as the axes, construct a parametric diagram to indicate the subsets of $\overline{\Lambda}$ associated with the different parametrically optimal extreme points and edges.

5-8 Consider the reduced cost rows:

$$\left(c_j^1 - z_j^1\right) = (3, 0, -4, -5, 0, 0)$$

$$\left(c_j^2 - z_j^2\right) = (-2, 0, 2, 2, 0, 0)$$

(a) Does there exist a $\lambda \in R^2$ such that the convex combination of the two rows is nonpositive and the first element of the convex combination is 0? If yes, specify λ.

(b) Does there exist a $\lambda \in R^2$ such that the convex combination is nonpositive and the third element is 0? If yes, specify λ.

(c) Does there exist a $\lambda \in R^2$ such that the convex combination is nonpositive and the fourth element is 0? If yes, specify λ.

5-9 Consider a firm that manufactures three products: x_1, x_2, and x_3. The firm's overall objective function (the more of z the better) has been estimated as

$$-2x_1 + 3x_2 \qquad = z$$

The following describes the limitations on the firm's operating environment:

$$\begin{aligned} x_2 \qquad &\leqq 4 \\ x_1 + x_2 \qquad &\leqq 6 \\ -x_1 \qquad + 2x_3 &\leqq 0 \\ x_1, x_2, x_3 &\geqq 0 \end{aligned}$$

The firm, however, is worrying about its objective function. The firm is not concerned about the x_2 term or the absence of the x_3 term. But the coefficient -2 for the x_1 term is based on incomplete information and represents the most pessimistic estimate. The most optimistic estimate is 4.

(a) Graphically determine all parametrically optimal extreme points.

(b) Which ones would a commercial LP code find?

(c) Employing a methodology of crashing down the list of parametrically optimal bases, compute all parametrically optimal extreme points.

5-10 Allowing the feasible region to be defined by

$$\begin{aligned} x_1 + x_2 &\leqq 3 \\ 3x_1 + 4x_2 &\leqq 12 \\ x_1, x_2 &\geqq 0 \end{aligned}$$

consider the parametric program in which:

$$\mathbf{c}^1 = (0, 2)$$

$$\mathbf{c}^2 = (1, 0)$$

$$\mathbf{c}^3 = (2, 2)$$

With λ_1 and λ_2 as the axes, construct a parametric diagram to indicate the subsets of $\bar{\Lambda}$ associated with the :

(a) Parametrically optimal bases.

(b) Parametrically optimal extreme points.

5-11C Let the objective rows of the following be generators of the criterion cone:

	x_1	x_2	x_3	x_4	x_5		
Objs	2	2	−1	4	1	max	
	3		1		1	max	
s.t.		1	5	1	2	$\leqq$	6
				5	5	$\leqq$	10
	4					$\leqq$	8
	3	3		2		$\leqq$	7
	5			2	1	$\leqq$	10

(a) Using a commercial LP code, solve for all parametrically optimal extreme points.

(b) Using an MOLP code, solve for all parametrically optimal extreme points.

CHAPTER 6

Utility Functions, Nondominated Criterion Vectors, and Efficient Points

A multiple objective linear program (MOLP) is written

$$\begin{aligned} &\max \{\mathbf{c}^1 \mathbf{x} = z_1\} \\ &\max \{\mathbf{c}^2 \mathbf{x} = z_2\} \\ &\quad \vdots \\ &\max \{\mathbf{c}^k \mathbf{x} = z_k\} \\ &\text{s.t.} \qquad \mathbf{x} \in \mathrm{S} \end{aligned}$$

or

$$\text{“max”} \{\mathbf{Cx} = \mathbf{z} \mid \mathbf{x} \in S\}$$

where

k is the number of objectives.
$\mathbf{c}^i$ is the *gradient* (vector of objective function coefficients) of the ith objective function.
z_i is the *criterion value* (objective function value, z-value) of the ith objective.
S is the feasible region.
"max" indicates that the purpose is to maximize all objectives simultaneously.
$\mathbf{C}$ is the $k \times n$ *criterion matrix* (matrix of objective function coefficients) whose rows are the gradients $\mathbf{c}^i$ of the k objective functions.
$\mathbf{z}$ is the *criterion vector* (objective function vector, $\mathbf{z}$-vector).

Since multiple objective problems rarely have points that simultaneously maximize all of the objectives, we are typically in a situation of trying to maximize each objective to the "greatest extent possible." To understand the difficulties that can be encountered in attempting to do this, we introduce the topics of utility functions, nondominated criterion vectors, and efficient points.

6.1 UTILITY FUNCTION SHAPES

Regardless of whether we know its mathematical representation or not, throughout the text we will assume that the decision maker has a *utility function* U: $R^k \to R$. The utility function U maps criterion vectors into the real line so that the greater the value along the real line, the more preferred the criterion vector. Because different utility function shapes have different implications for multiple criteria analysis, let us consider the following definitions.

Definition 6.1. A function U: $R^k \to R$ is *nondecreasing* iff for all $\mathbf{z}^1, \mathbf{z}^2 \in R^k$ such that $\mathbf{z}^1 \leqq \mathbf{z}^2$, $U(\mathbf{z}^1) \leqq U(\mathbf{z}^2)$.

Definition 6.2. A function U: $R^k \to R$ is *increasing* iff for all $\mathbf{z}^1, \mathbf{z}^2 \in R^k$ such that $\mathbf{z}^1 < \mathbf{z}^2$, $U(\mathbf{z}^1) < U(\mathbf{z}^2)$.

Definition 6.3. A function U: $R^k \to R$ is *concave* iff for all $\mathbf{z}^1, \mathbf{z}^2 \in R^k$

$$U(\lambda\mathbf{z}^1 + (1-\lambda)\mathbf{z}^2) \geqq \lambda U(\mathbf{z}^1) + (1-\lambda)U(\mathbf{z}^2)$$

for all $\lambda \in [0,1]$.

Definition 6.4. A function U: $R^k \to R$ is *strictly concave* iff for all $\mathbf{z}^1, \mathbf{z}^2 \in R^k$ such that $\mathbf{z}^1 \neq \mathbf{z}^2$,

$$U(\lambda\mathbf{z}^1 + (1-\lambda)\mathbf{z}^2) > \lambda U(\mathbf{z}^1) + (1-\lambda)U(\mathbf{z}^2)$$

for all $\lambda \in (0,1)$.

By replacing $U(\mathbf{z}^1) \leqq U(\mathbf{z}^2)$ by $U(\mathbf{z}^1) \geqq U(\mathbf{z}^2)$ in Definition 6.1, we obtain the definition of *nonincreasing*. By replacing $U(\mathbf{z}^1) < U(\mathbf{z}^2)$ by $U(\mathbf{z}^1) > U(\mathbf{z}^2)$ in Definition 6.2, we arrive at the definition of *decreasing*. By reversing the inequalities in Definitions 6.3 and 6.4, we have the definitions of *convex* and *strictly convex*, respectively. (One way to picture the difference between concavity and convexity is to note that the arc is on or above the chord with concavity, and the arc is on or below the chord with convexity.) Applying the terms of Definitions 6.1 to 6.4 when applicable, we have Example 1.

Example 1. To illustrate the notions of Definitions 6.1 through 6.4, consider the graphs of Fig. 6.1.

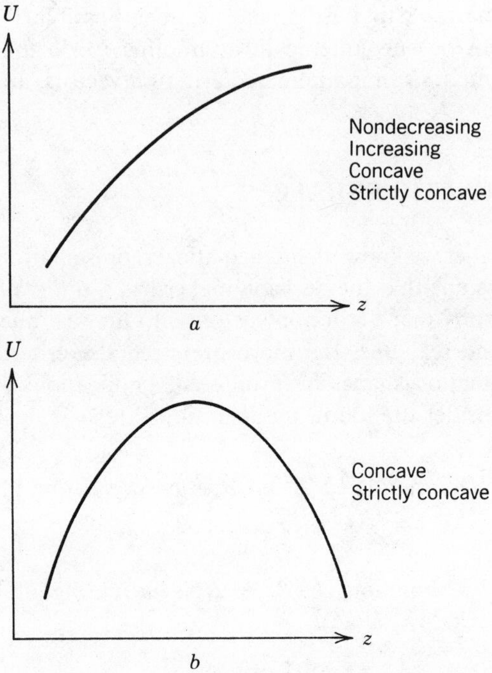

Figure 6.1. Graphs of Example 1.

Definition 6.5. A function U: $R^k \to R$ is *quasiconcave* iff for all $\mathbf{z}^1, \mathbf{z}^2 \in R^k$ such that $U(\mathbf{z}^1) \leqq U(\mathbf{z}^2)$,

$$U(\mathbf{z}^1) \leqq U(\lambda \mathbf{z}^1 + (1 - \lambda)\mathbf{z}^2)$$

for all $\lambda \in [0, 1]$.

Definition 6.6. A function U: $R^k \to R$ is *strictly quasiconcave* iff for all $\mathbf{z}^1, \mathbf{z}^2 \in R^k$ such that $U(\mathbf{z}^1) < U(\mathbf{z}^2)$,

$$U(\mathbf{z}^1) < U(\lambda \mathbf{z}^1 + (1 - \lambda)\mathbf{z}^2)$$

for all $\lambda \in (0, 1)$.

Definition 6.7. Let U: $R^k \to R$ be differentiable on R^k. Then, U is *pseudoconcave* iff for all $\mathbf{z}^1, \mathbf{z}^2 \in R^k$ such that $\nabla U(\mathbf{z}^1)(\mathbf{z}^2 - \mathbf{z}^1) \leqq 0$, we have $U(\mathbf{z}^2) \leqq U(\mathbf{z}^1)$.

By reversing *all* of the inequalities in Definitions 6.5, 6.6, and 6.7, we arrive at the definitions of *quasiconvex*, *strictly quasiconvex*, and *pseudoconvex*. Note that pseudoconcave and pseudoconvex functions are differentiable. Applying the terms of Definitions 6.1 to 6.7 when applicable, we have Example 2.

Example 2. To illustrate the notions of Definitions 6.1 through 6.7, consider the graphs of Fig. 6.2.

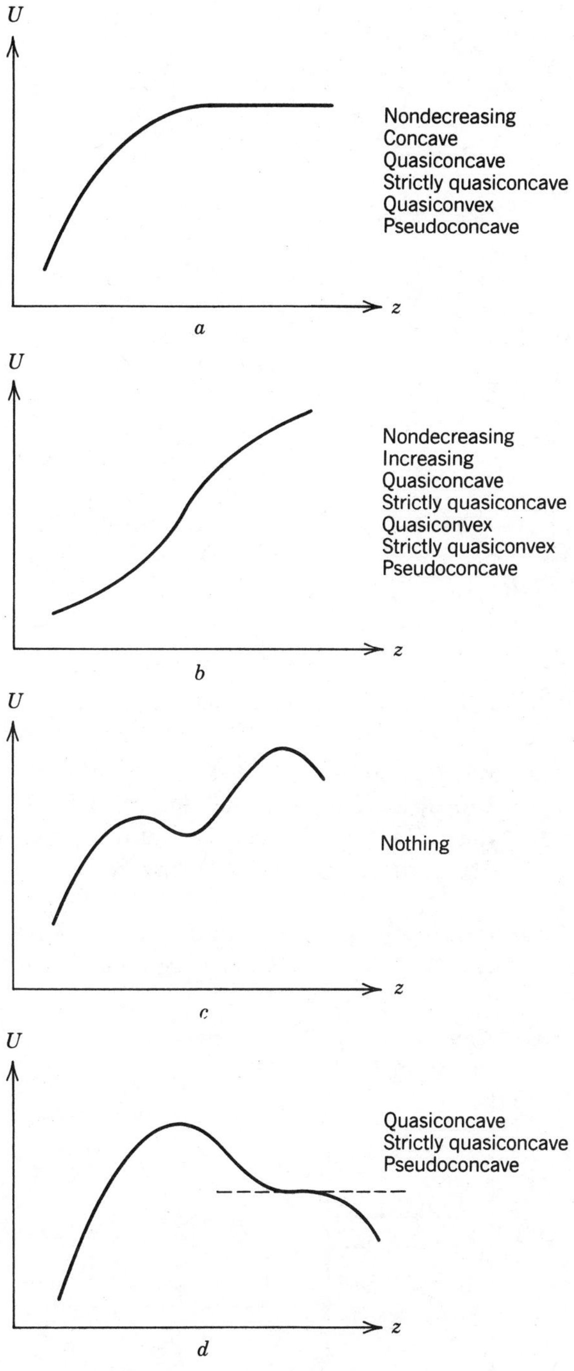

Figure 6.2. Graphs of Example 2.

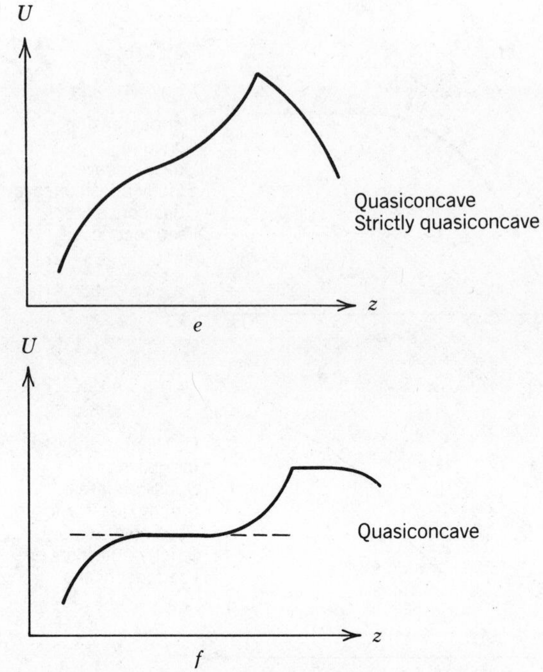

Figure 6.2. Graphs of Example 2 (cont.)

Quasiconcave functions rule out dips, as in Fig. 6.2*c*. The functions of Figs. 6.2*e* and *f* are not pseudoconcave because they are not differentiable. Figures 6.2*a* and *f* show that there can be more than one optimal point. With implications as indicated, we have the relationships among concave functions as illustrated in Fig. 6.3.

Every nonnegative linear combination of concave functions is concave. However, it is not true that every nonnegative linear combination of quasiconcave

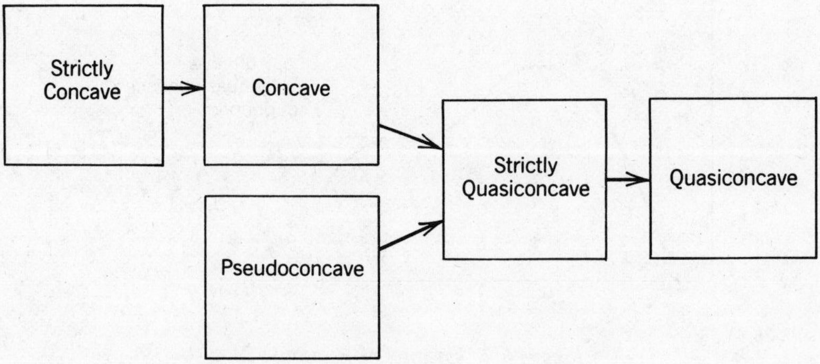

Figure 6.3. Concave function relationships.

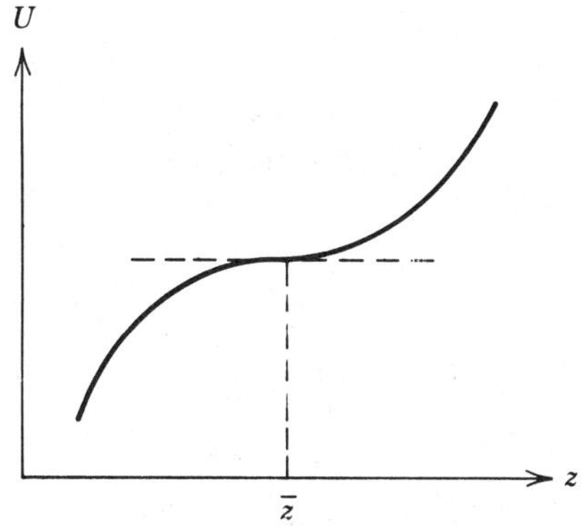

Figure 6.4. Graph of Example 3.

(strictly quasiconcave, pseudoconcave) functions is quasiconcave (strictly quasiconcave, pseudoconcave).

6.2 MONOTONICITY

With a maximization objective, we have *monotonicity* if more is always better than less. With a minimization objective, we have monotonicity if less is always better than more. A *coordinatewise increasing* utility function implies that all objectives are in maximization form and that monotonicity holds for each of them (regardless of the values at which the other criteria are held constant). Example 3 shows us that an inflection point does not necessarily destroy monotonicity.

Example 3. Consider the differentiable, strictly quasiconcave function U of Fig. 6.4. Although the partial derivative of U with respect to z at $\bar{z}$ is zero, U is coordinatewise increasing, with respect to z, because more is always better than less (as we move from left to right along the dashed line).

Let us consider objectives that have curves, as in Figs. 6.2a, d, e, and f. In Figs. 6.2d, e, and f, too much is as bad as too little. In Fig. 6.2a, only too little is undesirable. Nonmonotonic objectives such as these can, however, be remodeled monotonically using *deviational variables* (d^- for underachievement and d^+ for overachievement). The deviational variables are used to measure undesirable deviations from intermediate ideal amounts. Then, the originally nonmonotonic objectives can be rewritten monotonically as in Example 4.

Example 4. Consider the problem in which 68 to 72° is ideal for temperature and 56% or less is ideal for humidity. That is, below 68° or above 72° is undesirable, but only above 56% humidity is undesirable. Using deviational variables with the functions $\mathbf{c}^1\mathbf{x}$ and $\mathbf{c}^2\mathbf{x}$ for temperature and humidity, respectively, we obtain the formulation

$$\begin{array}{llll}
\min\{ & d_1^- + d_1^+ & & \} \\
\min\{ & & d_2^+ & \} \\
\text{s.t.} & \mathbf{c}^1\mathbf{x} + d_1^- & & \geqq 68 \\
& \mathbf{c}^1\mathbf{x} \quad - d_1^+ & & \leqq 72 \\
& \mathbf{c}^2\mathbf{x} & - d_2^+ & \leqq 56 \\
& \mathbf{A}\mathbf{x} & & \leqq \mathbf{b} \\
& & \mathbf{x}, d_1^-, d_1^+, d_2^+ & \geqq 0
\end{array}$$

Conversion of the objectives from minimization to maximization form yields the MOLP

$$\begin{array}{llll}
\max\{ & -d_1^- - d_1^+ & & \} \\
\max\{ & & -d_2^+ & \} \\
\text{s.t.} & \mathbf{c}^1\mathbf{x} + d_1^- & & \geqq 68 \\
& \mathbf{c}^1\mathbf{x} \quad - d_1^+ & & \leqq 72 \\
& \mathbf{c}^2\mathbf{x} & - d_2^+ & \leqq 56 \\
& \mathbf{A}\mathbf{x} & & \leqq \mathbf{b} \\
& & \mathbf{x}, d_1^-, d_1^+, d_2^+ & \geqq 0
\end{array}$$

The problem has now been transformed into one in which the decision maker's utility function is coordinatewise increasing in terms of the values of the new criterion functions.

Further discussions about the creative application of deviational variables are deferred until Chapter 10. Nevertheless, the point is that many initially nonmonotonic criteria can be remodeled so as to be coordinatewise increasing in the decision maker's utility function. Hence, the MOLP format

$$\begin{array}{ll}
\max\{\mathbf{c}^1\mathbf{x} = z_1\} \\
\max\{\mathbf{c}^2\mathbf{x} = z_2\} \\
\vdots \\
\max\{\mathbf{c}^k\mathbf{x} = z_k\} \\
\text{s.t.} \quad \mathbf{x} \in S
\end{array}$$

has broad applicability.

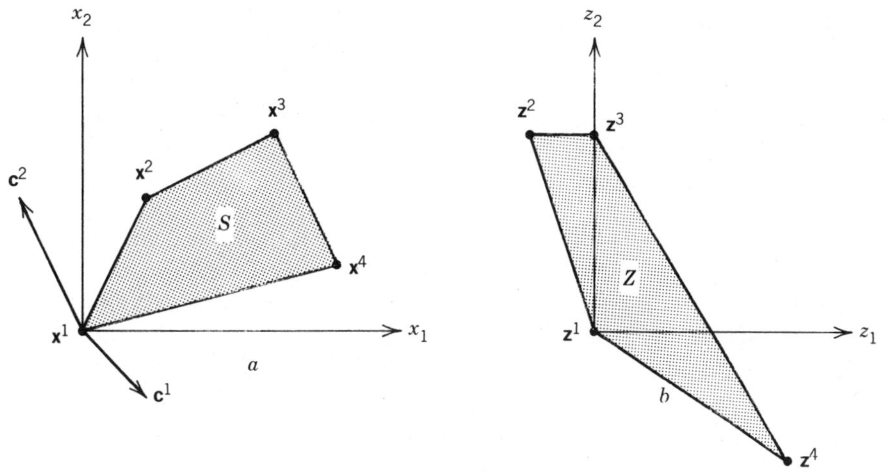

Figure 6.5. Graphs of Example 5.

6.3 FEASIBLE REGION IN CRITERION SPACE

In this book we will sometimes be graphing a multiple objective problem in decision space, and at other times, it will be more revealing to graph the problem in criterion space. To demonstrate the feasible region in criterion space, consider an MOLP. Whereas S denotes the feasible region in *decision space*, Z denotes the feasible region in *criterion space*. In set theoretic notation,

$$Z = \{\mathbf{z} \in R^k \mid \mathbf{z} = \mathbf{C}\mathbf{x}, \quad \mathbf{x} \in S\}$$

In other words, Z is the set of images of all points in S. Although S is confined to the nonnegative orthant of R^n, Z (the set of all feasible criterion vectors) is not necessarily confined to the nonnegative orthant of R^k.

Example 5. Consider the MOLP of Fig. 6.5a with $\mathbf{c}^1 = (1, -1)$ and $\mathbf{c}^2 = (-1, 2)$. The criterion vectors associated with the extreme points of S are as follows:

$$\begin{aligned} \mathbf{x}^1 &= (0,0) & \mathbf{z}^1 &= (0,0) \\ \mathbf{x}^2 &= (1,2) & \mathbf{z}^2 &= (-1,3) \\ \mathbf{x}^3 &= (3,3) & \mathbf{z}^3 &= (0,3) \\ \mathbf{x}^4 &= (4,1) & \mathbf{z}^4 &= (3,-2) \end{aligned}$$

In Fig. 6.5b, we observe that Z is convex and that the extreme points of Z are images of extreme points of S. Having portions of Z outside of the nonnegative orthant of R^k is a likely occurrence when the problem has objective functions with negative coefficients.

6.4 UTILITY FUNCTION APPROACH

In this book, we are assuming that, in practice, we will never know a mathematical representation of the DM's utility function U. However, for the sake of discussion, let us assume in this section that we are able to construct U. Then, the ideal way to solve an MOLP would be to solve the *utility function program*

$$\max \{ U(\mathbf{z}) | \mathbf{z} = \mathbf{Cx}, \quad \mathbf{x} \in S \}$$

Because of the difficulties in obtaining U and the almost certain nonlinearity of the utility function program, this method for solving a multiple criterion optimization problem is not seriously considered. However, the utility function approach is useful for conceptual reasons. For example, it enables us to define optimality in a multiple objective context. That is, if $(\mathbf{x}^0, \mathbf{z}^0)$ solves the utility function program, $\mathbf{x}^0$ is an *optimal point* and $\mathbf{z}^0$ is an *optimal criterion vector*. Also, the utility function method tells us how we would like to solve a multiple objective problem if we only could. But because we cannot, we must resort to the solution methods whose descriptions begin in Chapter 8. To graphically visualize the utility function approach, we have Examples 6 and 7.

Example 6. Consider the MOLP

$$\begin{aligned}
\max \{ x_1 \ + x_2 &= z_1 \} \\
\max \{ x_1 \qquad &= z_2 \} \\
\text{s.t.} \quad 4x_1 + 3x_2 &\leqq 12 \\
x_1, x_2 &\geqq 0 \\
\text{where } U &= 2z_1z_2
\end{aligned}$$

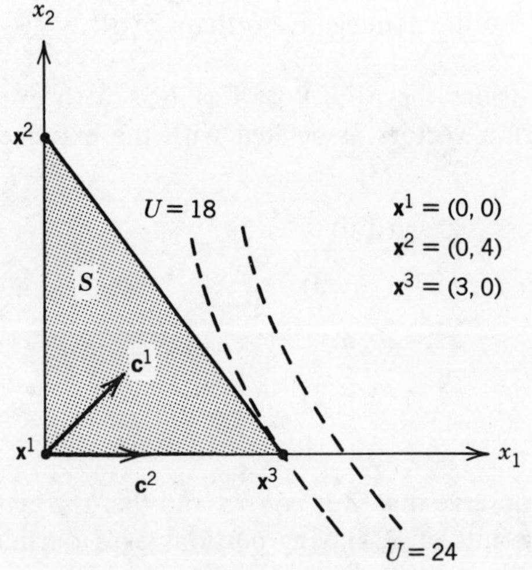

Figure 6.6. Graph of Example 6.

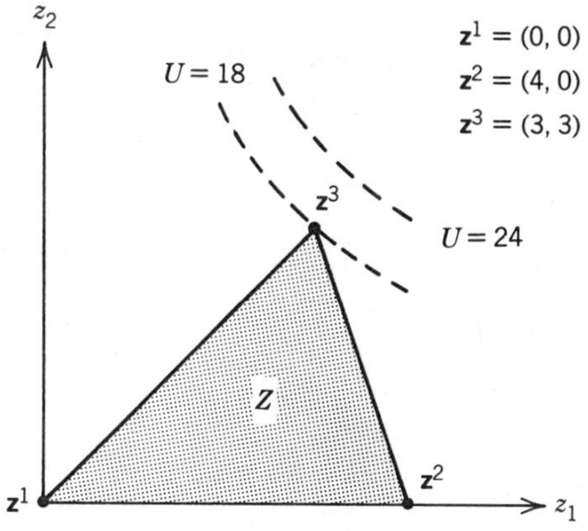

Figure 6.7. Graph of Example 7.

To solve the problem in *decision space*, we rewrite U in terms of the $\mathbf{x}$-variables to yield

$$U = 2(x_1)^2 + 2x_1x_2$$

Portraying *contours* of U in Fig. 6.6, we see that the optimal point is $\mathbf{x}^3 = (3, 0)$, the optimal criterion vector is (3, 3), and the optimal value of U is 18.

Example 7. Solving the MOLP of Example 6 in *criterion space*, we have Fig. 6.7, in which $\mathbf{z}^3 = (3, 3)$ is the optimal criterion vector and 18 is the optimal value of U. Since $\mathbf{x}^3 = (3, 0)$ is the inverse image of $\mathbf{z}^3$, $\mathbf{x}^3$ is the optimal point.

6.5 DOMINANCE

With criterion vectors, we have two forms of *dominance*.

Definition 6.8. Let $\mathbf{z}^1, \mathbf{z}^2 \in R^k$ be two criterion vectors. Then, $\mathbf{z}^1$ *dominates* $\mathbf{z}^2$ iff $\mathbf{z}^1 \geqq \mathbf{z}^2$ and $\mathbf{z}^1 \neq \mathbf{z}^2$ (i.e., $z_i^1 \geqq z_i^2$ for all i and $z_i^1 > z_i^2$ for at least one i).

If $\mathbf{z}^1$ dominates $\mathbf{z}^2$, no component of $\mathbf{z}^1$ is less than the corresponding component of $\mathbf{z}^2$, and at least one component of $\mathbf{z}^1$ is greater than its corresponding component of $\mathbf{z}^2$.

Definition 6.9. Let $\mathbf{z}^1, \mathbf{z}^2 \in R^k$ be two criterion vectors. Then, $\mathbf{z}^1$ *strongly dominates* $\mathbf{z}^2$ iff $\mathbf{z}^1 > \mathbf{z}^2$ (i.e., $z_i^1 > z_i^2$ for all i).

If $\mathbf{z}^1$ strongly dominates $\mathbf{z}^2$, each component of $\mathbf{z}^1$ is greater than its corresponding element of $\mathbf{z}^2$. Sometimes, the word "weakly" is used with the concept of Definition 6.8 to distinguish it from the concept of Definition 6.9. Note that if a criterion vector *strongly* dominates another, it *weakly* dominates it as well.

TABLE 6.1
Data for Example 8

Criterion Vector	*Criterion Values* z_1	z_2	z_3	*Dominated by*
$\mathbf{z}^1$	-1	3	4	$\mathbf{z}^2$ (strongly)
$\mathbf{z}^2$	2	4	6	
$\mathbf{z}^3$	2	2	5	$\mathbf{z}^2$ (weakly), $\mathbf{z}^4$ (weakly)
$\mathbf{z}^4$	3	2	5	
$\mathbf{z}^5$	8	3	-1	$\mathbf{z}^6$ (weakly)
$\mathbf{z}^6$	8	3	0	

Example 8. Consider the six criterion vectors in Table 6.1. Criterion vectors $\mathbf{z}^1$, $\mathbf{z}^3$, and $\mathbf{z}^5$ are dominated by the other vectors in the set as indicated. No form of dominance exists among $\mathbf{z}^2$, $\mathbf{z}^4$, and $\mathbf{z}^6$.

6.6 NONDOMINATED CRITERION VECTORS

A criterion vector is *nondominated* if it is not dominated by any other feasible criterion vector.

Definition 6.10. Let $\bar{\mathbf{z}} \in Z$. Then, $\bar{\mathbf{z}}$ is *nondominated* iff there does not exist another $\mathbf{z} \in Z$ such that $\mathbf{z} \geq \bar{\mathbf{z}}$ and $\mathbf{z} \neq \bar{\mathbf{z}}$. Otherwise, $\bar{\mathbf{z}}$ is a *dominated* criterion vector.

Let N be the *nondominated set* (i.e., the set of all nondominated criterion vectors). Our interest in N stems from Theorems 6.11 and 6.12.* Theorem 6.11 states that a criterion vector cannot be optimal unless it is nondominated. Theorem 6.12 states that there exists for each nondominated criterion vector a coordinatewise increasing utility function that makes it optimal.

Theorem 6.11. Let U: $R^k \to R$ be coordinatewise increasing. Then, if $\mathbf{z}^0 \in Z$ is optimal, $\mathbf{z}^0$ is nondominated.

Proof. Suppose $\mathbf{z}^0$ is dominated. Then, there exists a $\bar{\mathbf{z}} \in Z$ such that $\bar{\mathbf{z}} \geqq \mathbf{z}^0$, $\bar{\mathbf{z}} \neq \mathbf{z}^0$. Since U is coordinatewise increasing, this means $U(\bar{\mathbf{z}}) > U(\mathbf{z}^0)$, which contradicts the optimality of $\mathbf{z}^0$. Thus, $\mathbf{z}^0$ is nondominated. ◀

Theorem 6.12. Let $\bar{\mathbf{z}} \in Z$ be nondominated. Then, there exists a coordinatewise increasing utility function U: $R^k \to R$ such that $\bar{\mathbf{z}}$ is optimal.

Proof. Follows from the proof of Theorem 1 in Soland (1979). ◀

*Throughout the text, definitions, lemmas, theorems, and equations are numbered collectively, by chapter, in the order in which they appear.

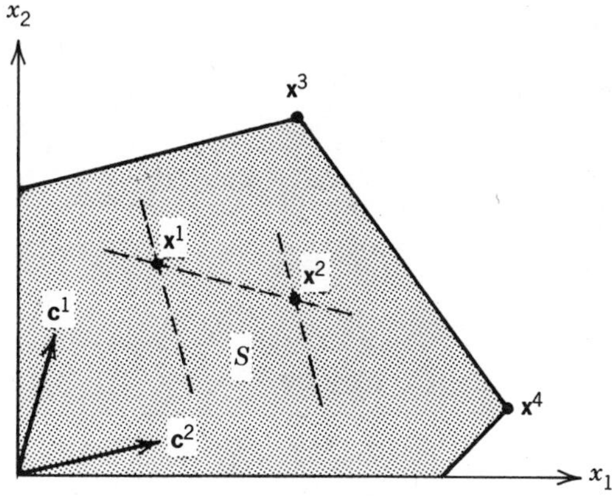

Figure 6.8. Graph of Example 9.

6.7 EFFICIENCY

Whereas the ideas of dominance refer to vectors in criterion space, the idea of efficiency refers to points in decision space. Recalling that the criterion vector of $\mathbf{x}$ in an MOLP is given by $\mathbf{Cx}$, we have the following.

Definition 6.13. A point $\bar{\mathbf{x}} \in S$ is *efficient* iff there does not exist another $\mathbf{x} \in S$ such that $\mathbf{Cx} \geqq \mathbf{C}\bar{\mathbf{x}}$ and $\mathbf{Cx} \neq \mathbf{C}\bar{\mathbf{x}}$. Otherwise, $\bar{\mathbf{x}}$ is *inefficient.*

A point $\bar{\mathbf{x}} \in S$ is efficient if its criterion vector is not dominated by the criterion vector of some other point in S. That is, from an efficient point, it is not possible to move feasibly so as to increase one of the objectives without necessarily decreasing at least one of the others. In other words, we are in a "take from Peter to pay Paul" situation. In other disciplines, the term efficiency is also known as *Pareto optimality*, *admissibility*, or *noninferiority*. Let us call the set of all efficient points the *efficient set*.

Example 9. In Fig. 6.8, the dashed lines are level curves of the objectives. Point $\mathbf{x}^1$ is inefficient because its criterion vector, for instance, is dominated by the criterion vector of $\mathbf{x}^2$. However, $\mathbf{x}^2$ is inefficient because its criterion vector, for instance, is dominated by the criterion vector of $\mathbf{x}^3$. In this problem, the efficient set is $\gamma(\mathbf{x}^3, \mathbf{x}^4)$.

A point is efficient if and only if it is the inverse image of a nondominated criterion vector, and a point is inefficient if and only if it is the inverse image of a dominated criterion vector. Therefore, with reference to Theorems 6.11 and 6.12, efficient points are *contenders* for optimality and inefficient points are *noncontenders* for optimality.

TABLE 6.2
Data for Example 10

Feasible Point	*Criterion Values* z_1	z_2	z_3	*Dominated by Criterion Vectors of*
$\mathbf{x}^1$	7	5	-1	
$\mathbf{x}^2$	2	6	-2	
$\mathbf{x}^3$	1	1	-2	$\mathbf{x}^1, \mathbf{x}^2, \mathbf{x}^4$
$\mathbf{x}^4$	5	2	-1	$\mathbf{x}^1$
$\mathbf{x}^5$	9	2	-6	

Normally, it is easier to show a point to be inefficient rather than efficient. To show a point $\mathbf{x}^1 \in S$ to be inefficient, it only takes a counterexample. That is, it only takes the identification of some other point $\mathbf{x}^2 \in S$ whose criterion vector dominates the criterion vector of $\mathbf{x}^1$. On the other hand, to show a point to be efficient involves an exhaustive test. It must be shown that none of the criterion vectors of other points in the feasible region dominates the criterion vector of the point in question.

Example 10. Assume that we have a multiple objective program whose feasible region consists of the five discrete points whose criterion vectors are given in Table 6.2. Because the criterion vectors of $\mathbf{x}^3$ and $\mathbf{x}^4$ are dominated, $\mathbf{x}^3$ and $\mathbf{x}^4$ are inefficient. Since the criterion vectors of $\mathbf{x}^1$, $\mathbf{x}^2$, and $\mathbf{x}^5$ are not dominated by the criterion vectors of any other points in the feasible region, $\mathbf{x}^1$, $\mathbf{x}^2$, and $\mathbf{x}^5$ are efficient.

A question might be asked about what happens to the notion of efficiency when the number of objectives is one. When $k = 1$, $E = \Theta$ because Definition 6.13 reduces to the following: A point $\bar{\mathbf{x}} \in S$ is efficient iff there does not exist an $\mathbf{x} \in S$ such that $\mathbf{c}^T\mathbf{x} > \mathbf{c}^T\bar{\mathbf{x}}$. Therefore, by finding an efficient point in a single objective linear program, we have found an optimal point, and by finding all efficient points in a single objective linear program, we have found all optimal points.

6.8 DETECTING EFFICIENCY USING DOMINATION SETS

We now discuss some concepts that have use in the graphical detection of efficiency. The concepts involve cones, and the material on cones that was presented in Section 2.4 should be reviewed before proceeding. To test for efficiency at a given point $\bar{\mathbf{x}} \in S$, the concept of a *domination set* is introduced by means of Definitions 6.14 and 6.15.

Definition 6.14. Let $\bar{\mathbf{x}} \in S$ and let $C^{\geq}$ be the *semi-positive polar cone* generated by the gradients of the k objective functions where

$$C^{\geq} = \{\mathbf{y} \in R^n \mid \mathbf{C}\mathbf{y} \geqq \mathbf{0}, \quad \mathbf{C}\mathbf{y} \neq \mathbf{0}\} \cup \{\mathbf{0} \in R^n\}$$

Definition 6.15. The domination set

$$D_{\bar{x}} = \{\bar{\mathbf{x}}\} \oplus C^{\geq}$$

is given by the set addition of $\{\bar{\mathbf{x}}\}$ and $C^{\geq}$.

Another way to write the domination set is $D_{\bar{x}} = \{\mathbf{x} \in R^n \mid \mathbf{x} = \bar{\mathbf{x}} + \mathbf{y}, \mathbf{C}\mathbf{y} \geqq \mathbf{0}, \mathbf{C}\mathbf{y} \neq \mathbf{0}\}$. The domination set $D_{\bar{x}}$ contains all points whose criterion vectors dominate the criterion vector of $\bar{\mathbf{x}} \in S$. Observe that in the specification of $C^{\geq}$, $\mathbf{C}\mathbf{y}$ dominates (but not necessarily strongly) the null vector $\mathbf{0} \in R^k$. Also, note that the set addition effects a translation of the semi-positive polar cone $C^{\geq}$ from the origin to the point in question. Theorem 6.16 states the importance of the domination set $D_{\bar{x}}$ in the detection of efficiency.

Theorem 6.16. Let $D_{\bar{x}}$ be the domination set at $\bar{\mathbf{x}} \in S$. Then, $\bar{\mathbf{x}}$ is efficient $\Leftrightarrow D_{\bar{x}} \cap S = \{\bar{\mathbf{x}}\}$.

Proof. $\Rightarrow$ Suppose $D_{\bar{x}} \cap S \neq \{\bar{\mathbf{x}}\}$. Then, there exists an $\hat{\mathbf{x}} \in (D_{\bar{x}} \cap S)$, $\hat{\mathbf{x}} \neq \bar{\mathbf{x}}$. With $\hat{\mathbf{x}} \in D_{\bar{x}}$ let us write $\hat{\mathbf{x}} = \bar{\mathbf{x}} + \mathbf{y}$ where $\mathbf{y} \in C^{\geq}$. Since $\mathbf{C}\mathbf{y} \geqq \mathbf{0}$, $\mathbf{C}\mathbf{y} \neq \mathbf{0}$, we have $\mathbf{C}\hat{\mathbf{x}} \geqq \mathbf{C}\bar{\mathbf{x}}$, $\mathbf{C}\hat{\mathbf{x}} \neq \mathbf{C}\bar{\mathbf{x}}$. But this is a contradiction because $\bar{\mathbf{x}}$ is efficient. Thus, if $\bar{\mathbf{x}}$ is efficient, $D_{\bar{x}} \cap S = \{\bar{\mathbf{x}}\}$.
$\Leftarrow$ If $D_{\bar{x}} \cap S = \{\bar{\mathbf{x}}\}$, this implies that if the criterion vector of $\tilde{\mathbf{x}}$ dominates the criterion vector of $\bar{\mathbf{x}}$, $\tilde{\mathbf{x}} \notin S$. Thus, the criterion vector of $\bar{\mathbf{x}}$ is nondominated, and hence $\bar{\mathbf{x}}$ is efficient. ◀

Theorem 6.16 provides a test for detecting efficient points that can be geometrically visualized. If the intersection of the domination set with the feasible region only contains $\bar{\mathbf{x}}$, then $\bar{\mathbf{x}}$ is efficient. If there are other points in the intersection, then $\bar{\mathbf{x}}$ is inefficient. It is helpful at this time to introduce the following notation

$$\begin{aligned} E &= \{\mathbf{x} \in S \mid \mathbf{x} \text{ is efficient}\} \\ E_x &= \{\mathbf{x} \in S \mid \mathbf{x} \text{ is an efficient extreme point of } S\} \\ E_\mu &= \{\mu(\mathbf{x}, \mathbf{v}) \subset S \mid \mu(\mathbf{x}, \mathbf{v}) \text{ is an unbounded efficient edge of } S\}. \end{aligned}$$

Examples 11 through 14 illustrate the domination set approach for detecting efficiency.

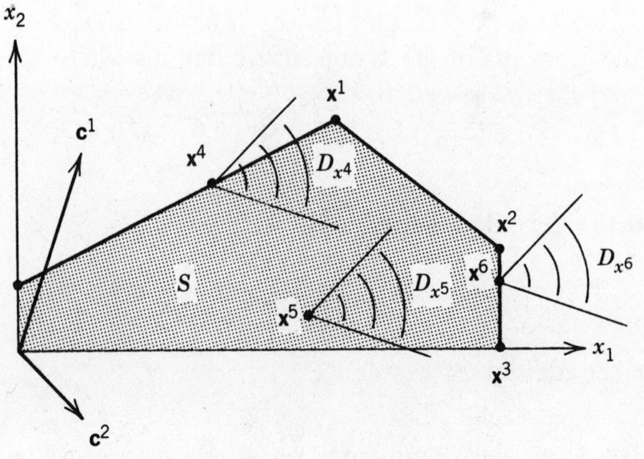

Figure 6.9. Graph of Example 11.

Example 11. Whereas $\mathbf{x}^4$ and $\mathbf{x}^5$ are inefficient because $D_{x^4} \cap S \neq \{\mathbf{x}^4\}$ and $D_{x^5} \cap S \neq \{\mathbf{x}^5\}$, $\mathbf{x}^6$ is efficient because $D_{x^6} \cap S = \{\mathbf{x}^6\}$. In Fig. 6.9,

$$E = \gamma(\mathbf{x}^1, \mathbf{x}^2) \cup \gamma(\mathbf{x}^2, \mathbf{x}^3)$$

$$E_x = \{\mathbf{x}^1, \mathbf{x}^2, \mathbf{x}^3\}$$

$$E_\mu = \varnothing$$

Example 12. In the MOLP of Fig. 6.10, the domination set approach identifies $\mathbf{x}^1$ as the only efficient point. Because both objectives are maximized at the

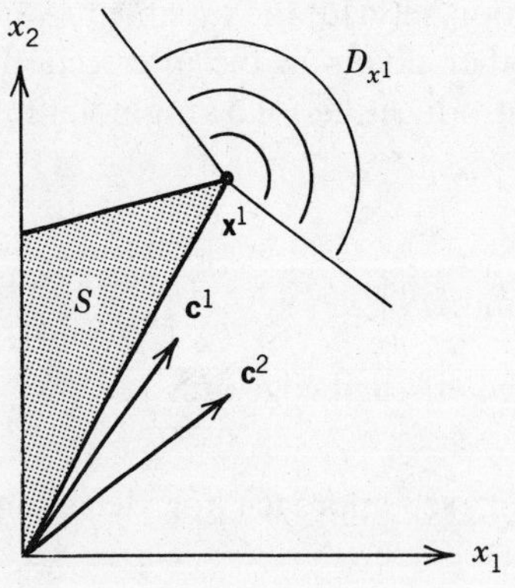

Figure 6.10. Graph of Example 12.

same point, the DM has the "best of both worlds." In this problem,

$$E = \{\mathbf{x}^1\}$$

$$E_x = \{\mathbf{x}^1\}$$

$$E_\mu = \varnothing$$

Example 13. In the MOLP of Fig. 6.11, $D_\mathbf{x} \cap S \neq \{\mathbf{x}\}$ for all $\mathbf{x} \in S$. Therefore, $E = E_x = E_\mu = \varnothing$. This MOLP does not have any optimal points.

Since the notion of efficiency is equivalent to that of optimality when there is only one objective, the domination set approach can be used to identify all optimal points in a single objective linear program. The tricky thing to note here is that the domination set in a single objective linear program is an open half-space, except for the fact that it contains one boundary point—the point in question.

Example 14. In the single objective LP of Fig. 6.12, $\mathbf{c}$ is normal to the unbounded edge $\mu(\mathbf{x}^1, \mathbf{v})$. Note that $D_{\mathbf{x}^2}$ is otherwise open except for $\mathbf{x}^2$. Since $D_{\mathbf{x}^2} \cap S = \{\mathbf{x}^2\}$, $\mathbf{x}^2$ is efficient. In a similar fashion, all other points along the $\mu(\mathbf{x}^1, \mathbf{v})$ edge are seen to be efficient. Thus, in this problem,

$$E = \Theta = \mu(\mathbf{x}^1, \mathbf{v})$$

$$E_x = \Theta_x = \{\mathbf{x}^1\}$$

$$E_\mu = \Theta_\mu = \{\mu(\mathbf{x}^1, \mathbf{v})\}$$

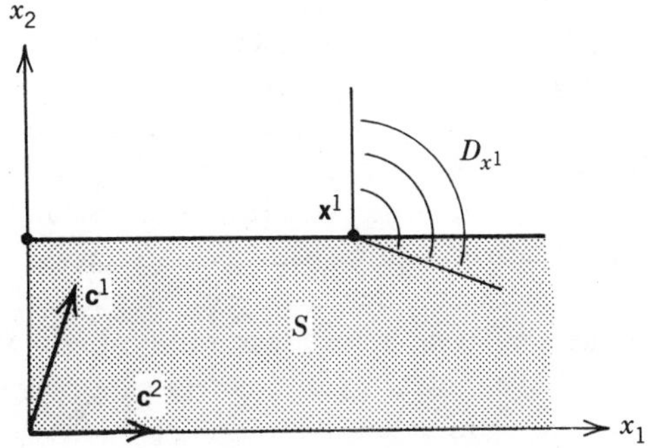

Figure 6.11. Graph of Example 13.

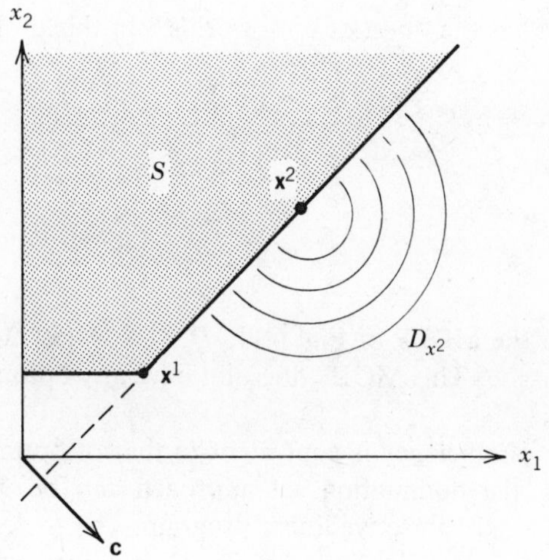

Figure 6.12. Graph of Example 14.

From the preceding examples we observe that the larger the domination set $D_{\bar{x}}$, the greater the chance $\bar{\mathbf{x}} \in S$ is inefficient. This is because the larger $D_{\bar{x}}$, the greater the opportunity that the intersection of $D_{\bar{x}}$ with S will contain more points than just $\bar{\mathbf{x}}$. Similarly, the smaller $D_{\bar{x}}$, the greater the chance $\bar{\mathbf{x}}$ is efficient.

6.9 NONCONCAVE UTILITY FUNCTIONS

There is a likelihood, perhaps even a strong likelihood, that the decision maker's coordinatewise increasing utility function is nonconcave. To illustrate, consider the following scenario.

A man would like to collect both stamps and coins. The more he has of each, the better. However, with regard to his collections, he must live within a limited budget and other restrictions. Also, he is the type of person who "likes to do things well or not at all." That is, he gets very little gratification from doing things half-way. Suppose his set of feasible criterion vectors is Z in Fig. 6.13.

It is conceivable that $\mathbf{z}^1$ (a larger stamp collection) and $\mathbf{z}^3$ (a larger coin collection) could be alternatively optimal. Criterion vectors in the middle of the $\gamma(\mathbf{z}^1, \mathbf{z}^3)$ edge such as $\mathbf{z}^2$ might be much less desirable because little satisfaction is drawn from either collection. With $\mathbf{z}^2$ of lower utility, the man's utility function contours would have to wave outwardly in this region, as shown in Fig. 6.13.

Since many people may exhibit behavior similar to that of our hypothetical stamp/coin collector, we should not be surprised to encounter nonconcave utility functions in practice. In fact, we might even expect nonconcave utility functions.

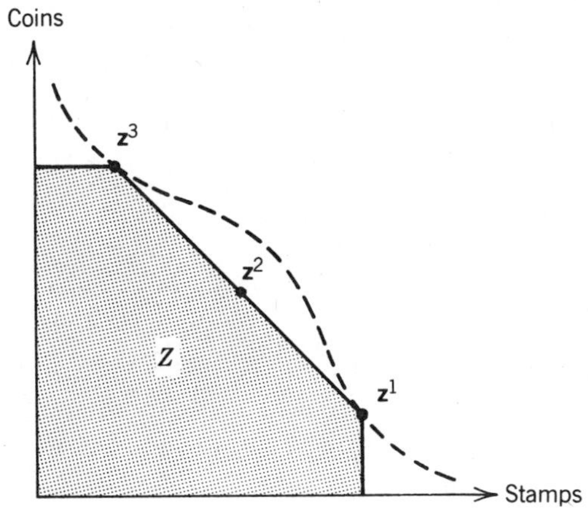

Figure 6.13. Stamp-coin illustration.

6.10 OPTIMALITY PECULIARITIES WITH MULTIPLE OBJECTIVES

In single objective linear programming, the efficient set E and the optimal set Θ are well behaved in that they are both convex (recall that in single objective LP, $E = \Theta$). However, when multiple objectives are present, E and Θ are not as well behaved. Figure 6.13 from the previous section shows that

1. The optimal set may be finite and disconnected (this means that more than one optimal point does not imply an infinite number of optimal points).
2. An MOLP may have more than one distinct optimal criterion vector (in LP there is only one optimal criterion value regardless of the number of optimal points).

Example 15 shows that

1. E can easily be nonconvex.
2. The optimal point can easily occur at a nonextreme point.
3. An MOLP may have local optima that are not global optima.

Example 16 shows that it is possible for the optimal set to be empty when E is nonempty.

Example 15. In Fig. 6.14, $E = \gamma(\mathbf{x}^1, \mathbf{x}^4)$. Although $\Theta_x = \Theta_\mu = \varnothing$, the problem has an optimal point at $\mathbf{x}^2$ (which is not extreme). Hence, Θ_x and Θ_μ cannot

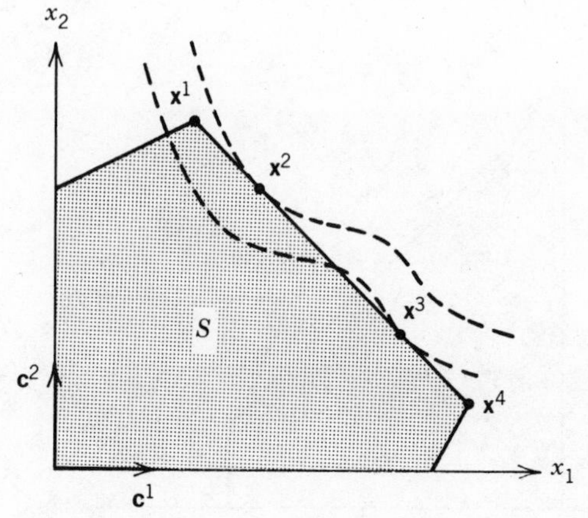

Figure 6.14. Graph of Example 15.

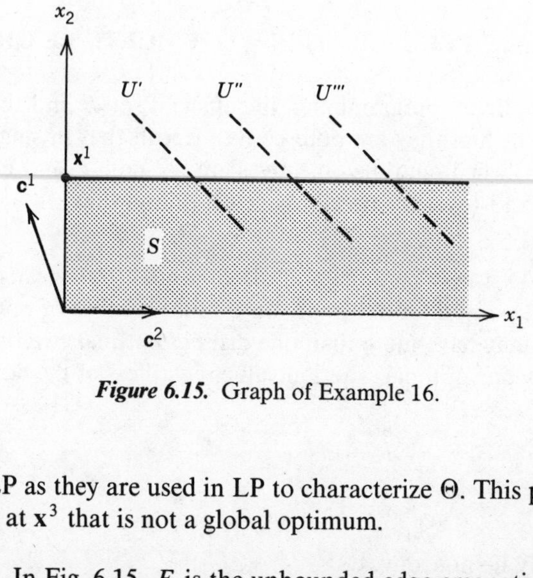

Figure 6.15. Graph of Example 16.

be used in MOLP as they are used in LP to characterize Θ. This problem also has a local optimum at $\mathbf{x}^3$ that is not a global optimum.

Example 16. In Fig. 6.15, E is the unbounded edge emanating from $\mathbf{x}^1$. With a utility function whose contours are as shown, there are no optimal points. Such a result is only possible when E is unbounded.

6.11 EFFICIENT EXTREME POINTS OF GREATEST UTILITY

In many problems, the efficient extreme point of greatest utility, if not optimal itself, will be close enough to an optimal point to suffice in its stead. However, in other problems, this might not be the case.

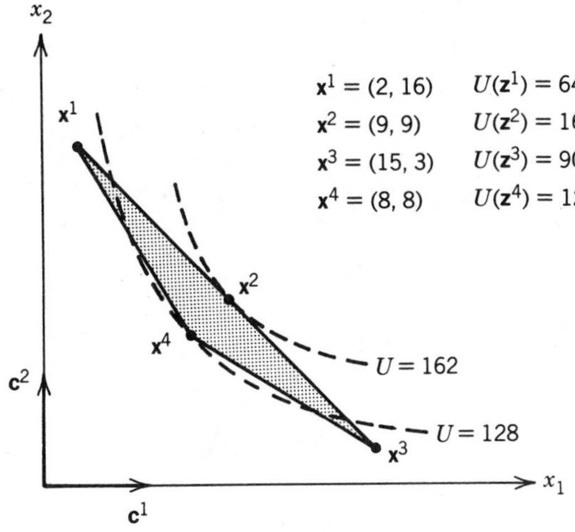

Figure 6.16. Graph of Example 17.

Example 17. Consider the following MOLP that is graphed in Fig. 6.16

$$
\begin{aligned}
\max\{\ x_1 \qquad\quad &= z_1\} \\
\max\{\qquad\quad x_2 &= z_2\} \\
\text{s.t.} \quad x_1 + x_2 &\leqq 18 \\
8x_1 + 6x_2 &\geqq 112 \\
5x_1 + 7x_2 &\geqq 96 \\
x_1, x_2 &\geqq 0 \\
\text{where } U &= 2z_1z_2
\end{aligned}
$$

in which

$$
\begin{aligned}
E &= \gamma(\mathbf{x}^1, \mathbf{x}^3) \\
E_x &= \{\mathbf{x}^1, \mathbf{x}^3\} \\
\Theta &= \{\mathbf{x}^2\}
\end{aligned}
$$

Because of the utility differences between $\mathbf{x}^1$ and $\mathbf{x}^2$ on one hand, and $\mathbf{x}^2$ and $\mathbf{x}^3$ on the other, neither efficient extreme point is likely to suffice as an approximation of $\mathbf{x}^2$. In the MOLP of this example, we thus have the situation in which the best extreme point (in utility terms) is the one that is inefficient, and the worst two extreme points of S are the ones that are efficient.

6.12 EFFECT OF NONLINEARITIES

In an MOLP, the efficient set E is always connected (proved in Theorem 9.23). However, in a multiple objective program that possesses nonlinearities, E may not be connected. A multiple objective program that possesses nonlinearities (either in the constraints, objectives, or both) is called a *nonlinear multiple objective program* and is written

$$\begin{aligned}
&\max \{f_1(\mathbf{x}) = z_1(\mathbf{x})\} \\
&\max \{f_2(\mathbf{x}) = z_2(\mathbf{x})\} \\
&\quad\vdots \\
&\max \{f_k(\mathbf{x}) = z_k(\mathbf{x})\} \\
&\text{s.t.} \qquad \mathbf{x} \in S
\end{aligned}$$

Since Definition 6.13 was predicted on an MOLP, we use the following statement of efficiency with nonlinear multiple objective programs.

Definition 6.17. A point $\bar{\mathbf{x}} \in S$ is *efficient* iff there does not exist another $\mathbf{x} \in S$ such that $\mathbf{z}(\mathbf{x}) \geqq \mathbf{z}(\bar{\mathbf{x}})$, $\mathbf{z}(\mathbf{x}) \neq \mathbf{z}(\bar{\mathbf{x}})$.

The domination set approach for detecting efficiency of Section 6.8 can be used as stated with multiple objective programs that have linear objectives and nonlinear constraints.

Example 18. In Fig. 6.17, the efficient set consists of the boundary segment $\mathbf{x}^1$ to $\mathbf{x}^2$ (but not including $\mathbf{x}^2$), along with the boundary segment $\mathbf{x}^3$ to $\mathbf{x}^4$ (inclusive). Point $\mathbf{x}^2$ is not efficient because its criterion vector is dominated by the criterion vector of $\mathbf{x}^3$.

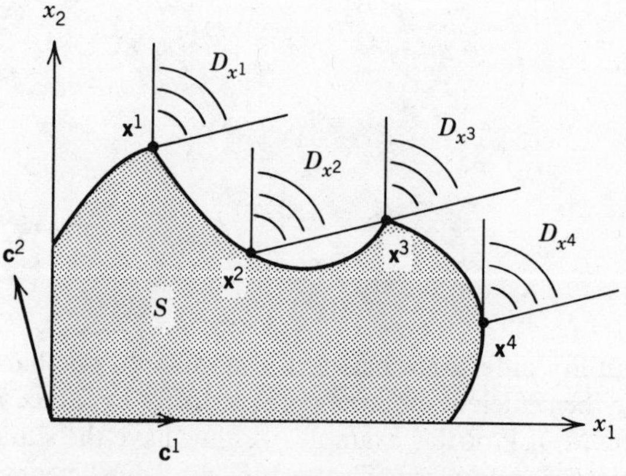

Figure 6.17. Graph of Example 18.

From Example 18, we see that the efficient set need not be connected in a nonlinear multiple objective program.

Readings

1. Arrow, K. J., E. W. Barankin, and D. Blackwell (1953). "Admissible Points of Convex Sets." In H. W. Kuhn and A. W. Tucker (eds.), *Contributions to the Theory of Games*, Princeton, New Jersey: Princeton University Press, pp. 87–91.
2. Charnes, A. and W. W. Cooper (1961). *Management Models and Industrial Applications of Linear Programming*, Vol. I, New York: John Wiley & Sons.
3. Charnes, A., W. W. Cooper, and J. P. Evans (1972). "Connectedness of the Efficient Extreme Points in Linear Multiple Objective Programs," College of Business Administration, University of North Carolina, Chapel Hill, North Carolina.
4. Koopmans, T. C. (1951). "Analysis of Production as an Efficient Combination of Activities." In T. C. Koopmans (ed.), *Activity Analysis of Production and Allocation*, Cowles Commission Monograph No. 13, New York: John Wiley & Sons, pp. 33–97.
5. Mangasarian, O. L. (1969). *Nonlinear Programming*, New York: McGraw-Hill.
6. Pareto, V. (1909). *Manuel d'Economie Politique*, Paris: Giard.
7. Soland, R. M. (1979). "Multicriteria Optimization: A General Characterization of Efficient Solutions," *Decision Sciences*, Vol. 10, No. 1, pp. 26–38.
8. von Neumann, J. and O. Morgenstern (1953). *Theory of Games and Economic Behavior*, 3rd Edition, Princeton, New Jersey: Princeton University Press.
9. Yu, P. L. (1974). "Cone Convexity, Cone Extreme Points and Nondominated Solutions in Decision Problems with Multiobjectives," *Journal of Optimization Theory and Applications*, Vol. 14, No. 3, pp. 319–377.

PROBLEM EXERCISES

6-1 With $S = \{\mathbf{x} \in R^5 \mid \mathbf{Ax} = \mathbf{b}, \mathbf{x} \geqq \mathbf{0}, \mathbf{b} \in R^2\}$

$$\mathbf{C} = \begin{bmatrix} 2 & 4 & -1 & 0 & 0 \\ 3 & 0 & 2 & 0 & 0 \\ 0 & -1 & -1 & 0 & 0 \\ 7 & 9 & 2 & 0 & 0 \end{bmatrix}$$

$$\mathbf{A} = \begin{bmatrix} 1 & 2 & 1 & 1 & 0 \\ 1 & 0 & 0 & 0 & 1 \end{bmatrix}$$

$$\mathbf{b} = \begin{bmatrix} 5 \\ 4 \end{bmatrix}$$

explicitly write out the MOLP formulation.

6-2 Which of the terms

nondecreasing (nonincreasing)
increasing (decreasing)

concave (convex)
strictly concave (strictly convex)
quasiconcave (quasiconvex)
strictly quasiconcave (strictly quasiconvex)
pseudoconcave (pseudoconvex)

apply to each of the following functions:

(a) $U = 2z + 5$

(b) $U = \sqrt{z}$ for $z \geqq 0$

(c) $U = -(z-2)^2 + 4$

(d) $U = -(z-2)^3 + 2$

(e) $U = \begin{cases} z & 0 \leqq z \leqq 1 \\ 1 & z \geqq 1 \end{cases}$

(f) $U = \begin{cases} -(z-2)^2 + 4 & 0 \leqq z \leqq 2 \\ 4 & 2 \leqq z \leqq 6 \\ -(z-6)^2 + 4 & z \geqq 6 \end{cases}$

6-3 Which of the terms of Problem 6-2 apply to the graphs of Fig. 6.18?

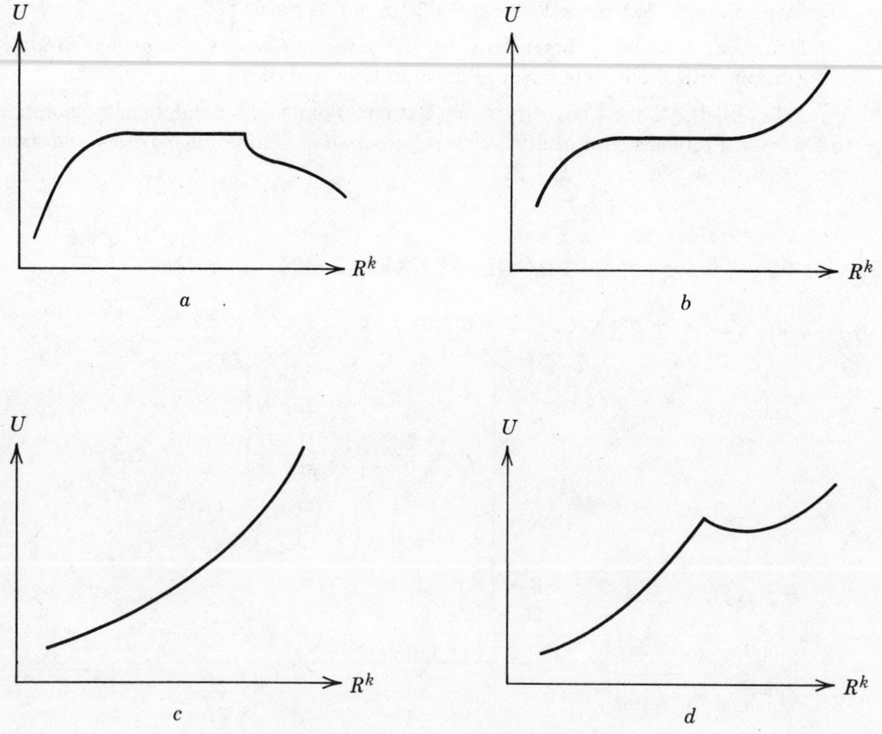

Figure 6.18. Graphs of Problem 6-3.

6-4 Consider the utility function:

$$U = \frac{\sqrt{z_1}}{(10 - z_2)} \quad \text{over} \quad Z = \{\mathbf{z} \in R^2 \,|\, z_1 \geqq 1, \;\; 0 \leqq z_2 \leqq 8\}$$

Coordinatewise, which of the terms of Problem 6-2 pertain to U when z_2 is fixed, and which pertain to U when z_1 is fixed.

6-5 Consider:

$$\begin{aligned}
&\max \{x_1 \quad\quad = z_1\} \\
&\max \{\quad\quad x_2 = z_2\} \\
&\text{s.t.} \quad\quad \mathbf{x} \in S \\
&\text{where } U = z_1 z_2
\end{aligned}$$

(a) In the nonnegative orthant, are the contours of U convex or concave?

(b) Is U convex or concave?

(c) Graph $S = \{\mathbf{x} \in R^2 \,|\, x_i \geqq 1, \; x_1 + x_2 \leqq 4\}$. Is U coordinatewise increasing over S?

(d) Graph $S = \{\mathbf{x} \in R^2 \,|\, x_i \geqq 0, \; x_1 + x_2 \leqq 4\}$. Is U coordinatewise increasing over S?

6-6 Graph the feasible region in criterion space $Z = \{\mathbf{z} \in R^k \,|\, \mathbf{Cx} = \mathbf{z}, \, \mathbf{x} \in S\}$ for:

(a)
$$\begin{aligned}
\max \{-x_1 + x_2 &= z_1\} \\
\max \{\quad\quad\; x_2 &= z_2\} \\
\text{s.t.} \quad x_1 + x_2 &\leqq 4 \\
-x_1 + x_2 &\leqq 2 \\
x_1, x_2 &\geqq 0
\end{aligned}$$

(b)
$$\begin{aligned}
&\max \{-x_1 + x_2 = z_1\} \\
&\max \{\quad\quad\; x_2 = z_2\} \\
&\max \{\; x_1 \quad\quad = z_3\} \\
&\text{s.t.} \quad\quad \text{same as (a)}
\end{aligned}$$

6-7 Graphically displaying S and Z, which extreme points of S map into extreme points of Z and which do not in:

$$\begin{aligned}
\max \{-x_1 + 2x_2 \quad\quad &= z_1\} \\
\max \{\; 3x_1 + \;\; x_2 \quad\quad &= z_2\} \\
\text{s.t.} \quad\quad x_2 \quad\quad &\leqq 3 \\
-x_1 \quad\quad + x_3 &\leqq 0 \\
x_1 \quad\quad + x_3 &\leqq 4 \\
x_1, x_2, x_3 &\geqq 0
\end{aligned}$$

6-8 In decision space, graphically solve the MOLP:

$$\begin{aligned}
\max\{\ & x_1 && = z_1\} \\
\max\{\ & && x_2 = z_2\} \\
\text{s.t.}\ & x_1 && \geqq 2 \\
& x_1 + x_2 && \leqq 6 \\
& -x_1 + x_2 && \leqq 0 \\
& x_1, x_2 && \geqq 0 \\
& \text{where } U = \frac{1 - z_1}{z_2 - 4}
\end{aligned}$$

by plotting level curves of the utility function.

6-9 **(a)** Is the U of Problem 6-8 coordinatewise increasing over S?
(b) Is the U of Problem 6-8 coordinatewise increasing over R^n?

6-10 Consider:

$$\begin{aligned}
\max\{\ & x_1 + 2x_2 = z_1\} \\
\max\{\ & 2x_1 + x_2 = z_2\} \\
\text{s.t.}\ & x_1 + x_2 \leqq 4 \\
& -x_1 + x_2 \leqq 2 \\
& -x_1 + x_2 \geqq -1 \\
& x_1, x_2 \geqq 0 \\
& \text{where } U = z_1 z_2
\end{aligned}$$

(a) By plotting level curves of U in *decision space*, graphically determine the optimal point and optimal criterion vector.

(b) By plotting level curves of U in *criterion space*, graphically determine the optimal criterion vector and the optimal point.

6-11 Design a coordinatewise increasing utility function and an MOLP example for which Θ is unbounded but $\Theta_x = \Theta_\mu = \emptyset$.

6-12 What is the distinction when applying the terms "nondominated" and "efficient"?

6-13 Consider a multiple objective problem that has the eight points of Table 6.3 as its feasible region. Determine which points are efficient and which are inefficient.

6-14 Determine which (if any) of the points in Table 6.4 are inefficient. Can any be determined to be efficient? Why?

6-15 Using the domination set approach, determine which of the points

$$\mathbf{x}^1 = (2,3)$$
$$\mathbf{x}^2 = (4,6)$$
$$\mathbf{x}^3 = (0,3)$$
$$\mathbf{x}^4 = (7,7)$$

TABLE 6.3
Data of Problem 6-13

Point	z_1	z_2	z_3	z_4
$\mathbf{x}^1$	18	42	−16	0
$\mathbf{x}^2$	18	43	−18	2
$\mathbf{x}^3$	17	43	−18	2
$\mathbf{x}^4$	12	44	−20	1
$\mathbf{x}^5$	12	44	−19	1
$\mathbf{x}^6$	36	42	−12	1
$\mathbf{x}^7$	18	43	−19	2
$\mathbf{x}^8$	12	43	−21	1

TABLE 6.4
Data of Problem 6-14

Point	z_1	z_2	z_3	z_4	z_5
⋮					
$\mathbf{x}^{23}$	4	−2	3	7	−5
$\mathbf{x}^{24}$	9	3	−5	0	4
$\mathbf{x}^{25}$	0	0	0	2	−1
$\mathbf{x}^{26}$	−6	3	4	5	1
$\mathbf{x}^{27}$	4	1	3	9	−2
$\mathbf{x}^{28}$	−2	−4	−1	−3	−2
⋮					

are efficient in:

$$\max\{-4x_1 + x_2 = z_1\}$$
$$\max\{x_1 - x_2 = z_2\}$$
$$\text{s.t.} \quad -x_1 + 2x_2 \leq 8$$
$$-x_1 + 2x_2 \geq 4$$
$$x_1, x_2 \geq 0$$

6-16 Using the domination set approach, determine the efficiency status of

$$\mathbf{x}^1 = (5,3)$$
$$\mathbf{x}^2 = (2,2)$$
$$\mathbf{x}^3 = (3,0)$$
$$\mathbf{x}^4 = (5,1)$$

with regard to the MOLP:

$$\begin{aligned} \max\{x_1 + x_2 &= z_1\} \\ \max\{\quad x_2 &= z_2\} \\ \text{s.t.} \quad x_2 &\leqq 2 \\ x_1 + x_2 &\leqq 6 \\ x_1, x_2 &\geqq 0 \end{aligned}$$

6-17 Specify E, E_x, and E_μ for:

(a)
$$\begin{aligned} \max\{x_1 \quad &= z_1\} \\ \text{s.t.} \quad x_1 + x_2 &\leqq 5 \\ x_1 \quad &\leqq 4 \\ x_1, x_2 &\geqq 0 \end{aligned}$$

(b)
$$\begin{aligned} \max\{x_1 \quad &= z_1\} \\ \max\{\quad x_2 &= z_2\} \\ \text{s.t.} \quad x_1 + x_2 &\leqq 5 \\ x_1 \quad &\leqq 4 \\ x_1, x_2 &\geqq 0 \end{aligned}$$

(c)
$$\begin{aligned} \max\{\quad x_1 + 2x_2 &= z_1\} \\ \max\{-2x_1 - 4x_2 &= z_2\} \\ \text{s.t.} \quad -x_1 + 2x_2 &\geqq 0 \\ -x_1 + 2x_2 &\leqq 4 \\ x_1, x_2 &\geqq 0 \end{aligned}$$

6-18 Describe the efficient set of the multiple objective program (that has two linear objectives and a nonlinear feasible region) of Fig. 6.19.

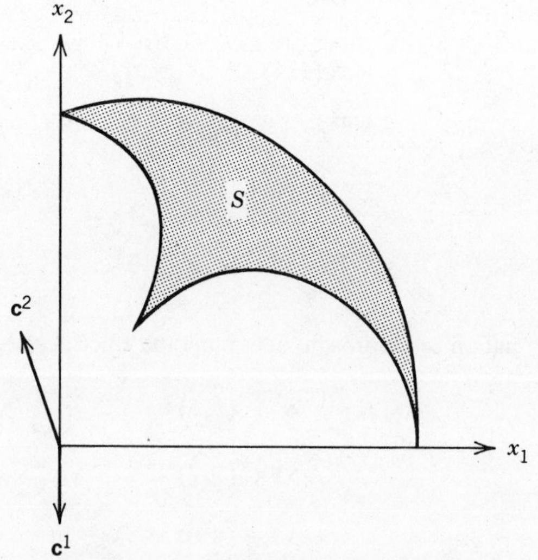

Figure 6.19. Graph of Problem 6-19.

CHAPTER 7

Point Estimate Weighted-Sums Approach

A frequently discussed method in MOLP is the *point estimate* weighted-sums approach. The method is as follows. Each objective $\mathbf{c}^i\mathbf{x}$ is multiplied by a strictly positive scalar weight λ_i. Then, the k weighted objectives are summed to form a *composite* (or *weighted-sums*) objective function. With $\mathbf{C}$ the $k \times n$ criterion matrix whose rows are the $\mathbf{c}^i$, the composite objective function is written $\boldsymbol{\lambda}^T\mathbf{C}\mathbf{x}$. Without loss of generality, we will assume that each *weighting vector* $\boldsymbol{\lambda} \in R^k$ is normalized so that its elements sum to one (in accordance with the L_1-norm).

Let Λ denote the set of all such weighting vectors where

$$\Lambda = \left\{ \boldsymbol{\lambda} \in R^k \mid \lambda_i > 0, \quad \sum_{i=1}^{k} \lambda_i = 1 \right\}$$

Then, by doing one's best to estimate a $\boldsymbol{\lambda} \in \Lambda$, it is hoped that the composite (or weighted-sums) LP

$$\max \{ \boldsymbol{\lambda}^T\mathbf{C}\mathbf{x} \mid \mathbf{x} \in S \}$$

will produce a solution that is optimal; or if not, one that is close enough to being optimal to be useful. Thus, the point estimate weighted-sums technique can be viewed as a method that experiments with strictly convex combinations of the objectives.

Example 1. Consider the MOLP

$$\begin{aligned} &\max \{ \mathbf{c}^1\mathbf{x} = z_1 \} \\ &\max \{ \mathbf{c}^2\mathbf{x} = z_2 \} \\ &\text{s.t.} \qquad \mathbf{x} \in S \end{aligned}$$

to which the weighting vector $\boldsymbol{\lambda} = (.8, .2)$ has been assigned, as in Fig. 7.1. When

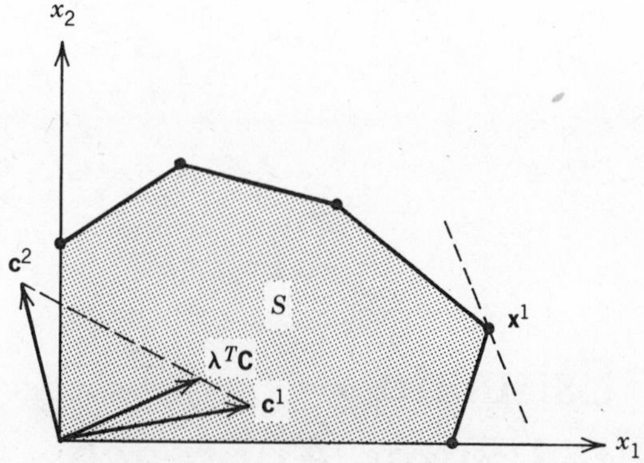

Figure 7.1. Graph of Example 1.

we solve the weighted-sums LP

$$\max \{ \lambda^T \mathbf{C}\mathbf{x} \mid \mathbf{x} \in S \}$$

extreme point $\mathbf{x}^1$, which is efficient, results. How good $\mathbf{x}^1$ is depends on the utility difference between $\mathbf{x}^1$ and an optimal point.

The point estimate weighted-sums method looks straightforward enough. By using the device of a weighting vector, we are able to convert an MOLP into a single criterion LP that can be solved using standardly available LP software. To make the method work, all that is needed is a "good" weighting vector. Therein, however, lies the difficulty.

7.1 FIGURE-OF-MERIT INTERPRETATION

Whereas the criterion values z_i of the original objectives $\mathbf{c}^i\mathbf{x}$ have physical meanings, no such physical meaning is easily ascribed to the value of the composite criterion function $\lambda^T \mathbf{C}\mathbf{x}$. Perhaps the best way to interpret the composite criterion function value is as a "figure of merit." What the point estimate weighted-sums method does is rank the points in the feasible region according to their figures of merit. The higher the figure of merit, supposedly, the more desirable the solution. Of course, the success of the figure-of-merit interpretation is directly dependent on the appropriateness of the weights assigned to the different objectives.

7.2 MATHEMATICAL MOTIVATION

Interest in the point estimate weighted-sums approach stems from the following two theorems.

Theorem 7.1. Let $\bar{\mathbf{x}} \in S$ maximize the weighted-sums LP $\max\{\bar{\lambda}^T C\mathbf{x} \mid \mathbf{x} \in S\}$ where $\bar{\lambda} \in \Lambda$. Then, $\bar{\mathbf{x}}$ is efficient.

Proof. Assume $\bar{\mathbf{x}}$ is inefficient. Then, there exists an $\hat{\mathbf{x}} \in S$ such that $C\hat{\mathbf{x}} \geqq C\bar{\mathbf{x}}$, $C\hat{\mathbf{x}} \neq C\bar{\mathbf{x}}$. Since $\bar{\lambda} \in \Lambda$ is strictly positive, $\bar{\lambda}^T C\hat{\mathbf{x}} > \bar{\lambda}^T C\bar{\mathbf{x}}$, which contradicts the fact that $\bar{\mathbf{x}}$ maximizes the weighted-sums LP. ◀

Theorem 7.2. Let $\bar{\mathbf{x}} \in S$ be efficient. Then, there exists a $\bar{\lambda} \in \Lambda$ such that $\bar{\mathbf{x}}$ is a maximal solution of $\max\{\bar{\lambda}^T C\mathbf{x} \mid \mathbf{x} \in S\}$.

Proof. See the proof of Theorem 9.6. ◀

By Theorem 7.1, all points that maximize the weighted-sums LP for any strictly positive weighting vector are efficient. By Theorem 7.2, every efficient point has associated with it at least one strictly positive weighting vector that causes the weighted-sums LP to have it as a maximal solution.

Suppose the weighted-sums LP has an unbounded objective function value for a particular weighting vector. All this means is that for such a weighting vector, there is no associated efficient point. This does not imply that the efficient set E is empty since there may be other strictly positive weighting vectors for which the weighted-sums LP is bounded.

What would happen if zero weights were allowed in the weighting vector? If one or more of the weights were zero, it is not guaranteed that all points maximizing the resultant composite LP are efficient, as in Example 2.

Example 2. In the MOLP of Fig. 7.2, $E = \gamma(\mathbf{x}^1, \mathbf{x}^3)$. If $\lambda = (0, 1)$ were the weighting vector, then the set of points that maximize the weighted-sums LP would be $\gamma(\mathbf{x}^2, \mathbf{x}^3)$, of which only $\mathbf{x}^3$ is efficient.

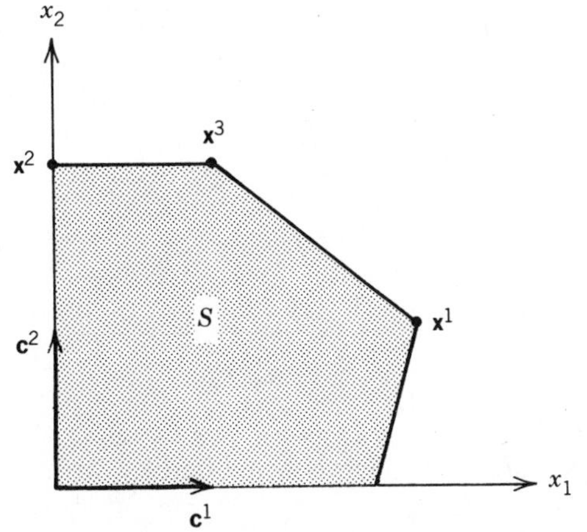

Figure 7.2. Graph of Example 2.

If we were to solve the composite LP of Example 2 using standardly available LP packages, the codes would pivot from the origin to $\mathbf{x}^2$ (which is not efficient) and then stop. We would not necessarily be told about $\mathbf{x}^3$ because none of the commercial packages as of this writing computes all maximizing extreme points. They only produce the first one they encounter and then terminate.

7.3 WHAT ARE THE WEIGHTS?

Assume that the DMs utility function $U\colon R^k \to R$ is differentiable where

$$U(z_1, z_2, \ldots, z_k) \equiv U(\mathbf{c}^1\mathbf{x}, \mathbf{c}^2\mathbf{x}, \ldots, \mathbf{c}^k\mathbf{x})$$

Consider an arbitrary point $\bar{\mathbf{x}} \in S$. By the chain rule, the gradient of U at $\bar{\mathbf{x}}$ is given by

$$\nabla_x U = \sum_{i=1}^{k} \left(\frac{\partial U}{\partial z_i} \right) \nabla_x z_i$$

evaluated at $\bar{\mathbf{x}}$ where $\nabla_x z_i$ is the gradient of the ith objective. Thus,

$$\nabla_x U = \sum_{i=1}^{k} \left(\frac{\partial U}{\partial z_i} \right) \mathbf{c}^i$$

Without loss of generality, let the first objective be such that $(\partial U/\partial z_1) > 0$ at $\bar{\mathbf{x}}$. With this property, the first objective is called the *reference criterion*. Since positive scaling does not affect the direction of a vector, the direction of the gradient of U at $\bar{\mathbf{x}}$ is given by

$$\sum_{i=1}^{k} w_i \mathbf{c}^i$$

where $w_i = (\partial U/\partial z_i)/(\partial U/\partial z_1)$ evaluated at $\bar{\mathbf{x}}$. Normalizing the w_i, we obtain

$$\sum_{i=1}^{k} \lambda_i \mathbf{c}^i$$

where

$$\lambda_i = \frac{w_i}{\sum_{j=1}^{k} w_j}$$

Hence, the weighted-sums approach is a method by which we attempt to apply weights to the various objectives so that the weighted-sums objective function gradient $\boldsymbol{\lambda}^T C$ points in the direction of the gradient of U.

7.4 DETERMINING THE WEIGHTS

A problem in determining the λ_i weights is that U is, in general, nonlinear. Thus, the quantities $(\partial U/\partial z_i)$ are likely to vary from point to point. To estimate the *locally relevant* λ_i at $\bar{\mathbf{x}}$, we consider the tangent hyperplane to the contour (indifference surface) of U at $\bar{\mathbf{x}}$. This equation is

$$\frac{\partial U}{\partial z_1}(\bar{z}_1 - \mathbf{c}^1\mathbf{x}) + \frac{\partial U}{\partial z_2}(\bar{z}_2 - \mathbf{c}^2\mathbf{x}) + \cdots + \frac{\partial U}{\partial z_k}(\bar{z}_k - \mathbf{c}^k\mathbf{x}) = 0$$

where $\bar{z}_i = \mathbf{c}^i\bar{\mathbf{x}}$. Dividing by $(\partial U/\partial z_1) > 0$, we obtain

$$1(\bar{z}_1 - z_1) + w_2(\bar{z}_2 - z_2) + \cdots + w_k(\bar{z}_k - z_k) = 0$$

where $z_i = \mathbf{c}^i\mathbf{x}$. The w_i are now recognized as the *marginal rates of substitution* of criterion value 1 for criterion value i. To estimate the w_i, let Δ_i be the amount by which the ith criterion must be increased to compensate for a loss in the first criterion of Δ_1 units holding all other criterion values constant. Thus, the quantity

$$\frac{\Delta_1}{\Delta_i}$$

evaluated at $\bar{\mathbf{x}}$ is an estimator of w_i with the property that

$$w_i = \lim_{\Delta_1 \to 0} \frac{\Delta_1}{\Delta_i}$$

With $\Delta_1 = 1$, we have

$$w_1 = 1$$

$$w_2 \text{ is estimated by } \frac{1}{\Delta_2}$$

$$w_3 \text{ is estimated by } \frac{1}{\Delta_3}$$

$$\vdots$$

$$w_k \text{ is estimated by } \frac{1}{\Delta_k}$$

Normalizing the estimated w_i at $\bar{\mathbf{x}}$, we have the estimated λ_i at $\bar{\mathbf{x}}$.

Example 3. Consider the two-objective MOLP whose feasible region in criterion space is graphed in Fig. 7.3. Let $\bar{\mathbf{x}}$, whose criterion vector is $\bar{\mathbf{z}}$, be the point at which locally relevant weights are to be estimated. To compensate for a loss of Δ_1 units in the first objective, we see that we must increase the second objective by Δ_2 units. This analysis leads to estimated weights that cause the composite LP gradient to point in a direction normal to the hypotenuse of the triangle. However, the correct weights at $\bar{\mathbf{x}}$ would cause the composite LP gradient to point in a direction normal to the tangent hyperplane at $\bar{\mathbf{x}}$.

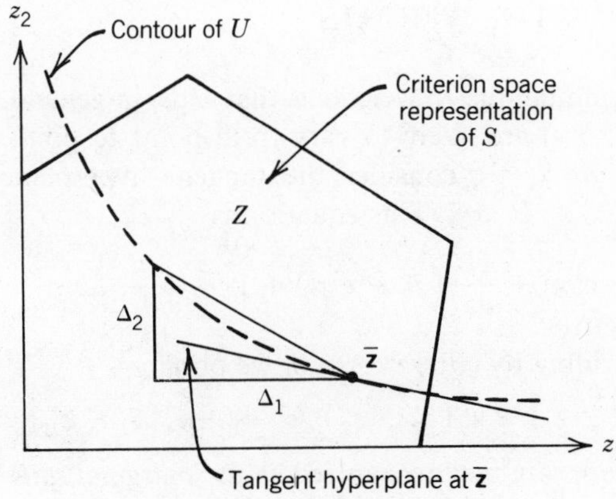

***Figure** 7.3.* Graph of Example 3.

Although, with the method of this section, Δ_1 can be made infinitesimally small in theory, in practice Δ_1 must be *meaningfully* large so that the decision maker will be able to respond successfully to the tradeoff questions. Thus, we should not expect perfection with this weight estimation technique.

7.5 CRITERION CONE

The criterion cone is an important concept in multiple objective programming. The size of the criterion cone is an indicator of the size of the efficient set and of the difficulty we are likely to have in solving the multiple objective problem. With regard to the MOLP

$$\begin{aligned}
&\max\{\mathbf{c}^1\mathbf{x} = z_1\}\\
&\max\{\mathbf{c}^2\mathbf{x} = z_2\}\\
&\qquad\vdots\\
&\max\{\mathbf{c}^k\mathbf{x} = z_k\}\\
&\text{s.t.} \qquad \mathbf{x} \in S
\end{aligned}$$

we have the two following definitions.

Definition 7.3. The *criterion cone* is the convex cone generated by the k objective function gradients $\{\mathbf{c}^1, \mathbf{c}^2, \ldots, \mathbf{c}^k\}$.

Definition 7.4. The *null vector condition* is in effect if there exists a strictly positive linear combination of the objective function gradients that produces the null vector. That is, the null vector condition holds if there exists an $\boldsymbol{\alpha} \in R^k$ with $\alpha_i > 0$ for all i such that

$$\sum_{i=1}^{k} \alpha_i \mathbf{c}^i = \mathbf{0} \in R^n$$

In all cases, the criterion cone is closed and convex. The origin $\mathbf{0} \in R^n$ is always a member of the criterion cone. The criterion cone need not be pointed. When pointed (the usual case), the origin is the cone's vertex. The only pointed criterion cone for which the null vector condition holds is $\{\mathbf{0} \in R^n\}$. Because the criterion cone is generated by the gradients of the objective functions, it is polyhedral with at most k extreme rays. Except in the instance when the criterion cone is $\{\mathbf{0} \in R^n\}$, the criterion cone is an unbounded set. The dimensionality of the criterion cone is given by the rank of its associated criterion matrix $\mathbf{C}$.

7.5.1 When the Null Vector Condition Does Not Hold

The following examples illustrate criterion cones with respect to which the null vector condition does not hold; this is the typical case in multiple objective linear programming.

Example 4. (Normal single objective linear programming.) The null vector condition does not hold in the LP

$$\begin{aligned} &\max \{\mathbf{c}^T\mathbf{x} = z\} \\ &\text{s.t.} \qquad \mathbf{x} \in S \end{aligned}$$

unless $\mathbf{c} = \mathbf{0} \in R^n$. When $\mathbf{c} \neq \mathbf{0}$, the criterion cone is the one-dimensional (unbounded) half-ray $\mu(\mathbf{0}, \mathbf{c})$.

Example 5. (When $r(\mathbf{C}) = 2$, but $k > 2$.) Consider the MOLP

$$\begin{aligned} &\max \{\mathbf{c}^1\mathbf{x} = z_1\} \\ &\max \{\mathbf{c}^2\mathbf{x} = z_2\} \\ &\max \{\mathbf{c}^3\mathbf{x} = z_3\} \\ &\text{s.t.} \qquad \mathbf{x} \in S \end{aligned}$$

that is graphed in Fig. 7.4. Of dimensionality 2, the criterion cone is as drawn. Gradients $\mathbf{c}^1$ and $\mathbf{c}^3$ are *essential* gradients because the criterion cone cannot be generated without them. In this example, $\mathbf{c}^2$ is a *nonessential* gradient because the criterion cone is the same with or without it.

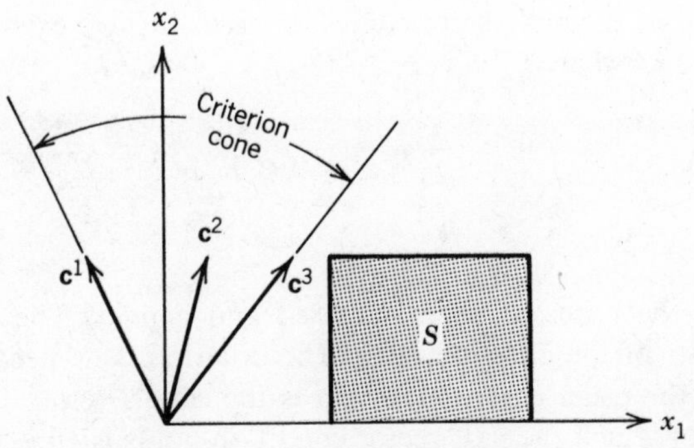

Figure 7.4. Graph of Example 5.

Example 6. (When the number of extreme rays is greater than $r(\mathbf{C})$.) Consider the MOLP

$$\begin{aligned} &\max \{ \mathbf{c}^1\mathbf{x} = z_1 \} \\ &\max \{ \mathbf{c}^2\mathbf{x} = z_2 \} \\ &\max \{ \mathbf{c}^3\mathbf{x} = z_3 \} \\ &\max \{ \mathbf{c}^4\mathbf{x} = z_4 \} \\ &\text{s.t.} \qquad \mathbf{x} \in S \end{aligned}$$

whose criterion cone is graphed in Fig. 7.5. As drawn, the criterion cone is of dimensionality 3 but has four extreme rays. Even though the set $\{\mathbf{c}^1, \mathbf{c}^2, \mathbf{c}^3, \mathbf{c}^4\}$ is linearly dependent, all gradients are essential.

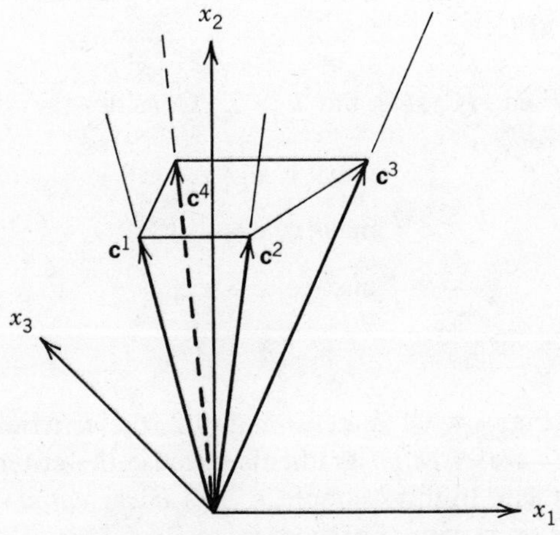

Figure 7.5. Graph of Example 6.

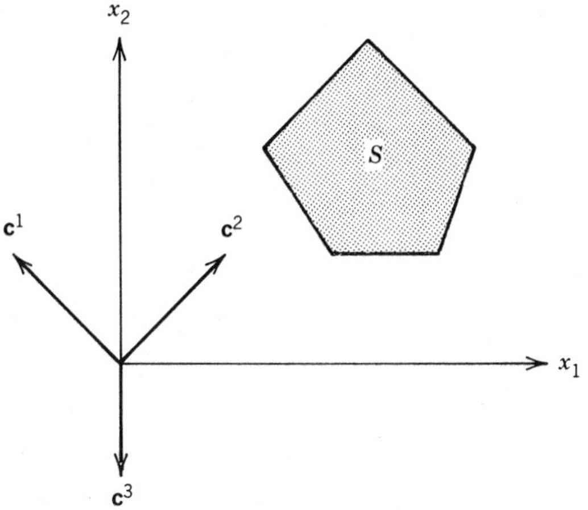

Figure 7.6. Graph of Example 7.

7.5.2 When the Null Vector Condition Holds

The following examples illustrate criterion cones with respect to which the null vector condition holds.

Example 7. (When $r(\mathbf{C}) > 0$.) Consider the MOLP that is graphed in Fig. 7.6. In this MOLP, the null vector condition holds because there exists a strictly positive $\boldsymbol{\alpha} \in R^3$ such that

$$\sum_{i=1}^{3} \alpha_i \mathbf{c}^i = \mathbf{0} \in R^n$$

Although $r(\mathbf{C}) = 2$, all three gradients are essential. Note that if the MOLP had consisted of any two of the three objectives, the null vector condition would not hold and the criterion cone would be the two-dimensional convex cone generated by the two gradients.

Example 8. (When the status of the null vector condition changes depending on which objectives are included in the problem.) Consider the objective function gradients that are graphed in Fig. 7.7. With $\mathbf{c}^1$ as the only gradient, the null vector condition does not hold and the criterion cone is the one-dimensional unbounded half-ray $\mu(\mathbf{0}, \mathbf{c}^1)$. Introducing $\mathbf{c}^2$ causes the null vector condition to hold. By further introducing $\mathbf{c}^3$, we return to the situation in which the null vector condition does not hold. With all three objectives in the problem, the criterion cone is the two-dimensional half-space $\{\mathbf{x} \in R^2 \mid \mathbf{c}^3\mathbf{x} \geqq 0\}$ that does not have a vertex. Although the half-space has two extreme rays, all three gradients are essential.

We now see what typically happens when we go from LP to MOLP. Instead of the usual one-dimensional half-ray emanating from the origin, the criterion cone

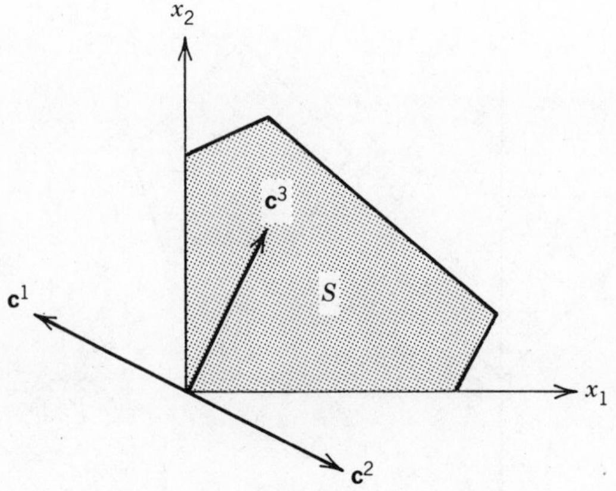

***Figure** 7.7.* Graph of Example 8.

expands into multiple dimensions as additional objectives are inserted into the problem.

7.6 RELATIVE INTERIOR OF THE CRITERION CONE

We now discuss the concept of the *relative interior* of a cone because this notion is instrumental to the material of Section 7.7. Let $\{\mathbf{v}^1, \mathbf{v}^2, \ldots, \mathbf{v}^k\}$ be a set of essential generators of the closed convex cone V. The *relative interior* of V (denoted $\text{rel}\, V$) consists of all strictly positive linear combinations of the $\mathbf{v}^i$. The *relative boundary* of V is given by the difference between V and its relative interior.

Example 9. With regard to the cone generated by $\{\mathbf{v}^1, \mathbf{v}^2\}$ in Fig. 7.8*a*, the relative interior of V is the interior of V, and the relative boundary of V is

$$\mu(\mathbf{0}, \mathbf{v}^1) \cup \mu(\mathbf{0}, \mathbf{v}^2)$$

With regard to the cone generated by $\{\mathbf{v}\}$ in Fig. 7.8*b*, the relative interior of V is V less the origin, and the relative boundary is the origin.

The only time $\text{rel}\, V = V$ is when the null vector condition holds. The significance of the relative interior of the criterion cone is that it is the set of all positive scalar multiples of the set of all strictly convex linear combinations of the $\mathbf{v}^i$. This leads to Corollary 7.5.

Corollary 7.5. Point $\bar{\mathbf{x}} \in S$ is efficient $\Leftrightarrow$ there exists a $\boldsymbol{\beta} \in R^n$ that is a member of the relative interior of the criterion cone such that $\bar{\mathbf{x}}$ maximizes the LP

$$\max\{\boldsymbol{\beta}^T\mathbf{x} \mid \mathbf{x} \in S\}$$

Proof. Since $\boldsymbol{\beta}$ is a member of the relative interior of the criterion cone, there exists an $\alpha \in R$, $\alpha > 0$ such that $\boldsymbol{\beta} = \alpha\boldsymbol{\lambda}^T\mathbf{C}$ for some $\boldsymbol{\lambda} \in \Lambda$. Thus, the corollary immediately results from Theorems 7.1 and 7.2. ◀

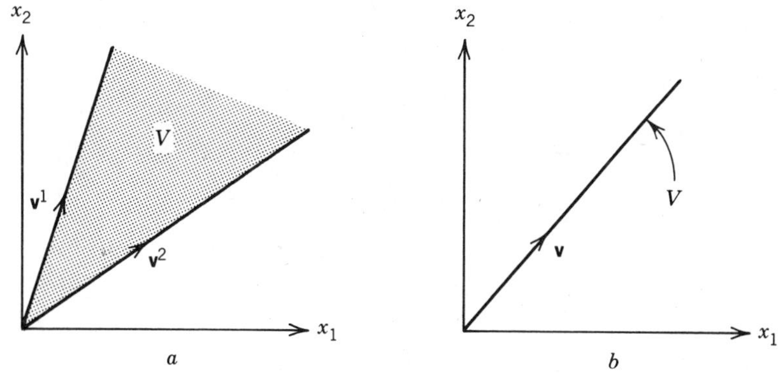

Figure 7.8. Graphs of Example 9.

The significance of Corollary 7.5 is that the efficient set is given by the union of all maximizing solutions of members of the family of LPs

$$\{\max [\boldsymbol{\beta}^T\mathbf{x} \mid \mathbf{x} \in S]\}$$

where $\boldsymbol{\beta}$ is a member of the relative interior of the criterion cone. If $\boldsymbol{\beta}$ were a member of the relative boundary of the criterion cone, a given maximizing solution of $\max\{\boldsymbol{\beta}^T\mathbf{x} \mid \mathbf{x} \in S\}$ might not be efficient.

7.7 DETECTING EFFICIENT POINTS USING COMPOSITE GRADIENTS

A method based upon the relative interior of the criterion cone that is useful for the graphical detection of efficiency is now discussed. Whereas the domination set approach of Section 6.8 is the best method for the graphical detection of efficiency in integer and nonlinear multiple objective programs, the method of this section is more convenient with MOLPs. The method is presented as follows.

A vector $\mathbf{v} \neq \mathbf{0}$ is said to *point into* the cone V when $\mathbf{v} \in V$. Also, a point $\bar{\mathbf{x}} \in S$ is said to *maximize the gradient* $\boldsymbol{\beta}$ when $\bar{\mathbf{x}}$ maximizes the LP max $\{\boldsymbol{\beta}^T\mathbf{x} \mid \mathbf{x} \in S\}$. Since, by Corollary 7.5, the efficient set is produced by maximizing all points in the relative interior of the criterion cone, we arrive at the following examples.

Example 10. The criterion cone of the MOLP of Fig. 7.9 is the nonnegative orthant. In this MOLP, the relative interior of the criterion cone is $\{\mathbf{x} \in R^2 \mid x_1, x_2 > 0\}$ and

$$E = \gamma(\mathbf{x}^2, \mathbf{x}^3)$$

$$E_x = \{\mathbf{x}^2, \mathbf{x}^3\}$$

$$E_\mu = \varnothing$$

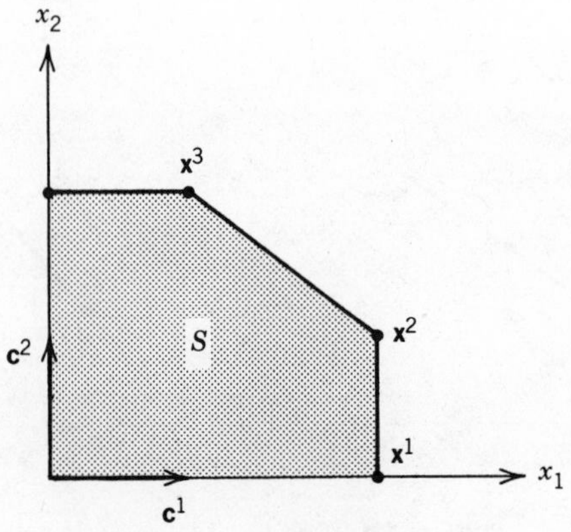

Figure 7.9. Graph of Example 10.

One might ask why all the points along the $\gamma(\mathbf{x}^1, \mathbf{x}^2)$ edge except for $\mathbf{x}^2$ are inefficient. It is because they do not maximize any vectors that point into the relative interior of the criterion cone. (Although they maximize $\mathbf{c}^1$, $\mathbf{c}^1$ does not point into the relative interior of the criterion cone.)

Example 11. In the MOLP of Fig. 7.10, the relative interior of the criterion cone is the interior of the convex cone generated by $\{\mathbf{c}^1, \mathbf{c}^2\}$ and

$$E = \{\mathbf{x}^1\}$$

$$E_x = \{\mathbf{x}^1\}$$

$$E_\mu = \varnothing$$

In this problem, $\mathbf{c}^2$ is a nonessential gradient and $\mathbf{x}^1$ maximizes all gradients that point into the relative interior of the criterion cone. Since $\mathbf{x}^1$ is the only efficient point, $\Theta = \{\mathbf{x}^1\}$.

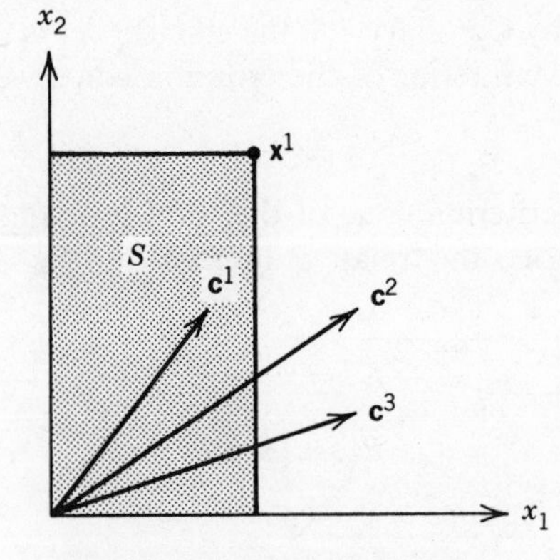

Figure 7.10. Graph of Example 11.

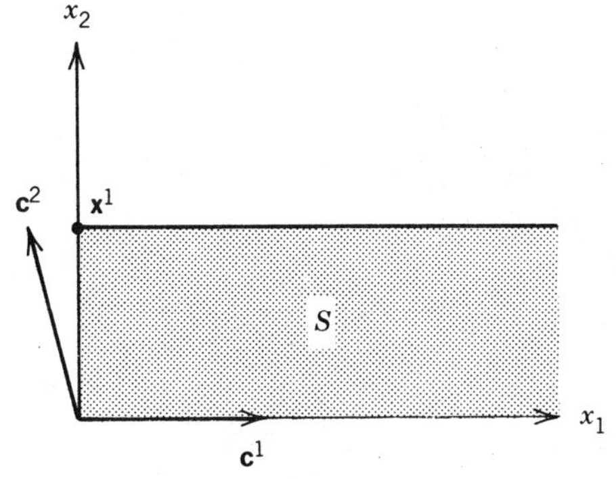

Figure 7.11. Graph of Example 12.

Example 12. Consider the MOLP

$$\max\{\ 4x_1 \qquad = z_1\}$$
$$\max\{-x_1 + 4x_2 = z_2\}$$
$$\text{s.t.} \qquad \mathbf{x} \in S$$

that is graphed in Fig. 7.11. Let $\boldsymbol{\lambda} \in \Lambda$ be the weighting vector used in the composite LP $\max\{\boldsymbol{\lambda}^T C\mathbf{x} \mid \mathbf{x} \in S\}$. For $\lambda_1 \in (0, \frac{1}{5})$, $\mathbf{x}^1$ uniquely maximizes the composite LP. For $\lambda_1 = \frac{1}{5}$, the unbounded edge $\mu(\mathbf{x}^1, \mathbf{v})$ where $\mathbf{v} = (1,0)$ maximizes the composite LP. And for $\lambda_1 \in (\frac{1}{5}, 1)$, the composite LP has an unbounded objective function value. Thus,

$$E = \mu(\mathbf{x}^1, \mathbf{v})$$
$$E_x = \{\mathbf{x}^1\}$$
$$E_\mu = \{\mu(\mathbf{x}^1, \mathbf{v})\}$$

Example 13. With the criterion gradients of the MOLP of Fig. 7.12 pointing in opposite directions, the null vector condition holds. In this problem,

$$E = S$$
$$E_x = \{\mathbf{x}^1, \mathbf{x}^2, \mathbf{x}^3\}$$
$$E_\mu = \varnothing$$

because the null vector is a member of the relative interior of the criterion cone.

Normally, the larger the criterion cone, the larger the efficient set. In Example 14, we have the typical case. As the criterion cone increases in size continuously, E increases in size by "quantum" amounts. Exceptions to this rule are illustrated in Examples 15 and 16.

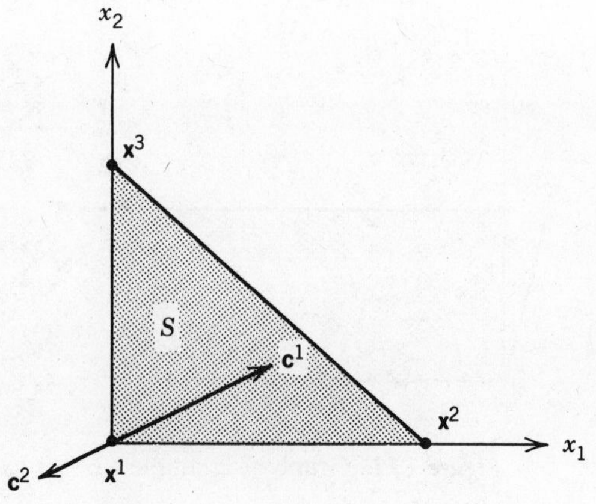

Figure 7.12. Graph of Example 13.

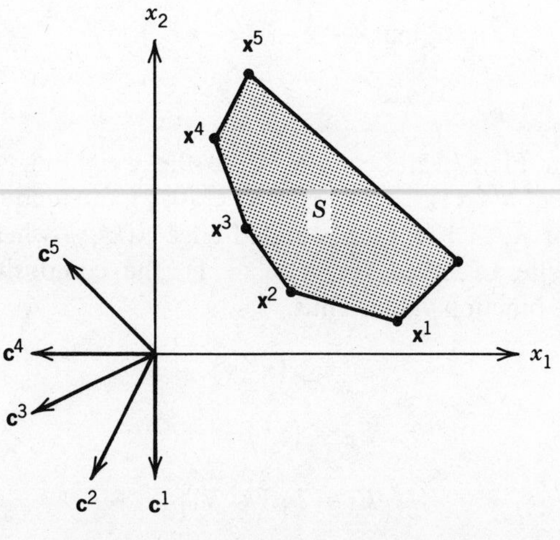

Figure 7.13. Graph of Example 14.

Example 14. Consider the criterion gradients and feasible region in Fig. 7.13. For the four criterion cones in which each is a superset of the previous, we have the resulting efficient sets in Table 7.1. The efficient sets form a series of nested supersets because the *relative interiors* of the criterion cones form a series of nested supersets.

Example 15. Consider the feasible region S in Fig. 7.14. With regard to

$$\max \{\mathbf{0}^T\mathbf{x} = z\}$$
$$\text{s.t.} \quad \mathbf{x} \in S$$

TABLE 7.1
Efficient Sets of Example 14

Generators of the Criterion Cone	*Efficient Set*
$\{\mathbf{c}^1, \mathbf{c}^2\}$	$\gamma(\mathbf{x}^1, \mathbf{x}^2)$
$\{\mathbf{c}^1, \mathbf{c}^3\}$	$\gamma(\mathbf{x}^1, \mathbf{x}^2) \cup \gamma(\mathbf{x}^2, \mathbf{x}^3)$
$\{\mathbf{c}^1, \mathbf{c}^4\}$	$\bigcup_{i=1}^{3} \gamma(\mathbf{x}^i, \mathbf{x}^{i+1})$
$\{\mathbf{c}^1, \mathbf{c}^5\}$	$\bigcup_{i=1}^{4} \gamma(\mathbf{x}^i, \mathbf{x}^{i+1})$

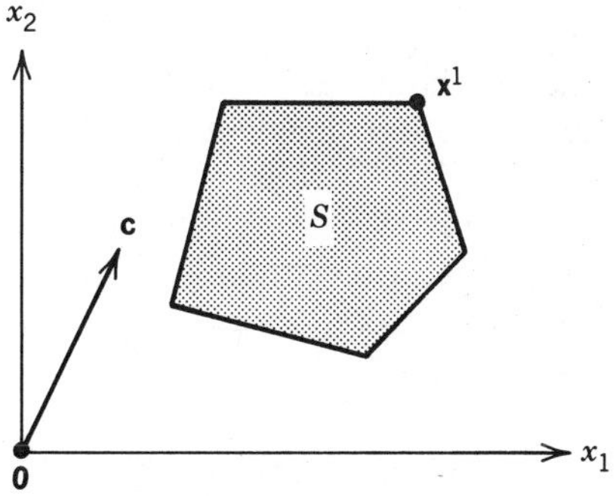

Figure 7.14. Graph of Example 15.

and

$$\max \{\mathbf{c}^T\mathbf{x} = z\}$$
$$\text{s.t.} \quad \mathbf{x} \in S$$

we have an instance in which the efficient set decreases as the criterion cone increases. With the criterion cone of the first LP being the singleton set $\{\mathbf{0} \in R^2\}$, $E = S$. With the criterion cone of the second LP being $\mu(\mathbf{0}, \mathbf{c})$, $E = \{\mathbf{x}^1\}$. The explanation is that the relative interior of the criterion cone of the first LP is *not* a subset of the relative interior of the criterion cone of the second LP.

Example 16. In the MOLP of Fig. 7.15, the efficient set disappears as the criterion cone increases. With only $\mathbf{c}^1$ in the problem, $E = \mu(\mathbf{x}^1, \mathbf{v})$ where $\mathbf{v} = (1, 0)$. With $\mathbf{c}^2$ also in the problem, $E = \varnothing$. This is because the relative interior of the criterion cone generated by $\mathbf{c}^1$ is disjoint from the relative interior of the criterion cone generated by $\{\mathbf{c}^1, \mathbf{c}^2\}$.

The key, then, when attempting to graphically visualize the efficient set using weighting vectors is to work with the *relative interior* of the criterion cone.

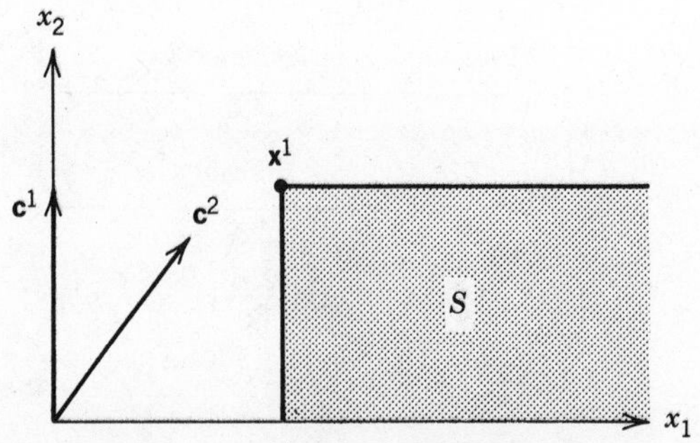

Figure 7.15. Graph of Example 16.

7.8 RELATIONSHIP BETWEEN THE CRITERION CONE AND DOMINATION SET

When the null vector condition holds, Stiemke's Theorem of the Alternative [see Mangasarian (1969)] states that there does not exist a $\mathbf{y} \in R^n$ such that $\mathbf{c}^i\mathbf{y} \geqq 0$ for all i and $\mathbf{c}^i\mathbf{y} > 0$ for at least one i. Thus, $C^{\geq} = \varnothing$ in Definition 6.14 and the domination set at $\bar{\mathbf{x}} \in S$ is simply $\{\bar{\mathbf{x}}\}$. Therefore, when the null vector condition holds, $E = S$. When the null vector condition does not hold, the domination set normally gets smaller (becomes a subset of itself) as the criterion cone gets larger (becomes a superset of itself).

Example 17. In Fig. 7.16*a*, the criterion cone is the cone generated by $\mathbf{c}^1$ and the domination set $D_{\bar{x}}$ is the open half-space as drawn, except that it also includes $\bar{\mathbf{x}}$. As the criterion cone generated by $\{\mathbf{c}^1, \mathbf{c}^2\}$ enlarges to supersets of itself in Figs. 7.16*b* and *c*, $D_{\bar{x}}$, whose generators are $\{\mathbf{y}^1, \mathbf{y}^2\}$, reduces to subsets of itself. Then, in Fig. 7.16*d*, $D_{\bar{x}} = \mu(\bar{\mathbf{x}}, \mathbf{c}^3)$ because the criterion cone is (the closed half-space) generated by $\{\mathbf{c}^1, \mathbf{c}^2, \mathbf{c}^3\}$.

In contrast to the behavior exhibited in Fig. 7.16, there are cases when the domination set does not become a subset of itself when the criterion cone becomes a superset of itself.

Example 18. The domination sets of Fig. 7.17 correspond to Example 16. With only $\mathbf{c}^1$ in the problem, the domination set is

$$\{\bar{\mathbf{x}}\} \cup \{\mathbf{x} \mid \mathbf{c}^1\mathbf{x} > \mathbf{c}^1\bar{\mathbf{x}}\}$$

With $\mathbf{c}^1$ and $\mathbf{c}^2$ in the problem, the domination set is generated by $\{\mathbf{y}^1, \mathbf{y}^2\}$ as drawn in Fig. 7.17*b*. Because of the inclusion of the extreme ray generated by $\mathbf{y}^1$ in the domination set of Fig. 7.17*b*, the domination set of Fig. 7.17*b* is *not* a subset of the domination set of Fig. 7.17*a*. Thus, the reason for the efficient set behavior in Example 16.

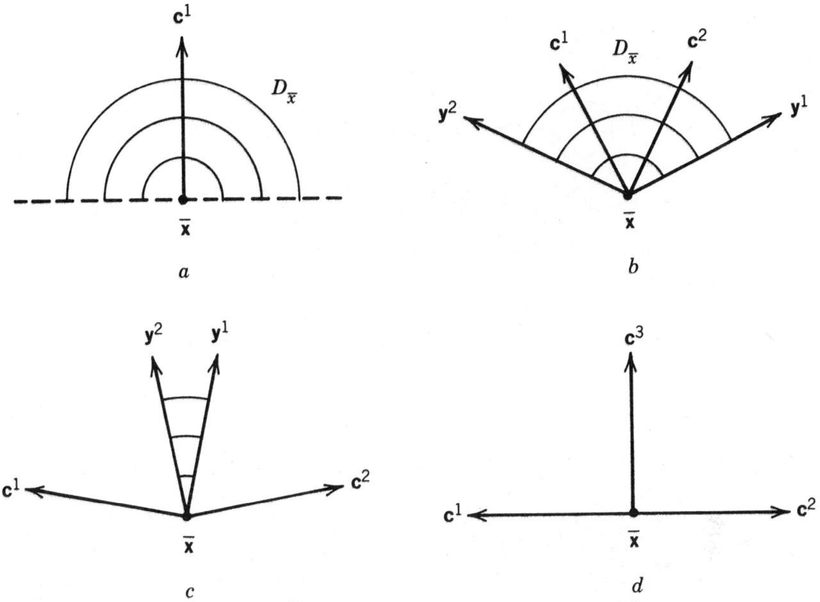

Figure 7.16. Graphs of Example 17.

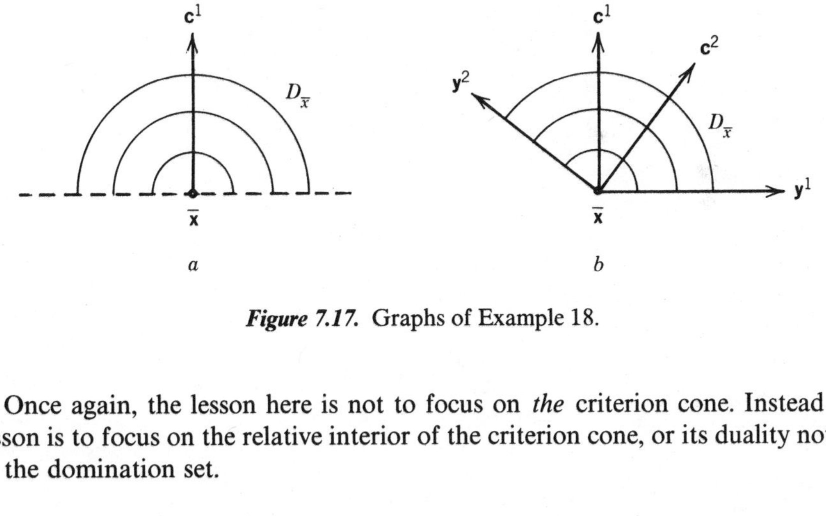

Figure 7.17. Graphs of Example 18.

Once again, the lesson here is not to focus on *the* criterion cone. Instead the lesson is to focus on the relative interior of the criterion cone, or its duality notion of the domination set.

7.9 EFFICIENT FACETS OF THE FEASIBLE REGION *S*

Let $F \subset S$ and let H be a supporting hyperplane of S. Then, F is a *facet* of S if and only if there exists an H such that $H \cap S = F$. F is said to be an *f-facet* of S if F is of dimensionality f. Extreme points are 0-facets and 1-facets are *edges*. In R^3, a 2-facet is referred to as a *face*.

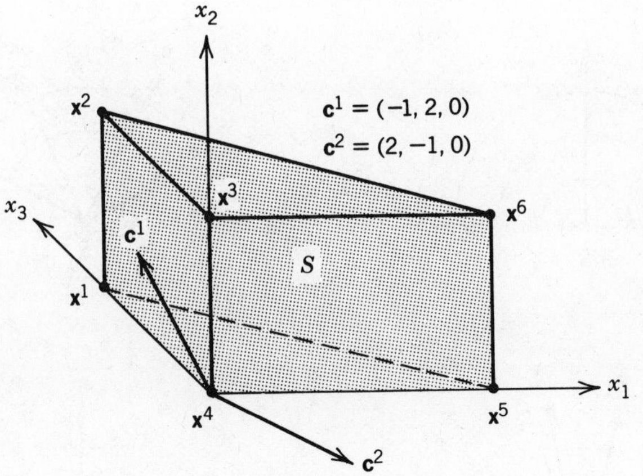

Figure 7.18. Graph of Example 19.

Example 19. In Fig. 7.18, S has six 0-facets

$$\{\mathbf{x}^1\} \qquad \{\mathbf{x}^3\} \qquad \{\mathbf{x}^5\}$$
$$\{\mathbf{x}^2\} \qquad \{\mathbf{x}^4\} \qquad \{\mathbf{x}^6\}$$

nine 1-facets (edges)

$$\gamma(\mathbf{x}^1, \mathbf{x}^2) \qquad \gamma(\mathbf{x}^2, \mathbf{x}^3) \qquad \gamma(\mathbf{x}^3, \mathbf{x}^6)$$
$$\gamma(\mathbf{x}^1, \mathbf{x}^4) \qquad \gamma(\mathbf{x}^2, \mathbf{x}^6) \qquad \gamma(\mathbf{x}^4, \mathbf{x}^5)$$
$$\gamma(\mathbf{x}^1, \mathbf{x}^5) \qquad \gamma(\mathbf{x}^3, \mathbf{x}^4) \qquad \gamma(\mathbf{x}^5, \mathbf{x}^6)$$

and five 2-facets (faces)

$$\overset{4}{\underset{i=1}{\gamma}} (\mathbf{x}^i)$$
$$\overset{6}{\underset{i=3}{\gamma}} (\mathbf{x}^i)$$
$$\gamma(\mathbf{x}^1, \mathbf{x}^2, \mathbf{x}^5, \mathbf{x}^6)$$
$$\gamma(\mathbf{x}^1, \mathbf{x}^4, \mathbf{x}^5)$$
$$\gamma(\mathbf{x}^2, \mathbf{x}^3, \mathbf{x}^6)$$

Let us define *maximal* and *maximally efficient* facets. The notion of a maximally efficient facet is useful because the efficient set is given by the union of all maximally efficient facets.

Definition 7.6. Let F be an f-facet of S. Then, F is a *maximal* facet iff there does not exist a g-facet G of S such that $F \subset G$ and $f < g$.

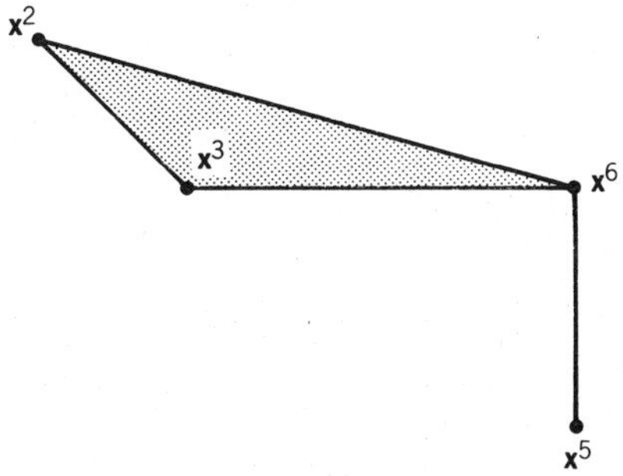

Figure 7.19. Efficient set of Example 20.

Definition 7.7. Let F be an efficient f-facet of S. Then, F is a *maximally efficient* facet iff there does not exist an efficient g-facet G of S such that $F \subset G$ and $f < g$.

Example 20. Consider the feasible region S in Fig. 7.18 along with the two objective function gradients $\mathbf{c}^1 = (-1, 2, 0)$ and $\mathbf{c}^2 = (2, -1, 0)$. With $\mathbf{c}^1$ and $\mathbf{c}^2$ in the $x_1 : x_2$ plane, E is displayed in Fig. 7.19. Consequently, there are four efficient 0-facets

$$\{\mathbf{x}^2\} \qquad \{\mathbf{x}^3\} \qquad \{\mathbf{x}^5\} \qquad \{\mathbf{x}^6\}$$

four efficient 1-facets

$$\gamma(\mathbf{x}^2, \mathbf{x}^3) \qquad \gamma(\mathbf{x}^3, \mathbf{x}^6)$$
$$\gamma(\mathbf{x}^2, \mathbf{x}^6) \qquad \gamma(\mathbf{x}^5, \mathbf{x}^6)$$

and one efficient 2-facet

$$\gamma(\mathbf{x}^2, \mathbf{x}^3, \mathbf{x}^6)$$

However, there are only two maximally efficient facets: the 1-facet $\gamma(\mathbf{x}^5, \mathbf{x}^6)$ and the 2-facet $\gamma(\mathbf{x}^2, \mathbf{x}^3, \mathbf{x}^6)$. Thus, we write E as

$$E = \gamma(\mathbf{x}^5, \mathbf{x}^6) \cup \gamma(\mathbf{x}^2, \mathbf{x}^3, \mathbf{x}^6)$$

As illustrated by $\gamma(\mathbf{x}^5, \mathbf{x}^6)$ in Example 20, a maximally efficient facet may not be a maximal facet of S. The "quantum" increases in the efficient set alluded to in connection with Example 14 occur as new facets of S become maximally efficient.

7.10 COMPUTING SUBSETS OF WEIGHTING VECTORS

Pertaining to each efficient facet is at least one $\bar{\lambda} \in \Lambda$ such that all points constituting the efficient facet maximize the weighted-sums LP $\max\{\bar{\lambda}^T C\mathbf{x} \mid \mathbf{x} \in S\}$. Examples 21, 22, and 23 illustrate the subsets of Λ pertaining to different efficient facets.

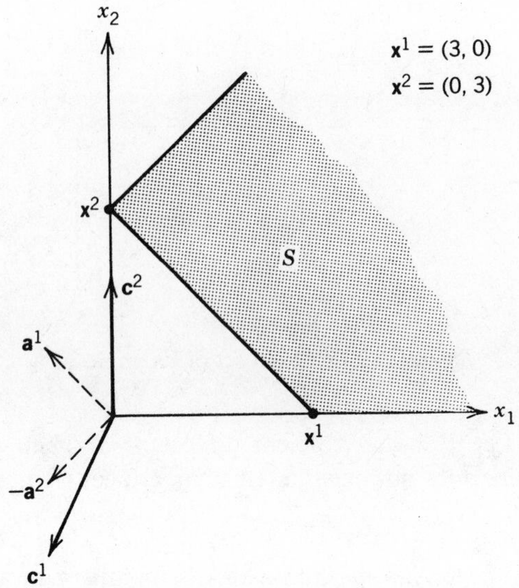

Figure 7.20. Graph of Example 21.

Example 21. Consider the MOLP

$$\begin{aligned} \max\{-x_1 - 2x_2 &= z_1\} \\ \max\{\qquad 2x_2 &= z_2\} \\ \text{s.t.} \quad -x_1 + x_2 &\leqq 3 \\ x_1 + x_2 &\geqq 3 \\ x_1, x_2 &\geqq 0 \end{aligned}$$

that is graphed in Fig. 7.20. In this problem,

$$E = \gamma(\mathbf{x}^1, \mathbf{x}^2) \cup \mu(\mathbf{x}^2, \mathbf{v})$$

where $\mathbf{v} = (1, 1)$.

There are two efficient 0-facets $\{\mathbf{x}^1\}$ and $\{\mathbf{x}^2\}$ and two efficient 1-facets $\gamma(\mathbf{x}^1, \mathbf{x}^2)$ and $\mu(\mathbf{x}^2, \mathbf{v})$. With $\mathbf{a}^1$ and $\mathbf{a}^2$ the gradients of the main constraints, the convex combinations of $\mathbf{c}^1$ and $\mathbf{c}^2$ that point in the directions $\mathbf{a}^1$ and $-\mathbf{a}^2$ are given by the weighting vectors $(\frac{2}{5}, \frac{3}{5})$ and $(\frac{2}{3}, \frac{1}{3})$. Consequently, we have the following subsets of Λ:

$$\begin{aligned} \Lambda_{x^1} &= \{\boldsymbol{\lambda} \in \Lambda \mid \lambda_1 \in [\tfrac{2}{3}, 1), \lambda_2 \in (0, \tfrac{1}{3}]\} \\ \Lambda_{\gamma(x^1, x^2)} &= \{\boldsymbol{\lambda} \in \Lambda \mid \lambda_1 = \tfrac{2}{3}, \lambda_2 = \tfrac{1}{3}\} \\ \Lambda_{x^2} &= \{\boldsymbol{\lambda} \in \Lambda \mid \lambda_1 \in [\tfrac{2}{5}, \tfrac{2}{3}], \lambda_2 \in [\tfrac{1}{3}, \tfrac{3}{5}]\} \\ \Lambda_{\mu(x^2, v)} &= \{\boldsymbol{\lambda} \in \Lambda \mid \lambda_1 = \tfrac{2}{5}, \lambda_2 = \tfrac{3}{5}\} \\ \Lambda_{\varnothing} &= \{\boldsymbol{\lambda} \in \Lambda \mid \lambda_1 \in (0, \tfrac{2}{5}), \lambda_2 \in (\tfrac{3}{5}, 1)\} \end{aligned}$$

Although there are only four efficient facets, there are five subsets of Λ. The extra

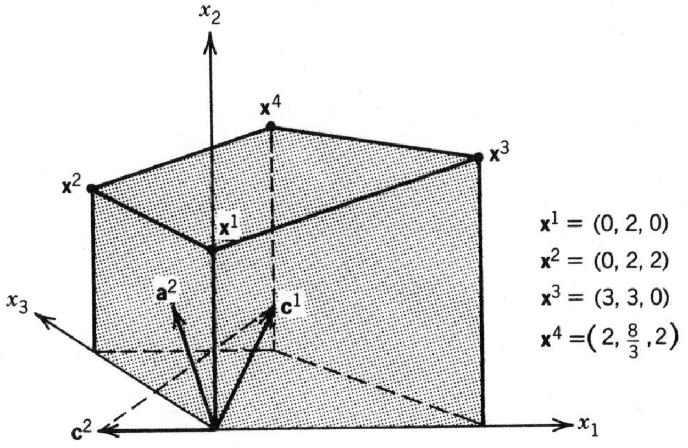

Figure 7.21. Graph of Example 22.

subset is $\Lambda_\varnothing$. The $\varnothing$ subscript indicates that the weighted-sums LP has an unbounded objective function value for all weighting vectors in $\Lambda_\varnothing$.

Example 22. Consider the MOLP

$$\begin{aligned}
\max\{\ \ & x_1 + 2x_2 && = z_1\} \\
\max\{ & -2x_1 && = z_2\} \\
\text{s.t.}\ \ & 2x_1 \qquad + x_3 && \leqq 6 \\
& -x_1 + 3x_2 && \leqq 6 \\
& \qquad\qquad\quad x_3 && \leqq 2 \\
& x_1, x_2, x_3 && \geqq 0
\end{aligned}$$

that is graphed in Fig. 7.21. In this problem, $E = \gamma(\mathbf{x}^1, \mathbf{x}^2, \mathbf{x}^3, \mathbf{x}^4)$. Since the weighting vector of the convex combination of $\mathbf{c}^1$ and $\mathbf{c}^2$ that points in the direction $\mathbf{a}^2$ is $(\frac{6}{11}, \frac{5}{11})$, we have

$$\Lambda_{\gamma(x^1, x^2)} = \{\boldsymbol{\lambda} \in \Lambda \mid \lambda_1 \in (0, \tfrac{6}{11}], \lambda_2 \in [\tfrac{5}{11}, 1)\}$$

$$\Lambda_{\gamma_{i=1}^4(x^i)} = \{\boldsymbol{\lambda} \in \Lambda \mid \lambda_1 = \tfrac{6}{11}, \lambda_2 = \tfrac{5}{11}\}$$

$$\Lambda_{x^3} = \{\boldsymbol{\lambda} \in \Lambda \mid \lambda_1 \in [\tfrac{6}{11}, 1), \lambda_2 \in (0, \tfrac{5}{11}]\}$$

We note that

$$\Lambda_{x^1} = \Lambda_{x^2} = \Lambda_{\gamma(x^1, x^2)}$$

and

$$\Lambda_{x^4} = \Lambda_{\gamma(x^1, x^3)} = \Lambda_{\gamma(x^2, x^4)} = \Lambda_{\gamma(x^3, x^4)} = \Lambda_{\gamma_{i=1}^4(x^i)}$$

Thus, different efficient facets may share the same subset of Λ. Also, it is possible for an efficient extreme point (e.g., $\mathbf{x}^4$) to have only one weighting vector associated with it.

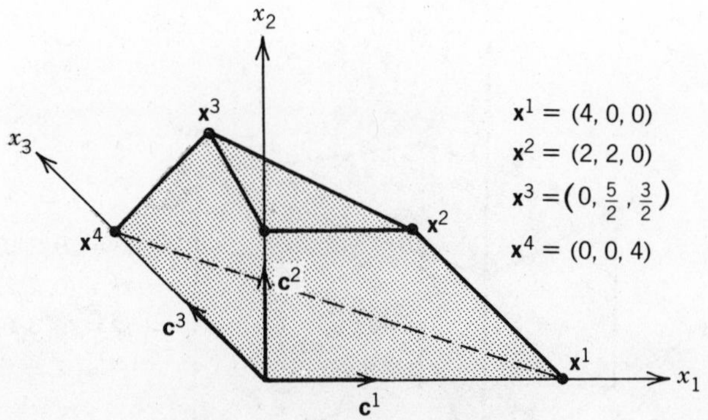

Figure 7.22. Graph of Example 23.

Example 23. Consider the MOLP

$$\begin{aligned} \max\{\, x_1 \qquad\qquad &= z_1\} \\ \max\{\qquad x_2 \qquad &= z_2\} \\ \max\{\qquad\qquad x_3 &= z_3\} \\ \text{s.t.}\quad x_1 + x_2 + x_3 &\leqq 4 \\ 3x_2 - x_3 &\leqq 6 \\ x_1, x_2, x_3 &\geqq 0 \end{aligned}$$

that is graphed in Fig. 7.22. In this problem, $E = \gamma(\mathbf{x}^1, \mathbf{x}^2, \mathbf{x}^3, \mathbf{x}^4)$. Consider Table 7.2. From the nonbasic $c_j - z_j$ columns of the simplex tableau at $\mathbf{x}^1$ as explained in Section 5.3, $\Lambda_{x^1} \subset \Lambda$ is defined by

$$\begin{aligned} -\lambda_1 + \lambda_2 &\leqq 0 \\ 2\lambda_1 + \lambda_2 &\geqq 1 \\ -\lambda_1 \qquad &\leqq 0 \end{aligned}$$

From the nonbasic $c_j - z_j$ columns of the tableaus at the other three efficient

TABLE 7.2
Simplex Tableau at $\mathbf{x}^1$

		x_1	x_2	x_3	s_4	s_5	
x_1	4	1	1	1	1	0	
s_5	6	0	3	−1	0	1	
z_1	4	0	−1	−1	−1	0	$c_j^1 - z_j^1$
z_2	0	0	1	0	0	0	$c_j^2 - z_j^2$
z_3	0	0	0	1	0	0	$c_j^3 - z_j^3$

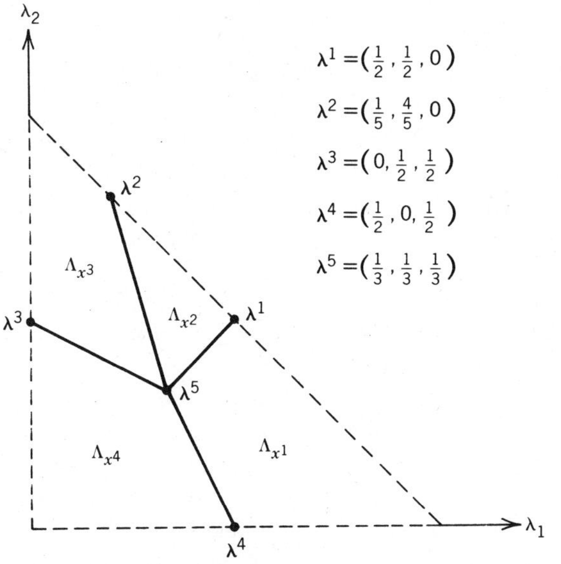

Figure 7.23. Parametric diagram of Example 23.

extreme points, $\Lambda_{x^2} \subset \Lambda$ is defined by

$$\begin{aligned} 7\lambda_1 + 2\lambda_2 &\geqq 3 \\ -\lambda_1 \qquad &\leqq 0 \\ -\lambda_1 + \lambda_2 &\geqq 0 \end{aligned}$$

$\Lambda_{x^3} \subset \Lambda$ is defined by

$$\begin{aligned} 7\lambda_1 + 2\lambda_2 &\leqq 3 \\ 3\lambda_1 + 2\lambda_2 &\leqq 3 \\ \lambda_1 + 2\lambda_2 &\geqq 1 \end{aligned}$$

and $\Lambda_{x^4} \subset \Lambda$ is defined by

$$\begin{aligned} 2\lambda_1 + \lambda_2 &\leqq 1 \\ \lambda_1 + 2\lambda_2 &\leqq 1 \\ \lambda_1 + \lambda_2 &\leqq 1 \end{aligned}$$

Using the sets of inequalities, we obtain the *parametric diagram* of Fig. 7.23.

In Fig. 7.23, the subset of Λ pertaining to the $\gamma(\mathbf{x}^2, \mathbf{x}^3)$ efficient edge is given by

$$\Lambda_{\gamma(x^2, x^3)} = \gamma(\boldsymbol{\lambda}^2, \boldsymbol{\lambda}^5) - \{\boldsymbol{\lambda}^2\}$$

The subset of Λ pertaining to the $\gamma_{i=1}^{4}(\mathbf{x}^i)$ efficient facet is given by

$$\Lambda_{\gamma_{i=1}^{4}(x^i)} = \{\boldsymbol{\lambda}^5\}$$

and so forth.

The way in which the different subsets of Λ are connected shows the way the different facets of E are connected.

Readings

1. Easton, A. (1973). *Complex Managerial Decisions Involving Multiple Objectives*, New York: John Wiley & Sons.
2. Eckenrode, R. T. (1965). "Weighting Multiple Criteria," *Management Science*, Vol. 12, No. 3, pp. 180–192.
3. Einhorn, H. J. and R. M. Hogarth (1975). "Unit Weighting Schemes for Decision Making," *Organizational Behavior and Human Performance*, Vol. 13, No. 2, pp. 171–192.
4. Gal, T. and H. Leberling (1977). "Redundant Objective Functions in Linear Vector Maximum Problems and Their Determination," *European Journal of Operational Research*, Vol. 1, No. 3, pp. 176–184.
5. Gale, D. (1960). *The Theory of Linear Economic Models*, New York: McGraw-Hill.
6. Geoffrion, A. M., J. S. Dyer, and A. Feinberg (1972). "An Interactive Approach for Multi-Criterion Optimization, with an Application to the Operation of an Academic Department" (Section 2), *Management Science*, Vol. 19, No. 4, pp. 357–368.
7. Gershon, M. (1984). "The Role of Weights and Scales in the Application of Multiobjective Decision Making," *European Journal of Operational Research*, Vol. 15, No. 2, pp. 244–250.
8. Hobbs, B. J. (1980). "A Comparison of Weighting Methods in Power Plant Siting," *Decision Sciences*, Vol. 11, No. 4, pp. 725–737.
9. Keeney, R. L. and H. Raiffa (1976). *Decisions with Multiple Objectives: Preferences and Value Tradeoffs*, New York: John Wiley & Sons.
10. Knoll, A. L. and A. Engelberg (1978). "Weighting Multiple Objectives: The Churchman-Ackoff Technique Revisited," *Computers and Operations Research*, Vol. 5, No. 3, pp. 165–177.
11. MacCrimmon, K. R. and J. K. Siu (1974). "Making Trade-Offs," *Decision Sciences*, Vol. 5, No. 4, pp. 680–704.
12. Saaty, T. L. (1977). "A Scaling Method for Priorities in Hierarchical Structures," *Journal of Mathematical Psychology*, Vol. 15, No. 3, pp. 234–281.
13. Saaty, T. L. (1980). *The Analytic Hierarchy Process*, New York: McGraw-Hill.
14. Schoemaker, P. J. H. and C. C. Waid (1982). "An Experimental Comparison of Different Approaches to Determining Weights in Additive Utility Models," *Management Science*, Vol. 28, No. 2, pp. 182–196.
15. Srinivasan, V. and A. D. Shocker (1973). "Estimating the Weights for Multiple Attributes in a Composite Criterion Using Pairwise Judgments," *Psychometrika*, Vol. 38, No. 4, pp. 473–493.

PROBLEM EXERCISES

7-1 Graphing the criterion cones generated by

(a)	**(b)**	**(c)**
$\mathbf{c}^1 = (1,2)$	$\mathbf{c}^1 = (0,-4,0)$	$\mathbf{c}^1 = (1,0)$
$\mathbf{c}^2 = (2,4)$	$\mathbf{c}^2 = (0,-4,1)$	$\mathbf{c}^2 = (0,1)$
$\mathbf{c}^3 = (3,3)$	$\mathbf{c}^3 = (0,-4,2)$	$\mathbf{c}^3 = (-1,0)$
$\mathbf{c}^4 = (3,1)$	$\mathbf{c}^4 = (-2,-4,0)$	$\mathbf{c}^4 = (0,-1)$
	$\mathbf{c}^5 = (-2,-4,2)$	

specify:

(i) The dimensionality of the criterion cone.

(ii) The number of extreme rays of the criterion cone.

(iii) Which gradients are essential and which are nonessential.

7-2 With essential objective function gradients, design an MOLP example in which $E = S$ and the null vector condition does not hold.

7-3 Using composite gradients, graphically determine E, E_x, and E_μ for:

(a)
$$\begin{aligned} \max\{ -x_1 &= z_1\} \\ \max\{ x_1 - x_2 &= z_2\} \\ \text{s.t.} \quad x_1 + x_2 &\geqq 2 \\ x_2 &\leqq 3 \\ x_1, x_2 &\geqq 0 \end{aligned}$$

(b)
$$\begin{aligned} \max\{ -x_1 &= z\} \\ \text{s.t.} \quad -x_1 + x_2 &\leqq 2 \\ x_1, x_2 &\geqq 0 \end{aligned}$$

(c)
$$\begin{aligned} \max\{-2x_1 + x_2 &= z_1\} \\ \max\{ x_1 - 2x_2 &= z_2\} \\ \text{s.t.} \quad x_1, x_2 &\geqq 0 \end{aligned}$$

(d)
$$\begin{aligned} \min\{ -2x_1 - x_2 &= z_1\} \\ \min\{ -x_1 - x_2 &= z_2\} \\ \text{s.t.} \quad x_2 &\leqq 2 \\ x_1, x_2 &\geqq 0 \end{aligned}$$

7-4 **(a)** Graph the MOLP:

$$\begin{aligned} \max\{ x_1 + x_2 &= z_1\} \\ \max\{-2x_1 - x_2 &= z_2\} \\ \text{s.t.} \quad x_1 &\leqq 6 \\ x_1 &\geqq 4 \\ x_2 &\leqq 3 \\ x_3 &\leqq 2 \\ x_1 + x_2 + x_3 &\geqq 6 \\ x_1, x_2, x_3 &\geqq 0 \end{aligned}$$

(b) Specify E, E_x, and E_μ.

(c) What λ-vector causes $\mathbf{x}^1 = (5, 3, 1)$ to maximize the composite LP?

(d) What λ-vector causes $\mathbf{x}^2 = (4, 2, 1)$ to maximize the composite LP?

7-5 Consider the MOLP:

$$\begin{aligned} \max\{-x_1 - x_2 &= z_1\} \\ \max\{-x_1 + x_2 &= z_2\} \\ \max\{ -x_3 &= z_3\} \\ \text{s.t.} \quad x_3 &\leqq 2 \\ x_1 + x_2 &\leqq 2 \\ x_1, x_2, x_3 &\geqq 0 \end{aligned}$$

(a) Specify all maximal facets of S.

(b) When the only objective gradients in the problem are

(i) $\mathbf{c}^1$

(ii) $\mathbf{c}^1$ and $\mathbf{c}^2$

(iii) $\mathbf{c}^1$, $\mathbf{c}^2$, and $\mathbf{c}^3$

write E as the union of maximally efficient facets.

7-6 Graphically determine the subsets of Λ associated with each efficient facet of the following MOLPs:

(a)
$$\begin{aligned}
\max\{\ 2x_1 - x_2 &= z_1\}\\
\max\{\ \qquad x_2 &= z_2\}\\
\text{s.t.}\quad x_1 + x_2 &\leqq 7\\
x_2 &\leqq 2\\
-x_1 + x_2 &\leqq 1\\
x_1, x_2 &\geqq 0
\end{aligned}$$

(b)
$$\begin{aligned}
\min\{\ x_1 - 2x_2 &= z_1\}\\
\min\{-x_1 \qquad &= z_2\}\\
\text{s.t.}\quad x_2 &\leqq 2\\
x_1, x_2 &\geqq 0
\end{aligned}$$

7-7 For the MOLP

$$\begin{aligned}
\max\{-2x_1 - x_2 &= z_1\}\\
\max\{\ x_1 + 2x_2 &= z_2\}\\
\text{s.t.}\quad x_1 &\leqq 4\\
x_2 + x_3 &\leqq 3\\
x_1, x_2, x_3 &\geqq 0
\end{aligned}$$

graphically determine:

(a) The dimensionality of each maximally efficient facet.

(b) The subset of Λ associated with each maximally efficient facet.

7-8 Consider the MOLP:

$$\begin{aligned}
\max\{4x_1 \qquad - x_3 &= z_1\}\\
\max\{\ x_1 + 5x_2 + x_3 &= z_2\}\\
\max\{\ x_1 \qquad + 4x_3 &= z_3\}\\
\text{s.t.}\quad x_1 + x_2 + x_3 &\leqq 3\\
x_1 &\leqq 2\\
x_1, x_2, x_3 &\geqq 0
\end{aligned}$$

(a) Graph the MOLP.

(b) Write E in terms of maximally efficient facets.

(c) From the tableaus at the different efficient extreme points, construct a parametric diagram.

7-9 Consider the MOLP:

$$
\begin{array}{lrcl}
\max\{ & x_1 + x_2 + x_3 & = & z_1\} \\
\max\{ & -x_1 + x_2 - 2x_3 & = & z_2\} \\
\max\{ & x_3 & = & z_3\} \\
\text{s.t.} & x_1 + x_2 + x_3 & \leqq & 5 \\
 & x_2 & \leqq & 3 \\
 & x_1, x_2, x_3 & \geqq & 0
\end{array}
$$

(a) Graph the MOLP.

(b) Write E in terms of maximally efficient facets.

(c) From the tableaus at the different efficient extreme points, construct a parametric diagram.

(d) Specify a set of generators for the subset of the criterion cone pertaining to each of the maximally efficient facets.

7-10C Consider the MOLP:

	x_1	x_2	x_3	x_4		
Objs	1	−2		3	max	
	−3	6		−9	max	
	5	1	2		max	
s.t.	−2		1		≦	6
	1	1	5		≦	6
	6	5	−3		≦	6
			−1	4	≦	8

Compute all extreme points efficient with respect to:

(a) Only the first objective function.

(b) The first two objective functions.

(c) All three objective functions.

7-11C Consider the MOLP:

	x_1	x_2	x_3	x_4	x_5		
Objs	5	−1	−1	−1	1	max	
	1	3	3		−1	max	
	−6	−2	−2	1		max	
		4	2		−3	max	
s.t.	3	−1	3		4	≦	6
	−1	2	−1	5	4	≦	9
	−1		5	4	2	≦	8
	5			−1	3	≦	7
	−1		−1	2		≦	7

Compute all extreme points efficient with respect to:

(a) The first two objective functions.

(b) The first three objective functions.

(c) All four objective functions.

7-12C Consider the MOLP:

	x_1	x_2	x_3	x_4	x_5	x_6	x_7	x_8	x_9	x_{10}		
	15	5	12	−8	2	7	−17	−1	−14	9	max	
Objs	−17	−19	−1	3	−4	−18	14	4	−16	−11	max	
	−3	3	13	8	15	18	17	−19	14	−19	max	
	−1			2	8	−4	7	9			max	
	18					11		6		3	≤	100
	4		19	15	1	11	13				≤	100
		8		3				11			≤	100
s.t.					12	2	5		3	4	≤	100
	13		4		9	7	3	13	12	3	≤	100
			4		19				8	9	≤	100
	8	3	18		3		2		2	5	≤	100
		1		9					13	19	≤	100

Compute all extreme points efficient with respect to:

(a) The first two objective functions.

(b) The first three objective functions.

(c) All four objective functions.

7-13C With regard to the MOLP of Problem 7-12C, solve the weighted-sums problem with the following weighting vectors:

(a) $\lambda = (.4, .2, .3, .1)$

(b) $\lambda = (.4, .3, .1, .2)$

(c) $\lambda = (.1, .2, .6, .1)$

(d) $\lambda = (.1, .2, .2, .5)$

CHAPTER 8

Optimal Weighting Vectors, Scaling, and Reduced Feasible Region Methods

This chapter is a conglomerate. In Sections 8.1, 8.2, and 8.3, we discuss reasons why optimal weighting vectors are difficult to estimate, and in some cases, difficult to apply. In Section 8.4, we discuss methods for normalizing and rescaling objective functions. In Sections 8.5 and 8.6, we discuss methods for reducing the feasible region until it hopefully "traps" an optimal solution.

8.1 OPTIMAL WEIGHTING VECTOR ESTIMATION DIFFICULTIES

Consider the multiple objective linear program

$$\begin{aligned} &\max \{\mathbf{c}^1\mathbf{x} = z_1\} \\ &\max \{\mathbf{c}^2\mathbf{x} = z_2\} \\ &\quad\vdots \\ &\max \{\mathbf{c}^k\mathbf{x} = z_k\} \\ &\text{s.t.} \quad \mathbf{x} \in S \end{aligned}$$

with reference to which

1. $\Lambda = \{\boldsymbol{\lambda} \in R^k \mid \lambda_i > 0,\ \sum_{i=1}^k \lambda_i = 1\}$.
2. $\Lambda_\Theta \subset \Lambda$ is the set of *optimal weighting vectors* in the sense that a member of the optimal set Θ maximizes the *composite* LP

   $$\max \{\boldsymbol{\lambda}^T C\mathbf{x} \mid \mathbf{x} \in S\}$$

   whenever $\boldsymbol{\lambda} \in \Lambda_\Theta$.

The idea of the composite (or weighted-sums) LP is to use it with a member of Λ_Θ in order to obtain an optimal solution. Unfortunately, the set of optimal

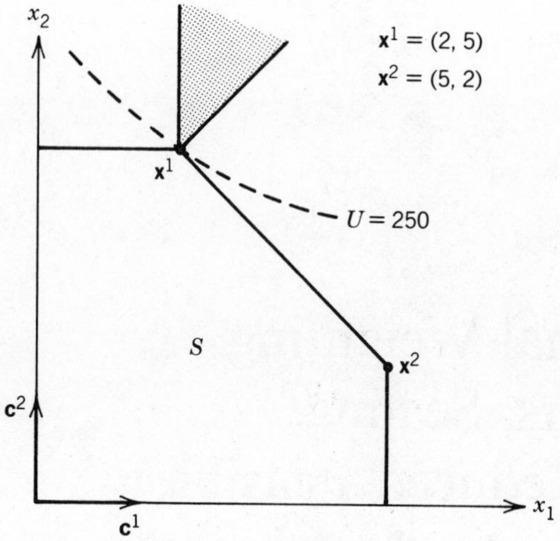

Figure 8.1. Graph of MOLP with $U = z_1(z_2)^3$.

weighting vectors Λ_Θ is more than just a function of the decision maker's preferences. It is also dependent on the relative lengths of the objective function gradients and the geometry of the feasible region.

8.1.1 Dependence on the Decision Maker's Preferences

Suppose a firm's operation is modeled as the MOLP

$$\begin{aligned} &\max \{x_1 \quad = z_1\} \\ &\max \{ \quad x_2 = z_2\} \\ &\text{s.t.} \qquad \mathbf{x} \in S \end{aligned}$$

that is graphed in Fig. 8.1. Under current policy, assume that the DM's utility function is $U = z_1(z_2)^3$. Thus, $\mathbf{x}^1$ is optimal. When we note that the extreme ray directions of the shaded "cone" emanating from $\mathbf{x}^1$ are $(0, 1)$ and $(1, 1)$, the set of $\boldsymbol{\lambda}$'s that cause the composite LP to have $\mathbf{x}^1$ as a maximizing point is

$$\Lambda_\Theta = \{\boldsymbol{\lambda} \in \Lambda \,|\, \lambda_1 \in (0, \tfrac{1}{2}],\quad \lambda_2 \in [\tfrac{1}{2}, 1)\}$$

Suppose there is a change in policy so that the utility function becomes $U = (z_1)^3 z_2$. In this case, as portrayed in Fig. 8.2, $\mathbf{x}^2$ is optimal with

$$\Lambda_\Theta = \{\boldsymbol{\lambda} \in \Lambda \,|\, \lambda_1 \in [\tfrac{1}{2}, 1),\quad \lambda_2 \in (0, \tfrac{1}{2}]\}$$

Thus, Λ_Θ is demonstrated to be a function of U.

8.1.2 Dependence on the Relative Lengths of the Objective Function Gradients

Holding the DM's utility function and feasible region S fixed, we show how the set of optimal weighting vectors Λ_Θ can change when the relative lengths of the

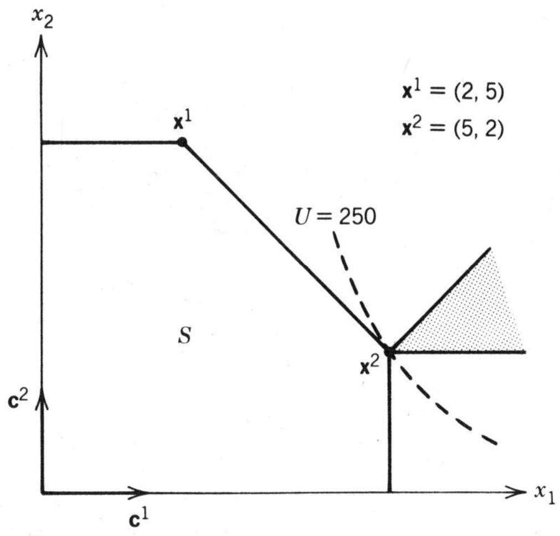

Figure 8.2. Graph of MOLP with $U = (z_1)^3 z_2$.

objective function gradients change. This can happen whenever the objectives are rescaled by different factors. Such might occur if we normalized the objective functions. The relative lengths of the objective function gradients might also change if we change the units in which the objectives are measured (i.e., from board-feet to cubic feet of timber production, or from barrels to tons of fuel oil consumption, etc.).

To illustrate, consider the MOLP

$$\begin{aligned} &\max\{2x_1 \qquad = z_1\} \\ &\max\{ \qquad 2x_2 = z_2\} \\ &\text{s.t.} \qquad \mathbf{x} \in S \end{aligned}$$

that is graphed in Fig. 8.3 in which $\mathbf{x}^2$ is optimal. In the above formulation, the two objective function gradients, $\mathbf{c}^1$ and $\mathbf{c}^2$, are of equal length and

$$\Lambda_\Theta = \{\boldsymbol{\lambda} \in \Lambda \mid \lambda_1 \in [\tfrac{1}{3}, \tfrac{2}{3}], \quad \lambda_2 \in [\tfrac{1}{3}, \tfrac{2}{3}]\}$$

Now, when we rescale the first objective by the factor $\frac{1}{4}$ and the second objective by the factor 2, the MOLP becomes

$$\begin{aligned} &\max\{\tfrac{1}{2}x_1 \qquad = z_1\} \\ &\max\{ \qquad 4x_2 = z_2\} \\ &\text{s.t.} \qquad \mathbf{x} \in S \end{aligned}$$

In this version of the MOLP, as seen in Fig. 8.3, the length of $\mathbf{c}^2$ is eight times the length of $\mathbf{c}^1$. For this version of the MOLP,

$$\Lambda_\Theta = \{\boldsymbol{\lambda} \in \Lambda \mid \lambda_1 \in [\tfrac{4}{5}, \tfrac{16}{17}], \quad \lambda_2 \in [\tfrac{1}{17}, \tfrac{1}{5}]\}$$

Since we have not changed the utility function or feasible region, this demon-

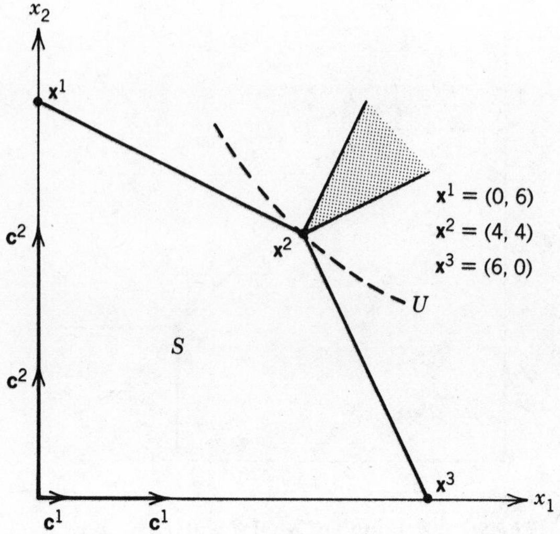

Figure 8.3. MOLP with different objective function gradient lengths.

strates that Λ_Θ is a function of the relative lengths of the objective function gradients.

8.1.3 Dependence on the Feasible Region

Holding the DM's utility function and the relative lengths of the objective function gradients fixed, we show how the set of optimal weighting vectors Λ_Θ can change when the feasible region S changes. To illustrate, consider Firm A

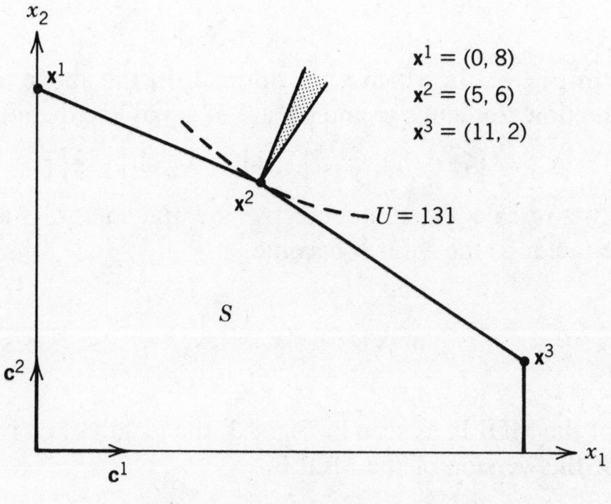

Figure 8.4. MOLP for Firm A.

whose operations are modeled with the MOLP

$$
\begin{aligned}
\max\{\ & x_1 & & = z_1\} \\
\max\{\ & & x_2 & = z_2\} \\
\text{s.t.}\quad & 2x_1 & +5x_2 & \leqq 40 \\
& 2x_1 & +3x_2 & \leqq 28 \\
& x_1 & & \leqq 11 \\
& & x_1, x_2 & \geqq 0
\end{aligned}
$$

Assume that the firm's utility function is $U = z_1 + z_1(z_2)^2 + z_2 - 60$.

If we graph Firm A's MOLP in Fig. 8.4, $\mathbf{x}^2$ is optimal and Firm A's set of optimal weighting vectors is

$$\Lambda_{\Theta}^{A} = \{\lambda \in \Lambda \mid \lambda_1 \in [\tfrac{2}{7}, \tfrac{2}{5}],\quad \lambda_2 \in [\tfrac{3}{5}, \tfrac{5}{7}]\}$$

Now consider Firm B, whose operations are modeled with the MOLP

$$
\begin{aligned}
\max\{\ & x_1 & & = z_1\} \\
\max\{\ & & x_2 & = z_2\} \\
\text{s.t.}\quad & 7x_1 & +5x_2 & \leqq 66 \\
& 5x_1 & +7x_2 & \leqq 66 \\
& & x_1, x_2 & \geqq 0
\end{aligned}
$$

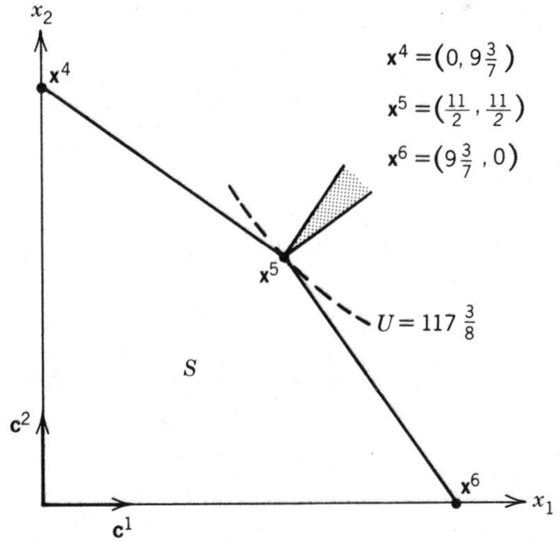

Figure 8.5. MOLP for Firm B.

Firm B, a competitor of Firm A, is in every way, except one, identical to Firm A. Firms A and B are roughly the same size, produce the same products, and have the same utility functions ($U = z_1 + z_1(z_2)^2 + z_2 - 60$). However, Firm B's feasible region is somewhat different from Firm A's because of some differences in their physical plants.

If we graph Firm B's MOLP in Fig. 8.5, $\mathbf{x}^5$ is optimal and Firm B's set of optimal weighting vectors is

$$\Lambda_\Theta^B = \{\boldsymbol{\lambda} \in \Lambda \mid \lambda_1 \in [\tfrac{5}{12}, \tfrac{7}{12}], \quad \lambda_2 \in [\tfrac{5}{12}, \tfrac{7}{12}]\}$$

Note in the illustration, that if any of Firm A's optimal weighting vectors are applied to Firm B, $\mathbf{x}^4$ would result (which is a disaster with $U = -50\frac{4}{7}$). Furthermore, if any of Firm B's optimal weighting vectors are applied to Firm A, $\mathbf{x}^3$ would result (which is also unfortunate with $U = -3$). Since the only thing that is different between Firm A and Firm B is the feasible region, we have demonstrated how Λ_Θ can be a function of the feasible region.

8.2 DIFFICULTIES CAUSED BY THE DEGREE TO WHICH OBJECTIVES ARE CORRELATED

Confounding the use of weighting vectors is the degree to which the objectives are *correlated*. Consider two objectives: i and j. A measure that suffices for the degree to which the ith and jth objectives are correlated is the angle

$$\alpha = \cos^{-1}\left(\frac{(\mathbf{c}^i)^T \mathbf{c}^j}{\|\mathbf{c}^i\|_2 \|\mathbf{c}^j\|_2}\right)$$

between the gradients $\mathbf{c}^i$ and $\mathbf{c}^j$. The smaller the angle, the more correlated the objectives.

If there is correlation among the objectives, strange things can happen. Assume the MOLP example of Fig. 8.6 in which the objective pertaining to $\mathbf{c}^1$ is the most important, the objective pertaining to $\mathbf{c}^2$ is the second most important, the objective pertaining to $\mathbf{c}^3$ is the least important, and the optimal set $\Theta = \{\mathbf{x}^1\}$.

In this MOLP, seemingly "good" weighting vectors exist that produce bad points, and there are seemingly "bad" weighting vectors that produce good points. Consider the seemingly good weighting vector

$$\boldsymbol{\lambda}^g = (.7, .2, .1)$$

Although consistent with the decision maker's ordinal ranking of the objectives, $\boldsymbol{\lambda}^g$ causes the composite LP to generate $\mathbf{x}^2$ that is not a good point. Now consider the weighting vector

$$\boldsymbol{\lambda}^b = (.0, .1, .9)$$

that is seemingly bad because of its inconsistency with the decision maker's ordinal ranking of the objectives. However, $\boldsymbol{\lambda}^b$ causes the composite LP to generate the optimal point $\mathbf{x}^1$. Note the awkwardness we would experience trying

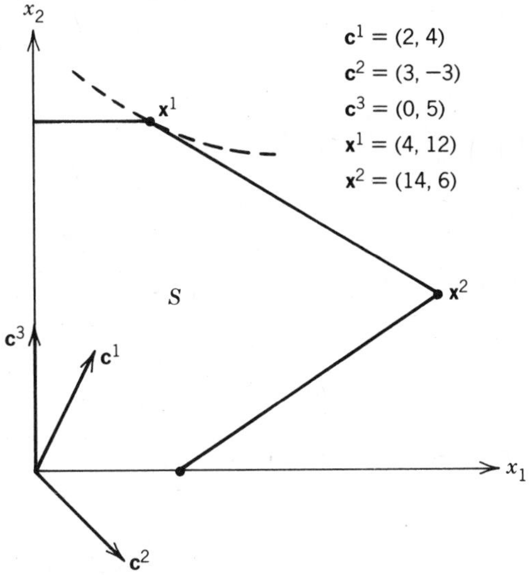

Figure 8.6. Graphical example of correlated objectives.

to explain to a client that we found his optimal solution by placing a zero weight on his most important objective.

Such counter-intuitive weighting vector behavior is a result of the fact that $\mathbf{c}^1$ and $\mathbf{c}^3$ are highly correlated. By placing a large enough weight on $\mathbf{c}^3$, it is not necessary to place any weight on $\mathbf{c}^1$ because $\mathbf{c}^1$ comes along free for the ride. It is because of this that $\boldsymbol{\lambda}^b$ was able to produce the optimal point.

8.3 ADDITIONAL OPTIMAL WEIGHTING VECTOR DIFFICULTIES

In addition to the previous considerations affecting Λ_Θ, the size of Λ_Θ can be expected to decrease as the problem size increases. This is because we expect the efficient set to increase as the problem size increases. This adds to the difficulty of trying to estimate an optimal weighting vector $\boldsymbol{\lambda} \in \Lambda_\Theta$.

Even if we were able to estimate an optimal weighting vector, we may have difficulty finding an optimal point. Assume the MOLP of Fig. 8.7. Let us consider two cases: (1) when the optimal point is nonextreme (at $\mathbf{x}^4$), and (2) when the optimal point is extreme (at $\mathbf{x}^5$). For both points, $(\frac{11}{25}, \frac{14}{25}) \in \Lambda_\Theta$.

When solving the composite LP with $\boldsymbol{\lambda} = (\frac{11}{25}, \frac{14}{25})$ by the simplex method, let us assume that the pivoting goes $\mathbf{x}^1$ to $\mathbf{x}^2$ to $\mathbf{x}^3$. When the optimal point is not extreme, as with $\mathbf{x}^4$, the composite LP is not able to compute it regardless of what weighting vector is used. When the optimal point is extreme, as with $\mathbf{x}^5$, and there are alternative optima, the composite LP may compute the wrong extreme point. This can happen even though, as in this illustration, we used a weighting vector

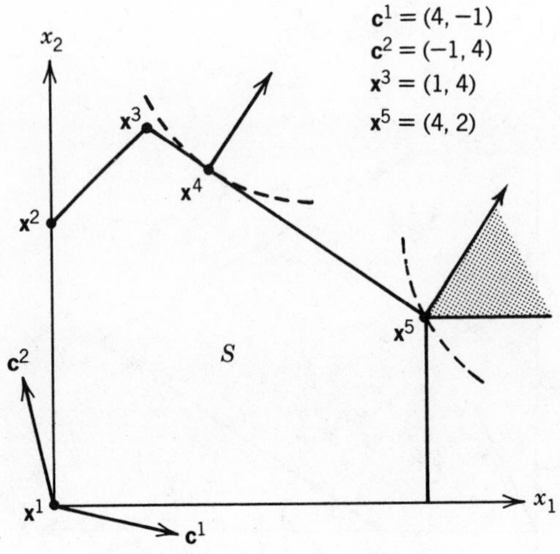

Figure 8.7. Graph of Section 8.3.

that caused the composite LP objective function gradient to point precisely in the local direction of the utility function gradient.

An additional difficulty with weighting vectors is that in many situations, the decision maker may not be willing to specify a weighting vector. For instance, a politician would not want to specify a weighting vector that would imply he is willing to trade off 300,000 jobs for a .5% reduction in the inflation rate. Consequently, care is imperative in the use of methods employing weighting vectors for a variety of reasons.

8.4 SCALING THE OBJECTIVE FUNCTIONS

Three philosophies are available for rescaling the objective functions: (a) normalization, (b) use of 10 raised to an appropriate power, and (c) the application of range equalization factors. Normalization was covered in Section 2.5.1. The use of 10 raised to an appropriate power and range equalization factors are discussed in Sections 8.4.1 and 8.4.2, respectively.

8.4.1 Use of 10 Raised to an Appropriate Power

Assume that our purpose is numerical (i.e., to bring all objective function coefficients into the same order of magnitude). Besides normalization, another option is to rescale each objective function by 10 raised to an appropriate power. Whereas normalization is likely to change the coefficients to unrecognizable numbers, the *recognition* of each coefficient is retained with 10 raised to an appropriate power because only the decimal point moves.

Example 1. Consider the MOLP

$$\max\{63400x_1 + 189600x_2 - 50400x_3 = z_1\}$$
$$\max\{\ 9.61x_1 - \quad 2.35x_2 + \ 6.78x_3 = z_2\}$$
$$\text{s.t.} \quad \mathbf{x} \in S$$

If we normalize each objective function using the L_1-norm, we have

$$\max\{.209x_1 + .625x_2 - .166x_3 = z_1\}$$
$$\max\{.513x_1 - .125x_2 + .362x_3 = z_2\}$$
$$\text{s.t.} \quad \mathbf{x} \in S$$

However, if we rescale the first objective by 10^{-5} and the second objective by 10^{-1}, we have

$$\max\{.634x_1 + 1.896x_2 - .504x_3 = z_1\}$$
$$\max\{.961x_1 - \ .235x_2 + .678x_3 = z_2\}$$
$$\text{s.t.} \quad \mathbf{x} \in S$$

In this way, we have brought all objective function coefficients into the same order of magnitude, yet we have retained the recognizability of the objective function coefficients.

8.4.2 Use of Range Equalization Factors

Suppose our purpose is to equalize the ranges of the criterion values over the efficient set. One way to do this is to multiply each objective by its representative *range equalization* factor

$$\pi_i = \frac{1}{R_i}\left[\sum_{j=1}^{k} \frac{1}{R_j}\right]^{-1}$$

where R_i is the *range width* of the ith criterion value over the efficient set.

Example 2. Since $E = \gamma(\mathbf{x}^1, \mathbf{x}^2)$ in the MOLP of Fig. 8.8, the range of z_1 over E is $[600, 800]$ and the range of z_2 is $[-24, 12]$. Thus, $R_1 = 200$ and $R_2 = 36$. Hence, $\pi_1 = .1525$ and $\pi_2 = .8475$.

After we scale the objectives by the π_i in Example 2, the range of z_1 over E is $[91.5, 122.0]$ and the range of z_2 is $[-20.3, 10.2]$. Suppose it is also desired to equalize the midpoints of the scaled ranges at a given value. This can be accomplished by adding K_i constant terms to the scaled objective functions. If the given value is 50, the range midpoints would be equalized with $K_1 = -56.75$

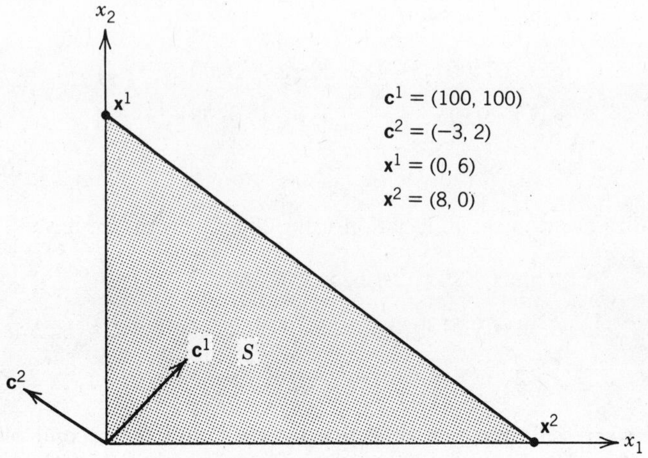

Figure 8.8. Graph of Example 2.

and $K_2 = 55.05$. Therefore, if

$$\max\{\mathbf{c}^1\mathbf{x} = z_1\}$$
$$\max\{\mathbf{c}^2\mathbf{x} = z_2\}$$
$$\text{s.t.} \quad \mathbf{x} \in S$$

were the starting MOLP of Example 2, the ending MOLP would be

$$\max\{\pi_1(\mathbf{c}^1\mathbf{x}) + K_1 = z_1\}$$
$$\max\{\pi_2(\mathbf{c}^2\mathbf{x}) + K_2 = z_2\}$$
$$\text{s.t.} \quad \mathbf{x} \in S$$

where π_1, π_2, K_1, and K_2 are as given above.

8.5 e-CONSTRAINT REDUCED FEASIBLE REGION METHOD

We now discuss a method that attempts to "trap" an optimal solution of the following MOLP in a reduced feasible region. Assume the MOLP

$$\max\{\mathbf{c}^1\mathbf{x} = z_1\}$$
$$\max\{\mathbf{c}^2\mathbf{x} = z_2\}$$
$$\vdots$$
$$\max\{\mathbf{c}^k\mathbf{x} = z_k\}$$
$$\text{s.t.} \quad \mathbf{x} \in \mathrm{S}$$

In this method, one of the objectives is selected for maximization subject to e_j lower bounds on the other objectives to form the ith *objective* **e**-*constraint program*

$$\max\{\mathbf{c}^i\mathbf{x} = z_i\}$$
$$\text{s.t.} \quad \mathbf{c}^j\mathbf{x} \geqq \mathrm{e}_j \quad j \neq i$$
$$\mathbf{x} \in S$$

Because of the conversion of $k-1$ of the objectives to constraints, the feasible region of the ith objective **e**-constraint program is a subset of the original feasible region S. Then, over the reduced feasible region, the selected objective is maximized.

Example 3. Consider the MOLP

$$\max\{\text{profit}\}$$
$$\max\{\text{market share}\}$$
$$\max\{\text{employment}\}$$
$$\text{s.t. } \mathbf{x} \in S$$

In hopes of generating a point close to an optimal point, suppose the DM subordinates the market share and employment objectives to constraints with RHSs of e_m and e_e, respectively. This results in the profit objective **e**-constraint program

$$\max\{\text{profit}\}$$
$$\text{s.t.} \quad \text{market share} \geqq e_m$$
$$\text{employment} \geqq e_e$$
$$\mathbf{x} \in S$$

Assume that the graph of the **e**-constraint program is as depicted in Fig. 8.9, in which $\mathbf{c}^p$, $\mathbf{c}^m$, and $\mathbf{c}^e$ are the gradients of the three objectives. In Fig. 8.9, we make the following observations: (a) although $\mathbf{x}^5$ is the point in S that maximizes profit, the point maximizing profit over the reduced feasible region is $\mathbf{x}^3$; (b) $\mathbf{x}^3$ is efficient because $E = \gamma(\mathbf{x}^1, \mathbf{x}^2) \cup \gamma(\mathbf{x}^2, \mathbf{x}^4) \cup \gamma(\mathbf{x}^4, \mathbf{x}^5)$; (c) although $\mathbf{x}^3$ is an extreme point of the reduced feasible region, it is not an extreme point of S; and (d) $\mathbf{x}^3$ is immediately sensitive to changes in e_e but not immediately sensitive to changes in e_m. Point (d) suggests opportunities for sensitivity analysis that are discussed in Section 8.5.3.

8.5.1 Operation of the e-Constraint Approach

There is no established protocol for employing the **e**-constraint approach. The method is ad hoc because it is never precisely clear how to configure the problem. Which objectives are to be converted to constraints at what RHS values is largely determined by the user's sense of experiment.

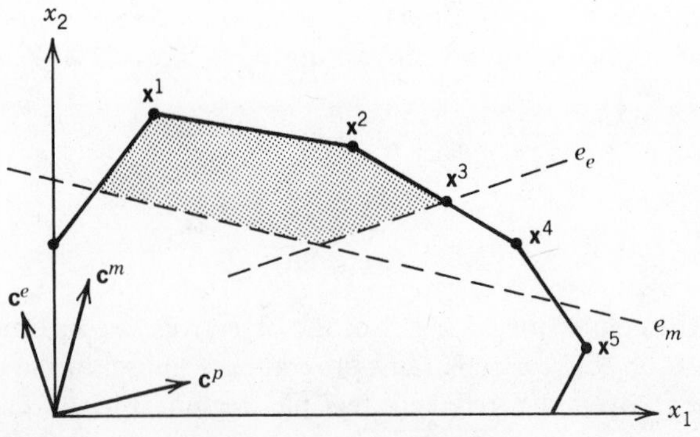

Figure 8.9. Graph of Example 3.

Basically, the idea is to configure the problem to produce a good candidate solution. Then, from the solution generated, it is hoped that another configuration will suggest itself that will lead to a better solution. Continuing to experiment in this fashion, the DM will sooner or later reach a point at which he or she will stop (perhaps by losing patience with the procedure). Then, from the series of solutions generated, the DM will pick the point with which he or she feels most comfortable as the final choice. Although the **e**-constraint approach has its weaknesses, the method has enjoyed some popularity because it is easy to understand and uses standardly available software.

With the **e**-constraint method, two features emerge that we will see more of in the rest of this book. The first is the *interactive* nature of the process. After each cycle of computation, the results are evaluated before deciding how to reconfigure the problem for the next cycle of computation. The second is the generation of a series of candidate solutions from which the final solution will eventually be selected.

8.5.2 Different Types of Outcomes

Normally, when $E \neq \varnothing$, the solution generated by an **e**-constraint LP is efficient, and when $E = \varnothing$, the **e**-constraint LP has an unbounded objective function value solution. However, disconcerting results are possible, as portrayed in Examples 4, 5, and 6.

Example 4. (When the **e**-constraint program is inconsistent.) Consider the MOLP of Fig. 8.10. In this problem, $E = \gamma(\mathbf{x}^1, \mathbf{x}^2)$. However, the **e**-constraint program with e_1 and e_2 as the RHSs of the first two objectives is inconsistent. This is because the shaded region in Fig. 8.10 does not intersect S. The inconsistency difficulty is remedied by downgrading the e_j values.

Example 5. (When the solution generated is inefficient even though $E \neq \varnothing$.) Consider the MOLP of Figure 8.11. In this problem, $E = \{\mathbf{x}^2\}$. Suppose the **e**-constraint program is formed in which the second objective is converted to a lower bound constraint whose RHS value is e_2 as indicated. If we use standardly

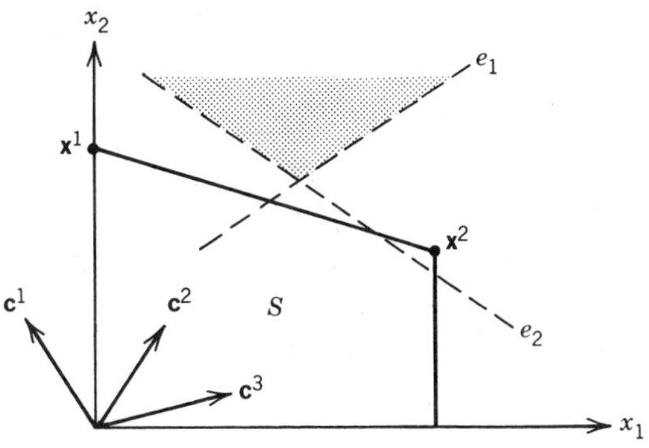

Figure 8.10. Graph of Example 4.

available LP software to maximize the first objective over the reduced feasible region, $\mathbf{x}^1$ results. Note that $\mathbf{x}^1$ is inefficient. The problem here is caused by alternative optima.

Example 6. (When a maximizing point is generated even though $E = \varnothing$.) Consider the MOLP of Figure 8.12. In this problem, $E = \varnothing$. Suppose the **e**-constraint program is formed in which the first objective is converted into a lower bound constraint whose RHS value is e_1 as indicated. Maximizing the second objective over the reduced feasible region results in $\mathbf{x}^1$, which is inefficient. As in Example 5, the **e**-constraint method has generated a point that is a noncontender for optimality.

Our conclusions are as follows. When $E \neq \varnothing$, the **e**-constraint program, when consistent, will have a bounded objective function value and in its set of maximizing points will be at least one efficient point. When $E = \varnothing$, the **e**-constraint program, when consistent, may have a bounded objective function value, but when it does, the set of maximizing points will be unbounded.

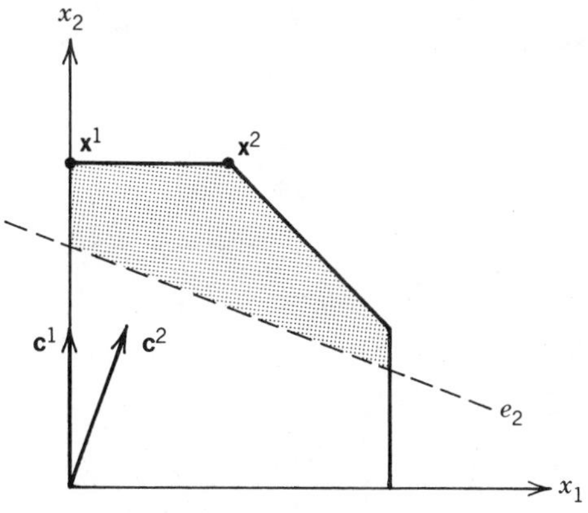

Figure 8.11. Graph of Example 5.

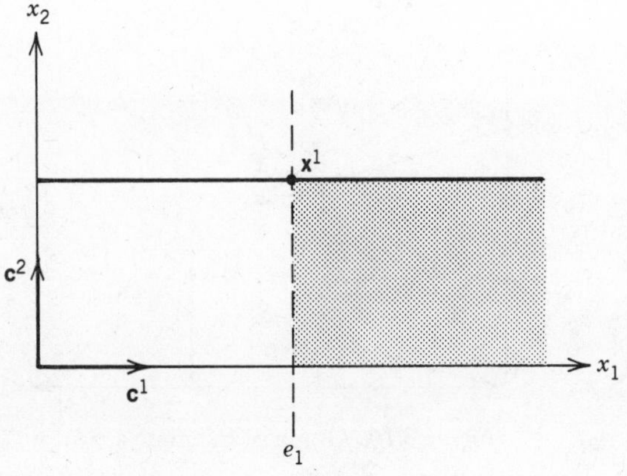

Figure 8.12. Graph of Example 6.

8.5.3 Sensitivity Analysis

Let $\bar{\mathbf{x}}$ be an efficient point and $\bar{\mathbf{z}}$ its nondominated criterion vector. Then, using an e-constraint program, we can compute the *nondominated partial* tradeoff rate

$$\left.\frac{\partial z_i}{\partial z_j}\right|_{\bar{\mathbf{x}}}$$

between the ith and jth objectives at $\bar{\mathbf{x}}$ holding all other objectives fixed at their values at $\bar{\mathbf{x}}$. Such an (i, j) nondominated partial tradeoff rate is given by the value of the dual variable associated with the jth objective constraint in the LP solution of

$$\begin{aligned} &\max \{ \mathbf{c}^i\mathbf{x} = z_i \} \\ \text{s.t.} \quad &\mathbf{c}^j\mathbf{x} \geqq \bar{z}_j \quad j \neq i \\ &\mathbf{x} \in S \end{aligned}$$

The partial tradeoff rate is "nondominated" in the sense that after a perturbation, the resulting criterion vector is still in the nondominated set. Thus, the e-constraint format is useful for generating locally relevant nondominated tradeoff information.

8.6 NEAR OPTIMALITY ANALYSIS

Another reduced feasible region approach is that of *near optimality analysis*. With reference to a weighted-sums LP, the near optimality analysis of the (efficient) point generated is performed as follows.

1. Solve the weighted-sums LP

$$\begin{aligned} &\max \{\bar{\boldsymbol{\lambda}}^T\mathbf{C}\mathbf{x} = z\} \\ \text{s.t.} \quad &\mathbf{x} \in S \end{aligned}$$

as usual to determine the maximizing "figure-of-merit" value z^*.

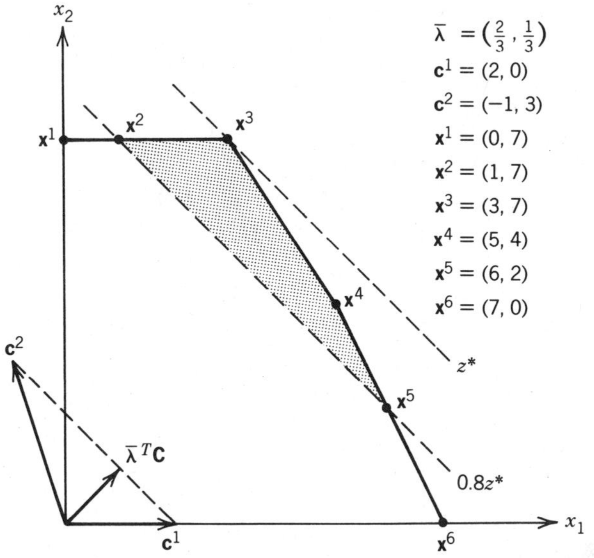

Figure 8.13. Graph of Example 7.

2. Selecting a $\bar{z}$ less than z^*, solve the null objective function/reduced feasible region LP

$$\max \{\mathbf{0}^T\mathbf{x}\}$$
$$\text{s.t.} \quad \mathbf{x} \in S$$
$$\bar{\lambda}^T\mathbf{C}\mathbf{x} \geqq \bar{z}$$

for all maximizing extreme points (i.e., all extreme points of the reduced feasible region created by the additional constraint $\bar{\lambda}^T\mathbf{C}\mathbf{x} \geqq \bar{z}$).
3. Compute the criterion vector $\mathbf{z}^i \in R^k$ associated with each extreme point $\mathbf{x}^i$ of the reduced feasible region.
4. Select the point $\mathbf{x}^i$ associated with the most preferred criterion vector as the final solution of the MOLP.

Although standardly available LP software can solve the weighted-sums LP of 1, special software that contains "all alternative optima" capabilities is required to compute all extreme points of the reduced feasible region in 2.

Example 7. Assume that solving the weighted-sums LP in Fig. 8.13 results in $\mathbf{x}^3$. If we relax the maximizing figure of merit z^* to $.8z^*$, the reduced feasible region is solved for all its extreme points $\mathbf{x}^2$, $\mathbf{x}^3$, $\mathbf{x}^4$, and $\mathbf{x}^5$. We note that all of these points are efficient since

$$E = \gamma(\mathbf{x}^1, \mathbf{x}^3) \cup \gamma(\mathbf{x}^3, \mathbf{x}^4) \cup \gamma(\mathbf{x}^4, \mathbf{x}^6)$$

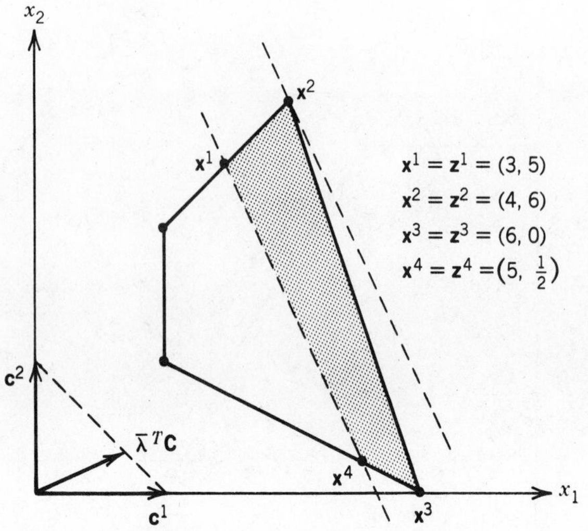

Figure 8.14. Graph of Example 8.

Then, we compute the criterion vectors

$$\mathbf{z}^2 = \begin{bmatrix} 2 \\ 20 \end{bmatrix} \quad \mathbf{z}^3 = \begin{bmatrix} 6 \\ 18 \end{bmatrix} \quad \mathbf{z}^4 = \begin{bmatrix} 10 \\ 7 \end{bmatrix} \quad \mathbf{z}^5 = \begin{bmatrix} 12 \\ 0 \end{bmatrix}$$

Selecting the most preferred criterion vector, say $\mathbf{z}^4$, results in $\mathbf{x}^4$ being identified as the final solution.

In Example 7, all extreme points of the reduced feasible region are efficient, but this is not always the case, as shown in Example 8.

Example 8. Consider Fig. 8.14. Although $\mathbf{x}^1$, $\mathbf{x}^2$, $\mathbf{x}^3$, and $\mathbf{x}^4$ are the extreme points of the reduced feasible region, only $\mathbf{x}^2$ and $\mathbf{x}^3$ are efficient since $E = \gamma(\mathbf{x}^2, \mathbf{x}^3)$.

As seen in Fig. 8.14, pairwise comparisons among the nondominated criterion vectors may not suffice for eliminating all inefficient extreme points. Since $\mathbf{z}^2$ dominates $\mathbf{z}^1$, $\mathbf{x}^1$ can be eliminated as inefficient. However, it is not possible to eliminate $\mathbf{x}^4$ by this method because $\mathbf{z}^4$ is not dominated by any of the other extreme point criterion vectors. Thus, inefficient points may be included among the points presented to the decision maker unless a more sophisticated method is employed.

Readings

1. Chankong, V. and Y. Y. Haimes (1983). *Multiobjective Decision Making: Theory and Methodology* (Sections 4.3, 6.3, and 8.2), New York: North-Holland.

2. Goicoechea, A., D. R. Hansen, and L. Duckstein (1982). *Multiobjective Decision Analysis with Engineering and Business Applications* (Sections 3.5 and 4.4), New York: John Wiley & Sons.
3. Lin, J. G. (1976). "Proper Equality Constraints and Maximization of Index Vectors," *Journal of Optimization Theory and Applications*, Vol. 20, No. 4, pp. 215–244.
4. Lin, J. G. (1977). "Proper Inequality Constraints and Maximization of Index Vectors," *Journal of Optimization Theory and Applications*, Vol. 21, No. 4, pp. 505–521.
5. Musselman, K. and J. Talavage (1980). "A Tradeoff Cut Approach to Multiple Objective Optimization," *Operations Research*, Vol. 28, No. 6, pp. 1424–1435.
6. van de Panne, C. (1971). *Linear Programming and Related Techniques*, Amsterdam: North-Holland.

PROBLEM EXERCISES

8-1 Consider the MOLP:

$$\begin{aligned} \max\{\quad x_1 \qquad &= z_1\} \\ \max\{-5x_1 + 5x_2 &= z_2\} \\ \text{s.t.} \qquad\quad \mathbf{x} &\in S \end{aligned}$$

(a) Specify the direction of the objective function gradient of the weighted-sums LP when $\lambda = (\frac{1}{2}, \frac{1}{2})$.

(b) After normalizing each objective according to the L_1-norm, specify the direction of the objective function gradient of the weighted-sums LP when $\lambda = (\frac{1}{2}, \frac{1}{2})$.

(c) Follow the same instructions as in (b), but after the objectives have been normalized according to the L_∞-norm.

8-2 Consider the MOLP:

$$\begin{aligned} \max\{2x_1 \qquad\qquad &= z_1\} \\ \max\{\ x_1 + 100x_2 &= z_2\} \\ \text{s.t.} \quad x_1 \qquad\qquad &\leqq 3 \\ x_2 &\leqq 3 \\ x_1 + \qquad x_2 &\leqq 4 \\ x_1, x_2 &\geqq 0 \end{aligned}$$

(a) With criterion weights of

$$\lambda_1 = .75$$
$$\lambda_2 = .25$$

what point is generated by the weighted-sums approach?

(b) Specify the subset of Λ pertaining to the solution point of (a).

(c) Normalize each objective according to the L_∞-norm. If the same criterion weights

$$\lambda_1 = .75$$
$$\lambda_2 = .25$$

are used, what point is generated by the weighted-sums approach?

(d) With respect to the "normalized" MOLP, specify the subset of Λ pertaining to the solution point of (c).

8-3 Rescale the objectives of

$$\begin{aligned} \max\ &\{300x_1 = z_1\} \\ \max\ &\{-2x_1 + 3x_2 = z_2\} \\ \text{s.t.}\ & -x_1 + 3x_2 \leqq 9 \\ & x_1 \leqq 3 \\ & x_1, x_2 \geqq 0 \end{aligned}$$

(a) Using the L_∞-norm.

(b) Using range equalization factors.

(c) Using range equalization factors with constant terms to center the equalized ranges at 10.

8-4 Consider the MOLP:

$$\begin{aligned} \max\ &\{-x_1 + x_2 = z_1\} \\ \max\ &\{x_2 = z_2\} \\ \max\ &\{x_1 + x_2 = z_3\} \\ \text{s.t.}\ & x_1 + x_2 \leqq 6 \\ & -x_1 + 2x_2 \leqq 6 \\ & x_1, x_2 \geqq 0 \end{aligned}$$

(a) Is there a point in S whose criterion vector is $\bar{\mathbf{z}} = (-1, 2, 3)$?

(b) Determine the set of points in S whose criterion vectors dominate $\bar{\mathbf{z}} = (-1, 2, 3)$.

8-5C Consider the MOLP:

	x_1	x_2	x_3	x_4		
Objs		5	3	5	max	
	4	1		2	max	
	−1		3		max	
s.t.	4		2		≦	5
		−1	3		≦	7
	−1	2	5	3	≦	8
	1				≦	7

(a) Compute all vertices of the reduced feasible region defined by

$$z_1 \geqq 7$$
$$z_2 \geqq 3$$
$$z_3 \geqq 2$$

(b) How many of the points generated in (a) are extreme points of S?

(c) How many of the vertices from (a) have their criterion vectors dominated by the criterion vectors of other vertices from (a)?

(d) How many vertices of the reduced feasible region are efficient?

8-6C Consider the MOLP:

	x_1	x_2	x_3	x_4	x_5		
Objs		−3	1	3	3	max	
	−3	−1	2	−2	−1	max	
	3	4	1	1	1	max	
s.t.	2		5	−2	8	$\leqq$	54
		1		2		$\leqq$	53
	5		4	4		$\leqq$	62
	4	6		4	2	$\leqq$	63

By following a domination set strategy, determine if

(a) $\mathbf{x}^a = (8, 3, 5, 0, 1)$ is efficient.

(b) $\mathbf{x}^b = (0, 5.25, 10.8, 0, 0)$ is efficient.

8-7C Consider the MOLP:

	x_1	x_2	x_3	x_4	x_5	x_6		
Objs				5	−5	−5	max	
	−4	−2		4	−3	1	max	
	6	2	6	6	4	2	max	
		−4	6	3	−5	2	max	
s.t.	4	6	4	−1	4		$\leqq$	9
		−1	6	4	4		$\leqq$	5
		4	3	6	2	6	$\leqq$	5
	5	2	1		−1		$\leqq$	5
		4	5		2	6	$\leqq$	6
			1	4	2		$\leqq$	9
	5		3	5			$\leqq$	5
	1		2			−1	$\leqq$	7

What are the nondominated partial tradeoffs:

(a) $\dfrac{\partial z_2}{\partial z_3}$ at $(\frac{1}{6}, 0, 0, \frac{5}{6}, 0, 0)$

(b) $\dfrac{\partial z_2}{\partial z_4}$ at $(\frac{1}{8}, 0, \frac{5}{12}, \frac{5}{8}, 0, 0)$

(c) $\dfrac{\partial z_4}{\partial z_1}$ at $(\frac{3}{8}, 0, 0, \frac{5}{8}, \frac{5}{8}, 0)$

(d) $\dfrac{\partial z_4}{\partial z_2}$ at $(\frac{3}{8}, 0, 0, \frac{5}{8}, \frac{5}{8}, 0)$

8-8C Consider the MOLP:

	x_1	x_2	x_3	x_4	x_5	x_6	x_7		
	4	4	2	−1	4	3	3	max	
Objs			4	4	3	3	3	max	
	−1	4	2		4	3	−1	max	
	3	2		4	−1	2		max	
	1			1	2			≤	7.0
							5	≤	6.2
		5				4		≤	9.6
s.t.				4		5		≤	6.0
	−1							≤	5.6
	5		3					≤	9.2

(a) With respect to $\lambda = (.1, .3, .4, .2)$, compute all the vertices of the $.95z^*$ near optimality region.

(b) Of the vertices generated in (a), what are the utility values of the vertices of least utility and of greatest utility when

$$U = \left(\frac{z_1}{10}\right)^{0.5}\left(\frac{z_2}{10}\right)^{1.5}\left(\frac{z_3}{10}\right)^{2}\left(\frac{z_4+1}{10}\right)^{1.5}$$

CHAPTER 9

Vector-Maximum Algorithms

A *vector-maximum* problem attempts to "maximize" a vector. Consider the MOLP

$$\max\left\{\mathbf{c}^1\mathbf{x} = z_1\right\}$$
$$\max\left\{\mathbf{c}^2\mathbf{x} = z_2\right\}$$
$$\vdots$$
$$\max\left\{\mathbf{c}^k\mathbf{x} = z_k\right\}$$
$$\text{s.t. } \mathbf{x} \in S = \{\mathbf{x} \in R^n \mid \mathbf{A}\mathbf{x} = \mathbf{b}, \quad \mathbf{x} \geqq \mathbf{0}, \quad \mathbf{b} \in R^m\}$$

where $\mathbf{A}$ is of full row rank. In a more compact format, an MOLP is sometimes written

$$\text{"max"}\ \{\mathbf{C}\mathbf{x} = \mathbf{z} \mid \mathbf{x} \in S\}$$

MOLPs are vector-maximum problems in the sense that their purpose is to find the point or points in S that "maximize" the criterion vector

$$\mathbf{z} = \begin{bmatrix} z_1 \\ z_2 \\ \vdots \\ z_k \end{bmatrix}$$

Unfortunately, the idea of "maximization," when applied to a vector, is insufficiently robust. There would be no problem if in each MOLP a point exists in S that simultaneously maximizes all objectives. Since this is rarely the case, the more generalized solution concept of *efficiency* is adopted.

Definition 9.1. $\bar{\mathbf{x}} \in S$ is *efficient* iff there does not exist another $\mathbf{x} \in S$ such that $\mathbf{C}\mathbf{x} \geq \mathbf{C}\bar{\mathbf{x}}$, $\mathbf{C}\mathbf{x} \neq \mathbf{C}\bar{\mathbf{x}}$.

That is, $\bar{\mathbf{x}} \in S$ is efficient if and only if there does not exist another $\mathbf{x} \in S$ such that $z_i(\mathbf{x}) \geqq z_i(\bar{\mathbf{x}})$ for all i and $z_i(\mathbf{x}) > z_i(\bar{\mathbf{x}})$ for at least one i. Thus, when we discuss the vector-maximum solution of an MOLP, we are talking about the specification or characterization of the efficient set. In this regard, it is often convenient to write an MOLP in vector-maximum format

$$\text{eff}\,\{\mathbf{Cx} = \mathbf{z} \mid \mathbf{x} \in S\}$$

where *eff* signifies that the set of all solutions to the problem is the efficient set.

There are several types of algorithms for characterizing the efficient set of a linear vector-maximum problem. There are vector-maximum algorithms that compute all efficient extreme points, algorithms that compute all efficient extreme points and unbounded efficient edges, and algorithms that compute all maximally efficient facets.

9.1 FOUNDATIONS

To provide a foundation for the study of vector-maximum algorithms, we have the following.

Theorem 9.2. (Tucker's Theorem of the Alternative.) Let **L**, **M**, and **N** be given $p^2 \times n$, $p^3 \times n$, and $p^4 \times n$ matrices with **L** nonvacuous. Then, either

$$\text{(i)} \quad \mathbf{Lx} \geqq \mathbf{0}, \quad \mathbf{Lx} \neq \mathbf{0}, \quad \mathbf{Mx} \geqq \mathbf{0}, \quad \mathbf{Nx} = \mathbf{0}$$

has a solution $\mathbf{x} \in R^n$

or

$$\text{(ii)} \quad \mathbf{L}^T\mathbf{y}^2 + \mathbf{M}^T\mathbf{y}^3 + \mathbf{N}^T\mathbf{y}^4 = \mathbf{0}$$

$$\mathbf{y}^2 > \mathbf{0} \quad \mathbf{y}^3 \geqq \mathbf{0}$$

has a solution $\mathbf{y}^2 \in R^{p^2}$, $\mathbf{y}^3 \in R^{p^3}$, and $\mathbf{y}^4 \in R^{p^4}$

but never both.

Theorem 9.3. (Motzkin's Theorem of the Alternative.) Let **K**, **M**, and **N** be given $p^1 \times n$, $p^3 \times n$, and $p^4 \times n$ matrices with **K** nonvacuous. Then, either

$$\text{(i)} \quad \mathbf{Kx} > 0, \quad \mathbf{Mx} \geqq \mathbf{0}, \quad \mathbf{Nx} = \mathbf{0}$$

has a solution $\mathbf{x} \in R^n$

or

$$\text{(ii)} \quad \mathbf{K}^T\mathbf{y}^1 + \mathbf{M}^T\mathbf{y}^3 + \mathbf{N}^T\mathbf{y}^4 = \mathbf{0}$$

$$\mathbf{y}^1 \geqq \mathbf{0} \quad \mathbf{y}^1 \neq \mathbf{0} \quad \mathbf{y}^3 \geqq \mathbf{0}$$

has a solution $\mathbf{y}^1 \in R^{p^1}$, $\mathbf{y}^3 \in R^{p^3}$, and $\mathbf{y}^4 \in R^{p^4}$

but never both.

For a reference about the above two theorems, see Mangasarian (1969, Chapter 2).

Theorem 9.4. Let $\bar{\mathbf{x}} \in S$ and $\mathbf{D}$ be an $n \times n$ diagonal matrix with

$$d_{jj} = \begin{cases} 1 & \text{if } \bar{x}_j = 0 \\ 0 & \text{otherwise} \end{cases}$$

Then, $\bar{\mathbf{x}} \in E \Leftrightarrow$ the system

$$\mathbf{Cu} \geqq \mathbf{0}, \quad \mathbf{Cu} \neq \mathbf{0}, \quad \mathbf{Du} \geqq \mathbf{0}, \quad \mathbf{Au} = \mathbf{0}$$

has no solution $\mathbf{u} \in R^n$.

Proof. $\Rightarrow$ Suppose $\mathbf{u}$ satisfies the system. Let $\hat{\mathbf{x}} = \bar{\mathbf{x}} + \alpha\mathbf{u}$. Then, there exists an $\bar{\alpha} > 0$ such that for all $\alpha \in [0, \bar{\alpha}]$, $\hat{\mathbf{x}} \in S$. But $\mathbf{C}\hat{\mathbf{x}} - \mathbf{C}\bar{\mathbf{x}} = \alpha\mathbf{Cu} \geqq \mathbf{0}$ and $\alpha\mathbf{Cu} \neq \mathbf{0}$. This implies that $\bar{\mathbf{x}}$ is not efficient.

$\Leftarrow$ Suppose the system is inconsistent and let $\mathbf{x} \in S$. Let $\mathbf{u} = \mathbf{x} - \bar{\mathbf{x}}$. Consequently, $\mathbf{Au} = \mathbf{0}$ and $\mathbf{Du} \geqq \mathbf{0}$. Hence, it is not true that $\mathbf{Cu} \geqq \mathbf{0}$, $\mathbf{Cu} \neq \mathbf{0}$. Thus, it is not true that $\mathbf{Cx} \geqq \mathbf{C}\bar{\mathbf{x}}$, $\mathbf{Cx} \neq \mathbf{C}\bar{\mathbf{x}}$. Since $\mathbf{x} \in S$ was arbitrary, $\bar{\mathbf{x}}$ is efficient. ◀

Theorem 9.5. Let $\bar{\mathbf{x}} \in S$ and $\mathbf{D}$ be defined as above. Then, $\bar{\mathbf{x}}$ is efficient $\Leftrightarrow$ there exist $\boldsymbol{\pi} \in R^k$, $\mathbf{y}^3 \in R^n$, and $\mathbf{y}^4 \in R^m$ such that

$$\mathbf{C}^T\boldsymbol{\pi} + \mathbf{D}^T\mathbf{y}^3 + \mathbf{A}^T\mathbf{y}^4 = \mathbf{0}$$

$$\boldsymbol{\pi} > \mathbf{0} \quad \mathbf{y}^3 \geqq \mathbf{0}$$

Proof. Tucker's Theorem of the Alternative and Theorem 9.4. ◀

Theorem 9.6. $\bar{\mathbf{x}} \in S$ is efficient $\Leftrightarrow$ there exists a

$$\boldsymbol{\lambda} \in \Lambda = \left\{ \boldsymbol{\lambda} \in R^k \,|\, \lambda_i > 0, \quad \sum_{i=1}^{k} \lambda_i = 1 \right\}$$

such that $\bar{\mathbf{x}}$ maximizes the *weighted-sums* (composite) LP

$$\max \{ \boldsymbol{\lambda}^T\mathbf{Cx} \,|\, \mathbf{x} \in S \}$$

Proof. $\Rightarrow$ Let $\bar{\mathbf{x}}$ be an arbitrary efficient point. Since $\bar{\mathbf{x}}$ is efficient, by Theorem 9.5 there exist $\boldsymbol{\pi} \in R^k$, $\mathbf{y}^3 \in R^n$, and $\mathbf{y}^4 \in R^m$ such that the system

$$\mathbf{C}^T\boldsymbol{\pi} + \mathbf{D}^T\mathbf{y}^3 + \mathbf{A}^T\mathbf{y}^4 = \mathbf{0}$$

$$\boldsymbol{\pi} > 0 \quad \mathbf{y}^3 \geqq \mathbf{0}$$

is consistent. Allowing

$$\alpha = \sum_{i=1}^{k} \pi_i > 0$$

we rewrite the above system as

$$(\mathbf{C}^T\boldsymbol{\lambda})\alpha + \mathbf{D}^T\mathbf{y}^3 + \mathbf{A}^T\mathbf{y}^4 = \mathbf{0}$$

$$\boldsymbol{\lambda} \in \Lambda \quad \mathbf{y}^3 \geqq \mathbf{0}$$

But by Tucker's Theorem of the Alternative, the system

$$(\boldsymbol{\lambda}^T\mathbf{C})\mathbf{u} \geqq 0$$

$$(\boldsymbol{\lambda}^T\mathbf{C})\mathbf{u} \neq 0$$

$$\mathbf{Du} \geqq \mathbf{0}$$

$$\mathbf{Au} = \mathbf{0}$$

has no solution. Now for any $\mathbf{x} \in S$, $\mathbf{D}(\mathbf{x} - \bar{\mathbf{x}}) \geqq \mathbf{0}$ and $\mathbf{A}(\mathbf{x} - \bar{\mathbf{x}}) = \mathbf{0}$. Hence, it is not true that $\boldsymbol{\lambda}^T\mathbf{C}(\mathbf{x} - \bar{\mathbf{x}}) \geqq 0$ and $\boldsymbol{\lambda}^T\mathbf{C}(\mathbf{x} - \bar{\mathbf{x}}) \neq 0$. This implies that

$$\boldsymbol{\lambda}^T\mathbf{Cx} \leqq \boldsymbol{\lambda}^T\mathbf{C}\bar{\mathbf{x}}$$

since $\boldsymbol{\lambda}^T\mathbf{Cu}$ is a scalar.

$\Leftarrow$Assume $\bar{\mathbf{x}} \in S$ is not efficient. Then there exists an $\mathbf{x} \in S$ such that $\mathbf{Cx} \geqq \mathbf{C}\bar{\mathbf{x}}$, $\mathbf{Cx} \neq \mathbf{C}\bar{\mathbf{x}}$. Since $\boldsymbol{\lambda} > \mathbf{0}$, this implies that $\boldsymbol{\lambda}^T\mathbf{Cx} > \boldsymbol{\lambda}^T\mathbf{C}\bar{\mathbf{x}}$ which contradicts the maximality of $\bar{\mathbf{x}}$. ◀

9.2 VECTOR-MAXIMUM THEORY

We now provide theory for the construction of a vector-maximum algorithm.

Theorem 9.7. If S has an efficient point, then at least one extreme point of S is efficient.

Proof. Follows from Theorem 9.6 and the fact that if a linear program has an optimal solution, it has an optimal extreme point. ◀

Definition 9.8. Let $\mathbf{C}_B$ denote the basic columns of criterion matrix $\mathbf{C}$, $\mathbf{C}_N$ the nonbasic columns, and $\mathbf{N}$ the nonbasic columns of the constraint matrix $\mathbf{A}$. Then, let $\mathbf{W}$ denote the $k \times (n - m)$ *reduced cost* matrix where $\mathbf{W} = \mathbf{C}_N - \mathbf{C}_B\mathbf{B}^{-1}\mathbf{N}$.

Definition 9.9. $\mathbf{B}$ is an *efficient basis* iff $\mathbf{B}$ is an optimal basis of the weighted-sums LP for some $\boldsymbol{\lambda} \in \Lambda$.

Since the reduced cost row of the weighted-sums LP is given by $\boldsymbol{\lambda}^T\mathbf{W}$, basis $\mathbf{B}$ is efficient if and only if the system

$$\boldsymbol{\lambda}^T\mathbf{W} \leqq \mathbf{0}$$

$$\boldsymbol{\lambda} > \mathbf{0}$$

is consistent.

Theorem 9.10. Let $\mathbf{x} \in S$ be the extreme point associated with efficient basis $\mathbf{B}$. Then, $\mathbf{x}$ is efficient.

Proof. Since there exists a $\boldsymbol{\lambda} \in \Lambda$ for which $\mathbf{B}$ is an optimal basis in the weighted-sums LP, by Theorem 9.6, $\mathbf{x}$ is efficient. ◀

Theorem 9.11. Let $\mathbf{x} \in S$ be an efficient extreme point. Then, there exists an efficient basis $\mathbf{B}$ associated with $\mathbf{x}$.

Proof. By Theorem 9.6, there exists a $\boldsymbol{\lambda} \in \Lambda$ that causes $\mathbf{x}$ to maximize the weighted-sums LP. Then, since $\mathbf{x}$ is an extreme point, there exists an optimal basis $\mathbf{B}$ associated with $\mathbf{x}$. ◀

Therefore, by finding all efficient bases, from Theorems 9.10 and 9.11, we find all efficient extreme points.

Definition 9.12. Bases $\overline{\mathbf{B}}$ and $\hat{\mathbf{B}}$ are *adjacent* iff one can be obtained from the other in one pivot.

Definition 9.13. Let $\mathbf{B}$ be an efficient basis. Then, x_j is an *efficient nonbasic variable* w.r.t. $\mathbf{B}$ iff there exists a $\boldsymbol{\lambda} \in \Lambda$ such that

$$\boldsymbol{\lambda}^T \mathbf{W} \leqq \mathbf{0}$$

$$\boldsymbol{\lambda}^T \mathbf{w}^j = 0$$

where $\mathbf{w}^j$ is the jth column of $\mathbf{W}$.

Definition 9.14. Let $\mathbf{B}$ be an efficient basis and x_j an efficient nonbasic entering variable. Then, any feasible pivot from $\mathbf{B}$ (including any with a negative pivot element whose associated basic variable is degenerate) is an *efficient pivot* w.r.t. $\mathbf{B}$ and x_j.

Theorem 9.15. Let $\mathbf{B}$ be an efficient basis. Then, any efficient pivot from $\mathbf{B}$ yields an adjacent efficient basis $\hat{\mathbf{B}}$.

Proof. Let x_j be the entering variable associated with an efficient pivot from $\mathbf{B}$. Then, there exists a $\boldsymbol{\lambda} \in \Lambda$ such that

$$\boldsymbol{\lambda}^T \mathbf{W}_B \leqq \mathbf{0}$$

$$\boldsymbol{\lambda}^T \mathbf{w}_B^j = 0$$

Since the reduced costs do not change when pivoting into the basis a nonbasic variable whose $c_j - z_j$ element is zero, we have

$$\boldsymbol{\lambda}^T \mathbf{W}_{\hat{B}} \leqq \mathbf{0}$$

$$\boldsymbol{\lambda}^T \mathbf{w}_{\hat{B}}^j = 0$$

Thus, $\hat{\mathbf{B}}$ is an adjacent efficient basis. ◀

Theorem 9.16. Let $\bar{\mathbf{B}}$ and $\hat{\mathbf{B}}$ be adjacent efficient bases such that one can be obtained from the other by means of an efficient pivot. Let $\bar{\mathbf{x}}$ and $\hat{\mathbf{x}}$ be the extreme points associated with $\bar{\mathbf{B}}$ and $\hat{\mathbf{B}}$, respectively. Then, the edge $\gamma(\bar{\mathbf{x}}, \hat{\mathbf{x}})$ is efficient.

Proof. Follows from the fact that we pivot into the basis a nonbasic variable whose composite LP reduced cost is zero. ◀

Unbounded efficient edges are detected by the existence of efficient nonbasic variables in whose columns positive pivot elements cannot be found.

Consider an efficient basis **B**. To determine computationally whether a variable nonbasic relative to **B** is efficient, we have the subproblem test of Theorem 9.17 in which **e** is the *sum vector* of ones.

Theorem 9.17. Let x_j be a nonbasic variable w.r.t. efficient basis **B**. Let **W** be the reduced cost matrix associated with **B**. Then, all feasible pivots (even those with negative pivot elements) that can be made from the x_j entering column are efficient pivots ⇔ the *subproblem*

$$\begin{aligned}
&\max \{\mathbf{e}^T\mathbf{v}\} \\
&\text{s.t.} \quad -\mathbf{W}\mathbf{y} + \mathbf{w}^j\delta + \mathbf{I}\mathbf{v} = \mathbf{0} \\
&\qquad\qquad \mathbf{0} \leqq \mathbf{y} \in R^{n-m} \\
&\qquad\qquad 0 \leqq \delta \in R \\
&\qquad\qquad \mathbf{0} \leqq \mathbf{v} \in R^k
\end{aligned}$$

has an optimal objective function value of zero.

Proof. From Definition 9.13, x_j is an efficient nonbasic variable w.r.t. **B** iff

$$\begin{aligned}
&\min \{\mathbf{0}^T\boldsymbol{\lambda}\} \\
&\text{s.t.} \quad \mathbf{W}^T\boldsymbol{\lambda} \leqq \mathbf{0} \\
&\qquad (\mathbf{w}^j)^T\boldsymbol{\lambda} = 0 \\
&\qquad \mathbf{I}\boldsymbol{\lambda} \geqq \mathbf{e} \\
&\qquad \boldsymbol{\lambda} \geqq \mathbf{0}
\end{aligned}$$

has an optimal objective function value of zero. With the first two sets of constraints equivalent to

$$-\mathbf{W}^T\boldsymbol{\lambda} \geqq \mathbf{0}$$

$$(\mathbf{w}^j)^T\boldsymbol{\lambda} \geqq 0$$

we have the dual

$$\max \{\mathbf{e}^T\mathbf{v}\}$$

$$\text{s.t.} \qquad -\mathbf{W}\mathbf{y} + \mathbf{w}^j\delta + \mathbf{I}\mathbf{v} + \mathbf{I}\mathbf{t} = \mathbf{0}$$

$$\mathbf{0} \leqq \mathbf{y} \in R^{n-m}$$

$$0 \leqq \delta \in R$$

$$\mathbf{0} \leqq \mathbf{v}, \mathbf{t} \in R^k$$

The slack vector **t**, however, is not necessary because if there exists a $t_i > 0$, we can increase the value of the objective function by setting $t_i = 0$. ◀

We note that the subproblem is always consistent. Thus, if x_j is an inefficient nonbasic variable, the subproblem has an optimal objective function value that is positive-unbounded. Highlighting the results of Theorem 9.17, we have

(i) x_j is *efficient* iff subproblem is *bounded.*
(ii) x_j is *inefficient* iff subproblem is *unbounded.*

Since the subproblem only has as many constraints as rows of the criterion matrix, the subproblem is usually "small" because k is usually much less than m. In the case of a nondegenerate efficient extreme point, the effect of Theorem 9.17 is to determine which emanating edges are efficient.

Definition 9.18. Let $\bar{\mathbf{B}}$ and $\hat{\mathbf{B}}$ be efficient bases. If one can be obtained from the other by performing only efficient pivots, $\bar{\mathbf{B}}$ and $\hat{\mathbf{B}}$ are said to be *connected.*

Noting that an optimal pivot in a single objective LP is an efficient pivot, we prove Theorem 9.19 using known results from linear and parametric programming.

Theorem 9.19. All efficient bases are connected.

Proof. It suffices to show that two arbitrary efficient bases, $\bar{\mathbf{B}}$ and $\hat{\mathbf{B}}$, are connected. Let $\bar{\boldsymbol{\lambda}}, \hat{\boldsymbol{\lambda}} \in \Lambda$ be weighting vectors for which $\bar{\mathbf{B}}$ and $\hat{\mathbf{B}}$ are respectively optimal in the composite LP. Allowing $\bar{\mathbf{B}}$ to be the starting basis, consider the parametric linear programming problem whose objective function is

$$(\boldsymbol{\lambda}^+)^T\mathbf{C} = \bar{\boldsymbol{\lambda}}^T\mathbf{C} + \Phi(\hat{\boldsymbol{\lambda}}^T\mathbf{C} - \bar{\boldsymbol{\lambda}}^T\mathbf{C})$$

$$\text{where} \quad \Phi \in [0, 1]$$

After performing a sequence of parametric programming pivots, we arrive at basis $\tilde{\mathbf{B}}$ that is optimal for

$$\max \left\{ \hat{\boldsymbol{\lambda}}^T\mathbf{C}\mathbf{x} \,|\, \mathbf{x} \in S \right\}$$

The pivots are efficient because $\boldsymbol{\lambda}^+ \in \Lambda$ for all $\Phi \in [0,1]$ since $\boldsymbol{\lambda}^+ = \Phi\hat{\boldsymbol{\lambda}} +$

$(1 - \Phi)\bar{\lambda}$. If $\tilde{\mathbf{B}} = \hat{\mathbf{B}}$, we are done. If $\tilde{\mathbf{B}} \neq \hat{\mathbf{B}}$, we know, since $\tilde{\mathbf{B}}$ and $\hat{\mathbf{B}}$ are both optimal bases for $\max\{\hat{\lambda}^T\mathbf{C}\mathbf{x} \mid \mathbf{x} \in S\}$, that $\hat{\mathbf{B}}$ can be obtained from $\tilde{\mathbf{B}}$ by performing only optimal (i.e., efficient) pivots. Thus, in an efficient pivot sense, $\bar{\mathbf{B}}$ and $\hat{\mathbf{B}}$ are connected. ◀

Theorem 9.20. Let $\mu(\mathbf{x}, \mathbf{v})$ be an unbounded efficient edge of S. Then, $\mathbf{x}$ is an efficient extreme point and $\mathbf{x}$ has associated with it an efficient basis B.

Proof. Because S is contained in the nonnegative orthant, unbounded edge $\mu(\mathbf{x}, \mathbf{v})$ contains an extreme point and that extreme point is $\mathbf{x}$. Since $\mu(\mathbf{x}, \mathbf{v})$ is efficient, $\mathbf{x}$ is efficient. By Theorem 9.11, there exists an efficient basis $\mathbf{B}$ associated with $\mathbf{x}$. ◀

Theorem 9.21. By applying the subproblem test of Theorem 9.17 and pivoting among all efficient bases, we are able to identify all efficient extreme points and all unbounded efficient edges of S.

Proof. Follows from Theorem 9.19 (all efficient bases are connected) and from Theorem 9.20 (each unbounded efficient edge can be identified from some efficient basis). ◀

Definition 9.22. Two efficient extreme points of S are *edge-connected* if they are connected by means of a path of efficient edges of S.

Theorem 9.23. All efficient extreme points of S are edge-connected.

Proof. Follows from Theorems 9.19 and 9.16. ◀

9.3 A VECTOR-MAXIMUM ALGORITHM

A vector-maximum algorithm for computing all efficient extreme points and all unbounded efficient edges possesses the following three phases.

Phase I: Obtain an extreme point or terminate problem with $S = \emptyset$ (inconsistent) if none exist.

Phase II: Obtain an *initial* efficient basis or terminate problem with $E = \emptyset$ (efficient set is empty) if none exist.

Phase III: Compute all efficient extreme points and all unbounded efficient edges.

Phase I is the ordinary Phase I from single objective linear programming. For finding an initial efficient basis in Phase II, several approaches are discussed in Section 9.6. Obtaining an efficient basis is necessary because Phase III cannot get started without one. In Phase III, we apply the same *bookkeeping system/master lists/crashing* strategy described in Chapter 4. The difference is that the four master lists are defined as follows.

L_B Contains all coded efficient bases.

L_x Contains all coded efficient extreme points of the corresponding entries in L_B.

L_e Contains the coded efficient extreme points from which emanate unbounded efficient edges.

L_μ Contains the coded directions of the unbounded efficient edges emanating from the corresponding entries in L_e.

To illustrate the structure of an algorithm for computing all efficient extreme points and all unbounded efficient edges, we have Fig. 9.1. By "unprocessed" in Block 10, we mean bases in L_B to which we have not yet crashed. The different efficient extreme points are printed out in Block 12 as they are encountered to save storage.

9.4 WEAK EFFICIENCY

In addition to (regular) efficiency, there is the more relaxed notion called *weak-efficiency* (w-efficiency).

Definition 9.24. $\bar{\mathbf{x}} \in S$ is w-efficient iff there does not exist another $\mathbf{x} \in S$ such that $\mathbf{Cx} > \mathbf{C}\bar{\mathbf{x}}$.

In this connection, let

$$E^w = \text{set of all } w\text{-efficient points.}$$

$$E_x^w = \text{set of all } w\text{-efficient extreme points.}$$

$$E_\mu^w = \text{set of all unbounded } w\text{-efficient edges.}$$

and

$$\bar{\Lambda} = \left\{ \boldsymbol{\lambda} \in R^k \mid \lambda_i \geqq 0, \quad \sum_{i=1}^{k} \lambda_i = 1 \right\}$$

We note that $E \subset E^w$. Furthermore, with only minor modifications, the theory and algorithm of Sections 9.2 and 9.3 can be used to compute E_x^w and E_μ^w (in place of E_x and E_μ). With regard to w-efficiency, we have the following results.

Theorem 9.25. $\bar{\mathbf{x}} \in S$ is w-efficient $\Leftrightarrow$ there exists a $\boldsymbol{\lambda} \in \bar{\Lambda}$ such that $\bar{\mathbf{x}}$ maximizes the weighted-sums LP

$$\max \{ \boldsymbol{\lambda}^T \mathbf{Cx} \mid \mathbf{x} \in S \}$$

Proof. Similar to the proof of Theorem 9.6, except that we use Motzkin's Theorem of the Alternative (Theorem 9.3). ◀

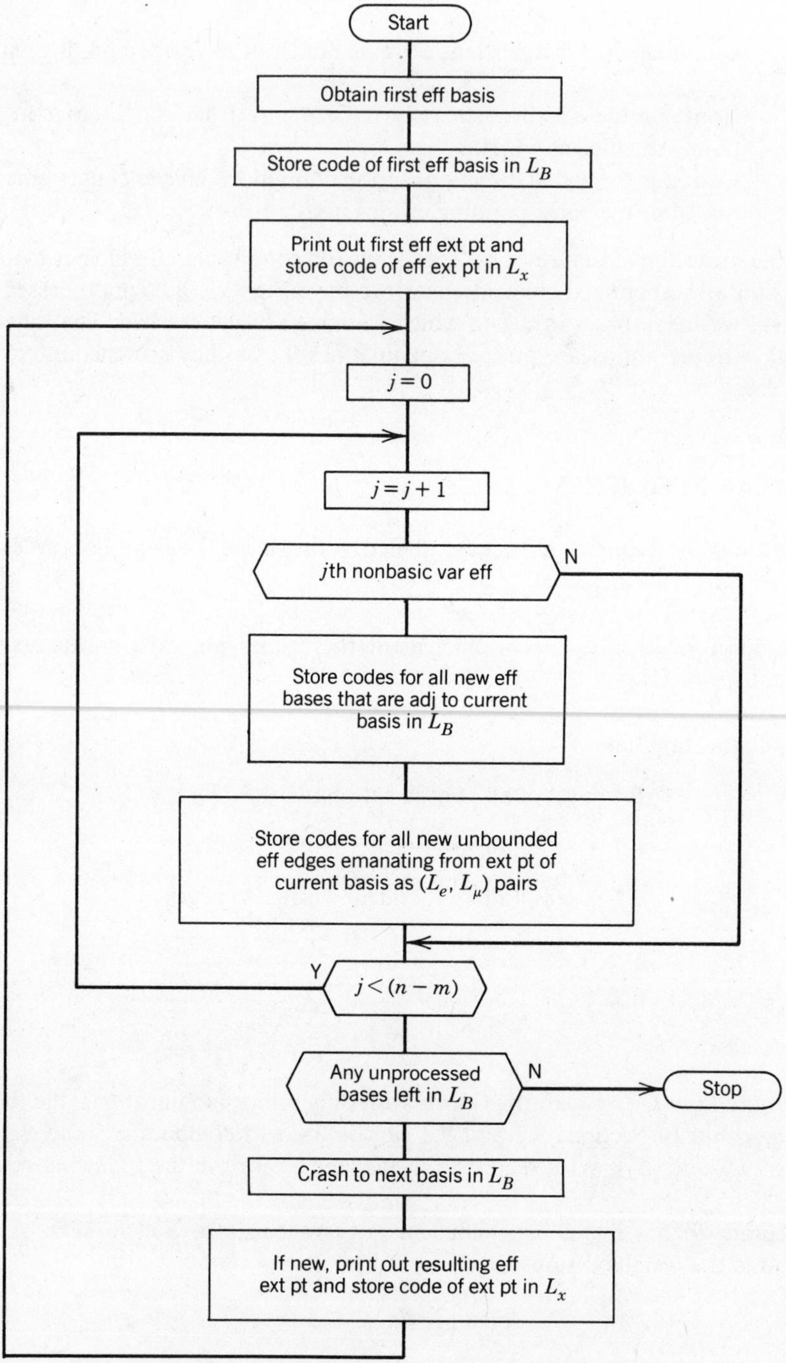

Figure 9.1. Flowchart of a vector-maximum algorithm.

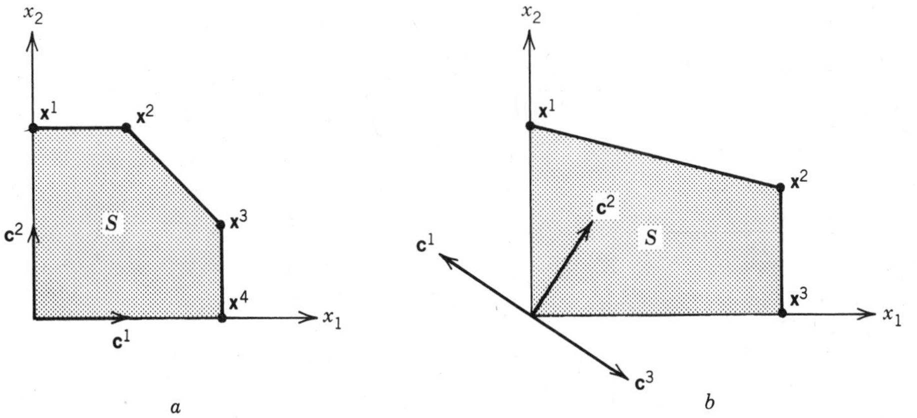

Figure 9.2. Graphs of Example 1.

Since zero weights are allowed in the weighting vectors, all points that maximize a given weighted-sums objective function are w-efficient. In many (probably most) problems, $E = E^w$. Illustrating that E^w can be different from E, we have Example 1.

Example 1. In Fig. 9.2a, $E = \gamma(\mathbf{x}^2, \mathbf{x}^3)$, but $E^w = \gamma(\mathbf{x}^1, \mathbf{x}^2) \cup \gamma(\mathbf{x}^2, \mathbf{x}^3) \cup \gamma(\mathbf{x}^3, \mathbf{x}^4)$. In Fig. 9.2$b$, $E = \gamma(\mathbf{x}^1, \mathbf{x}^2) \cup \gamma(\mathbf{x}^2, \mathbf{x}^3)$ and $E^w = S$.

Definition 9.26. $\mathbf{B}$ is a w-efficient basis iff $\mathbf{B}$ is an optimal basis of the weighted-sums LP for some $\boldsymbol{\lambda} \in \overline{\Lambda}$.

Let $\mathbf{W}$ be the reduced cost matrix associated with $\mathbf{B}$. Then, $\mathbf{B}$ is a w-efficient basis iff the system

$$\begin{aligned} \boldsymbol{\lambda}^T \mathbf{W} &\leqq \mathbf{0} \\ \boldsymbol{\lambda} &\geqq \mathbf{0} \end{aligned}$$

is consistent.

Theorem 9.27. Let x_j be a nonbasic variable w.r.t. w-efficient basis $\mathbf{B}$. Then, all feasible pivots that can be made from the x_j entering column are w-efficient pivots $\Leftrightarrow$ the subproblem

$$\begin{aligned} &\max \{r\} \\ &\text{s.t.} \qquad -\mathbf{W}\mathbf{y} + \mathbf{w}^j \delta + \mathbf{e} r \leqq \mathbf{0} \\ &\qquad\qquad\qquad \mathbf{0} \leqq \mathbf{y} \in R^{n-m} \\ &\qquad\qquad\qquad 0 \leqq \delta, r \in R \end{aligned}$$

has an optimal objective function value of zero.

Proof. Nonbasic variable x_j is w-efficient w.r.t. $\mathbf{B}$ iff

$$\begin{aligned} \min\ &\{\mathbf{0}^T\boldsymbol{\lambda}\} \\ \text{s.t.}\ \ &\mathbf{W}^T\boldsymbol{\lambda} \leqq \mathbf{0} \\ &(\mathbf{w}^j)^T\boldsymbol{\lambda} \geqq 0 \\ &\mathbf{e}^T\boldsymbol{\lambda} = 1 \\ &\boldsymbol{\lambda} \geqq \mathbf{0} \end{aligned}$$

has an optimal objective function value of zero. Thus, x_j is a w-efficient nonbasic variable iff

$$\begin{aligned} &\max\ \{r\} \\ &\text{s.t.} \qquad -\mathbf{W}\mathbf{y} + \mathbf{w}^j\delta + \mathbf{e}r \leqq \mathbf{0} \\ &\qquad\qquad \mathbf{0} \leqq \mathbf{y} \in R^{n-m} \\ &\qquad\qquad 0 \leqq \delta \in R \\ &\qquad\qquad r \in R \text{ unrestricted} \end{aligned}$$

has an optimal objective function value of zero. Because $(\mathbf{y}, \delta, r) = (\mathbf{0}, 0, 0)$ is a feasible solution, the subproblem can be written with $r \geqq 0$ as in the theorem. ◀

Theorem 9.28. By pivoting among all w-efficient bases and applying the subproblem test of Theorem 9.27, we are able to identify all w-efficient extreme points and all unbounded w-efficient edges of S.

Proof. Similar to that of Theorem 9.21. ◀

Other than for the slightly different subproblem, the Phase III algorithm for finding all w-efficient extreme points and all unbounded w-efficient edges is the same as the Phase III algorithm for finding all efficient extreme points and all unbounded efficient edges.

9.5 CLASSIFICATION OF MOLPs

Allowing E to be the efficient set, consider the following mutually exclusive and collectively exhaustive classification of multiple objective linear programs.

1. $S = \emptyset$.
2. $S \neq \emptyset$, $E = \emptyset$, and at least one objective is bounded.
3. $S \neq \emptyset$, $E = \emptyset$, and all objectives are unbounded.
4. $S \neq \emptyset$, E contains only one point.
5. $S \neq \emptyset$, E is bounded and contains an infinite number of points.
6. $S \neq \emptyset$, E is unbounded, and at least one objective is bounded.
7. $S \neq \emptyset$, E is unbounded, and all objectives are unbounded.

Classification 1 is inconsistency, and examples of classifications 2–5 are found in earlier chapters. Classifications 6 and 7 are illustrated by Examples 4 and 5 of this chapter.

9.6 FINDING AN INITIAL EFFICIENT BASIS

To start Phase III, we must be at an efficient basis. We now discuss methods for use in Phase II to find an *initial* efficient extreme point. Should the initial efficient extreme point be characterized by an inefficient basis, a method such as described by Ecker and Hegner (1978) would have to be utilized for moving to an efficient basis.

9.6.1 Subproblem Test for Extreme Point Efficiency

To determine if a given extreme point of S is efficient, we have the subproblem test of Theorem 9.29 that can be formulated from the tableau of the point in question.

Theorem 9.29. Let $\mathbf{x}$ be an extreme point of S, $(\mathbf{B}^{-1}\mathbf{N})_D$ the rows of $(\mathbf{B}^{-1}\mathbf{N})$ associated with degenerate basic variables, and d the number of degrees of degeneracy at $\mathbf{x}$. Then, $\mathbf{x}$ is efficient $\Leftrightarrow$ the subproblem test for *extreme point efficiency*

$$\begin{aligned} &\max\{\mathbf{e}^T\mathbf{v}\} \\ &\text{s.t.} \quad -\mathbf{W}\mathbf{y} + \mathbf{I}\mathbf{v} = \mathbf{0} \\ &(\mathbf{B}^{-1}\mathbf{N})_D\mathbf{y} + \mathbf{I}\mathbf{s} = \mathbf{0} \\ &\mathbf{0} \leqq \mathbf{y} \in R^{n-m} \\ &\mathbf{0} \leqq \mathbf{v} \in R^{k} \\ &\mathbf{0} \leqq \mathbf{s} \in R^{d} \end{aligned}$$

has a bounded objective function value of zero.

Proof. If we partition $\mathbf{A}$ and $\mathbf{C}$ into basic and nonbasic parts, by Theorem 9.4 $\mathbf{x}$ is efficient iff the system

$$\begin{aligned} \mathbf{C}_B\mathbf{u}_B + \mathbf{C}_N\mathbf{u}_N &\geqq \mathbf{0} \\ \mathbf{C}_B\mathbf{u}_B + \mathbf{C}_N\mathbf{u}_N &\neq \mathbf{0} \\ \mathbf{D}\mathbf{u} &\geqq \mathbf{0} \\ \mathbf{B}\mathbf{u}_B + \mathbf{N}\mathbf{u}_N &= \mathbf{0} \end{aligned}$$

is inconsistent. If we let $\mathbf{u}_B = -\mathbf{B}^{-1}\mathbf{N}\mathbf{u}_N$ and note that for the degenerate basic variables $\mathbf{u}_D = (-\mathbf{B}^{-1}\mathbf{N})_D\mathbf{u}_N \geqq \mathbf{0}$, $\mathbf{x}$ is efficient iff the system

$$\begin{aligned} \mathbf{C}_B\mathbf{B}^{-1}\mathbf{N}\mathbf{u}_N - \mathbf{C}_N\mathbf{u}_N &\leqq \mathbf{0} \\ \mathbf{C}_B\mathbf{B}^{-1}\mathbf{N}\mathbf{u}_N - \mathbf{C}_N\mathbf{u}_N &\neq \mathbf{0} \\ (\mathbf{B}^{-1}\mathbf{N})_D\mathbf{u}_N &\leqq \mathbf{0} \\ \mathbf{u}_N &\geqq \mathbf{0} \end{aligned}$$

is inconsistent. In other words, $\mathbf{x}$ is efficient iff the system

$$\begin{aligned} -\mathbf{W}\mathbf{u}_N + \mathbf{I}\mathbf{v} &= \mathbf{0} \\ (\mathbf{B}^{-1}\mathbf{N})_D\mathbf{u}_N &\leqq \mathbf{0} \\ \mathbf{u}_N &\geqq \mathbf{0} \end{aligned}$$

does not possess a solution such that $\mathbf{v} \geq \mathbf{0}$, $\mathbf{v} \neq \mathbf{0}$. Since the above system is consistent, we have the subproblem test for extreme point efficiency as stated in the theorem. ◀

Theorem 9.30. Let $\mathbf{x}$ be an extreme point of S. Then, $\mathbf{x}$ is w-efficient $\Leftrightarrow$ the subproblem test for *extreme point w-efficiency*.

$$\begin{aligned} &\max\{r\} \\ &\text{s.t.} \quad -\mathbf{W}\mathbf{y} + \mathbf{e}r \qquad \leqq \mathbf{0} \\ &(\mathbf{B}^{-1}\mathbf{N})_D\mathbf{y} \qquad + \mathbf{I}\mathbf{s} = \mathbf{0} \\ &\qquad \mathbf{0} \leqq \mathbf{y} \in R^{n-m} \\ &\qquad 0 \leqq r \in R \\ &\qquad \mathbf{0} \leqq \mathbf{s} \in R^d \end{aligned}$$

has a bounded objective function value of zero.

Proof. Similar to that of Theorem 9.29. ◀

In Theorems 9.29 and 9.30, if $\mathbf{x}$ is inefficient, the subproblem tests have positive-unbounded objective function values.

9.6.2 Methods for Finding an Efficient Extreme Point

The following six methods are possible approaches for finding an initial efficient extreme point.

1. Weighted-sums.
2. Weighted-sums with subproblem testing.
3. Lexicographic maximization.
4. Lexicographic maximization with subproblem testing.
5. Ecker-Kouada method.
6. Benson's method.

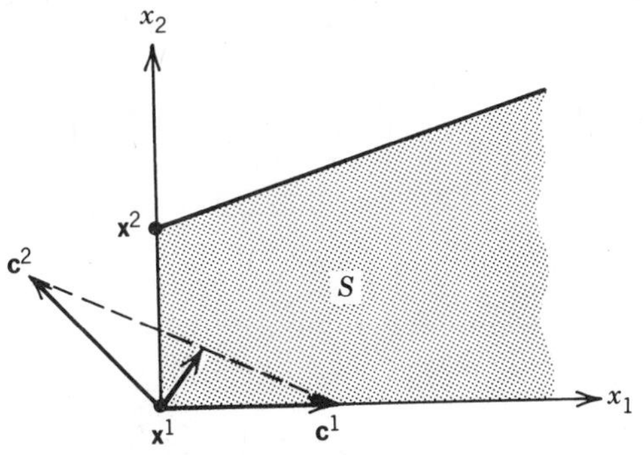

Figure 9.3. Graph of Example 2.

Weighted-Sums. In this method we select a $\boldsymbol{\lambda} \in \Lambda = \{\boldsymbol{\lambda} \in R^k \mid \lambda_i > 0, \sum_{i=1}^k \lambda_i = 1\}$ and solve

$$\max \{ \boldsymbol{\lambda}^T \mathbf{C}\mathbf{x} \mid \mathbf{x} \in S \}$$

If a solution is obtained, by Theorem 9.6 we know that it is efficient. If $S \neq \varnothing$ and bounded, the method is *fail-safe*—it will produce an efficient extreme point. However, as seen in Example 2, the method is not fail-safe when S is unbounded.

Example 2. In Fig. 9.3, E is the unbounded edge emanating from $\mathbf{x}^2$. Suppose that we select a $\boldsymbol{\lambda} \in \Lambda$ such that the composite gradient points in the direction indicated. In maximizing the weighted-sums LP, the pivoting would likely be from $\mathbf{x}^1$ to $\mathbf{x}^2$, at which point an unbounded objective function value would be indicated. Curiously, the weighted-sums method terminates nonproductively, sitting on top of efficient extreme point $\mathbf{x}^2$, but does not realize it.

Weighted-Sums with Subproblem Testing. Since the weighted-sums method may well encounter efficient extreme points before the weighted-sums objective function is maximized, time may be saved by subproblem testing extreme points along the way. Consider Example 3.

Example 3. In Fig. 9.4, extreme points $\mathbf{x}^2, \mathbf{x}^3, \ldots, \mathbf{x}^7$ are efficient. In maximizing the weighted-sums LP, the pivoting would likely proceed through $\mathbf{x}^2$, $\mathbf{x}^3$, $\mathbf{x}^4$, and $\mathbf{x}^5$ (all of which are efficient) before terminating at $\mathbf{x}^6$. However, with subproblem testing, the method would end at $\mathbf{x}^2$.

The advantages of the weighted-sums method with subproblem testing are (a) it may save time, and (b) it decreases the probability of the weighted-sums approach failing when S is unbounded. Because of the time consumed in performing the subproblem tests, the degree to which total time is saved depends a great deal on implementation—that is, when to commence subproblem testing and how often to apply it.

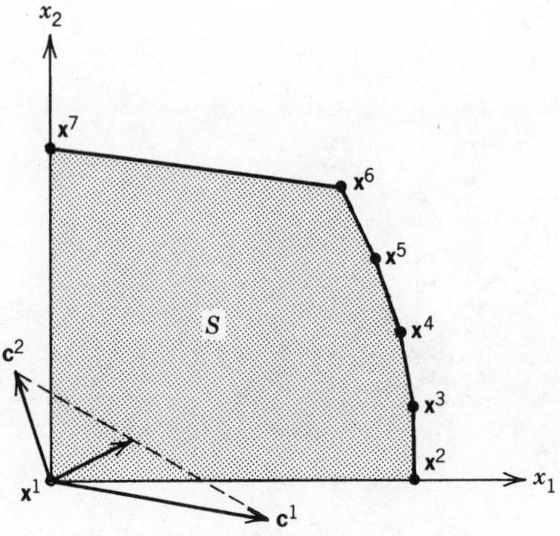

Figure 9.4. Graph of Example 3.

Lexicographic Maximization. The lexicographic maximization process constructs the following *recursively* defined reduced feasible regions:

$$S_0 = S$$

$$S_1 = \left\{\mathbf{y} \in S \,|\, \mathbf{c}^{j_1}\mathbf{y} = \max\left[\mathbf{c}^{j_1}\mathbf{x} : \mathbf{x} \in S_0\right]\right\}$$

$$\vdots$$

$$S_h = \text{reduced feasible region after } h \text{ maximizations}$$

$$\vdots$$

$$S_k = \left\{\mathbf{y} \in S \,|\, \mathbf{c}^{j_k}\mathbf{y} = \max\left[\mathbf{c}^{j_k}\mathbf{x} : \mathbf{x} \in S_{k-1}\right]\right\}$$

The process begins by maximizing an objective function $\mathbf{c}^{j_1}\mathbf{x}$ that is bounded over S_0. By constraining S_0 to the maximal value of the j_1th criterion, S_1 is obtained. Then, a second objective function $\mathbf{c}^{j_2}\mathbf{x}$ that is bounded over S_1 is maximized. By constraining S_1 to the maximal value of the j_2th criterion, S_2 is obtained. The process continues in this fashion until we obtain either $S_k \neq \varnothing$, or for some ℓ $(1 \leqq \ell \leqq k)$, $S_\ell, S_{\ell+1}, \ldots, S_k = \varnothing$. The latter occurs when all remaining objective functions are unbounded over $S_{\ell-1}$. In support of the lexicographic maximization approach, we have the following theory.

Theorem 9.31. Let $E \neq \varnothing$, $\bar{\mathbf{x}}$ be inefficient, $T = \{\mathbf{x} \in S \,|\, \mathbf{Cx} \geq \mathbf{C}\bar{\mathbf{x}}\}$, and E_T the efficient set of eff$\{\mathbf{Cx} = \mathbf{z} \,|\, \mathbf{x} \in T\}$. Then, $E_T \neq \varnothing$ and $E_T \subset E$.

Proof. Assume $E_T = \varnothing$. This implies that max$\{\boldsymbol{\lambda}^T\mathbf{Cx} \,|\, \mathbf{x} \in T\}$ is unbounded for each $\boldsymbol{\lambda} \in \Lambda$, which in turn implies that $E = \varnothing$, a contradiction. Hence, $E_T \neq \varnothing$. Let $\hat{\mathbf{x}} \in E_T$. If $\hat{\mathbf{x}} \notin E$, there exists a $\tilde{\mathbf{x}} \in S$ such that $\mathbf{C}\tilde{\mathbf{x}} \geq \mathbf{C}\hat{\mathbf{x}} \geq \mathbf{C}\bar{\mathbf{x}}$, $\mathbf{C}\tilde{\mathbf{x}} \neq \mathbf{C}\hat{\mathbf{x}}$. Since this contradicts $\hat{\mathbf{x}} \in E_T$, $\hat{\mathbf{x}} \in E$. Since $\hat{\mathbf{x}}$ was arbitrary, $E_T \subset E$. ◀

Theorem 9.32. Let $E \neq \varnothing$. Then, the criterion vector of each inefficient point is dominated by the criterion vector of an efficient point.

Proof. Let $\bar{\mathbf{x}} \notin E$ and $T = \{\mathbf{x} \in S \mid \mathbf{C}\mathbf{x} \geq \mathbf{C}\bar{\mathbf{x}}\}$. Since $E \neq \varnothing$, by Theorem 9.31, $E_T \neq \varnothing$ and $E_T \subset E$. Let $\hat{\mathbf{x}} \in E_T$. Since $\hat{\mathbf{x}} \in T$, $\mathbf{C}\hat{\mathbf{x}} \geqq \mathbf{C}\bar{\mathbf{x}}$. If $\mathbf{C}\hat{\mathbf{x}} = \mathbf{C}\bar{\mathbf{x}}$, then $\boldsymbol{\lambda}^T\mathbf{C}\hat{\mathbf{x}} = \boldsymbol{\lambda}^T\mathbf{C}\bar{\mathbf{x}}$ for all $\boldsymbol{\lambda} \in \Lambda$, which implies that $\hat{\mathbf{x}} \notin E$, a contradiction. Thus, $\mathbf{C}\hat{\mathbf{x}} \neq \mathbf{C}\bar{\mathbf{x}}$ and the theorem is proved. ◀

Theorem 9.33. Extreme points of S_ℓ, $1 \leqq \ell \leqq k$, are extreme points of S.

Proof. From linear programming, we know that the set of extreme points of the optimal set of $\max\{\mathbf{c}^{j_i}\mathbf{x} \mid \mathbf{x} \in S\}$ are extreme points of S. Thus, if $\bar{\mathbf{x}}$ is an extreme point of S_1, $\bar{\mathbf{x}}$ is an extreme point of S. Similarly, if $\bar{\mathbf{x}}$ is an extreme point of S_2, $\bar{\mathbf{x}}$ is an extreme point of S_1, and so forth. In this way, the theorem is proved. ◀

Theorem 9.34. Let $E \neq \varnothing$. If $S_\ell \neq \varnothing$ for some ℓ, $1 \leqq \ell \leqq k$, then S_ℓ contains an efficient extreme point of S.

Proof. Suppose all extreme points of S_ℓ are inefficient. This implies that $\max\{\boldsymbol{\lambda}^T\mathbf{C}\mathbf{x} \mid \mathbf{x} \in S\}$ is unbounded for all $\boldsymbol{\lambda} \in \Lambda$, which in turn implies that $E = \varnothing$, a contradiction. Since all extreme points of S_ℓ are extreme points of S, S_ℓ contains an efficient extreme point of S. ◀

Theorem 9.35. Let $S_k \neq \varnothing$. Then, (i) $S_k \subset E$, and (ii) every extreme point of S_k is an efficient extreme point of S.

Proof. (i) Suppose $\mathbf{x} \in S_k$ is inefficient. Then, there exists an $\bar{\mathbf{x}} \in S$ such that $\mathbf{C}\bar{\mathbf{x}} \geq \mathbf{C}\mathbf{x}$, $\mathbf{C}\bar{\mathbf{x}} \neq \mathbf{C}\mathbf{x}$. However, by the construction of the S_ℓ, $1 \leqq \ell \leqq k$, no such $\bar{\mathbf{x}}$ exists. Thus, $S_k \subset E$.

(ii) Since $S_k \subset E$ and all extreme points of S_k are extreme points of S, every extreme point of S_k is an efficient extreme point of S. ◀

Examples 4 and 5 respectively illustrate the successful and unsuccessful application of the lexicographic maximization process.

Example 4. In Fig. 9.5, $E = \mu(\mathbf{x}^1, \mathbf{v})$ where $\mathbf{v} = (0, 1)$. In this problem, lexicographic maximization locates an efficient extreme point as follows.

- **a.** Tries to maximize $\mathbf{c}^1\mathbf{x}$ over S_0 but cannot.
- **b.** Maximizes $\mathbf{c}^2\mathbf{x}$ over S_0 to form S_1. In this case, $j_1 = 2$. (First maximization stage ends at $\mathbf{x}^1$.)
- **c.** Tries to maximize $\mathbf{c}^1\mathbf{x}$ over S_1, but cannot.
- **d.** Maximizes $\mathbf{c}^3\mathbf{x}$ over S_1 to form $S_2 = \{\mathbf{x}^1\}$. In this case, $j_2 = 3$. (Second maximization stage ends at $\mathbf{x}^1$.)
- **e.** Maximizes $\mathbf{c}^1\mathbf{x}$ over S_2 to form $S_3 = \{\mathbf{x}^1\}$. In this case, $j_3 = 1$. We conclude with $\mathbf{x}^1$ as an efficient extreme point.

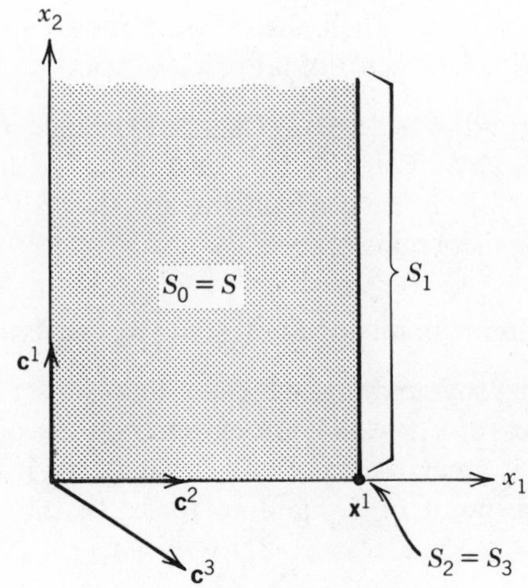

Figure 9.5. Graph of Example 4.

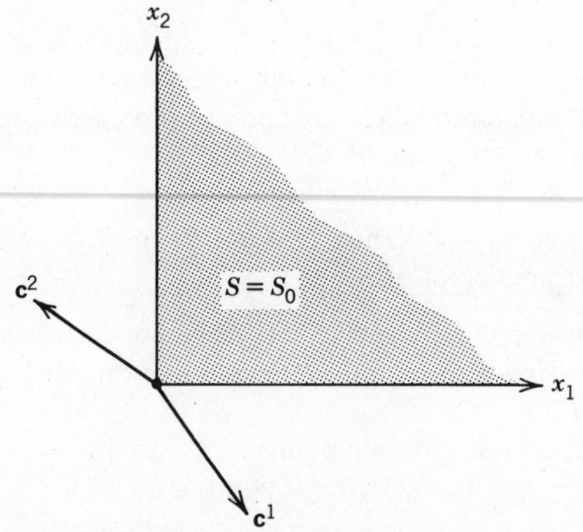

Figure 9.6. Graph of Example 5.

Example 5. In Fig. 9.6, E is the boundary of the nonnegative orthant. In this problem, lexicographic maximization fails because it cannot find an objective bounded over S with which to get started.

With regard to Theorem 9.34, if S is bounded, why use lexicographic maximization? Why not maximize one of the objectives, compute all maximizing extreme points, and perform pairwise comparisons until an efficient extreme point is identified? As seen in Example 6, this may not work.

Example 6. Consider Fig. 9.7*a*, in which $\mathbf{c}^1$ and $\mathbf{c}^3$ are in the $x_1 : x_3$ plane. Figure 9.7*b* portrays Z, the feasible region in criterion space. With regard to the second objective, $\mathbf{x}^1$, $\mathbf{x}^2$, and $\mathbf{x}^3$ are maximizing extreme points, and $\mathbf{z}^1$, $\mathbf{z}^2$, and $\mathbf{z}^3$

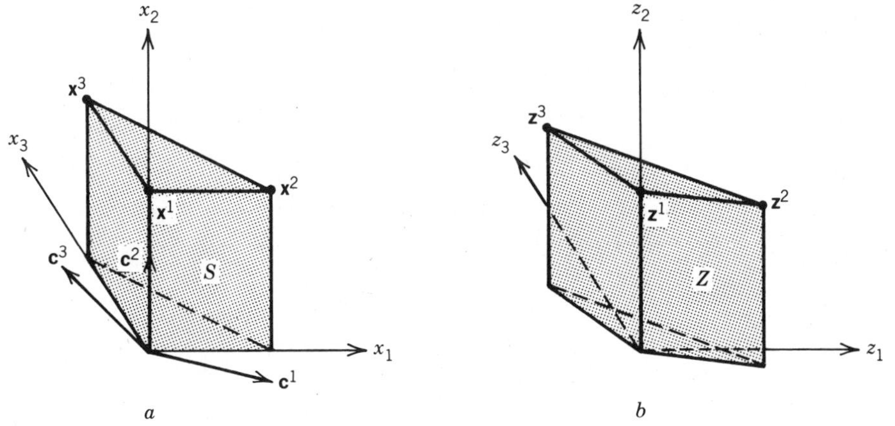

Figure 9.7. Graphs of Example 6.

are their associated criterion vectors. In pairwise comparing $\mathbf{z}^1$, $\mathbf{z}^2$, and $\mathbf{z}^3$, none can be eliminated as dominated. However, $\mathbf{z}^1$ is dominated because $E = \gamma(\mathbf{x}^2, \mathbf{x}^3)$.

In summary, lexicographic maximization is fail-safe for finding an efficient extreme point except when all objectives are unbounded over S.

Lexicographic Maximization with Subproblem Testing. To find an efficient extreme point sooner and enhance the ability of lexicographic maximization when all objectives are unbounded over S, subproblem testing can be employed. One possibility is to subproblem test at the starting extreme point and after each maximization stage. Another possibility is to subproblem test each encountered extreme point. Although operating with good reliability, lexicographic maximization with subproblem testing is still not fail-safe when all objectives are unbounded over S and $E \neq \varnothing$.

Ecker-Kouada Method. From Ecker and Kouada (1975), we have Theorem 9.36.

Theorem 9.36. Let $\mathbf{x}^0 \in S$ and (9.37) be the *E–K program*

$$\begin{aligned} &\max \{\mathbf{e}^T\mathbf{s}\} \\ &\text{s.t.} \quad \mathbf{Cx} = \mathbf{Is} + \mathbf{Cx}^0 \\ &\qquad \mathbf{Ax} = \mathbf{b} \\ &\qquad \mathbf{0} \leqq \mathbf{x} \in R^n \\ &\qquad \mathbf{0} \leqq \mathbf{s} \in R^k \end{aligned} \tag{9.37}$$

Then, (i) if $(\bar{\mathbf{x}}, \bar{\mathbf{s}})$ is an optimal solution of (9.37), $\bar{\mathbf{x}} \in E$, and (ii) if (9.37) has a positive-unbounded objective function value, $E = \varnothing$.

Proof. (i) Suppose $\bar{\mathbf{x}} \notin E$. Then, there exists an $\hat{\mathbf{x}}$ such that $\mathbf{C}\hat{\mathbf{x}} \geq \mathbf{C}\bar{\mathbf{x}}$, $\mathbf{C}\hat{\mathbf{x}} \neq \mathbf{C}\bar{\mathbf{x}}$. With $\mathbf{C}\hat{\mathbf{x}} - \mathbf{Cx}^0 = \mathbf{I}\hat{\mathbf{s}}$ and $\mathbf{C}\bar{\mathbf{x}} - \mathbf{Cx}^0 = \mathbf{I}\bar{\mathbf{s}}$, we have $\mathbf{I}\hat{\mathbf{s}} \geq \mathbf{I}\bar{\mathbf{s}}$, $\mathbf{I}\hat{\mathbf{s}} \neq \mathbf{I}\bar{\mathbf{s}}$. Since $(\hat{\mathbf{x}}, \hat{\mathbf{s}})$ contradicts the optimality of $(\bar{\mathbf{x}}, \bar{\mathbf{s}})$, $\bar{\mathbf{x}} \in E$.

(ii) The dual of (9.37) is

$$\begin{aligned} &\min \{(\mathbf{Cx}^0)^T\mathbf{p} + \mathbf{b}^T\mathbf{y}\} \\ &\text{s.t.} \quad \mathbf{C}^T\mathbf{p} + \mathbf{A}^T\mathbf{y} \geqq \mathbf{0} \\ &\quad -\mathbf{Ip} \geqq \mathbf{e} \\ &\quad \mathbf{p}, \mathbf{y} \text{ unrestricted} \end{aligned}$$

Allowing $\boldsymbol{\pi} = -\mathbf{p}$ and $\mathbf{y}^4 = -\mathbf{y}$, we have

$$\begin{aligned} &\min \left\{ -(\mathbf{Cx}^0)^T\boldsymbol{\pi} - \mathbf{b}^T\mathbf{y}^4 \right\} \\ &\text{s.t.} \quad \mathbf{C}^T\boldsymbol{\pi} + \mathbf{Iy}^3 + \mathbf{A}^T\mathbf{y}^4 = \mathbf{0} \\ &\quad \boldsymbol{\pi} > \mathbf{0} \quad \mathbf{y}^3 \geqq \mathbf{0} \end{aligned} \tag{9.38}$$

Suppose $E \neq \varnothing$ and let $\mathbf{x} \in E$. Then, by Theorem 9.5, there exist $\boldsymbol{\pi} \in R^k$, $\mathbf{y}^3 \in R^n$, and $\mathbf{y}^4 \in R^m$ such that

$$\begin{aligned} \mathbf{C}^T\boldsymbol{\pi} + \mathbf{D}^T\mathbf{y}^3 + \mathbf{A}^T\mathbf{y}^4 = \mathbf{0} \\ \boldsymbol{\pi} > \mathbf{0} \quad \mathbf{y}^3 \geqq \mathbf{0} \end{aligned}$$

Thus, there exists a $\mathbf{y}^3$ in (9.38) such that (9.38) has a feasible solution. Therefore, it is not possible for (9.37) to have a positive-unbounded objective function value. ◀

Thus, the E–K program of (9.37) provides us with a means for finding an efficient point when $E \neq \varnothing$ and tells us that E is empty when no efficient points exist. The unfortunate aspect of the Ecker-Kouada method is that when it generates an efficient point, there is no guarantee that the point will be extreme in S.

Benson's Method. From Benson (1981), we have the following procedure for finding an initial efficient extreme point.

Step 1: Let $S \neq \varnothing$. Find any $\mathbf{x}^0 \in S$.

Step 2: Solve

$$\begin{aligned} &\min \{ -\mathbf{z}^T\mathbf{Cx}^0 + \mathbf{u}^T\mathbf{b}\} \\ &\text{s.t.} \quad \mathbf{z}^T\mathbf{C} - \mathbf{u}^T\mathbf{A} + \mathbf{w}^T = -\mathbf{e}^T\mathbf{C} \\ &\quad \mathbf{w}, \mathbf{z} \geqq \mathbf{0} \end{aligned}$$

If no optimal solutions exist, the MOLP does not have any efficient points. Stop. Otherwise, let $(\mathbf{z}^0, \mathbf{u}^0, \mathbf{w}^0)$ be an optimal solution and go to Step 3.

Step 3: Let $\bar{\boldsymbol{\lambda}} = (\mathbf{z}^0 + \mathbf{e})$. Obtain an efficient extreme point by solving $\max \{\bar{\boldsymbol{\lambda}}^T\mathbf{Cx} \mid \mathbf{x} \in S\}$.

The validity of the procedure is shown in Benson (1981). Although the procedure involves the solution of two LPs, it is a fail-safe method for obtaining an initial efficient extreme point.

9.7 SUBPROBLEM TESTS FOR NONBASIC VARIABLE EFFICIENCY

In Phase III, we locate all efficient extreme points by pursuing each efficient pivot in each efficient nonbasic variable column in each *efficient tableau* (where an efficient tableau is one whose basis matrix **B** is efficient). Four subproblem tests for detecting nonbasic variable efficiency are now reviewed. Note that if the extreme point of an efficient tableau is nondegenerate, the subproblem test for nonbasic variable efficiency is a test for detecting emanating edge efficiency. The four subproblem tests discussed are (1) the Evans-Steuer test, (2) Isermann's test, (3) Ecker's method, and (4) the Zionts-Wallenius routine.

9.7.1 Evans-Steuer Test

Theorem 9.17 and its proof should be reviewed. The Evans-Steuer test is the subproblem test of Theorem 9.17. The test was introduced in Evans and Steuer (1973) and is as follows. To determine the efficiency status of the jth nonbasic variable, we solve the *Evans-Steuer* subproblem

$$\begin{aligned} &\max \{\mathbf{e}^T\mathbf{v}\} \\ &\text{s.t.} \quad -\mathbf{W}\mathbf{y} + \mathbf{w}^j\delta + \mathbf{I}\mathbf{v} = \mathbf{0} \\ &\qquad\qquad\qquad \mathbf{0} \leqq \mathbf{y} \in R^{n-m} \\ &\qquad\qquad\qquad 0 \leqq \delta \in R \\ &\qquad\qquad\qquad \mathbf{0} \leqq \mathbf{v} \in R^k \end{aligned}$$

where **W** is the $c_j - z_j$ reduced cost matrix of an efficient basis and $\mathbf{w}^j$ is the jth column of **W**. If the optimal objective function value is zero, the jth nonbasic variable is efficient. If the optimal objective function value is positive-unbounded, the jth nonbasic variable is inefficient. To determine all efficient nonbasic variables, the Evans-Steuer procedure solves one such problem for each column of **W**.

9.7.2 Isermann's Test

This test was introduced by Isermann (1977). For this test, consider

$$\begin{aligned} &\min \{\mathbf{e}^T\boldsymbol{\lambda}\} \\ &\text{s.t.} \quad -\mathbf{W}^T\boldsymbol{\lambda} \geqq \mathbf{0} \\ &\qquad -(\mathbf{W}^J)^T\boldsymbol{\lambda} = \mathbf{0} \\ &\qquad\qquad \boldsymbol{\lambda} \geqq \mathbf{e} \end{aligned}$$

where $\mathbf{W}^J$ is the matrix of columns of **W** corresponding to the nonbasic indices in an index set J. Clearly, if the above program has an optimal solution, (a) there exists a strictly positive weighting vector that causes the basis pertaining to **W** to be an optimal basis in the corresponding weighted-sums LP, and (b) each nonbasic variable x_j, $j \in J$, is efficient. Taking the dual, we obtain the *Isermann subproblem*

$$\begin{aligned} &\max\{\, \mathbf{e}^T\mathbf{v}\} \\ &\text{s.t.} \quad -\mathbf{W}\mathbf{y} + \mathbf{W}^J\boldsymbol{\delta} + \mathbf{I}\mathbf{v} = \mathbf{e} \\ &\qquad\qquad \mathbf{0} \leqq \mathbf{y} \in R^{n-m} \\ &\qquad\qquad \mathbf{0} \leqq \boldsymbol{\delta} \in R^{|J|} \\ &\qquad\qquad \mathbf{0} \leqq \mathbf{v} \in R^{k} \end{aligned}$$

where $|J|$ is the number of elements in J. Thus, in Isermann's subproblem test, each nonbasic variable x_j, $j \in J$, is efficient if and only if the subproblem has an optimal solution.

Although the Isermann subproblem is similar to the Evans-Steuer subproblem, particularly when $|J| = 1$, Isermann's method of implementation for determining all efficient nonbasic variables offers opportunities for being more economical. Indices pertaining to nonnull, nonpositive columns of **W** are never candidates for J because such columns are known to pertain to inefficient nonbasic variables without testing. To start, let us solve the Isermann subproblem with $J = \{r\}$ where r is a nonbasic index whose efficiency status is not known. Then, if the subproblem has an optimal solution (i.e., x_r is an efficient nonbasic variable), Isermann's method seeks to enlarge J utilizing the three following observations in order to reduce the number of subproblems that have to be solved to classify all nonbasic variables.

Observation 1. Consider the optimal tableau of the Isermann subproblem with $J = \{r\}$. Let y_j be in the optimal subproblem basis. Then, x_j is an efficient nonbasic variable, and we can enlarge J by each such j with the current optimal subproblem basis remaining optimal. (*Rationale*: The reduced costs of any such new columns of $\mathbf{W}^J$ would be zero.)

Observation 2. Let y_j be nonbasic in the optimal tableau of the Isermann subproblem with $J \neq \varnothing$. Consider dropping the sign restriction on y_j. Then, if we can bring y_j into the basis in exchange for a v_i and still be at an optimal subproblem solution, we perform such a pivot. For such a j, x_j is an efficient nonbasic variable and we can enlarge J to $J \cup \{j\}$. (*Rationale*: Removing the sign restriction is equivalent to having solved the subproblem with J initially equal to $J \cup \{j\}$.)

Observation 3. Let y_j be nonbasic in an optimal tableau of the Isermann subproblem with $J \neq \varnothing$. Then, if the reduced cost of y_j in the optimal subproblem tableau is zero, J can be enlarged to $J \cup \{j\}$. (*Rationale*: The current optimal basis would be optimal for the subproblem had J been initially equal to $J \cup \{j\}$ because the reduced cost of the new column of $\mathbf{W}^J$ would be zero.)

In Isermann's method, we first enlarge J using Observation 1. Then, we enlarge J as much as possible in accordance with Observation 2. From the final Observation 2 (optimal) tableau, we enlarge J to the greatest extent possible using Observation 3. The resulting index set J is a *maximal* index set of efficient nonbasic variables. Because we keep repeating Isermann's method with status unknown starting indices r until all nonbasic indices are classified, a given $\mathbf{W}$ is likely to produce several maximal index sets.

9.7.3 Ecker's Method

This method was introduced by Ecker and Kouada (1978). For this method, consider the *linear system*

$$\mathbf{W}^T(\boldsymbol{\lambda} + \mathbf{e}) + \mathbf{y} = \mathbf{0}$$
$$\mathbf{0} \leqq \boldsymbol{\lambda} \in R^k$$
$$\mathbf{0} \leqq \mathbf{y} \in R^{n-m}$$

in which $\mathbf{W}$ is the $c_j - z_j$ reduced cost matrix of an efficient basis. Then, x_j is an efficient nonbasic variable if and only if there exists a solution of the linear system $(\bar{\boldsymbol{\lambda}}, \bar{\mathbf{y}}) \in R^{k+n-m}$ such that $\bar{y}_j = 0$.

Rather than solve a series of subproblems, as in the Evans-Steuer and Isermann methods, in Ecker's method we apply a routine on simplex tableaus of the *rearranged* linear system

$$\mathbf{W}^T\boldsymbol{\lambda} + \mathbf{y} = -\mathbf{W}^T\mathbf{e}$$
$$\mathbf{0} \leqq \boldsymbol{\lambda} \in R^k$$
$$\mathbf{0} \leqq \mathbf{y} \in R^{n-m}$$

For instance, the initial tableau of the rearranged system is given by

		$\lambda_1 \quad \cdots \quad \lambda_k$	$y_1 \quad \cdots \quad y_{n-m}$
y_1 $\vdots$ y_{n-m}	$-\mathbf{W}^T\mathbf{e}$	$\mathbf{W}^T$	$\mathbf{I}$

Note that the solution of the initial tableau may not be feasible. With regard to simplex tableaus of the rearranged system, Ecker's routine utilizes the following observations.

Observation 1. If a y_j row of a tableau has a positive left constant and nonpositive nonbasic entries, x_j is an inefficient nonbasic variable. (*Rationale*: Since all variables are nonnegative, the only way we can have equality with the equation of this row is for the basic variable (y_j) to be positive.)

Observation 2. If a y_j row of a tableau has a positive left constant and nonpositive nonbasic entries, that row can be dropped from the tableau. (*Rationale*: Every nonnegative solution that satisfies all other rows of the tableau satisfies the y_j row. Thus, the y_j row is redundant (nonconstraining) and can be dropped.)

Observation 3. If a y_j can be made nonbasic in a feasible tableau, x_j is an efficient nonbasic variable. (*Rationale*: When y_j is nonbasic, its value is zero.)

Ecker's routine to determine the set of efficient nonbasic indices J is as follows.

Step 1: Let $L = \{1, \ldots, n - m\}$ and $J = \varnothing$.

Step 2: For each $j \in L$ such that the y_j row has a positive left constant and nonpositive nonbasic entries, drop the row and set $L = L - \{j\}$.

Step 3: Perform pivots (if necessary) on the tableau to obtain a nonnegative constant column.

Step 4: For each $j \in L$ such that y_j is nonbasic or basic at a zero value, set $J = J \cup \{j\}$ and $L = L - \{j\}$.

Step 5: For each $j \in L$ such that y_j is currently basic but can be made nonbasic in one pivot, set $J = J \cup \{j\}$ and $L = L - \{j\}$.

Step 6: If $L = \varnothing$, stop. Otherwise, select a $j \in L$ and set $L = L - \{j\}$. Add to the current tableau an objective row to minimize y_j. Over the course of the minimization, check for Steps 2, 4, and 5 after each pivot.

Step 7: If y_j has a minimum value of zero, set $J = J \cup \{j\}$. Go to Step 6.

9.7.4 Zionts–Wallenius Routine

Zionts and Wallenius (1980) prescribe a routine for classifying directions as to whether they are dominated or nondominated. The routine is valid provided that there are no nondominated directions in the set of directions to be classified that can be formed by a nonnegative linear combination of other directions in the set.

Consider the reduced cost matrix **W** at an efficient basis of an MOLP. The only such nonnegative linear combination that can occur in **W** is when one column is a positive scalar multiple of another column. Because scalar multiple duplicates can be easily identified and set aside from the analysis, the Zionts-Wallenius routine can be used for determining which of the remaining columns pertain to efficient nonbasic variables.

With respect to an efficient basis, let x_h be a nonbasic variable and $(w_{1h}, w_{2h}, \ldots, w_{kh})^T$ its associated column in the reduced cost matrix after the removal of scalar multiples. Then, x_h is an *efficient nonbasic variable* iff the *null objective function* program

$$\min \{\mathbf{0}^T\mathbf{u}\}$$

$$\text{s.t.} \quad \sum_{j \neq h} w_{ij}u_j \geqq w_{ih} \qquad 1 \leqq i \leqq k$$

$$\text{all } u_j \geqq 0$$

is inconsistent. If we write the dual, x_h is an efficient nonbasic variable iff the subproblem

$$\max \left\{ \sum_{i=1}^{k} w_{ih} y_i \right\}$$

$$\text{s.t.} \quad \sum_{i=1}^{k} w_{ij} y_i \leqq 0 \qquad j \neq h$$

$$\text{all } y_i \geqq 0$$

has a positive-unbounded objective function value. (Alternately, x_h is an inefficient nonbasic variable iff the optimal objective function value is zero.)

Rather than solving an individual subproblem for each nonbasic variable, the Zionts-Wallenius routine resolves the efficiency status of all nonbasic variables in one extended run. With regard to simplex tableaus (with basic columns suppressed) of the *linear system*

$$\sum_{i=1}^{k} w_{ij} y_i + v_j = 0 \quad \text{for all } j \text{ after scalar multiple removal}$$

$$\text{all } y_i, v_j \geqq 0$$

the algorithm is based upon the following observations.

Observation 1. There is complete degeneracy in each simplex tableau of the linear system.

Observation 2. If v_h is negative in an otherwise feasible solution of the linear system, x_h is an efficient nonbasic variable. (*Rationale*: $v_h < 0$ implies $\sum_{i=1}^{k} w_{ih} y_i > 0$, which makes the subproblem objective function value positive.)

Observation 3. Let v_h be nonbasic in a simplex tableau of the linear system. Then, if all elements in the v_h column are nonnegative, x_h is an efficient nonbasic variable. (*Rationale*: v_h can take on negative values in an otherwise feasible solution of the linear system.)

Observation 4. Let v_h be basic in a simplex tableau of the linear system. Then, if all elements in the v_h row are nonpositive, x_h is an inefficient nonbasic variable. (*Rationale*: The equation of the row of the tableau can be written

$$v_h = -\sum_{j=1}^{k} \alpha_{hj} y_j$$

where the α_{hj} are the (nonpositive) row elements and the y_j are nonbasic variables. Since v_h can never be negative, x_h is inefficient.)

Observation 5. Let v_h be nonbasic in a simplex tableau of the linear system. If there exists a row that is nonnegative except for a negative element in the v_h column, x_h is an inefficient nonbasic variable. (*Rationale*: A pivot on the negative element leads to the condition of Observation 4.)

Observation 6. Let x_h be an inefficient nonbasic variable and v_h be basic in a simplex tableau of the linear system. Then, the constraint in which v_h is basic can be disregarded from further consideration. (*Rationale*: Since $(w_{1h}, w_{2h}, \ldots, w_{kh})^T$ is dominated by a nonnegative linear combination of the other columns of W, its presence in the null objective function program at the beginning of this subsection is unneeded in determining the efficiency status of the other x_h nonbasic variables.)

Observation 7. A nonbasic x_h variable whose column has been removed from **W** is efficient (inefficient) if the nonbasic x_h variable pertaining to its positive scalar multiple counterpart remaining in **W** is efficient (inefficient).

From the seven observations, we have the following Zionts-Wallenius routine for classifying x_h nonbasic variables.

Step 1: Remove all columns of **W** that are positive scalar multiple duplicates of other columns of **W**.

Step 2: Let all basic variable columns be suppressed in the simplex tableaus of the resulting linear system. In the initial tableau, designate each v_j as *status unknown*.

Step 3: Designate as inefficient each basic v_j whose row has all nonpositive coefficients.

Step 4: Delete all inefficient basic v_j rows (all but one if the resulting tableau would otherwise have no rows).

Step 5: For any row whose only negative coefficient is in a status unknown v_j column, change the status of the v_j to inefficient.

Step 6: For any status unknown v_j column having all coefficients nonnegative, change the status of v_j to efficient.

Step 7: Consider any column containing only one positive element. If the positive element is in a status unknown v_j row, designate the basic v_j as efficient.

Step 8: Interchange rows if necessary so that the top row is one with a status unknown v_j basic variable. If no such v_j row exists, go to Step 10.

Step 9: Pivot into the basis a variable that has a positive element in the (new) top row so that the pivot element is not in the top row. Then, go to Step 3.

Step 10: If there are no status unknown v_j columns, stop. Otherwise, perform an iteration to make one of the status unknown v_j variables basic in the top row interchanging rows if necessary. Then, go to Step 3.

To illustrate the algorithm, consider the reduced cost matrix

$$\mathbf{W} = \begin{bmatrix} 2 & -1 & 1 & 1 & -2 & -1 & 4 \\ -1 & 2 & 1 & 2 & -2 & -3 & -2 \\ 1 & -2 & -2 & -1 & 1 & -4 & 2 \end{bmatrix}$$

in which the first, fourth, and seventh are the only nondominated column directions.

1st Iteration (Steps 1, 3, 4, and 9)

	y_1	y_2	y_3	
v_1	2	−1	1	
v_2	−1	2	−2	
v_3	[1]	1	−2	
v_4	1	2	−1	
v_5	−2	−2	1	
Ineff v_6	−1	−3	−4	X

2nd Iteration (Steps 5 and 9)

	Ineff v_3	y_2	y_3
v_1	−2	−3	5
v_2	1	3	−4
y_1	1	1	−2
v_4	−1	1	[1]
v_5	2	0	−3

3rd Iteration (Steps 7, 8, and 9)

	Ineff v_3	y_2	v_4
Eff v_1	3	−8	−5
v_2	−3	7	4
y_1	−1	3	2
y_3	−1	1	1
v_5	−1	3	3

	Ineff v_3	y_2	v_4
v_2	−3	7	4
Eff v_1	3	−8	−5
y_1	−1	[3]	2
y_3	−1	1	1
v_5	−1	3	3

4th Iteration (Steps 3, 4, 6, 8, and 9)

	Ineff v_3	y_1	Eff v_4	
Ineff v_2	$-\frac{2}{3}$	$-\frac{7}{3}$	$-\frac{2}{3}$	X
Eff v_1	$\frac{1}{3}$	$\frac{8}{3}$	$\frac{1}{3}$	
y_2	$-\frac{1}{3}$	$\frac{1}{3}$	$\frac{2}{3}$	
y_3	$-\frac{2}{3}$	$-\frac{1}{3}$	$\frac{1}{3}$	
v_5	0	−1	1	

	Ineff v_3	y_1	Eff v_4
v_5	0	−1	1
Eff v_1	$\frac{1}{3}$	$\frac{8}{3}$	[$\frac{1}{3}$]
y_2	$-\frac{1}{3}$	$\frac{1}{3}$	$\frac{2}{3}$
y_3	$-\frac{2}{3}$	$-\frac{1}{3}$	$\frac{1}{3}$

5th Iteration (Steps 3, 4, and 10)

	Ineff v_3	y_1	Eff v_1	
Ineff v_5	−1	−9	−3	X
Eff v_4	1	8	3	
y_2	−1	−5	−2	
y_3	−1	−3	−1	

Since the seventh column was removed because it is a positive scalar multiple of the first column, the nonbasic x_h variables associated with the first, fourth, and seventh columns of $\mathbf{W}$ are efficient.

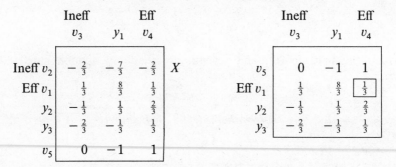

The Zionts-Wallenius routine has other uses. It can be used to determine which of a finite number of points are extreme points of the convex polyhedron defined by the points (provided that there are no duplicates in the set of points to be classified). To demonstrate, let $\{\mathbf{x}^1, \ldots, \mathbf{x}^q\}$ be the set of points defining a convex polyhedron (assuming $\mathbf{x}^i \neq \mathbf{x}^j$ for all $i \neq j$). Point $\mathbf{x}^h \in R^n$ is an extreme point of the polyhedron iff

$$\begin{aligned}
&\min \{ \mathbf{0}^T \boldsymbol{\mu} \} \\
\text{s.t.} \quad & \sum_{j \neq h} x_i^j \mu_j = x_i^h \quad 1 \leqq i \leqq n \\
& \sum_{j \neq h} \mu_j = 1 \\
& \text{all } \mu_j \geqq 0
\end{aligned}$$

is inconsistent. If we write the dual, $\mathbf{x}^h$ is an extreme point of the polyhedron iff the subproblem

$$\begin{aligned}
&\max \left\{ y_0 + \sum_{i=1}^{n} x_i^h y_i \right\} \\
\text{s.t.} \quad & y_0 + \sum_{i=1}^{n} x_i^j y_i \leqq 0 \quad j \neq h \\
& \text{all } y\text{'s unrestricted}
\end{aligned}$$

has a positive-unbounded objective function value. (Alternately, $\mathbf{x}^h$ is nonextreme iff the optimal objective function value is zero.)

After adding v_j variables, we have the *linear system*

$$\begin{aligned}
y_0 + \sum_{i=1}^{n} x_i^j y_i + v_j = 0 \quad \text{all } j \\
\text{all } v_j \geqq 0 \\
\text{all } y\text{'s unrestricted}
\end{aligned}$$

To deal with the unrestricted variables, we iteratively pivot each unrestricted variable into the basis and then eliminate the row associated with the variable. Along with the restriction that $\mathbf{x}^i \neq \mathbf{x}^j$ for all $i \neq j$, this mandates a new Step 1 for the Zionts-Wallenius routine when applied to the extreme point identification problem. The new Step 1 is as follows.

Step 1: Remove all duplicates from the set of points defining the polyhedron. Iteratively pivot into the basis each unrestricted variable and then remove the row associated with the variable. (*Rationale*: The rows of the unrestricted variables become in effect nonconstraining.)

With regard to the rest of the steps, efficiency is interpreted as extreme and inefficiency as nonextreme.

To illustrate the Zionts-Wallenius routine as applied to the extreme point problem, consider the convex polyhedron defined by

$$\{\mathbf{x}^1, \mathbf{x}^2, \mathbf{x}^3, \mathbf{x}^4, \mathbf{x}^5\} = \{(0,0), (4,0), (4,2), (3,1), (2,1)\}$$

of which $\mathbf{x}^4$ and $\mathbf{x}^5$ are not extreme.

Removal of Unrestricted Variables (Step 1)

	y_0	y_1	y_2
v_1	[1]	0	0
v_2	1	4	0
v_3	1	4	2
v_4	1	3	1
v_5	1	2	1

	v_1	y_1	y_2	
y_0	1	0	0	X
v_2	-1	[4]	0	
v_3	-1	4	2	
v_4	-1	3	1	
v_5	-1	2	1	

	v_1	v_2	y_2	
y_1	$-\frac{1}{4}$	$\frac{1}{4}$	0	X
v_3	0	-1	[2]	
v_4	$-\frac{1}{4}$	$-\frac{3}{4}$	1	
v_5	$-\frac{1}{2}$	$-\frac{1}{2}$	1	

	v_1	v_2	v_3	
y_2	0	$-\frac{1}{2}$	$\frac{1}{2}$	X
v_4	$-\frac{1}{4}$	$-\frac{1}{4}$	$-\frac{1}{2}$	
v_5	$-\frac{1}{2}$	0	$-\frac{1}{2}$	

1st Iteration (Steps 3, 4, 6, and 10)

		Eff		
	v_1	v_2	v_3	
Ineff v_4	$-\frac{1}{4}$	$-\frac{1}{4}$	$-\frac{1}{2}$	X
Ineff v_5	[$-\frac{1}{2}$]	0	$-\frac{1}{2}$	

2nd Iteration (Steps 6 and 7)

	Ineff	Eff	Eff
	v_5	v_2	v_3
Eff v_1	-2	0	1

9.8 COMPUTING ALL MAXIMALLY EFFICIENT FACETS

We have primarily been interested in the set of all efficient extreme points. Now, we turn our attention to the computation of which subgroups of efficient extreme points define the different maximally efficient facets. Our interest in maximally efficient facets is that the (entire) efficient set is given by the union of all maximally efficient facets.

Example 7. Assume Fig. 9.8 shows the portion of the surface of the feasible region that is efficient. Then, the efficient set can be characterized by the following

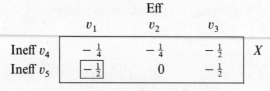

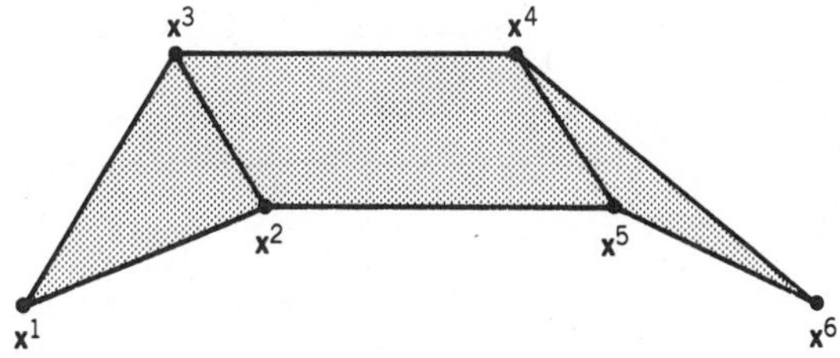

Figure 9.8. Graph of Example 7.

subsets of efficient extreme points:

$$\{\mathbf{x}^1, \mathbf{x}^2, \mathbf{x}^3\}$$

$$\{\mathbf{x}^2, \mathbf{x}^3, \mathbf{x}^4, \mathbf{x}^5\}$$

$$\{\mathbf{x}^4, \mathbf{x}^5, \mathbf{x}^6\}$$

Although procedures for computing (characterizing) all maximally efficient facets have been proposed by Yu and Zeleny (1975); Gal (1977); and Ecker, Hegner, and Kouada (1980), we shall review the method of Isermann (1977) in this section. Isermann's method for finding all maximally efficient facets is an extension of his method for classifying all nonbasic variables (Section 9.7.2). Assume S is bounded. For notation, let

I_x Be an index set denoting the set of all efficient bases.

To compute all maximally efficient facets using Isermann's method, we must perform the following operations.

1. For each $i \in I_x$, let $T^i = J^{i,1} \cup J^{i,2} \cup \cdots \cup J^{i,l}$ where the $J^{i,j}$ are the maximal index sets of efficient nonbasic variables resulting from classifying the nonbasic variables (as in Section 9.7.2) at efficient basis i.
2. For each T^i, we must do the following. Letting g denote the number of indices in T^i, we form the directed graph $\Gamma(T^i)$ whose nodes are the $\binom{g}{h}$ combinations of indices in T^i where $h = g, g-1, \ldots, 1$. The $\binom{g}{g}$ node is the source and the $\binom{g}{1}$ nodes are the sinks. Since each node of $\Gamma(T^i)$ represents a potential maximal index set at $\mathbf{B}^i$, the graph must be *adjusted* to reflect the maximal index sets $J^{i,1}, \ldots, J^{i,l}$ that are known. In adjusting $\Gamma(T^i)$ for a maximal index set that is known, the respective index set and all predecessor and all successor nodes are deleted. If a given node is known not to correspond to a maximal index set, the graph is adjusted deleting the respective node and all of its predecessors. When the adjusted graph has no nodes, all maximal index sets of efficient nonbasic variables have been

TABLE 9.1
MOLP of Example 8

	x_1	x_2	x_3	x_4	x_5		
	1	3	-2		1	max	
Objs	3	-1		3	1	max	
	1		2		3	max	
	2	4			3	$\leq$	27
			2	5	4	$\leq$	35
s.t.	5					$\leq$	26
				2		$\leq$	24
	5	5	2			$\leq$	36

found. To resolve a nonempty adjusted graph, we strategically invoke Isermann's subproblem test until no nodes remain.

3. Let D^i denote the indices of the basic variables at $i \in I_x$. Now, let us form the index sets.

$$Q^{i,j} = J^{i,j} \cup D^i \quad \text{for all } i,\, j$$

Let us also form the minimal number of minimal index sets $U^1, \ldots, U^\alpha$ that subsume all of the $Q^{i,j}$. The $U^1, \ldots, U^\alpha$ are minimal index sets in the sense (a) that for each $Q^{i,j}$, there exists an ℓ such that $Q^{i,j} \subset U^\ell$; (b) that for each U^ℓ, there exist i, j such that $Q^{i,j} = U^\ell$; and (c) $U^m \not\subset U^n$ for any $m \neq n$. Now for each $\ell \in \{1, \ldots, \alpha\}$, we form the index set

$$I_x^\ell = \{ i \in I_x \,|\, D^i \subset U^\ell \}$$

Thus, the ℓth maximally efficient facet is characterized by I_x^ℓ.

Example 8. Consider the MOLP of Table 9.1. This problem has 11 efficient extreme points and four maximally efficient facets. Which efficient extreme points define which maximally efficient facets are shown in the graph of Fig. 9.9 in which

$\mathbf{z}^1 = (20.25, 14.25, .00)$ $\quad$ $\mathbf{z}^7 = (11.20, 34.60, 5.20)$

$\mathbf{z}^2 = (19.80, 17.40, .90)$ $\quad$ $\mathbf{z}^8 = (-1.26, 20.26, 34.04)$

$\mathbf{z}^3 = (9.31, 8.56, 26.25)$ $\quad$ $\mathbf{z}^9 = (5.20, 36.60, 5.20)$

$\mathbf{z}^4 = (14.06, 30.58, 13.80)$ $\quad$ $\mathbf{z}^{10} = (.73, 22.85, 31.80)$

$\mathbf{z}^5 = (9.12, 9.87, 26.25)$ $\quad$ $\mathbf{z}^{11} = (-34.80, .60, 35.20)$

$\mathbf{z}^6 = (10.73, 28.85, 21.80)$

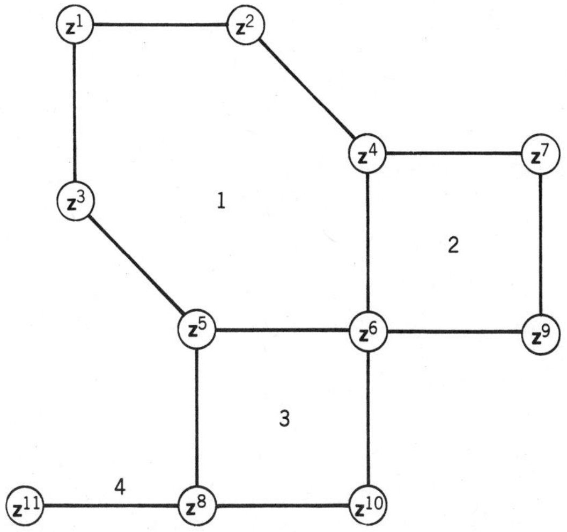

Figure 9.9. Graph of efficient extreme points and maximally efficient facets of Example 8.

9.9 VECTOR-MAXIMUM APPROACH FOR SOLVING AN MOLP

Consider the following approach for solving an MOLP. A vector-maximum algorithm is used to compute all efficient extreme points. The thought is that the decision maker, by reviewing the list of nondominated criterion vectors associated with the efficient extreme points, would be able to identify his or her efficient extreme point of greatest utility. It is hoped this extreme point would be optimal, or close enough to being optimal, to terminate the decision process.

However, there are difficulties with the vector-maximum approach with regard to (1) CPU time requirements, (2) the number of efficient extreme points, and (3) the possibility that the best efficient extreme point is an unacceptable approximation of an optimal point.

Because of their complexity, vector-maximum algorithms easily encounter CPU time limitations with most of the computer time consumed by the efficiency evaluation of the nonbasic variables at each efficient basis. For instance, in a $5 \times 25 \times 100$ MOLP with all constraints slack, 100 nonbasic variables would have to be evaluated at each efficient basis. With the prospect that such an MOLP may have more than 1,000 efficient extreme points, it is seen that the vector-maximum approach not only has voracious computational requirements, but it presents the decision maker with "information overload" problems. Even with procedures for only generating a subset of the set of all efficient extreme points (Section 9.10) and techniques for interactively searching through the generated extreme points (Chapter 11), it is not reasonable to recommend the vector-maximum approach for large problems. Furthermore, the possibility exists that the decision maker's optimal solution may lie in the middle of a large efficient facet. To deal with such a possibility, we would want to have at our disposal methods for searching the relative interior of efficient facets. This is easier said than done. Topics related to the exploration of the relative interior of efficient facets are discussed in Section 11.8 and Chapter 14.

9.10 CONTRACTING THE CRITERION CONE

On one hand, we have the vector-maximum problem

$$\text{eff}\,\{\mathbf{Cx} = \mathbf{z} \mid \mathbf{x} \in S\} \tag{9.39}$$

that requires no information about the decision maker's preferences. On the other hand, we have the *weighted-sums* LP

$$\max\,\{\boldsymbol{\lambda}^T\mathbf{Cx} \mid \mathbf{x} \in S\}$$
$$\boldsymbol{\lambda} \in \Lambda, \quad \boldsymbol{\lambda} \text{ fixed} \tag{9.40}$$

that requires very specific information about the decision maker's preferences. With the vector-maximum problem, we are usually inundated with efficiency information. With the weighted-sums LP, we usually do not generate enough efficiency information.

Consider the *interval criterion weights* problem, in which we solve the (infinite) family of weighted-sums LPs

$$\{\max\,[\boldsymbol{\lambda}^T\mathbf{Cx} \mid \mathbf{x} \in S] : \boldsymbol{\lambda} \in \tilde{\Lambda}\}$$

where

$$\tilde{\Lambda} = \left\{\boldsymbol{\lambda} \in R^k \mid \lambda_i \in \text{rel}\,[\ell_i, \mu_i],\ \sum_{i=1}^{k} \lambda_i = 1\right\}$$

for the union of all maximizing points. When all $\ell_i = 0$ and all $\mu_i = 1$, the interval criterion weights problem is the vector-maximum problem (9.39) because the union of all maximizing points is the efficient set. When all $\ell_i = \mu_i$, the interval criterion weights problem reduces to the weighted-sums LP (9.40). In this way, the vector-maximum problem (9.39) and the weighted-sums problem (9.40) can be viewed as *polar extremes* on the interval criterion weights continuum. Since each polar extreme has its disadvantages, problems somewhere in-between on the continuum may be more appropriate. Such problems can be created using *subinterval* weights. They can be constructed by regulating the ℓ_i and μ_i so that not all weighting vectors are allowed. In this way, by manipulating the ℓ_i and μ_i, we arrive at a strategy for computing subsets of the efficient set thereby controlling the amount of efficiency information generated.

Since the interval criterion weights problem involves, in general, an infinite number of weighted-sums problems, it cannot be solved as stated. However, it can be solved by transforming it into the *interval criterion weights* vector-maximum problem

$$\text{eff}\,\{\mathbf{Dx} \mid \mathbf{x} \in S\}$$
$$\text{where } \mathbf{D} \text{ is } q \times n$$

whose criterion cone is a subset of the criterion cone of $\text{eff}\,\{\mathbf{Cx} \mid \mathbf{x} \in S\}$.

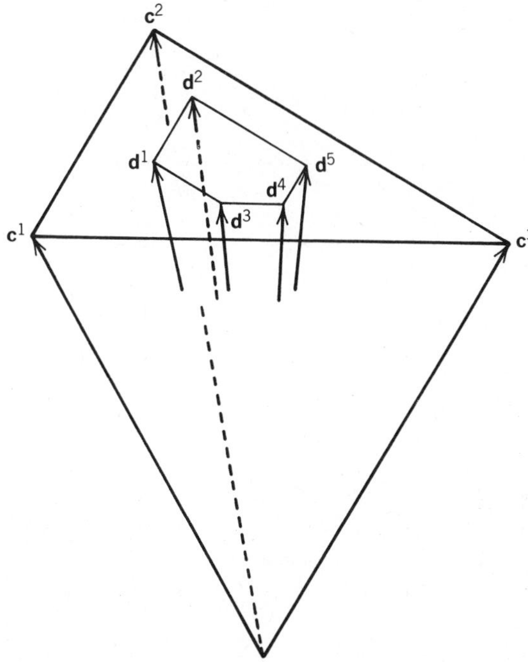

Figure 9.10. Interval criterion weights criterion cone.

Let us call the criterion cone of eff$\{\mathbf{Cx} \mid \mathbf{x} \in S\}$ the **C**-*cone* (or *original* criterion cone) and the criterion cone of eff$\{\mathbf{Dx} \mid \mathbf{x} \in S\}$ the **D**-*cone* (or *interval criterion weights* cone). In the interval criterion weights vector-maximum problem, the **D**-cone is a subset of the **C**-cone, as in Fig. 9.10.

Figure 9.10 is typical in the sense that the **D**-cone, although smaller, has more generators than the **C**-cone. We note, of course, that the narrower the $[\ell_i, \mu_i]$ interval widths, the smaller the **D**-cone.

9.10.1 Interval Criterion Weights Criterion Cone

Consider a convex combination weighting vector that, when applied to the generators of the **C**-cone, specifies a generator of the **D**-cone. Such a convex combination weighting vector is called a *critical vector*. With a critical vector for each generator of the interval criterion weights cone, criterion matrix **D** can be constructed.

Example 9. To see how the $\mathbf{d}^j$ generators of the interval criterion weights cone are determined, consider the cross-section of the three-dimensional **C**-cone generated by $\mathbf{c}^1$, $\mathbf{c}^2$, and $\mathbf{c}^3$ in Fig. 9.11, along with

$$\lambda_1 \in \text{rel}\,[.2, .7]$$
$$\lambda_2 \in \text{rel}\,[.2, .5]$$
$$\lambda_3 \in \text{rel}\,[.1, .8]$$

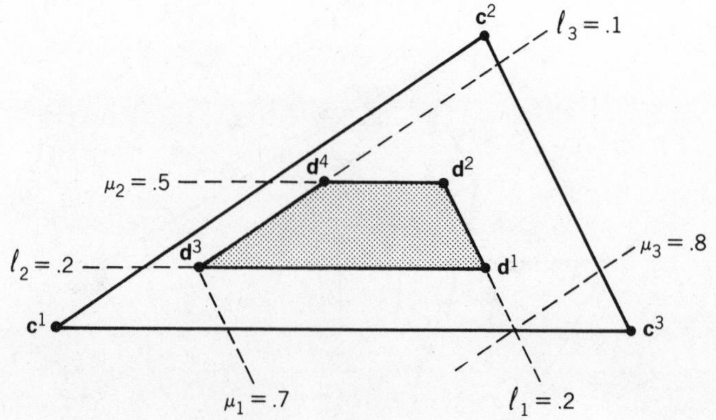

Figure 9.11. Graph of Example 9.

As seen from Fig. 9.11, each $\mathbf{d}^j$ is defined by $k - 1$ endpoints of the $[\ell_i, \mu_i]$ intervals. Generator $\mathbf{d}^1$ is defined by $\ell_1 = .2$ and $\ell_2 = .2$. This implies that $(.2, .2, .6)$ is the critical vector for $\mathbf{d}^1$ because

$$.2\mathbf{c}^1 + .2\mathbf{c}^2 + .6\mathbf{c}^3 = \mathbf{d}^1$$

Since $\mathbf{d}^2$ is defined by $\ell_1 = .2$ and $\mu_2 = .5$, $(.2, .5, .3)$ is the critical vector for $\mathbf{d}^2$, and so forth.

As described by Steuer (1976 and 1978), the q rows of $\mathbf{D}$ are constructed according to the following two stages.

Stage I: Obtain all q critical vectors. Form the $q \times k$ matrix $\mathbf{T}$ whose ith row is the ith critical vector.

Stage II: Premultiply criterion matrix $\mathbf{C}$ by $\mathbf{T}$ to obtain $\mathbf{D}$ (i.e., $\mathbf{TC} = \mathbf{D}$).

To illustrate Stage I, let us again consider the

$$[.2, .7]$$
$$[.2, .5]$$
$$[.1, .8]$$

intervals of Example 9. A systematic way to determine all critical vectors is to utilize *endpoint tables* to search all combinations of $k - 1$ of the interval endpoints. There are k endpoint tables and each is of size $2^{k-1} \times k$ as in Table 9.2. In Table 9.2,

1. The arrows refer to the weighting vector components that are *residually* determined.
2. OR signifies that the residual weight is outside of its $[\ell_i, \mu_i]$ range.
3. The circled numbers refer to critical vectors.
4. DUP means that the row is a duplicate.

TABLE 9.2
Endpoint Tables for Intervals of Example 9

		↓	
.2	.2	.6	①
.2	.5	.3	②
.7	.2	.1	③
.7	.5		OR

	↓		
.2		.1	OR
.2		.8	OR
.7	.2	.1	DUP
.7		.8	OR

↓			
.7	.2	.1	DUP
	.2	.8	OR
.4	.5	.1	④
	.5	.8	OR

Storing the critical weights identified in Table 9.2 in the *premultiplication* matrix **T**, we have

$$\mathbf{T} = \begin{bmatrix} .2 & .2 & .6 \\ .2 & .5 & .3 \\ .7 & .2 & .1 \\ .4 & .5 & .1 \end{bmatrix}$$

Performing the matrix multiplication for Stage II, we obtain

$$\begin{aligned} \mathbf{D} &= \mathbf{TC} \\ &= \begin{bmatrix} .2 & .2 & .6 \\ .2 & .5 & .3 \\ .7 & .2 & .1 \\ .4 & .5 & .1 \end{bmatrix} \mathbf{C} \end{aligned}$$

A difficulty with the interval criterion weights approach is that the number of generators of the **D**-cone increases rapidly with k (the original number of objectives). From Fig. 9.11, we can see that the largest q can be is 6 when $k = 3$. Computational experience regarding the number of **D**-cone generators using randomly generated intervals with larger values of k is given in Table 9.3.

From Table 9.3, we see that a practical upper limit for k when using the interval criterion weights vector-maximum problem is about 6. This is because of the increased work in detecting nonbasic variable efficiency. Each row of **D** adds a simplex row to the Evans-Steuer and Isermann tests and a simplex column to the Ecker and Zionts-Wallenius tests. Thus, after about 40 or 50 **D**-matrix rows, the subproblem computational burden is so large that the usefulness of the **D**-matrix technique becomes questionable.

Throughout our interval criterion weights discussions, we have implicitly assumed that all k original objective function gradients were linearly independent. What happens if the set of original objective function gradients is linearly dependent? A small proportion of the rows of **D** may be nonessential generators of the interval criterion weights criterion cone. However, an attempt to identify and remove such rows is probably more work than it is worth.

Table 9.3
Computational Experience Regarding the Number of D-Cone Generators

k	*Average* $[\ell_i, \mu_i]$ *Width*	*Average Number of* ***D****-Cone Generators*
4	.5	7.85
4	.25	11.00
4	.125	11.00
4	.0625	11.37
6	.5	20.15
6	.25	42.55
6	.125	55.40
6	.0625	55.35
8	.5	42.05
8	.25	116.75
8	.125	256.15
8	.0625	256.00

9.10.2 Enveloping Reduced Criterion Cone

To extend the range of applicability of interval criterion weights, we now consider the *enveloping reduced* criterion cone. The purpose of the enveloping reduced criterion cone (called the **E**-*cone*) is to encase the **D**-cone in a slightly larger cone that has, at most, k extreme rays. When the k original objective function gradients are linearly *independent*, the **E**-cone will have k extreme rays and be the smallest proportional version of the **C**-cone that contains the **D**-cone. When the k original objective function gradients are linearly *dependent*, the **E**-cone may have fewer than k generators and may not be shaped proportionally as the **C**-cone.

With the **E**-cone, we solve the *enveloping reduced* vector-maximum problem

$$\text{eff}\,\{\mathbf{Ex} \mid \mathbf{x} \in S\}$$

Since **E** has, at most, k rows, we do not have the nonbasic variable subproblem testing computational burden that we have with the **D**-cone. However, because the **E**-cone is a superset of the **D**-cone, we can expect to generate more efficiency information with eff $\{\mathbf{Ex} \mid \mathbf{x} \in S\}$.

The critical (weighting) vectors for the **E**-cone are determined rapidly by forming the intervals

$$[\hat{\ell}_1, 1]$$
$$[\hat{\ell}_2, 1]$$
$$\vdots$$
$$[\hat{\ell}_k, 1]$$

where

$$\hat{\ell}_i = \max\left\{\ell_i, \left(1 - \sum_{j \neq i} \mu_j\right)\right\}$$

Clearly, the ith weight cannot be less than $\hat{\ell}_i$. When using the $[\hat{\ell}_i, 1]$ intervals with endpoint tables, we need only consider the first row of each table. This is because all other rows are OR or DUPs due to the fact that all rows except the first in each endpoint table contain at least one 1. Hence **E** contains, at most, k rows.

Example 10. Let

$$\mathbf{C} = \begin{bmatrix} 2 & 2 & 8 \\ 9 & 12 & 8 \\ 17 & 2 & 8 \end{bmatrix}$$

and

$$\lambda_1 \in \text{rel}[.2, .3]$$
$$\lambda_2 \in \text{rel}[.1, .7]$$
$$\lambda_3 \in \text{rel}[.2, .4]$$

From the intervals, we obtain $\hat{\ell}_1 = .2$, $\hat{\ell}_2 = .3$, and $\hat{\ell}_3 = .2$. From the first rows of the endpoint tables, we obtain the three critical vectors for the **E**-cone, as in Table 9.4. Forming the premultiplication matrix **T**, we have

$$\mathbf{E} = \mathbf{TC}$$

$$= \begin{bmatrix} .2 & .3 & .5 \\ .2 & .6 & .2 \\ .5 & .3 & .2 \end{bmatrix} \begin{bmatrix} 2 & 2 & 8 \\ 9 & 12 & 8 \\ 17 & 2 & 8 \end{bmatrix}$$

$$= \begin{bmatrix} 11.6 & 5 & 8 \\ 9.2 & 8 & 8 \\ 7.1 & 5 & 8 \end{bmatrix}$$

As shown in Fig. 9.12, the **E**-cone is the smallest proportional version of the **C**-cone that contains the **D**-cone (where the **D**-cone is indicated by the cross-hatched area.)

TABLE 9.4
Endpoint Table First Rows of Example 10

.2	.3	.5 ↓	①
.2	.6 ↓	.2	②
.5 ↓	.3	.2	③

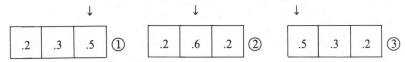

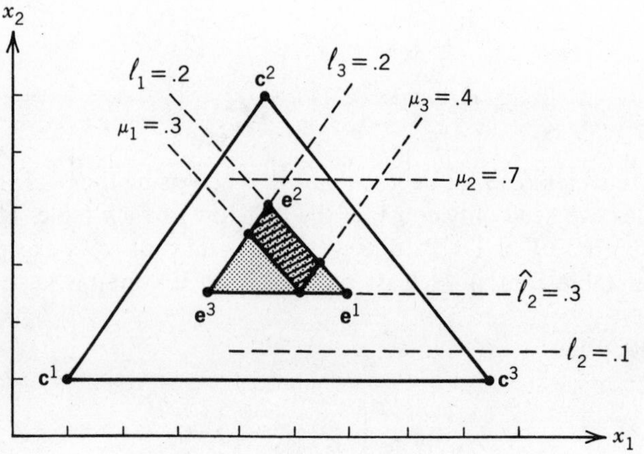

Figure 9.12. Graph of Example 10.

9.10.3 Some Specific E-Cone Contractions

Consider a **C**-cone whose $k \times n$ criterion matrix is **C**. Consider the $2k + 1$ rays in **C** given by $\boldsymbol{\lambda}^i\mathbf{C}$ where

$$\left.\begin{aligned} \boldsymbol{\lambda}^1 &= (1,0,\ldots,0) \\ &\vdots \\ \boldsymbol{\lambda}^k &= (0,0,\ldots,1) \end{aligned}\right\} \quad \text{Extreme ray convex combinations}$$

$$\left.\begin{aligned} \boldsymbol{\lambda}^{k+1} &= (1/k^2, r,\ldots,r) \\ &\vdots \\ \boldsymbol{\lambda}^{2k} &= (r,r,\ldots,1/k^2) \end{aligned}\right\} \quad \text{Off-center convex combinations}$$

$$\boldsymbol{\lambda}^{2k+1} = (1/k,1/k,\ldots,1/k)\} \quad \text{Center convex combination}$$

TABLE 9.5
Subset E-Cone $[\ell_i, \mu_i]$ Interval Criterion Weights

i	$\boldsymbol{\lambda}^i$ *Is an Extreme Ray Convex Combination*	$\boldsymbol{\lambda}^{k+i}$ *Is an Off-Center Convex Combination*	$\boldsymbol{\lambda}^{2k+1}$ *Is the Center Convex Combination*
1	$[0,1]$	$[p/h,1]$	$[p/k,1]$
2	$[0,1]$	$[p/h,1]$	$[p/k,1]$
$\vdots$			
i	$[p,1]$	$[0,1]$	$[p/k,1]$
$\vdots$			
k	$[0,1]$	$[p/h,1]$	$[p/k,1]$

and $r = (k+1)/k^2$. Suppose that about each $\boldsymbol{\lambda}^i\mathbf{C}$ ray (centered as well as possible), we wish to construct an **E**-cone whose cross-sectional volume is $(1/k)$th the cross-sectional volume of the **C**-cone. The interval criterion weights that yield the critical vectors (via the first rows of the endpoint tables) for the subset **E**-cones are presented in Table 9.5 (which comes from Steuer (1977)).

Example 11. Let $k = 7$. Then, the interval criterion weights from Table 9.5 pertaining to $\boldsymbol{\lambda}^4$ are

$$\begin{array}{c} [0,1] \\ [0,1] \\ [0,1] \\ [p,1] \\ [0,1] \\ [0,1] \\ [0,1] \end{array}$$

where $p = 1 - k^{-(1/(k-1))}$. Via the first rows of the seven endpoint tables, we obtain the seven critical vectors for the subset **E**-cone. Let the critical vectors be stored in the rows of the premultiplication matrix

$$\mathbf{T}_1 = \begin{bmatrix} & & & p & & & 1-p \\ & & & p & & 1-p & \\ & & & p & 1-p & & \\ & & & 1 & & & \\ & & 1-p & p & & & \\ & 1-p & & p & & & \\ 1-p & & & p & & & \end{bmatrix}$$

Example 12. Let $\mathbf{C}_2 = \mathbf{T}_1\mathbf{C}$ be the criterion matrix of the subset **E**-cone of Example 11. Now, let us construct an **E**-cone that is a subset of the cone defined by $\mathbf{C}_2$. Let this second subset **E**-cone be centered about the second off-center convex combination of the rows of $\mathbf{C}_2$. From Table 9.5, the interval criterion weights pertaining to $\boldsymbol{\lambda}^9$ are

$$\begin{array}{c} [p/h,1] \\ [0,1] \\ [p/h,1] \\ [p/h,1] \\ [p/h,1] \\ [p/h,1] \\ [p/h,1] \end{array}$$

where $h = k - 1$ from which we obtain

$$\mathbf{T}_2 = \frac{p}{h}\begin{bmatrix} 1 & 0 & 1 & 1 & 1 & 1 & \frac{h}{p}-5 \\ 1 & 0 & 1 & 1 & 1 & \frac{h}{p}-5 & 1 \\ 1 & 0 & 1 & 1 & \frac{h}{p}-5 & 1 & 1 \\ 1 & 0 & 1 & \frac{h}{p}-5 & 1 & 1 & 1 \\ 1 & 0 & \frac{h}{p}-5 & 1 & 1 & 1 & 1 \\ 1 & \frac{h}{p}-6 & 1 & 1 & 1 & 1 & 1 \\ \frac{h}{p}-5 & 0 & 1 & 1 & 1 & 1 & 1 \end{bmatrix}$$

The generators of the second subset **E**-cone are specified by the rows of $\mathbf{C}_3 = \mathbf{T}_2\mathbf{C}_2 = \mathbf{T}_2\mathbf{T}_1\mathbf{C}$.

9.11 ADBASE PROGRAM

ADBASE is a Fortran program for enumerating efficient extreme points and unbounded efficient edges. ADBASE employs the Phase III procedure of Fig. 9.1 along with the Evans-Steuer subproblem test for detecting nonbasic variable efficiency. ADBASE incorporates a number of features, including the ability to reduce the criterion cone via the specification of interval criterion weights. Thus, ADBASE is a program that can solve the family of weighted-sums problems

$$\{\max[\boldsymbol{\lambda}^T(\mathbf{Cx} + \boldsymbol{\alpha}) \mid \mathbf{x} \in S] : \boldsymbol{\lambda} \in \tilde{\Lambda}\}$$

where

$$S = \{\mathbf{x} \geqq \mathbf{0} \mid \mathbf{A}_k\mathbf{x} \leqq \mathbf{b}_k, \quad \mathbf{A}_e\mathbf{x} = \mathbf{b}_e, \quad \mathbf{A}_s\mathbf{x} \geqq \mathbf{b}_s\}$$

$$\tilde{\Lambda} = \left\{\boldsymbol{\lambda} \in R^k \mid \lambda_i \in \text{rel}[\ell_i, \mu_i], \quad \sum_{i=1}^{k} \lambda_i = 1\right\}$$

for all maximizing extreme points and unbounded maximizing edges. Described in terms of the parameters that must be set prior to execution, we have

1. **NUMB**: Number of problems to be solved.
2. **IFASE2**: Phase II switch

   = 1 Lexicographic maximization.

   = 2 Lexicographic maximization with test for extreme point efficiency after each maximization stage.

   = 3 Lexicographic maximization with test for efficiency at each extreme point encountered.

   = 4 Weighted-sums (with equal weights).

   = 5 Weighted-sums with test for efficiency at each extreme point encountered.

3. **IFASE3**: Phase III switch

= 0	Do not perform Phase III (i.e., terminate with first efficient extreme point located by Phase II).
= 1	Find all efficient extreme points.
= 2	Find all efficient extreme points and all unbounded efficient edges.

4. **IPRINT(*)**: Vector of on/off switches for various output options. If a given element of **IPRINT** is set to 0, the option is ignored. If set to 1, the option is performed. The most important **IPRINT** options are:

IPRINT(1)	Prints whether or not an efficient extreme point exists.
IPRINT(2)	Prints all criterion and nonzero basic structural variable values for each generated efficient basis and unbounded efficient edge.
IPRINT(3)	Prints all criterion and nonzero basic structural variable values for each generated efficient extreme point and unbounded efficient edge.
IPRINT(7)	Prints the numbers of efficient bases, efficient extreme points, and unbounded efficient edges.
IPRINT(11)	Prints out header card data, nonzero coefficients, and criterion weight interval bounds (if specified) for each of the problems to be solved.
IPRINT(12)	Writes to a file the criterion vector of each generated efficient extreme point.
IPRINT(13)	Prints the reduced cost matrix for each efficient basis (only operational if **IPRINT(2)** = 1).

5. **IPUFMT**: Format in which the criterion values that are written to a file are expressed:

= 1	F10.1
= 2	F10.2
= 3	F10.3
= 4	F10.4
= 5	F10.5
= 6	F10.6

9.11.1 ADBASE Input Format

Each problem to be solved by **ADBASE** must have its own data deck. If more than one problem is to be solved, the data decks for the problems to be solved are simply stacked, one after the other, in the input stream. A data deck may consist of as many as five different types of cards. The formats of the different types of cards are given in Fig. 9.13.

In the problem title card, we specify a literal title for the problem to be solved. In the last field of the header card, the **NGRAYS** parameter specifies the maximum number of reduced criterion cone generators allowed. In the next to the last

Problem Title Card

Title of Problem
2–72

Header Card

Problem Number	Number of Objectives	Number of Variables	Number of Slack Constraints	Number of Equality Constraints	Number of Surplus Constraints
1–8	9–16	17–24	25–32	33–40	41–48

`IFASEO` Option	`NGRAYS` Parameter
49–56	57–64

Count Cards

Number of Nonzero Coefficients
1–8

Nonzero Coefficient Cards

Row	Column	Nonzero Coefficient Value	. . .	Row	Column	Nonzero Coefficient Value
1–3	4–6	7–20		61–63	64–66	67–80

Interval Criterion Weight Cards
(Optional)

Objective Index i	ℓ_i Lower Bound	μ_i Upper Bound
1–10	11–30	31–50

Figure 9.13. Types of cards that comprise an `ADBASE` input deck.

(**IFASE0**) field, we have three Phase 0 (criterion cone contraction) options:

= 0	No criterion cone contraction.
= 1	Construction of the interval criterion weights criterion cone.
= 2	Construction of the enveloping reduced criterion cone.

If **IFASE0** = 1 and the number of generators of the interval criterion weights cone exceeds **NGRAYS**, **ADBASE** abandons the interval criterion weights cone and instead computes the enveloping reduced criterion cone.

Each count card specifies the number of nonzero coefficients in the section that follows. Each data deck has eight count cards that specify the number of nonzero $\mathbf{A}_k$, $\mathbf{b}_k$, $\mathbf{A}_e$, $\mathbf{b}_e$, $\mathbf{A}_s$, $\mathbf{b}_s$, C, and $\boldsymbol{\alpha}$ coefficients, respectively. All eight count cards must be included even if, for some of them, no nonzero coefficients follow.

The nonzero coefficient cards specify the values of the nonzero $\mathbf{A}_k$, $\mathbf{b}_k$, $\mathbf{A}_e$, $\mathbf{b}_e$, $\mathbf{A}_s$, $\mathbf{b}_s$, C, and $\boldsymbol{\alpha}$ coefficients. Each card has four 20-column partitions, and each partition consists of three subfields. The first two are three-column subfields that specify the row and column coordinates of the nonzero coefficient within $\mathbf{A}_k$, $\mathbf{b}_k$, $\mathbf{A}_e$, $\mathbf{b}_e$, $\mathbf{A}_s$, $\mathbf{b}_s$, C, and $\boldsymbol{\alpha}$. The last is a 14-column subfield that specifies the nonzero coefficient value. For the $\mathbf{b}_k$, $\mathbf{b}_e$, $\mathbf{b}_s$, and $\boldsymbol{\alpha}$ nonzero coefficients, the second three-column subfield is left blank because a column designation is unnecessary.

Interval criterion weight cards are included *if and only if* **IFASE0** has been set to 1 or 2 in the header card. They specify the ℓ_i and μ_i interval criterion weight bounds.

The five different types of cards are used to structure the data of an MOLP in **ADBASE** *input format* as follows:

Problem title card

Header card

Count card for $\mathbf{A}_k$

}Nonzero coefficient cards (if any) for $\mathbf{A}_k$

Count card for $\mathbf{b}_k$

}Nonzero coefficient cards (if any) for $\mathbf{b}_k$

Count card for $\mathbf{A}_e$

}Nonzero coefficient cards (if any) for $\mathbf{A}_e$

Count card for $\mathbf{b}_e$

}Nonzero coefficient cards (if any) for $\mathbf{b}_e$

Count card for $\mathbf{A}_s$

}Nonzero coefficient cards (if any) for $\mathbf{A}_s$

Count card for $\mathbf{b}_s$

}Nonzero coefficient cards (if any) for $\mathbf{b}_s$

TABLE 9.6
MOLP of Example 13

	x_1	x_2	x_3	x_4	x_5	x_6	x_7	x_8	α		
Objs			-1	-1						max	
					-1	-1				max	
							-1		-10	max	
								-1	18	max	
	2	4								$\leqq$	24
	4	3								$\leqq$	28
	1	1								$\leqq$	8
s.t.	1		1	-1						$=$	5
		1			1	-1				$=$	5.75
	5	-3					1	-1		$=$	26
	2	1								$\geqq$	9
	1	2								$\geqq$	7

all $x_i \geqq 0$

```
AN MOLP THAT HAS 8 EFFICIENT BASES AND 6 EFFICIENT EXTREME POINTS
     13        4         8         3         3         2         0        40
      6
1 1   2.0D0             1 2  4.0D0             2 1  4.0D0             2 2  3.0D0
3 1   1.0D0             3 2  1.0D0
      3
1    24.0D0             2   28.0D0             3    8.0D0
     10
1 1   1.0D0             1 3  1.0D0             1 4-1.0D0              2 2  1.0D0
2 5   1.0D0             2 6-1.0D0              3 1  5.0D0             3 2-3.0D0
3 7   1.0D0             3 8-1.0D0
      3
1     5.0D0             2    5.75D0            3   26.0D0
      4
1 1   2.0D0             1 2  1.0D0             2 1  1.0D0             2 2  2.0D0
      2
1     9.0D0             2     7.0D0
      6
1 3  -1.0D0             1 4-1.0D0              2 5-1.0D0              2 6-1.0D0
3 7  -1.0D0             4 8-1.0D0
      2
3   -10.0D0             4   18.0D0
```

Figure 9.14. MOLP of Example 13 in ADBASE input format.

Count card for **C**

}Nonzero coefficient cards (if any) for **C**

Count card for **α**-constants

}Nonzero coefficient cards (if any) for **α**-constants

Interval criterion weights cards (if any)

Example 13. In Fig. 9.14 is the data deck for the MOLP of Table 9.6 in ADBASE input format. Note that no interval criterion weights are present because IFASE0 = 0 in column 56 of the header card. We also note that the coefficient of x_1 in the first equality constraint has (1, 1) as its row and column indices. This is because the coefficients in $\mathbf{A}_k$, $\mathbf{b}_k$, $\mathbf{A}_e$, $\mathbf{b}_e$, $\mathbf{A}_s$, $\mathbf{b}_s$, **C**, and **α** are indexed independently.

9.11.2 Versatility of ADBASE

Because it is a vector-maximum algorithm that can accommodate interval criterion weights, ADBASE can solve a range of linear optimization problems. In addition to being able to solve an MOLP for all efficient extreme points and all unbounded efficient edges, ADBASE can solve the following.

1. *Single objective LPs for all alternative optima.*

$$\max\left\{\mathbf{c}^T\mathbf{x} + \alpha = z \mid \mathbf{x} \in S\right\}$$

By specifying one objective and setting IFASE0 = 0 and IFASE3 = 2, all optimal extreme points and all unbounded optimal edges are computed.

2. *Preemptive goal programming problems.* Store the negative of the deviational variable expression for the highest priority in the first row of **C**, the negative of the expression for the second highest priority in the second row of **C**, and so forth. Then, we set IFASE0 = 0, IFASE2 = 1 (to make use of the Phase II lexicographic capabilities), and IFASE3 = 0 to obtain a *lex min* solution of the preemptive goal program.
3. *MOLPs with interval criterion weights.*

$$\begin{aligned}
&\max\{\mathbf{c}^1\mathbf{x} + \alpha_1 = z_1\}\\
&\max\{\mathbf{c}^2\mathbf{x} + \alpha_2 = z_2\}\\
&\quad\vdots\\
&\max\{\mathbf{c}^k\mathbf{x} + \alpha_k = z_k\}\\
&\text{s.t.} \qquad\qquad \mathbf{x} \in S
\end{aligned}$$

where

$$\begin{aligned}
\lambda_1 &\in \operatorname{rel}[\ell_1, \mu_1]\\
\lambda_2 &\in \operatorname{rel}[\ell_2, \mu_2]\\
&\ \ \vdots\\
\lambda_k &\in \operatorname{rel}[\ell_k, \mu_k]
\end{aligned}$$

All extreme points and unbounded edges that are efficient with respect to the reduced criterion cone corresponding to the interval criterion weights are computed by including appropriate interval criterion weights cards and setting **IFASE0** = 1 or 2 and **IFASE3** = 2.

4. *Point estimate weighted-sums problems.*

$$\max\left\{\boldsymbol{\lambda}^T(\mathbf{Cx} + \boldsymbol{\alpha}) \mid \mathbf{x} \in S, \quad \boldsymbol{\lambda} \text{ fixed}\right\}$$

With $\boldsymbol{\lambda} \in \Lambda$ specified in the interval criterion weights cards such that $\ell_i = \mu_i$ for all i, all optimal extreme points and all unbounded optimal edges of the weighted-sums program are computed by setting **IFASE0** = 1 or 2 and **IFASE3** = 2.

9.11.3 ADBASE Six-Field Format

When **ADBASE** writes the efficient extreme point criterion vectors to a file (i.e., when **IPRINT(12)** = 1), the criterion vectors are written in **ADBASE** *six-field format*, as in Fig. 9.15.

If **ADBASE** generated 25 efficient extreme points and each criterion vector was of length four, the output file would contain 26 card images. If the criterion vectors, however, had eight components, there would be a second card for each criterion vector. Each second card would have a 2 in column 10, the seventh criterion value in columns 11–20, and the eighth criterion value in columns 21–30. In this case, the output file would contain 51 card images.

Header Card

Problem Number 1–5	Number of Vector Components 6–10	Number of Criterion Vectors 11–15

Criterion Vector Cards

Problem Number 1–4	Extreme Point Number 5–8	Vector Card Number 9–10	First Criterion Value 11–20	. . .	Sixth Criterion Value 61–70

Figure 9.15. Criterion vector output file in **ADBASE** six-field format.

9.11.4 An ADBASE Example

Using ADBASE with

```
NUMB       = 1
IFASE2     = 2
IFASE3     = 1
IPRINT(1)  = 1
IPRINT(3)  = 1
IPRINT(7)  = 1
IPRINT(11) = 1
IPRINT(12) = 1
IPUFMT     = 5
```

let us solve the interval criterion weights MOLP whose data deck in ADBASE input format is given in Fig. 9.16. In Fig. 9.16, note that the MOLP does not have any equality constraints, surplus constraints, or α-constants. The hardcopy output for the problem is given in Fig. 9.17 and the output file (because IPRINT(12) = 1) is given in Fig. 9.18.

```
THIS INTERVAL CRITERION WGHTS MOLP HAS 28 EFF BASES AND 22 EFF EXT PTS
   7400        4       7       7       0       0       1     40
     24
 1   1  7.0D0        1  3   6.0D0        1  4  2.0D0        1  5  5.0D0
 2   1  4.0D0        2  6   7.0D0        2  7  9.0D0        3  1  5.0D0
 3   4  6.0D0        4  3   9.0D0        4  7  4.0D0        5  1  2.0D0
 5   2  8.0D0        5  3   5.0D0        6  1  1.0D0        6  2  2.0D0
 6   3  2.0D0        6  4   6.0D0        6  5  5.0D0        6  6  8.0D0
 6   7  5.0D0        7  3   5.0D0        7  5  3.0D0        7  6  7.0D0
     7
 1    100.0D0        2    100.0D0        3    100.0D0        4    100.0D0
 5    100.0D0        6    100.0D0        7    100.0D0
     0
     0
     0
     0
     24
 1   1 -4.0D0        1  2  -2.0D0        1  3  1.0D0        1  4  2.0D0
 1   5 -4.0D0        1  6  -3.0D0        1  7  2.0D0        2  2 -3.0D0
 2   3 -4.0D0        2  5   5.0D0        2  6 -2.0D0        2  7  1.0D0
 3   1  5.0D0        3  2   5.0D0        3  4 -2.0D0        3  5  3.0D0
 3   7  5.0D0        4  1   3.0D0        4  2 -3.0D0        4  3  5.0D0
 4   5  2.0D0        4  6   2.0D0        4  7 -4.0D0        2  4  4.0D0
     0
       1   0.000D0             0.600D0
       2   0.100D0             0.500D0
       3   0.200D0             0.700D0
       4   0.300D0             1.000D0
```

Figure 9.16. Data input deck for the ADBASE example.

```
THIS INTERVAL CRITERION WGHTS MOLP HAS 28 EFF BASES AND 22 EFF EXT PTS

IPORIG=  4
N1 = 7
IK = 7
IE = 0
IS = 0
IFASE0= 1
NGRAYS=40
IFASE2= 2
IFASE3= 1

A(   1,  1)=       7.000000
A(   1,  3)=       6.000000
A(   1,  4)=       2.000000
A(   1,  5)=       5.000000
A(   2,  1)=       4.000000
A(   2,  6)=       7.000000
A(   2,  7)=       9.000000
A(   3,  1)=       5.000000
A(   3,  4)=       6.000000
A(   4,  3)=       9.000000
A(   4,  7)=       4.000000
A(   5,  1)=       2.000000
A(   5,  2)=       8.000000
A(   5,  3)=       5.000000
A(   6,  1)=       1.000000
A(   6,  2)=       2.000000
A(   6,  3)=       2.000000
A(   6,  4)=       6.000000
A(   6,  5)=       5.000000
A(   6,  6)=       8.000000
A(   6,  7)=       5.000000
A(   7,  3)=       5.000000
A(   7,  5)=       3.000000
A(   7,  6)=       7.000000

B(   1)=    100.000000
B(   2)=    100.000000
B(   3)=    100.000000
B(   4)=    100.000000
B(   5)=    100.000000
B(   6)=    100.000000
B(   7)=    100.000000

C(   1,  1)=      -4.000000
C(   1,  2)=      -2.000000
C(   1,  3)=       1.000000
C(   1,  4)=       2.000000
C(   1,  5)=      -4.000000
C(   1,  6)=      -3.000000
C(   1,  7)=       2.000000
C(   2,  2)=      -3.000000
C(   2,  3)=      -4.000000
C(   2,  4)=       4.000000
C(   2,  5)=       5.000000
C(   2,  6)=      -2.000000
C(   2,  7)=       1.000000
C(   3,  1)=       5.000000
C(   3,  2)=       5.000000
C(   3,  4)=      -2.000000
C(   3,  5)=       3.000000
C(   3,  7)=       5.000000
```

Figure 9.17. Hardcopy output for the **ADBASE** example.

```
C(  4,  1)=     3.000000
C(  4,  2)=    -3.000000
C(  4,  3)=     5.000000
C(  4,  5)=     2.000000
C(  4,  6)=     2.000000
C(  4,  7)=    -4.000000

WRANGE( 1,1)=   0.0         WRANGE( 1,2)=   0.600000
WRANGE( 2,1)=   0.100000    WRANGE( 2,2)=   0.500000
WRANGE( 3,1)=   0.200000    WRANGE( 3,2)=   0.700000
WRANGE( 4,1)=   0.300000    WRANGE( 4,2)=   1.000000



                         PROBLEM NO. 7400

                 4 REDUCED CRITERION CONE GENERATORS
                  AN EFFICIENT EXTREME POINT EXISTS

               LISTING OF ALL EFFICIENT EXTREME POINTS

EXTREME                                          VALUES OF NONZERO
 POINT       CRITERION VALUES              BASIC STRUCTURAL VARIABLES

   1     Z( 1)=      32.222222                X(  1)=       1.111111
         Z( 2)=       6.666667                X(  3)=      11.111111
         Z( 3)=     -20.000000                X(  4)=      12.777778
         Z( 4)=      58.888889

   2     Z( 1)=      24.444444                X(  3)=      11.111111
         Z( 2)=      11.111111                X(  4)=      11.111111
         Z( 3)=     -15.555556                X(  5)=       2.222222
         Z( 4)=      60.000000

   3     Z( 1)=      11.819066                X(  1)=       6.177043
         Z( 2)=       3.599222                X(  3)=       7.392996
         Z( 3)=      60.311284                X(  4)=       6.201362
         Z( 4)=      22.033074                X(  7)=       8.365759

                                    .
                                    .
                                    .

  28     Z( 1)=     -55.447942                X(  1)=       8.474576
         Z( 2)=     -46.004843                X(  3)=       6.779661
         Z( 3)=      42.372881                X(  6)=       9.443099
         Z( 4)=      78.208232

TOTAL NUMBER OF EFFICIENT BASES          =28
TOTAL NUMBER OF EFFICIENT EXTREME POINTS =22
```

Figure 9.17. *Continued*

9.12 COMPUTATIONAL EXPERIENCE

Tables 9.7 to 9.11 report computational experience concerning the generation of efficient extreme points in multiple objective linear programming. The test problems were randomly generated as follows: Starting with 50% zero density, (30% in Table 9.8), the remaining elements of the **A**-matrix were drawn from the uniform distribution of integers $[-1, \ldots, 10]$. All constraints were $\leqq$ and all RHS

```
 7400    4    22
7400    1 1   32.22222    6.66667 -20.00000  58.88889
7400    2 1   24.44444   11.11111 -15.55556  60.00000
7400    3 1   11.81907    3.59922  60.31128  22.03307
7400    4 1  -80.00000  100.00000  60.00000  40.00000
7400    6 1  -15.55556  -11.11111  20.00000  68.88889
7400    7 1  -16.54846  -13.00236  10.16548  74.46809
7400    8 1  -19.34426   24.59016  63.27869  35.08197
7400    9 1  -15.61644   17.12329  86.02740  20.13699
7400   14 1  -44.89796   10.20408  30.61224  69.38776
7400   15 1  -26.03175  -18.09524  20.00000  75.87302
7400   16 1  -27.89318   -6.48580  34.54854  64.90038
7400   17 1  -26.41366  -17.48893 116.80582   3.49779
7400   18 1  -52.97297   60.81081 105.94595   8.91892
7400   19 1  -29.76372  -47.52187  44.91101  68.26074
7400   21 1  -87.02703    8.10811  68.64865  58.37838
7400   22 1  -96.92308   38.46154 120.76923   6.15385
7400   23 1  -58.10324  -41.41657  43.81753  77.19088
7400   24 1  -26.98413  -57.14286  23.80952  82.53968
7400   25 1  -71.59236   11.78344 153.37580 -16.75159
7400   26 1  -29.62106  -51.00516  43.58655  69.80964
7400   27 1  -99.87097  -30.32258 116.83871  28.64516
7400   28 1  -55.44794  -46.00484  42.37288  78.20823
```

Figure 9.18. Output file in **ADBASE** six-field format for the **ADBASE** example.

elements were 100. The **A**-matrix sizes in the tables refer to the number of constraints by the number of structural variables. Except in Table 9.10, the elements of **C** were drawn from the uniform distribution of integers [0, ..., 10]. In the first three columns of Table 9.11, we also have randomly generated $[\ell_i, \mu_i]$ interval criterion weights. The quantities reported are

E_x	Average number of efficient extreme points per MOLP.
CPU time	Average CPU time (in seconds) per MOLP.
Calls to crash	Average number of calls to the crashing routine per MOLP.

Table 9.7
Controlling for Number of Variables

Objectives	*4*	*4*	*4*
***A**-Matrix*	*10 × 10*	*10 × 25*	*10 × 50*
Sample Size	*10*	*10*	*10*
E_x	22.4	47.3	76.6
CPU time (secs)	.9	5.2	23.5
Calls to crash	26.3	52.4	87.1
Crashing pivots	51.2	104.9	202.9
Calls to subproblem	273.0	1,337.1	4,404.5
Subproblem pivots	526.2	2,356.2	6,279.8

TABLE 9.8
Controlling for Number of Constraints

Objectives	*3*	*3*	*3*
A-Matrix	*10 × 40*	*20 × 40*	*40 × 40*
Sample Size	*10*	*10*	*5*
E_x	65.7	226.6	535.2
CPU time (secs)	9.6	44.3	155.4
Calls to crash	65.0	225.6	534.2
Crashing pivots	159.7	602.0	1,530.4
Calls to subproblem	2,640.0	9,064.0	21,408.0
Subproblem pivots	3,636.5	17,532.9	50,903.2

TABLE 9.9
Controlling for Number of Objectives

Objectives	*2*	*4*	*6*
A-Matrix	*20 × 20*	*20 × 20*	*20 × 20*
Sample Size	*10*	*10*	*3*
E_x	14.3	209.8	1,078.7
CPU time (secs)	.6	24.0	220.7
Calls to crash	13.3	212.2	1,077.7
Crashing pivots	13.3	509.5	2,568.7
Calls to subproblem	286.0	4,264.0	21,573.3
Subproblem pivots	290.3	11,579.3	89,279.3

TABLE 9.10
Controlling for Size of Criterion Cone

Objectives	*4*	*4*	*4*
A-Matrix	*15 × 30*	*15 × 30*	*15 × 30*
Criterion Cone	*[4, ..., 10]*	*[0, ..., 10]*	*[−10, ..., 10]*
Sample Size	*10*	*10*	*5*
E_x	25.7	70.7	105.6
CPU time (secs)	2.4	7.7	12.7
Calls to crash	24.7	70.3	110.0
Crashing pivots	54.4	170.5	283.1
Calls to subproblem	771.0	2,139.0	3,330.0
Subproblem pivots	847.2	3,429.0	6,007.1

TABLE 9.11
Controlling for Interval Criterion Weight Widths

Objectives	*4*	*4*	*4*	*4*
A-Matrix	*30 × 30*	*30 × 30*	*30 × 30*	*30 × 30*
Interval Width	*.1*	*.2*	*.4*	*1.0*
Sample Size	*10*	*10*	*10*	*3*
E_x w.r.t. intervals	7.0	17.3	62.1	1,471.7
CPU time (secs)	1.3	2.9	10.8	367.3
Calls to crash	6.0	16.3	61.1	1,470.7
Crashing pivots	10.8	31.8	132.8	3,859.7
Calls to subproblem	210.0	519.0	1,863.0	44,150.0
Subproblem pivots	131.4	511.1	2,974.0	144,170.0

Crashing pivots	Average number of MOLP pivots taken during crashing activities per MOLP.
Calls to subproblem	Average number of nonbasic variables tested for efficiency per MOLP.
Subproblem pivots	Average number of subproblem pivots performed per MOLP.

In Tables 9.7 to 9.11, all computations were performed using the **ADBASE** code on the University of Georgia IBM 3081D.

In Table 9.7, we note that the number of efficient extreme points grows at a rate that is less than the growth in the number of variables. In Table 9.8, the number of efficient extreme points grows more significantly with the number of constraints. In Table 9.9, we see an explosive growth in the number of efficient extreme points as the number of objectives increases. Holding the number of objectives constant, we see how the number of efficient extreme points increases as the criterion cone becomes larger (opens up) in Table 9.10. Relative to the fourth column of Table 9.11, we observe the ability of interval criterion weights to control the number of efficient extreme points generated.

Although no studies have been reported on the topic, there is suspicion that real-world MOLPs have fewer efficient extreme points than randomly generated problems. The reason for this probably has something to do with the fact that real-world problems usually have structure whereas randomly generated ones do not.

If there are no degenerate efficient extreme points (as in Table 9.11), the number of calls to the crashing routine will be one less than the number of efficient extreme points. When there are degenerate efficient extreme points, the difference will usually be greater than one. From the computational experience, there are about two to three master problem pivots per call to crash. Also from the computational experience, we see that it takes about two to three pivots, on the average, to resolve a nonbasic variable's efficiency status. As seen from the number of subproblem calls and pivots, the bulk of the computation involved in generating efficient extreme points takes place in the subproblem evaluation of the nonbasic variables.

9.13 MINIMUM CRITERION VALUES OVER THE EFFICIENT SET

To determine the maximal value of the ith criterion over the efficient set, we simply maximize the ith objective over the feasible region. To determine the minimal value of the ith criterion over the efficient set, we must solve the following problem

$$\min\left\{\mathbf{c}^i\mathbf{x} = z_i \,|\, \mathbf{x} \in E\right\}$$

Unfortunately, this problem cannot be solved in a straightforward manner because the efficient set E is not known explicitly. Moreover, for most MOLPs, E is nonconvex.

9.13.1 Payoff Tables

A *payoff table* is of the form of Table 9.12, where the rows are the criterion vectors resulting from individually maximizing each of the objectives. Unless special measures are taken when there are alternative optima, there is no guarantee that all row criterion vectors will be nondominated.

The z_i^* entries along the main diagonal form the vector of maximal criterion values (over the efficient set). The minimum value in the ith column of the payoff table is an estimate of the minimum criterion value of the ith objective over E. If the minimum column value occurs in a row whose criterion vector is dominated, the minimum column value may *underestimate* the minimum over E. Otherwise, the minimum column value will correctly specify the minimum criterion value over E or *overestimate* it.

Example 14. The MOLP of Table 9.13 has 12 efficient extreme points. For this MOLP, we obtain the payoff table of Table 9.14. Whereas the payoff table column minimums are -2.75, -1.25, -5.33, and -6.67, the criterion value minimums over the efficient set are -7.50, -7.50, -5.33, and -7.20, respectively. Thus, in this problem, three out of four column minimums incorrectly

TABLE 9.12
Payoff Table

	z_1	z_2		z_k
$\mathbf{z}^1$	z_1^*	z_{12}		z_{1k}
$\mathbf{z}^2$	z_{21}	z_2^*	$\ddots$	z_{2k}
$\mathbf{z}^k$	z_{k1}	z_{k2}		z_k^*

TABLE 9.13
MOLP of Example 14

	x_1	x_2	x_3	x_4	x_5	x_6		
	3	−2			−2		max	
Objs		−2	−2		−2	1	max	
	−2	2		2		2	max	
	1			−2	1	−2	max	
		2				1	≦	10
	2		1	5			≦	9
s.t.	3		4	5	4	2	≦	8
	3	1	4		4	1	≦	10
	3	4				3	≦	10
			3	5	4		≦	5

TABLE 9.14
Payoff Table of Example 14

	z_1	z_2	z_3	z_4
$\mathbf{z}^1$	8.00	.00	−5.33	2.67
$\mathbf{z}^2$	.00	3.33	6.67	−6.67
$\mathbf{z}^3$	−2.75	−1.25	7.75	−5.00
$\mathbf{z}^4$	8.00	.00	−5.33	2.67

specify the minimums over the efficient set. For the second objective, for instance, 57.7% of the range of the criterion value over the efficient set is below the payoff table column minimum. In this MOLP, six of the problem's 12 efficient extreme points have one or more criterion value components below their associated payoff table column minimums.

Although the payoff table approach is an easy way to obtain some information about the ranges of the criterion values over the efficient set, it is an unreliable method for obtaining the minimum values. Computational experience has shown that in the vast majority of MOLPs, one or more of the minimum criterion values over E would be incorrectly specified if we were to use the minimum column values from the payoff table.

9.13.2 Properties of the Minimum Criterion Value Problem

A point $\bar{\mathbf{x}} \in E$, at which the ith criterion attains its minimum value over the efficient set, solves the weighted-sums LP

$$\max\{\, \boldsymbol{\lambda}^T C\mathbf{x}\}$$
$$\text{s.t.} \quad A\mathbf{x} = \mathbf{b}$$
$$\mathbf{x} \geqq \mathbf{0}$$

for some $\bar{\boldsymbol{\lambda}} \in \Lambda$. Then, from duality theory, we know that there exists a $\bar{\boldsymbol{\mu}}$ that solves

$$\min\ \{\boldsymbol{\mu}^T\mathbf{b}\}$$
$$\text{s.t.}\quad \mathbf{A}^T\boldsymbol{\mu} \geqq \mathbf{C}^T\bar{\boldsymbol{\lambda}}$$

such that $\bar{\boldsymbol{\lambda}}^T\mathbf{C}\bar{\mathbf{x}} = \bar{\boldsymbol{\mu}}^T\mathbf{b}$. Hence, the problem of finding the ith criterion value minimum over E is given by the mathematical program

$$\begin{aligned} \min\ \{\mathbf{c}^i\mathbf{x} = z_i\} & \\ \text{s.t.} \qquad \mathbf{A}\mathbf{x} &= \mathbf{b} \\ \mathbf{x} &\geqq \mathbf{0} \\ \mathbf{A}^T\boldsymbol{\mu} \quad -\mathbf{C}^T\boldsymbol{\lambda} &\geqq \mathbf{0} \\ \boldsymbol{\lambda} &\geqq \boldsymbol{\delta} \\ \boldsymbol{\mu}^T\mathbf{b} \quad -\boldsymbol{\lambda}^T\mathbf{C}\mathbf{x} &= 0 \end{aligned}$$

where $\boldsymbol{\delta} \in R^k$ is a sufficiently small, strictly positive vector. This problem is difficult to solve because it is large and because the $\boldsymbol{\mu}^T\mathbf{b} - \boldsymbol{\lambda}^T\mathbf{C}\mathbf{x} = 0$ constraint is nonlinear.

Concerning the minimum criterion value problem, from Isermann and Steuer (1985), we know the following: (1) A criterion value minimum over the efficient set occurs at an extreme point. (2) Let $\mathbf{x}^g$ be an efficient extreme point at which the ith criterion achieves its maximum value, and let $\mathbf{x}^h$ be an efficient extreme point at which the ith criterion attains its minimum over E. Then, a path of efficient edges exists connecting $\mathbf{x}^g$ with $\mathbf{x}^h$ such that z_i is nonincreasing along the path from $\mathbf{x}^g$ to $\mathbf{x}^h$. (3) With regard to a given objective, it is possible for a point to be a local minimum over E and yet not be a global minimum over E.

Example 15. In the MOLP of Fig. 9.19, $E = \gamma(\mathbf{x}^1, \mathbf{x}^2) \cup \gamma(\mathbf{x}^2, \mathbf{x}^3) \cup \gamma(\mathbf{x}^3, \mathbf{x}^4)$. With $\mathbf{x}^2$ maximizing and $\mathbf{x}^4$ minimizing the second objective over E, $\gamma(\mathbf{x}^2, \mathbf{x}^3) \cup \gamma(\mathbf{x}^3, \mathbf{x}^4)$ is, with respect to z_2, an edge-connected nonincreasing path in E connecting $\mathbf{x}^2$ with $\mathbf{x}^4$. We also note that $\mathbf{x}^1$ is a local minimum over E with regard to the second objective, but it is not a global minimum.

9.13.3 A Simplex-Based Algorithm

As a simplex-based procedure to solve for the minimum criterion value of the ith objective over the efficient set, we have the following algorithm. Lexicographically, maximize each objective to construct a payoff table. Let z_i^m be the

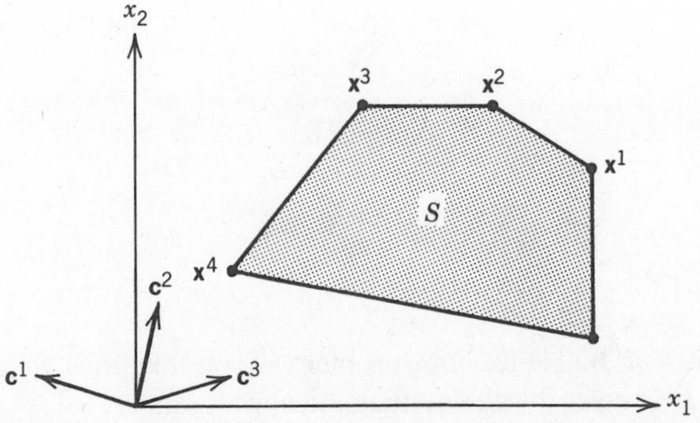

Figure 9.19. Graph of Example 15.

minimum criterion value in the ith column of the payoff table. Add $\mathbf{c}^i\mathbf{x} \leqq z_i^m$ to the constraint set. Start with the extreme point corresponding to z_i^m. Search the $\mathbf{c}^i\mathbf{x} = z_i^m$ facet of the (reduced) feasible region for an extreme point from which emanates an efficient edge that reduces the ith criterion value. If no such edge exists, the current value of z_i^m is the minimum value and we are done. If such an edge exists, pivot along the edge to its other extreme point. Assign to z_i^m the (lower) value of the ith criterion at the new extreme point. With the new value for z_i^m, add $\mathbf{c}^i\mathbf{x} \leqq z_i^m$ to the constraint set and repeat.

We lexicographically maximize each of the objectives so that each row of the payoff table will be a criterion vector of an efficient extreme point. We note that each $\mathbf{c}^i\mathbf{x} = z_i^m$ facet intersects every efficient nonincreasing edge-connected path between $\mathbf{x}^g$ and $\mathbf{x}^h$ (see Section 9.13.2) so that another iteration is required unless we are done. Hence, the algorithm terminates at a minimum over the efficient set.

Readings

1. Arrow, K. J., E. W. Barankin, and D. Blackwell (1953). "Admissible Points of Convex Sets." In H. W. Kuhn and A. W. Tucker (eds.), *Contributions to the Theory of Games*, Princeton, New Jersey: Princeton University Press.

2. Benson, H. P. (1981). "Finding an Initial Efficient Extreme Point for a Linear Multiple Objective Program," *Journal of the Operational Research Society*, Vol. 32, No. 6, pp. 495–498.

3. Benson, H. P. (1984). "Optimization Over the Efficient Set," *Journal of Mathematical Analysis and Applications*, Vol. 98, No. 2, pp. 562–580.

4. Bitran, G. R. (1980). "Linear Multiple Objective Programs with Interval Coefficients," *Management Science*, Vol. 26, No. 7, pp. 694–706.

5. Bitran, G. R. and T. L. Magnanti (1979). "The Structure of Admissible Points with Respect to Cone Dominance," *Journal of Optimization Theory and Applications*, Vol. 29, No. 4, pp. 573–614.

6. Charnes, A., W. W. Cooper, and J. P. Evans (1972). "Connectedness of the Efficient Extreme Points in Linear Multiple Objective Programs," College of Business Administration, University of North Carolina, Chapel Hill, North Carolina.

7. Dessouky, M. I., M. Ghiassi, and W. J. Davis (1979). "Determining the Worst Value of an Objective Function Within the Nondominated Solutions in Multiple Objective Linear Programming," Department of Mechanical and Industrial Engineering, University of Illinois, Urbana, Illinois.

8. Dinkelbach, W. and W. Dürr (1972). "Effizienzaussagen bei Ersatzprogrammen zum Vektormaximumproblem." In R. Henn, H. P. Künzi, and H. Schubert (eds.), *Operations Research Verfahren XII*, Verlag Anton Hain, Meisenheim, pp. 117–123.

9. Duesing, E. C. (1978). "Polyhedral Convex Sets and the Economic Analysis of Production," Ph.D Dissertation, University of North Carolina, Chapel Hill.

10. Ecker, J. G. and N. S. Hegner (1978). "On Computing an Initial Efficient Extreme Point," *Journal of the Operational Research Society*, Vol. 29, No. 10, pp. 1005–1007.

11. Ecker, J. G., N. S. Hegner, and I. A. Kouada (1980). "Generating All Maximal Efficient Faces for Multiple Objective Linear Programs," *Journal of Optimization Theory and Applications*, Vol. 30, No. 3, pp. 353–381.

12. Ecker, J. G. and I. A. Kouada (1975). "Finding Efficient Points for Linear Multiple Objective Programs," *Mathematical Programming*, Vol. 8, No. 3, pp. 375–377.

13. Ecker, J. G. and I. A. Kouada (1978). "Finding All Efficient Extreme Points for Multiple Objective Linear Programs," *Mathematical Programming*, Vol. 14, No. 2, pp. 249–261.

14. Ecker, J. G. and N. E. Shoemaker (1982). "Selecting Subsets from the Set of Nondominated Vectors in Multiple Objective Linear Programming," *SIAM Journal on Control and Optimization*, Vol. 19, No. 4, pp. 505–515.

15. Evans, J. P. and R. E. Steuer (1973). "A Revised Simplex Method for Multiple Objective Programs," *Mathematical Programming*, Vol. 5, No. 1, pp. 54–72.

16. Evans, J. P. and R. E. Steuer (1973). "Generating Efficient Extreme Points in Linear Multiple Objective Programming: Two Algorithms and Computing Experience." In J. L. Cochrane and M. Zeleny (eds.), *Multiple Criteria Decision Making*, Columbia, South Carolina: University of South Carolina Press, pp. 349–365.

17. Gal, T. (1977). "A General Method for Determining the Set of All Efficient Solutions to a Linear Vectormaximum Problem," *European Journal of Operational Research*, Vol. 1, No. 5, pp. 307–322.

18. Gal, T. and H. Leberling (1981). "Relaxation Analysis in Linear Vectorvalued Maximization," *European Journal of Operational Research*, Vol. 8, No. 3, pp. 274–282.

19. Geoffrion, A. M. (1968). "Proper Efficiency and the Theory of Vector Maximization," *Journal of Mathematical Analysis and Applications*, Vol. 22, No. 3, pp. 618–630.

20. Hartley, R. (1978). "On Cone-Efficiency, Cone-Convexity and Cone-Compactness," *SIAM Journal on Applied Mathematics*, Vol. 34, No. 2, pp. 211–222.

21. Isermann, H. (1977). "The Enumeration of the Set of All Efficient Solutions for a Linear Multiple Objective Program," *Operational Research Quarterly*, Vol. 28, No. 3, pp. 711–725.

22. Isermann, H. (1982). "Linear Lexicographic Optimization," *OR Spektrum*, Vol. 4, No. 4, pp. 223–228.

23. Isermann, H. (1984). "Operating Manual for the EFFACET Multiple Objective Linear Programming Package," Fakultät für Wirtschaftswissenschaften, Universität Bielefeld, West Germany.

24. Isermann, H., and R. E. Steuer (1985). "Payoff Tables and Minimum Criterion Values Over the Efficient Set," College of Business Administration, University of Georgia, Athens, Georgia.

25. Jahn, J. (1985). "Some Characterizations of the Optimal Solutions of a Vector Optimization Problem," *OR Spektrum*, Vol. 7, No. 1, pp. 7–17.

26. Lin, J. G. (1976). "Maximal Vectors and Multi-Objective Optimization," *Journal of Optimization Theory and Applications*, Vol. 18, No. 1, pp. 41–64.

27. Mangasarian, O. L. (1969). *Nonlinear Programming*, New York: McGraw-Hill.

28. Naccache, P. H. (1978). "Connectedness of the Set of Non-Dominated Outcomes in Multicriteria Optimization," *Journal of Optimization Theory and Applications*, Vol. 25, No. 3, pp. 459–467.

29. Philip, J. (1972). "Algorithms for the Vector-Maximization Problem," *Mathematical Programming*, Vol. 2, No. 2, pp. 207–229.

30. Rhode, R. and R. Weber (1984). "The Range of the Efficient Frontier in Multiple Objective Linear Programming," *Mathematical Programming*, Vol. 28, No. 1, pp. 84–95.

31. Salukvadze, M. E. (1974). "On the Existence of Solutions in Problems of Optimization under Vector-Valued Criteria," *Journal of Optimization Theory and Applications*, Vol. 13, No. 2, pp. 203–217.

32. Seiford, L. and P. L. Yu (1979). "Potential Solutions of Linear Systems: The Multi-Criteria Multiple Constraint Levels Program," *Journal of Mathematical Analysis and Applications*, Vol. 69, No. 2, pp. 283–303.

33. Steuer, R. E. (1976). "Multiple Objective Linear Programming with Interval Criterion Weights," *Management Science*, Vol. 23, No. 3, pp. 305–316.

34. Steuer, R. E. (1977). "An Interactive Multiple Objective Linear Programming Procedure," *TIMS Studies in the Management Sciences*, Vol. 6, pp. 225–239.

35. Steuer, R. E. (1978). "Vector-Maximum Gradient Cone Contraction Techniques," *Lecture Notes in Economics and Mathematical Systems*, No. 155, Berlin: Springer-Verlag, pp. 462–481.

36. Steuer, R. E. (1983). "Operating Manual for the ADBASE Multiple Objective Linear Programming Package," College of Business Administration, University of Georgia, Athens, Georgia.

37. Tamura, K. and S. Miura (1977). "On Linear Vector Maximization Problems," *Journal of the Operations Research Society of Japan*, Vol. 20, No. 3, pp. 139–149.

38. Weistroffer, H. R. (1983). "On Pessimistic Values in Multiple Objectives Optimization," Department of Mathematical Sciences, Virginia Commonwealth University, Richmond, Virginia.

39. Wendell, R. E. and D. N. Lee (1977). "Efficiency in Multiple Objective Optimization Problems," *Mathematical Programming*, Vol. 12, No. 3, pp. 406–414.

40. Yu, P. L. (1974). "Cone Convexity, Cone Extreme Points, and Nondominated Solutions in Decision Problems with Multiobjectives," *Journal of Optimization Theory and Applications*, Vol. 14. No. 3, pp. 319–377.

41. Yu, P. L. and G. Leitmann (1974). "Compromise Solutions, Domination Structures, and Salukvadze's Solution," *Journal of Optimization Theory and Applications*, Vol. 13, No. 3, pp. 362–378.

42. Yu, P. L. and M. Zeleny (1975). "The Set of All Non-Dominated Solutions in Linear Cases and a Multicriteria Simplex Method," *Journal of Mathematical Analysis and Applications*, Vol. 49, No. 2, pp. 430–468.

43. Zeleny, M. (1973). "Compromise Programming." In Cochrane, J. L. and M. Zeleny (eds.), *Multiple Critieria Decision Making*, Columbia, South Carolina: University of South Carolina Press, pp. 262–301.

44. Zeleny, M. (1974). "Linear Multiobjective Programming," *Lecture Notes in Economics and Mathematical Systems*, No. 95, Berlin: Springer-Verlag.

45. Zeleny, M. (1976). "Multicriteria Simplex Method: A Fortran Routine," *Lecture Notes in Economics and Mathematical Systems*, No. 123, Berlin: Springer-Verlag, pp. 323–345.

46. Zionts, S. and J. Wallenius (1980). "Identifying Efficient Vectors: Some Theory and Computational Results," *Operations Research*, Vol. 28, No. 3, pp. 785–793.

PROBLEM EXERCISES

9-1 Graphing the feasible region defined by $\mathbf{x}^1 = (0,0,0)$, $\mathbf{x}^2 = (2,4,0)$, $\mathbf{x}^3 = (2,4,2)$, and $\mathbf{x}^4 = (4,0,0)$, specify E and E^w when $\mathbf{c}^1 = (1,0,0)$, $\mathbf{c}^2 = (-2,1,0)$, and $\mathbf{c}^3 = (0,0,1)$.

9-2 Consider the MOLP:

$$\begin{aligned}
\max\{ \; x_1 \qquad\quad &= z_1\} \\
\max\{ \qquad\quad x_2 &= z_2\} \\
\text{s.t.} \qquad\quad x_2 &\leqq 4 \\
x_1 + 2x_2 &\leqq 12 \\
2x_1 + x_2 &\leqq 12 \\
x_1 \qquad &\leqq 4 \\
x_1, x_2 &\geqq 0
\end{aligned}$$

Specify the coordinates of the Phase II efficient extreme point located by:

(a) The weighted-sums method (with equal weights).

(b) Lexicographic maximization.

(c) Lexicographic maximization with the order of the objectives reversed.

9-3 For the MOLP

$$\begin{aligned}
\max\{ \; 12x_1 + 6x_2 &= z_1\} \\
\max\{ \; -x_1 \qquad &= z_2\} \\
\text{s.t.} \quad -2x_1 + x_2 &\leqq 4 \\
-x_1 + 3x_2 &\leqq 15 \\
x_2 &\leqq 8 \\
x_1, x_2 &\geqq 0
\end{aligned}$$

graphically determine the first efficient extreme point detected by using:

(a) Weighted-sums method (with equal weights).

(b) Weighted-sums method (with equal weights) with subproblem testing for efficiency at each extreme point visited.

(c) Lexicographic maximization.

9-4 For the MOLP

$$\begin{aligned} &\max\{-2x_1 + 4x_2 = z_1\} \\ &\max\{\ \ x_1 - 2x_2 = z_2\} \\ &\text{s.t.} \quad -x_1 + x_2 \leqq 2 \\ &\qquad\qquad x_2 \geqq 1 \\ &\qquad\qquad x_1, x_2 \geqq 0 \end{aligned}$$

graphically determine the first efficient extreme point detected by using:

(a) Weighted-sums method (with equal weights).

(b) Lexicographic maximization.

(c) Lexicographic maximization with subproblem testing for efficiency after each maximization stage.

(d) Lexicographic maximization with subproblem testing for efficiency at each extreme point visited.

(e) Ecker-Kouada method.

9-5 Using lexicographic maximization, specify S_1, S_2, and S_3 for the MOLP:

$$\begin{aligned} &\max\{\ x_1 = z_1\} \\ &\max\{2x_1 + x_2 = z_2\} \\ &\max\{\ x_1 + x_2 + x_3 = z_3\} \\ &\text{s.t.} \quad x_1 \leqq 3 \\ &\qquad x_1 + x_2 \leqq 5 \\ &\qquad x_2 \leqq 4 \\ &\qquad x_3 \leqq 2 \\ &\qquad x_1, x_2, x_3 \geqq 0 \end{aligned}$$

9-6 Using the origin as the reference point, consider the MOLP:

$$\begin{aligned} &\max\{\ 2x_1 + x_2 = z_1\} \\ &\max\{-x_1 + 3x_2 = z_2\} \\ &\text{s.t.} \quad x_2 \leqq 2 \\ &\qquad -x_1 + 2x_2 \leqq 2 \\ &\qquad x_1, x_2 \geqq 0 \end{aligned}$$

(a) Graphically determine the efficient point that would be found by the Ecker-Kouada method.

(b) Formulate the Ecker-Kouada program that will compute the efficient point of (a).

9-7C Consider the MOLP:

$$\begin{aligned}
\max\{\ 3x_1 + \ x_2 &= z_1\} \\
\max\{-x_1 + 2x_2 &= z_2\} \\
\text{s.t.} \quad 3x_1 + 2x_2 &\geqq 6 \\
x_1 \qquad &\leqq 10 \\
x_2 &\leqq 3 \\
x_1, x_2 &\geqq 0
\end{aligned}$$

(a) Use the computer to determine the extreme point found by Phase I.

(b) Using the Phase I extreme point as the reference point, use the computer to determine the efficient point found by the Ecker-Kouada method.

9-8C Let $\mathbf{x}^0 = (3,1)$. Applying Benson's method, what is the $\bar{\lambda}$ in Step 3 that is used to compute the efficient extreme point of:

$$\begin{aligned}
\max\{\ 4x_1 - \quad x_2 &= z_1\} \\
\max\{-x_1 + 100x_2 &= z_2\} \\
\text{s.t.} \qquad x_2 &\leqq 5 \\
x_1, x_2 &\geqq 0
\end{aligned}$$

9-9 For the following criterion matrix and interval weights, compute the **D** criterion matrix of the interval criterion weights vector-maximum problem.

$$\mathbf{C} = \begin{bmatrix} -6 & 2 & 0 \\ 1 & 5 & 7 \\ -3 & -1 & 4 \\ 2 & 0 & 1 \end{bmatrix} \qquad \begin{aligned} \lambda_1 &\in \text{rel}[.0,.2] \\ \lambda_2 &\in \text{rel}[.1,.3] \\ \lambda_3 &\in \text{rel}[.2,.6] \\ \lambda_4 &\in \text{rel}[.3,.5] \end{aligned}$$

9-10 Consider the following criterion matrix and interval weights.

$$\mathbf{C} = \begin{bmatrix} -3 & 3 & 1 \\ 1 & 5 & 2 \\ -3 & -1 & 2 \\ 2 & 0 & 1 \end{bmatrix} \qquad \begin{aligned} \lambda_1 &\in \text{rel}[.0,.2] \\ \lambda_2 &\in \text{rel}[.1,.4] \\ \lambda_3 &\in \text{rel}[.2,.6] \\ \lambda_4 &\in \text{rel}[.3,.5] \end{aligned}$$

(a) Graphing the $x_3 = 3$ cross-section of the **C**-cone, can any of the rows of **C** be eliminated as nonessential?

(b) Compute criterion matrix **D**.

(c) Graphing the $x_3 = 3$ cross-section of the **D**-cone, can any of the rows of **D** be eliminated as nonessential?

(d) Compute criterion matrix **E**.

9-11 For the following criterion matrix and "mixed" interval weights, compute the **E** criterion matrix of the enveloping reduced vector-maximum problem.

$$\mathbf{C} = \begin{bmatrix} 1 & 3 & 0 \\ 7 & 1 & -1 \\ -4 & 0 & 0 \\ 3 & 1 & 1 \end{bmatrix} \qquad \begin{array}{l} \lambda_1 \in \text{rel}[.0,.4] \\ \lambda_2 \in \text{rel}[.1,.1] \\ \lambda_3 \in \text{rel}[.3,.5] \\ \lambda_4 \in \text{rel}[.2,.3] \end{array}$$

9-12 Consider the following criterion matrix and interval weights.

$$\mathbf{C} = \begin{bmatrix} 0 & 0 & 10 \\ 10 & 0 & 10 \\ 5 & 10 & 10 \end{bmatrix} \qquad \begin{array}{l} \lambda_1 \in \text{rel}[.3,.4] \\ \lambda_2 \in \text{rel}[.2,.5] \\ \lambda_3 \in \text{rel}[.3,.5] \end{array}$$

Graphically determine the critical $\boldsymbol{\lambda}$-vectors pertaining to the generators (a) of the **D**-cone and (b) of the **E**-cone.

9-13 Using the Evans-Steuer method, determine which columns of **W** pertain to efficient nonbasic variables when:

$$\mathbf{W} = \begin{bmatrix} -1 & -4 & -1 & 2 & 1 & -1 \\ -6 & 3 & 0 & -4 & -5 & -2 \\ 2 & 5 & -2 & -1 & 5 & 1 \end{bmatrix}$$

9-14 Using Ecker's method, determine which columns of **W** pertain to efficient nonbasic variables when:

$$\mathbf{W} = \begin{bmatrix} -4 & 0 & 4 & 1 & 2 & -4 \\ 0 & -2 & -3 & 0 & -1 & -2 \\ 1 & 3 & 1 & -2 & 1 & 4 \end{bmatrix}$$

9-15 Using the Zionts-Wallenius routine, determine which columns of **W** pertain to efficient nonbasic variables when:

$$\mathbf{W} = \begin{bmatrix} 2 & -1 & -1 & -1 & 2 & -2 & 4 \\ -2 & -2 & -4 & 3 & -1 & 3 & -4 \\ -2 & -1 & 1 & -2 & 4 & 0 & -4 \end{bmatrix}$$

9-16 Let Z be the convex polyhedron defined by the following criterion vectors:

$$\mathbf{z}^1 = (5,4)$$
$$\mathbf{z}^2 = (4,5)$$
$$\mathbf{z}^3 = (0,6)$$
$$\mathbf{z}^4 = (0,0)$$
$$\mathbf{z}^5 = (8,3)$$

Let it be given that $\mathbf{z}^1$ is a nondominated extreme point of **Z**. With respect to $\mathbf{z}^1$, which of the other $\mathbf{z}^i$ constitute nondominated adjacent extreme points of Z? Use the Zionts-Wallenius routine.

9-17C Determine which extreme points and unbounded edges are efficient, and which are not, in the MOLP:

$$\begin{aligned}
\max \{ -10x_1 + 2x_2 &= z_1 \} \\
\max \{ x_1 - x_2 &= z_2 \} \\
\text{s.t.} \quad x_1 + x_2 + x_3 &\geqq 2 \\
x_1 + x_3 &\geqq 1 \\
x_1, x_2, x_3 &\geqq 0
\end{aligned}$$

9-18C Solve the following MOLP for all efficient extreme points.

	x_1	x_2	x_3	x_4	x_5	x_6	x_7		
Objs	2	−1		2	−1		2	max	
	2		1	4	−1	−1	−1	max	
	1	1	2	2	−1	−1		max	
		3		−1	−1	3	−1	max	
s.t.	2	3	3					≦	95
		4		5	3			≦	87
					4	3		≦	73
	3					2	−1	≦	54
	5		3				5	≦	67
		2		5			−1	≦	64

9-19C Solve the following MOLP for all efficient extreme points.

	x_1	x_2	x_3	x_4	x_5	x_6	x_7	x_8		
Objs		−1		−1	4			2	max	
	−1		−1		2	2	3	2	max	
	3			3	1	1	1		max	
	−1	3	−1	3	2	1	−1		max	
			2	4	−1		2	−1	max	
	1	3	2	4		3			max	
s.t.	5	5		6	2	1			≦	66
	1	6	3				6		≦	90
	3	2				5			≦	96
	5			2	4				≦	64
			2	4			3	6	≦	77
		6	6	6		4	3		≦	86
			1					5	≦	73
	5			4			2	6	≦	60

9-20C How many bases and extreme points are efficient in the following MOLP?

	x_1	x_2	x_3	x_4	x_5	x_6	x_7	x_8		
	−3	−5	3	3	−5	6	4	5	max	
Objs	−4	−3	4	−6	6	−1		−1	max	
	−2	−2	−4		2	−6	6	−2	max	
	2	1		2		1	4	7	≦	8
	4		5	4	3	5		8	≦	7
		1		3	8	2	−3	2	≦	7
	−3			−1	−3		3	8	≦	7
s.t.	−2			6	−3	5		7	≦	5
	7	−2		−1	4		7		≦	7
	−1		5		3	4	4	4	≦	6
	3		−1		4			−1	≦	6

9-21C How many bases, extreme points, and unbounded edges are efficient in the following MOLP?

	x_1	x_2	x_3	x_4	x_5	x_6	x_7	x_8		
		−7		−3	6	5	7	1	max	
Objs	−3	−2	2	−6	5	6	−6	5	max	
	−2	2	6	−1	3	2	2	−1	max	
	2		3		5			2	≦	8
	−3		6			6	5	−1	≦	6
	3	3		−3	5		4		≦	6
	4	−2		2	6	−1	6		≦	7
s.t.		−3				−2		−3	≦	7
	−3	−3	3	−1	4	3	1	−3	≦	10
		2	4	−2		−3	7	−1	≦	10
	1		8			2	4	2	≦	9

9-22C **(a)** Compute all efficient extreme points and maximally efficient facets of:

	x_1	x_2	x_3	x_4	x_5		
		2	2	−2	1	max	
Objs	−1	−2	1	2	2	max	
	−1			1	1	max	
		2			−2	max	
				3		≦	24
	3	3	4			≦	27
s.t.	4	1	4	2		≦	40
				4	2	≦	38
					2	≦	27

(b) Display the graph of all efficient extreme points and maximally efficient facets.

9-23C **(a)** Compute all efficient extreme points, efficient bases, and maximally efficient facets of:

	x_1	x_2	x_3	x_4	x_5	x_6		
Objs	-1	-1	-2	3		-2	max	
	2			-2		2	max	
	-1	3	2		-2	3	max	
s.t.			4	4	2		$\leq$	21
	4		5	5			$\leq$	32
	2		3	3	3	2	$\leq$	23
	2				5	3	$\leq$	30
		4		5	4	4	$\leq$	35
		3	5		3	5	$\leq$	36

(b) Display the graph of all efficient bases.
(c) Display the graph of all efficient extreme points and maximally efficient facets.

9-24C How many bases and extreme points would be efficient if we were to solve the following MOLP with:

(a) Only the first two objectives?
(b) The first three objectives?
(c) All four objectives?

	x_1	x_2	x_3	x_4	x_5		
Objs	3	1			3	max	
		4	4	3		max	
	-3	-5	-4	-3	-3	max	
		6				max	
s.t.				3	5	$\leq$	8
	-1		5	1	-1	$\leq$	9
	1					$\leq$	10
	2			5		$\leq$	8
		3		4	5	$\leq$	8

9-25C Solve the MOLP of this problem for all efficient extreme points. How many extreme points are efficient w.r.t. to the **D**-cone formed by the interval criterion weights?

(a) $\lambda_1 \in \text{rel}\,[.0, .6]$ **(b)** $\lambda_1 \in \text{rel}\,[.4, .6]$
$\lambda_2 \in \text{rel}\,[.0, .6]$ $\lambda_2 \in \text{rel}\,[.0, .2]$
$\lambda_3 \in \text{rel}\,[.0, .6]$ $\lambda_3 \in \text{rel}\,[.2, .4]$
$\lambda_4 \in \text{rel}\,[.0, .6]$ $\lambda_4 \in \text{rel}\,[.0, .2]$

	x_1	x_2	x_3	x_4	x_5	x_6	x_7	x_8	x_9	x_{10}		
Objs		2	3		2				3		max	
	3		2			1	1	3	−1	4	max	
	−1		3	3	3	4	2	1	−1	4	max	
	−1			4	−1	3		2	−1	3	max	
s.t.					1	2					≤	51
		8	6	8							≤	56
	2		5		1				3	5	≤	81
			5	7		1	3				≤	77
	7		5		8		2	1			≤	81
		7		4	3				6		≤	93
		1			4	6		3		7	≤	77
			8		7					6	≤	76
		8	2				5	8	3		≤	64
	5	3	8				6	1			≤	100

9-26C Do (a) and (b) of Problem 9-25C but w.r.t. to the **E**-cone (enveloping reduced criterion cone).

9-27C Consider the MOLP:

	x_1	x_2	x_3	x_4	x_5	x_6	x_7	x_8		
Objs	−2	4	1		6	−6	−7	−2	max	
	3	−4	−3	4	−1		−2	5	max	
	5	6	8	−6	−7			−3	max	
s.t.			4		1	−3	6	3	≤	6
	−1	7	1	8	8	7		3	≤	10
	1				2		2	5	≤	5
	5	7	−3	5	2		8	5	≤	6
	4			1		−2	−3		≤	8
	2	−1			4		8	−3	≤	10
		4		3					≤	10
	2	8			1	−3	−2	1	≤	10

(a) Individually maximize each of the objective functions. From the results, construct the payoff table.

(b) What are the criterion value ranges for the different objectives as observed from the payoff table?

(c) Solve the MOLP for all efficient extreme points.

(d) What are the criterion value ranges for the different objectives over the efficient set?

(e) How many of the efficient extreme points have one or more criterion values below the minimums seen in the payoff table?

9-28C Consider the MOLP:

	x_1	x_2	x_3	x_4	x_5	x_6		
Objs	5	6	2	−3	−2	4	max	
	3	−4	5	−1	5		max	
	−1		−5	1	−4	1	max	
	1		1	−1	−3	3	max	
s.t.			3	−1	3	6	≤	6
	3	6	2	1		6	≤	5
	5	5		5	3		≤	9
		6	6				≤	8
	2			4		4	≤	8
	6	1				4	≤	5
	5		5	4			≤	9

(a) Individually maximize each of the objective functions. From the results, construct the payoff table.

(b) What are the criterion value ranges for the different objectives as observed from the payoff table?

(c) Solve the MOLP for all efficient extreme points.

(d) What are the criterion value ranges for the different objectives over the efficient set?

9-29 Consider the MOLP:

$$\begin{aligned}
&\max\{\quad x_1 + 2x_2 = z_1\}\\
&\max\{-2x_1 + x_2 = z_2\}\\
&\max\{\quad x_1 - 2x_2 = z_3\}\\
&\text{s.t.} \qquad \mathbf{x} \in S
\end{aligned}$$

where S is the convex hull of $\mathbf{x}^1 = (1,0)$, $\mathbf{x}^2 = (0,3)$, $\mathbf{x}^3 = (4,4)$, and $\mathbf{x}^4 = (6,0)$.

(a) What is the efficient set?

(b) What are the criterion value minimums as seen in the payoff table?

(c) What are the criterion value minimums over the efficient set?

CHAPTER 10

Goal Programming

Initially conceived as an application of single objective linear programming by Charnes and Cooper (1955, 1961), goal programming (GP) gained popularity in the 1960s and 70s from the works of Ijiri (1965), Lee (1972), and Ignizio (1976). GP is now an important area of multiple criteria optimization. The idea of goal programming is to establish a goal level of achievement for each criterion. GP is ideal for criteria with respect to which *target* (or threshold) values of achievement are of significance. Goal programming is distinguished from linear programming by:

1. The conceptualization of objectives as *goals*.
2. The assignment of *priorities* and/or *weights* to the achievement of the goals.
3. The presence of *deviational* variables d_i^+ and d_i^- to measure *overachievement* and *underachievement* from target (or threshold) levels t_i.
4. The minimization of weighted-sums of deviational variables to find solutions that best satisfy the goals.

Usually, a point that satisfies all the goals is not feasible. Thus, we try to find a feasible point that achieves the goals "as closely as possible." The way in which such points are found using priority and/or weighting structures defines goal programming.

For a conventional treatment of goal programming, the reader is referred to sources such as Lee (1972), Ignizio (1976 and 1983), and Schniederjans (1984). However, in this chapter, we take what might be considered a more critical look at goal programming. Our purpose is to study goal programming in such a way that we will not be left with a naive understanding of GP's strengths and weaknesses. The chapter also relates GP to vector-maximization and shows how the concept of the deviational variable can be integrated into the practice of MOLP in general.

10.1 GOALS AND UTOPIAN SETS

A multiple objective problem may have four types of goal criteria, as portrayed in Fig. 10.1:

1. Greater than or equal to.
2. Less than or equal to.
3. Equality.
4. Range.

The t_i are target values (a) on or above which, (b) on or below which, (c) at which, or (d) between which we wish to reside.

A goal programming problem with, for instance, one of each type of goal criterion is expressed as

$$\begin{aligned} &\text{goal}\ \{\mathbf{c}^1\mathbf{x} = z_1\} && (z_1 \geqq t_1) \\ &\text{goal}\ \{\mathbf{c}^2\mathbf{x} = z_2\} && (z_2 \leqq t_2) \\ &\text{goal}\ \{\mathbf{c}^3\mathbf{x} = z_3\} && (z_3 = t_3) \\ &\text{goal}\ \{\mathbf{c}^4\mathbf{x} = z_4\} && (z_4 \in [t_4^\ell, t_4^\mu]) \\ &\text{s.t.} \quad \mathbf{x} \in S \end{aligned}$$

The information in parentheses on the left specifies the values of the z_i to be achieved (if possible) in relation to stipulated t_i target values.

Example 1. Consider the GP

$$\begin{aligned} &\text{goal}\ \{\mathbf{c}^1\mathbf{x} = z_1\} && (z_1 \geqq t_1) \\ &\text{goal}\ \{\mathbf{c}^2\mathbf{x} = z_2\} && (z_2 \in [t_2^\ell, t_2^\mu]) \\ &\text{s.t.} \quad \mathbf{x} \in S \end{aligned}$$

whose decision space representation is given in Fig. 10.2*a*. In Fig. 10.2*a* where $\mathbf{c}^1 = (1, \frac{1}{2})$, $\mathbf{c}^2 = (\frac{1}{2}, 1)$, $\mathbf{x}^1 = (4, 1)$, and $\mathbf{x}^2 = (0, 5)$, the cross-hatched area is the *utopian set* in *decision space*. This is the set of points in R^n at which all goals are simultaneously satisfied. The criterion space representation of the GP is given in Fig. 10.2*b*. In Fig. 10.2*b* where $\mathbf{z}^1 = (4\frac{1}{2}, 3)$ and $\mathbf{z}^2 = (2\frac{1}{2}, 5)$, the cross-hatched area is the *utopian set* in *criterion space*. This is the set of criterion vectors in R^k that simultaneously satisfy all goals. Since there are no points in Fig. 10.2 that feasibly satisfy all goals simultaneously, our goal programming endeavor is to find the point in S whose criterion vector "best" compares with the utopian set in criterion space.

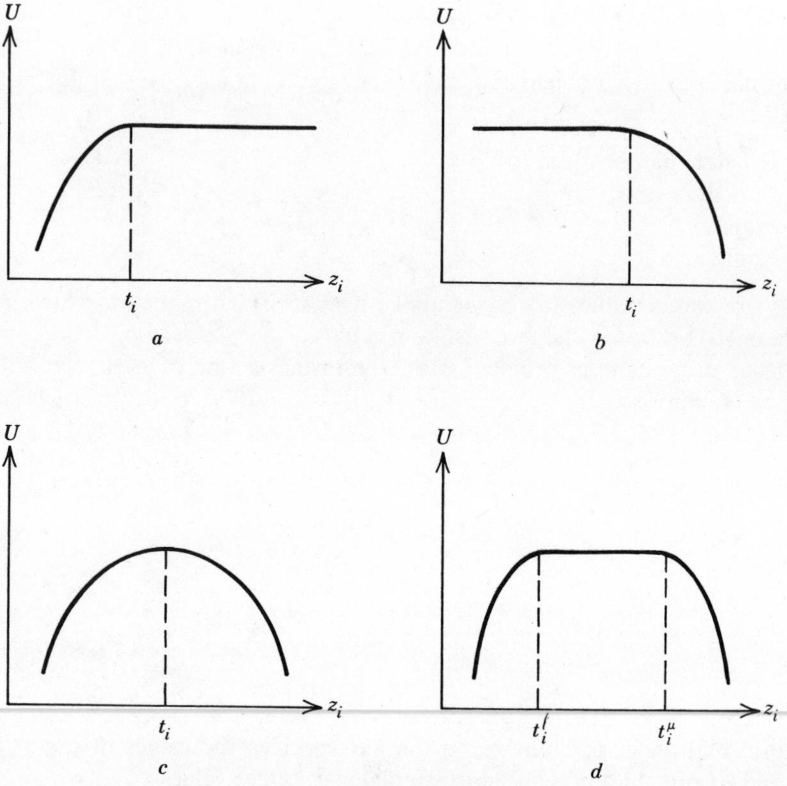

Figure 10.1. Goal criteria utility function shapes.

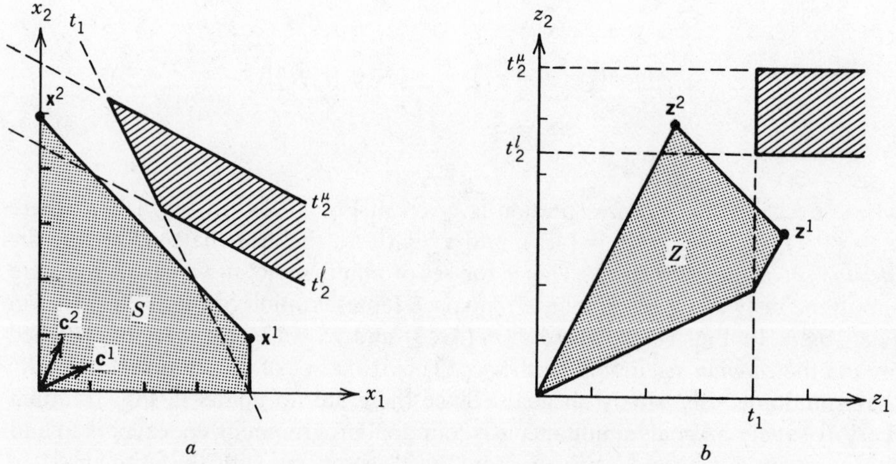

Figure 10.2. Graphs of Example 1.

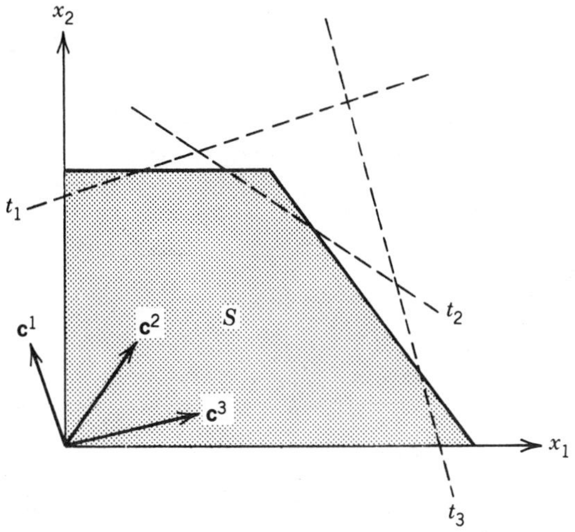

Figure 10.3. Graph of Example 2.

Example 2. Consider the GP

$$\begin{aligned} &\text{goal}\,\{\mathbf{c}^1\mathbf{x} = z_1\} && (z_1 = t_1) \\ &\text{goal}\,\{\mathbf{c}^2\mathbf{x} = z_2\} && (z_2 = t_2) \\ &\text{goal}\,\{\mathbf{c}^3\mathbf{x} = z_3\} && (z_3 = t_3) \\ &\text{s.t.} \quad \mathbf{x} \in S \end{aligned}$$

of Fig. 10.3. In this problem, the utopian set in decision space is empty, but the utopian set in criterion space consists of the point (t_1, t_2, t_3). Although the utopian set in *decision space* does not exist, there is no problem because the goal programming endeavor is to find the point in S whose criterion vector best compares with the utopian set in *criterion space*.

10.2 ARCHIMEDIAN GP

In goal programming there are two basic models: the *Archimedian* model and the *preemptive* model. With the Archimedian model, we generate candidate solutions by computing points in S whose criterion vectors are closest, in a weighted L_1-metric sense, to the utopian set in criterion space. With the preemptive model, we generate solutions whose criterion vectors are most closely related, in a lexicographic sense, to points in the utopian set. Archimedian goal programming is the subject of this section. Preemptive GP is discussed in Section 10.4. Consider the GP

$$\begin{array}{ll} \text{goal}\,\{\mathbf{c}^1\mathbf{x} = z_1\} & (z_1 \geqq t_1) \\ \text{goal}\,\{\mathbf{c}^2\mathbf{x} = z_2\} & (z_2 = t_2) \\ \text{goal}\,\{\mathbf{c}^3\mathbf{x} = z_3\} & (z_3 \in [t_3^\ell, t_3^\mu]) \\ \text{s.t.} \quad \mathbf{x} \in S & \end{array}$$

The *Archimedian formulation* of the above GP is

$$\begin{array}{lllll} \min\,\{w_1^- d_1^- + w_2^+ d_2^+ + w_2^- d_2^- + w_3^- d_3^- + w_3^+ d_3^+\} & & & & \\ \text{s.t.} \quad \mathbf{c}^1\mathbf{x} + d_1^- & & & \geqq t_1 & \\ \quad \mathbf{c}^2\mathbf{x} & -d_2^+ + d_2^- & & = t_2 & \\ \quad \mathbf{c}^3\mathbf{x} & & +d_3^- & \geqq t_3^\ell & \text{goal constraints} \\ \quad \mathbf{c}^3\mathbf{x} & & -d_3^+ & \leqq t_3^\mu & \\ & & & \mathbf{x} \in S & \\ & & & \text{all } d\text{'s} \geqq 0 & \end{array}$$

about which we make the following observations.

1. The w's in the objective function are positive *penalty* weights.
2. Each goal gives rise to a *goal constraint*, except range goals that give rise to two.
3. Only deviational variables associated with *undesirable* deviations need be employed in the formulation.
4. The Archimedian objective function is a weighted-sum of the undesirable deviational variables.
5. Archimedian GPs can be solved using conventional LP software.

The goal constraints are *soft* constraints in that they do not restrict the original feasible region S. In effect, they *augment* the feasible region by casting S into higher dimensional space, thereby creating the augmented (or Archimedian) GP feasible region.

The w's allow us to penalize undesirable deviations from goal with different degrees of severity, as shown in Fig. 10.4. In fact, more elaborate means of penalizing deviations from goal can be employed as in Fig. 10.5 using piecewise linear programming [see Charnes and Cooper (1980)].

10.2.1 Contours of Archimedian Objective Functions

Since the Archimedian objective function is a weighted-sum of the undesirable deviational variables, Archimedian GP is a *type* of L_1-metric optimization. Let us now study the objective function contours of Archimedian goal programs.

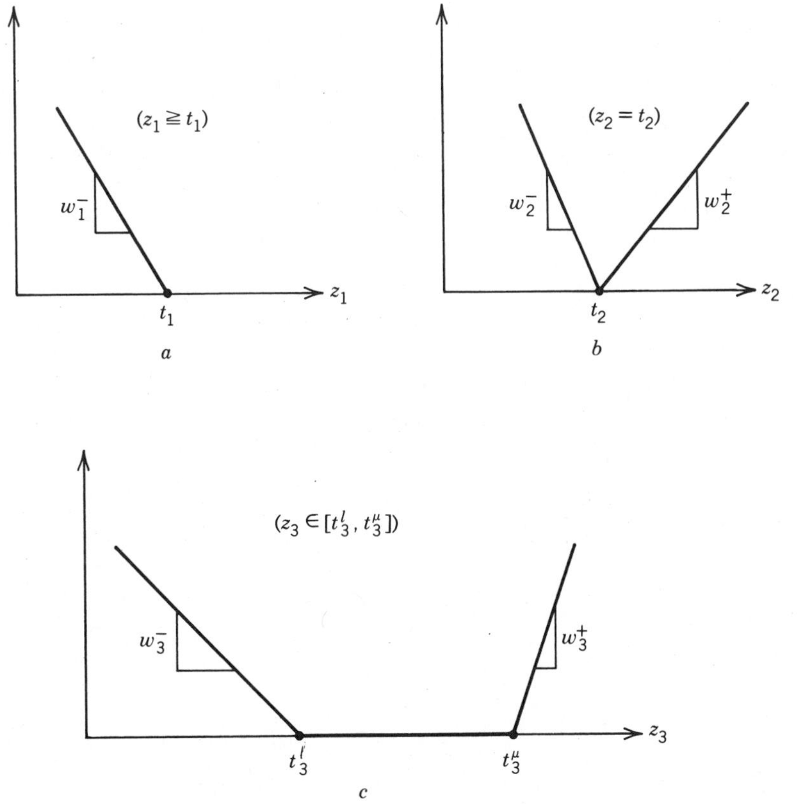

Figure 10.4. Effect of penalty weights.

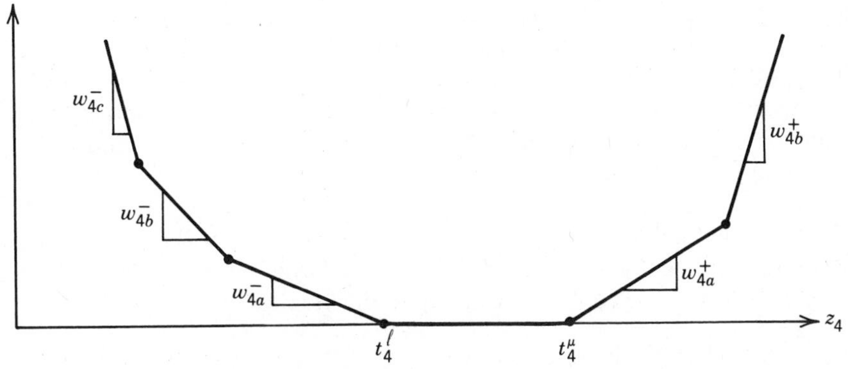

Figure 10.5. Piecewise linear penalties.

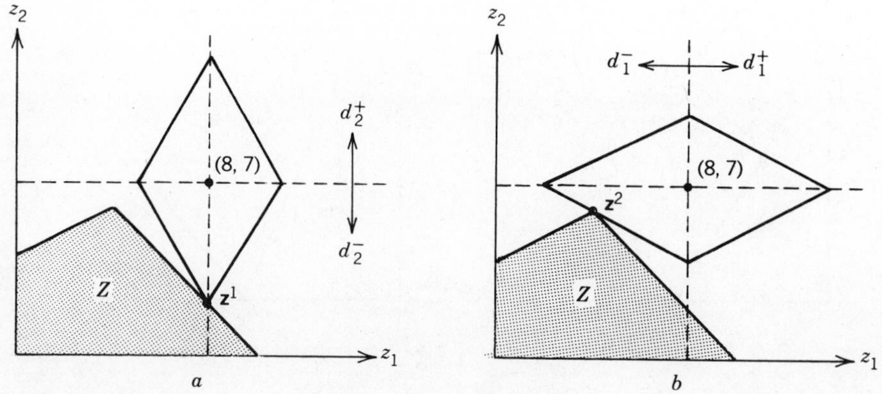

Figure 10.6. Graphs of Example 3.

Example 3. Consider the GP

$$\begin{aligned} &\text{goal}\{\ x_1 \qquad\quad = z_1\} \qquad (z_1 = 8) \\ &\text{goal}\{\qquad\quad x_2 = z_2\} \qquad (z_2 = 7) \\ &\text{s.t.} \quad -x_1 + 2x_2 \leqq 8 \\ &\qquad\quad x_1 + x_2 \leqq 10 \\ &\qquad\quad x_1, x_2 \geqq 0 \end{aligned}$$

The utopian set in criterion space is the point (8,7) and the Archimedian objective is

$$\min\{w_1^+ d_1^+ + w_1^- d_1^- + w_2^+ d_2^+ + w_2^- d_2^-\}$$

Let $w_1^+ = w_1^-$ and $w_2^+ = w_2^-$. With $w_1^+ > w_2^+$, the contours of the Archimedian objective function are *horizontally* compressed diamonds centered at (8,7). Thus, $\mathbf{z}^1 = (8, 2)$ is the criterion vector solution of the Archimedian GP because $\mathbf{z}^1$ is the point of intersection between the smallest diamond centered at (8,7) and Z, as shown in Fig. 10.6*a*. With $w_1^+ < w_2^+$, the diamond contours are *vertically* compressed yielding $\mathbf{z}^2 = (4, 6)$ as the criterion vector solution of the Archimedian GP as shown in Fig. 10.6*b*.

When the utopian set in criterion space is more than an individual point, the contours of the Archimedian objective function will not be diamond-shaped exactly as in Fig. 10.6. To illustrate, consider the utopian sets created by

1. Two range criteria.
2. One greater than or equal to criterion, and one equality criterion.
3. Two greater than or equal to criteria.

With equal weights, contour shapes for the above cases are shown in Figs. 10.7*a*, *b*, and *c*, respectively.

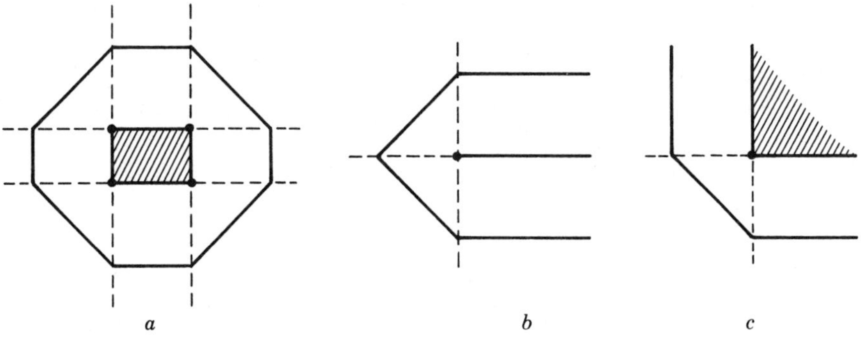

Figure 10.7. Archimedian contour shapes.

10.2.2 Archimedian GP Solutions

This section presents two more Archimedian GP examples. Then, the types of solutions that can be generated by the Archimedian approach are discussed.

Example 4. The objective of the Archimedian GP of Fig. 10.8 is

$$\min \{ w_1^+ d_1^+ + w_2^- d_2^- \}$$

Depending on the weights, either (a) $\mathbf{x}^1$ (uniquely), (b) $\gamma(\mathbf{x}^1, \mathbf{x}^2)$, or (c) $\mathbf{x}^2$ (uniquely) is generated as the solution of the GP.

Example 5. The solution set of the GP of Fig. 10.9 is the region $\gamma(\mathbf{x}^1, \mathbf{x}^2, \mathbf{x}^3, \mathbf{x}^4)$. It can be argued that the points in $\gamma(\mathbf{x}^1, \mathbf{x}^2, \mathbf{x}^3, \mathbf{x}^4)$ are not merely candidate solutions but that they are optimal solutions of the multiple goal program because they are in the utopian set.

When using conventional LP software with the Archimedian approach, we are only able to generate extreme points of the Archimedian GP feasible region (i.e.,

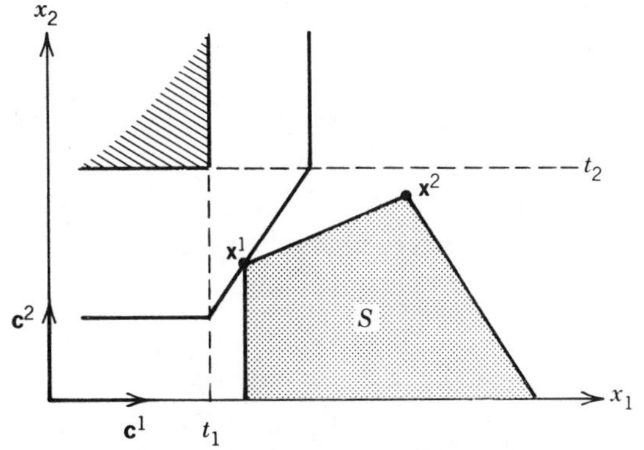

Figure 10.8. Graph of Example 4.

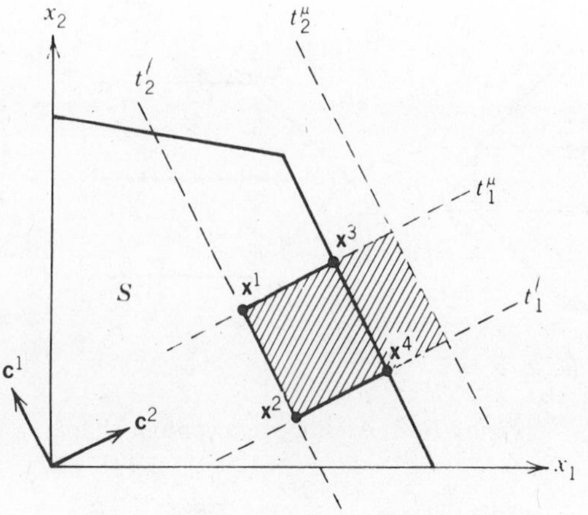

Figure 10.9. Graph of Example 5.

S after it has been augmented by the goal constraints). Thus, only a finite number of different points can be generated and they manifest themselves in decision space as either

1. Extreme points of S.
2. Nonextreme boundary points of S such as (a) the inverse image of $\mathbf{z}^1$ in Fig. 10.6 and (b) $\mathbf{x}^3$ and $\mathbf{x}^4$ in Fig. 10.9.
3. Interior points of S such as $\mathbf{x}^1$ and $\mathbf{x}^2$ in Fig. 10.9.

If the decision maker's most preferred point lies at other than an extreme point of the Archimedian GP feasible region, it cannot be generated unless we get into the business of manipulating the t_i target values (Section 10.8).

10.3 OCMM MODELS

Illustrative of some Archimedian GP models that have been used in practice are the manpower planning models that have been developed by Charnes, Cooper, Niehaus (1972) and others at the US Navy Office of Civilian Manpower Management (OCMM). They are multiperiod Archimedian goal programming models that employ *embedded* Markov matrices. A simple example of an OCMM model is described. Assume six job categories and two time periods (three discrete points in time). For variables, let

$x_i(t) \sim$ the number of people in job category i at time t.
$h_i(t) \sim$ the number of people to be added to job category i at time t.
$f_i(t) \sim$ the number of people to be reduced from job category i at time t.
$d_i^-(t), d_i^+(t) \sim$ deviations from manpower requirements for job category i at time t.

Table 10.1
An OCMM Model

	x(0)	x(1)	x(2)	h(1)	h(2)	f(1)	f(2)	$d^+(1)$	$d^+(2)$	$d^-(1)$	$d^-(2)$		RHS
Obj								α	β	γ	δ	min	
		I						**−I**		**I**		=	1st Period Manpower Requirements
			I						**−I**		**I**	=	2nd Period Manpower Requirements
s.t.	**I**											=	Initial Pop
	−M	**I**		**−I**		**I**						=	**0**
		−M	**I**		**−I**		**I**					=	**0**
		$\mathbf{c}^1$		$\mathbf{c}^2$		$\mathbf{c}^3$						$\leqq$	1st Period Budget
			$\mathbf{c}^1$		$\mathbf{c}^2$		$\mathbf{c}^3$					$\leqq$	2nd Period Budget

Matrix information for the OCMM model is given in Table 10.1, in which $\boldsymbol{\alpha}$, $\boldsymbol{\beta}$, $\boldsymbol{\gamma}$, and $\boldsymbol{\delta}$ are row vectors of penalty weights, the **M**'s are Markov matrices, and the **c**'s are row vectors of cost coefficients. The Markov matrices are typically constructed from historical records. The elements in these matrices capture the likelihood that a person will transit from one job category to another, one period to the next. Transitions may occur because of promotions, demotions, and reassignments. For the ith job category in the tth time period, the constraints involving the Markov matrices are given by

$$-\sum_j m_{ij}(t)x_j(t-1) + x_i(t) - h_i(t) + f_i(t) = 0$$

where $m_{ij}(t)$ is the probability that a person in job category j will transit to job category i at time t.

Because people may leave the organization, the rows of the Markov matrices will typically sum to less than one. With six job categories and two time periods, the OCMM model of Table 10.1 has 32 constraints and 66 variables. If the model were extended to 10 job categories and five time periods, it would have 115 constraints and 260 variables. Also, 100 penalty weights would have to be assigned.

10.4 PREEMPTIVE GP

In preemptive (*lexicographic*) goal programming, the goals are grouped according to priorities. The goals at the highest priority level are considered to be infinitely more important than goals at the second priority level, and the goals at the second priority level are considered to be infinitely more important than goals at the third priority level, and so forth. To illustrate, let us consider the preemptive GP

$$\begin{array}{ll} \text{goal}\,\{\mathbf{c}^1\mathbf{x} = z_1\} & P_1(z_1 \leqq t_1) \\ \text{goal}\,\{\mathbf{c}^2\mathbf{x} = z_2\} & P_2(z_2 \geqq t_2) \\ \text{goal}\,\{\mathbf{c}^3\mathbf{x} = z_3\} & P_3(z_3 = t_3) \\ \text{s.t.} \quad \mathbf{x} \in S & \end{array}$$

in which P_j identifies the goals at priority level j. The P_j also serve as *priority factors* where $P_j \ggg P_{j+1}$ with "$\ggg$" meaning very much larger. Using priority factors, some authors would write the above GP as follows

$$\begin{array}{llr} \min\,\{P_1(d_1^+) + P_2(d_2^-) + P_3(d_3^+ + d_3^-)\} & & \\ \text{s.t.} \quad \mathbf{c}^1\mathbf{x} - d_1^+ & & \leqq t_1 \\ \qquad \mathbf{c}^2\mathbf{x} \qquad + d_2^- & & \geqq t_2 \\ \qquad \mathbf{c}^3\mathbf{x} & - d_3^+ + d_3^- & = t_3 \\ & & \mathbf{x} \in S \\ & & \text{all } d\text{'s} \geqq 0 \end{array}$$

However, we write the preemptive GP in the following lexicographic format:

$$\begin{aligned}
&\text{lex min}\,\{d_1^+, d_2^-, (d_3^+ + d_3^-)\} \\
&\text{s.t.} \quad \mathbf{c}^1\mathbf{x} - d_1^+ \leqq t_1 \\
&\qquad \mathbf{c}^2\mathbf{x} + d_2^- \geqq t_2 \\
&\qquad \mathbf{c}^3\mathbf{x} - d_3^+ + d_3^- = t_3 \\
&\qquad \mathbf{x} \in S \\
&\qquad \text{all } d\text{'s} \geqq 0
\end{aligned}$$

To solve the above *lex min* problem using conventional LP software, as many as three optimization stages may be required. In the first stage, we solve

$$\begin{aligned}
&\min\,\{d_1^+\} \\
&\text{s.t.} \quad \mathbf{c}^1\mathbf{x} - d_1^+ \leqq t_1 \\
&\qquad \mathbf{x} \in S \\
&\qquad d_1^+ \geqq 0
\end{aligned}$$

If this problem has alternative optima, we form and then solve the second-stage problem

$$\begin{aligned}
&\min\,\{d_2^-\} \\
&\text{s.t.} \quad \mathbf{c}^1\mathbf{x} \leqq t_1 + (d_1^+)^* \\
&\qquad \mathbf{c}^2\mathbf{x} + d_2^- \geqq t_2 \\
&\qquad \mathbf{x} \in S \\
&\qquad d_2^- \geqq 0
\end{aligned}$$

where $(d_1^+)^*$ is the optimal value of d_1^+ from stage one. If the second-stage problem has alternative optima, we form and then solve the third-stage problem

$$\begin{aligned}
&\min\,\{(d_3^+ + d_3^-)\} \\
&\text{s.t.} \quad \mathbf{c}^1\mathbf{x} \leqq t_1 + (d_1^+)^* \\
&\qquad \mathbf{c}^2\mathbf{x} \geqq t_2 - (d_2^-)^* \\
&\qquad \mathbf{c}^3\mathbf{x} - d_3^+ + d_3^- = t_3 \\
&\qquad \mathbf{x} \in S \\
&\qquad d_3^+, d_3^- \geqq 0
\end{aligned}$$

where $(d_2^-)^*$ is the optimal value of d_2^- from stage two. Any solution to the third stage lexicographically minimizes the preemptive GP.

In general, we may not have to solve as many optimization stages as there are priority levels. We can cease our progression through the optimization stages as

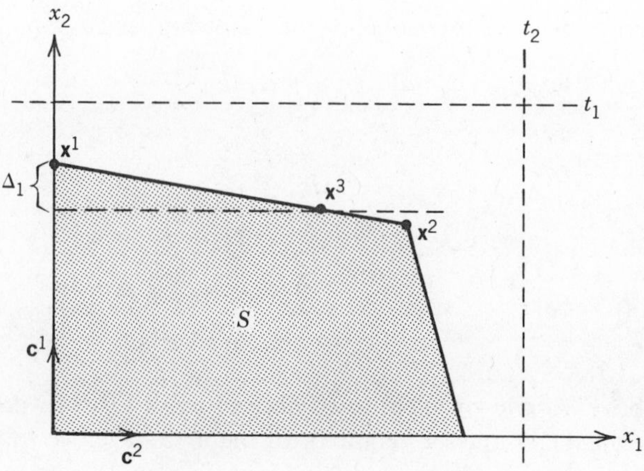

Figure 10.10. Graph of Example 6.

soon as an optimization stage is encountered that has a unique solution. Thus, a less than desirable consequence of the preemptive approach is that lower order goals may not get a chance to influence the GP generated solution.

We note that the different lexicographic optimization stages cannot be solved simultaneously. They must be formed and then solved sequentially because each subsequent stage needs optimality information from the previous stage. Hence, solving a preemptive GP is a dynamic process.

Preemptive GP has been criticized as inflexible because higher priority levels brook no compromise with lower priority levels. One suggestion for alleviating this inflexibility is to use *relaxation quantities.*

Example 6. Consider the GP of Fig. 10.10, in which both $\mathbf{c}^1$ and $\mathbf{c}^2$ pertain to greater than or equal to criteria. If $\mathbf{c}^1$ is at the first priority level and $\mathbf{c}^2$ at the second, $\mathbf{x}^1$ results. However, $\mathbf{x}^2$ may be preferred because little of the first criterion has to be given up in order to gain a large amount of the second. To deal with such a situation, we may wish to use relaxation quantities. With the relaxation quantity Δ_1 of Fig. 10.10, we have the modified second-stage lexicographic problem

$$\begin{aligned} &\min \{d_2^-\} \\ &\text{s.t.} \quad \mathbf{c}^1\mathbf{x} \qquad\quad \geqq t_1 - (d_1^-)^* - \Delta_1 \\ &\qquad \mathbf{c}^2\mathbf{x} + d_2^- \geqq t_2 \\ &\qquad\qquad\quad \mathbf{x} \in S \\ &\qquad\qquad\quad d_2^- \geqq 0 \end{aligned}$$

that yields $\mathbf{x}^3$. As can be imagined, this approach may or may not work successfully because of the ad hoc nature of selecting the Δ_i.

10.5 LEXICOGRAPHIC SIMPLEX METHOD

Rather than solve a preemptive GP via a sequence of LPs as in Section 10.4, a preemptive GP can be solved in one job-step using the *lexicographic simplex method*. To illustrate the lexicographic simplex method, consider the GP

$$\begin{aligned}
&\text{goal}\{\quad x_2 = z_1\} \qquad P_1(z_1 \geqq 5)\\
&\text{goal}\{-x_1 - x_2 = z_2\} \qquad P_2(z_2 \geqq 4)\\
&\text{goal}\{\quad x_3 = z_3\} \qquad P_3(z_3 \geqq 3)\\
&\text{s.t.}\quad x_2 \leqq 2\\
&\qquad x_3 \leqq 3\\
&\qquad x_1, x_2, x_3 \geqq 0
\end{aligned}$$

This converts to

$$\begin{aligned}
&\text{lex min}\{d_1^-, d_2^-, d_3^-\}\\
&\text{s.t.}\quad x_2 + d_1^- \geqq 5\\
&\qquad -x_1 - x_2 + d_2^- \geqq 4\\
&\qquad x_3 + d_3^- \geqq 3\\
&\qquad x_2 \leqq 2\\
&\qquad x_3 \leqq 3\\
&\qquad \text{all vars} \geqq 0
\end{aligned}$$

With a $c_j - z_j$ reduced cost row for the objective function of each P_j lexicographic level, we obtain the first-stage lexicographic minimizing tableau of Table 10.2 (in which the basic variable columns are suppressed). Because of alternative optima, we proceed to the second lexicographic stage.

TABLE 10.2
Minimizing Tableau of First Lexicographic Stage

		x_1	x_3	s_1	s_2	s_3	s_4	
d_1^-	3			−1			−1	
d_2^-	6	−1			−1		1	
d_3^-	3		1			−1		
x_2	2						1	
s_5	3		[1]					
P_1	3			1			1	$c_j - z_j$ rows
P_2	6	1			1		(−1)	
P_3	3		(−1)			1		

Table 10.3
Final Lexicographic Tableau

		x_1	s_1	s_2	s_3	s_4	s_5	
d_1^-	3		-1			-1		
d_2^-	6	-1		-1		1		
d_3^-	0				-1		-1	
x_2	2					1		
x_3	3						1	
P_1	3		1			1		$c_j - z_j$ rows
P_2	6	1		1		-1		
P_3	0				1		1	

Consider the circled element in the P_2 reduced cost row of Table 10.2. Although it may appear desirable to bring s_4 into the basis, it is not because there is a positive reduced cost above it in the same column. This means that with respect to s_4, the second lexicographic stage only can be further minimized by increasing the first-priority lexicographic stage. Since there are no other candidates to enter the basis (remember that we are only looking for negative reduced costs because we are minimizing), $(x_1, x_2, x_3) = (0, 2, 0)$ not only minimizes the first-stage problem, but it also minimizes the second-stage problem as well. Because of alternative optima, we proceed to the third lexicographic stage. Pivoting x_3 into the basis (because there are no positive elements above the circled element in the P_3 reduced cost row), we obtain $(x_1, x_2, x_3) = (0, 2, 3)$ as the lex min solution of the GP in Table 10.3. At this point, the first goal is underachieved by two units, the second goal is underachieved by one unit, and the third goal is achieved.

10.6 GOAL EFFICIENCY

Recall that in an MOLP $\bar{\mathbf{x}} \in S$ is efficient if and only if there does not exist another $\mathbf{x} \in S$ such that $\mathbf{c}^i\mathbf{x} \geqq \mathbf{c}^i\bar{\mathbf{x}}$ for all i and $\mathbf{c}^i\mathbf{x} > \mathbf{c}^i\bar{\mathbf{x}}$ for at least one i. Because the $\mathbf{c}^i\mathbf{x}$ of the goal criteria do not meet the monotonicity assumption (Section 6.2), this definition is not valid in GP. However, because each undesirable deviational variable is to be minimized, we have the following definition of efficiency for goal programming.

Definition 10.1. Let $\hat{\mathbf{d}}$ be the vector of undesirable deviational variables associated with $\hat{\mathbf{x}} \in S$. Then, $\hat{\mathbf{x}}$ is *goal-efficient* (in a minimization sense) if there does not exist another $\mathbf{x} \in S$ whose $\mathbf{d}$ satisfies $\mathbf{d} \leqq \hat{\mathbf{d}}$, $\mathbf{d} \neq \hat{\mathbf{d}}$.

Let $E^g \subset S$ be the set of all goal-efficient points. Goal-efficient points are of interest because in GP the decision maker's optimal point is goal-efficient.

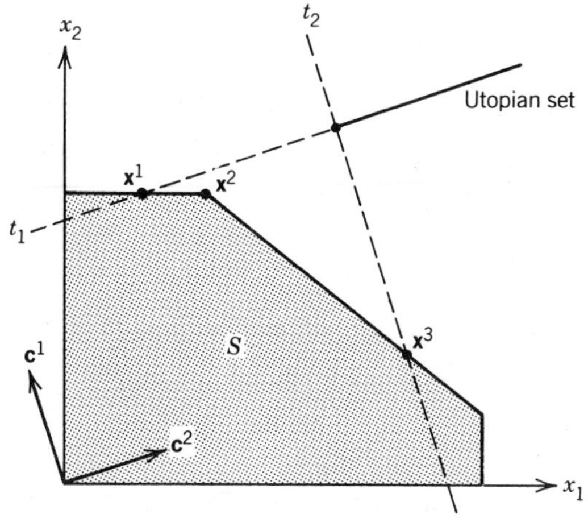

Figure 10.11. Graph of Example 7.

Example 7. Consider the GP

$$\begin{array}{lll} \text{goal}\,\{\mathbf{c}^1\mathbf{x} = z_1\} & & (z_1 = t_1) \\ \text{goal}\,\{\mathbf{c}^2\mathbf{x} = z_2\} & & (z_2 \geqq t_2) \\ \text{s.t.} & \mathbf{x} \in S & \end{array}$$

of Fig. 10.11. With the vector of undesirable deviational variables

$$\mathbf{d} = \begin{bmatrix} d_1^+ \\ d_1^- \\ d_2^- \end{bmatrix}$$

the goal-efficient set is $E^g = \gamma(\mathbf{x}^1, \mathbf{x}^2) \cup \gamma(\mathbf{x}^2, \mathbf{x}^3)$. Hence, the decision maker's optimal point is a member of $\gamma(\mathbf{x}^1, \mathbf{x}^2) \cup \gamma(\mathbf{x}^2, \mathbf{x}^3)$.

In Example 7, we again see the difficulty locating the DM's optimal point using goal programming. With the Archimedian approach, either $\mathbf{x}^1$, $\mathbf{x}^2$, or $\mathbf{x}^3$ would be generated. With the preemptive approach, either $\mathbf{x}^1$ or $\mathbf{x}^3$ would be generated depending on which way we ordered the priorities. Had the DM's optimal point, for instance, been in the middle of the $\gamma(\mathbf{x}^2, \mathbf{x}^3)$ line segment, we would not be doing a very good job coming close to it.

We can use a vector-maximum code (see Chapter 9) to solve for the goal-efficient set E^g by making a minimization objective out of each undesirable deviational variable. In the case of Example 7, the vector-maximum formulation would be

$$\begin{aligned}
&\min\{\quad d_1^+ \quad\quad\quad\}\\
&\min\{\quad\quad\quad d_1^- \quad\}\\
&\min\{\quad\quad\quad\quad\quad d_2^-\}\\
&\text{s.t.}\quad \mathbf{c}^1\mathbf{x} - d_1^+ + d_1^- \quad\quad = t_1\\
&\quad\quad\quad \mathbf{c}^2\mathbf{x} \quad\quad\quad\quad + d_2^- \geqq t_2\\
&\quad\quad\quad\quad\quad\quad\quad\quad\quad \mathbf{x} \in S\\
&\quad\quad\quad\quad\quad\quad d_1^+, d_1^-, d_2^- \geqq 0
\end{aligned}$$

10.7 SENSITIVITY ISSUES

In practice, goal programs typically have large numbers of deviational variables. With possibly different units of measure, assigning weights is often very difficult. Consequently, in Archimedian GP, we initially try to apply a reasonable set of penalty weights with plans to do sensitivity experiments with other sets of weights to see if better solutions can be located. Many times the strategy works satisfactorily. Many times it does not. When it does not, this is what might happen. The initial set of weights produces an unsatisfactory solution. Then, some of the weights are shifted and the problem resolved. Sometimes, the new solution will be the same as the old because both sets of weights pertain to the same augmented GP feasible region vertex. Sometimes, even though we may feel that we have made intelligent changes in the weights, a poorer solution results. Sometimes, despite the fact that we may have made only slight changes in the weights, a drastically different solution results because we may have jumped to a vertex on the opposite side of a large goal-efficient facet. The net effect is that often a user will come away from an Archimedian GP experience feeling frustrated about his or her ability to *control* the movement of the solution as the weights are varied.

In preemptive GP, users frequently address sensitivity concerns by *rotating priorities*. If there are r priority levels, there are $r!$ different ways of rotating the priorities. Normally, a user will select a small number of these possibilities and then resolve the problem for each of them. The result is usually a group of some of the most different goal-efficient points in S.

10.8 INTERACTIVE GP

Another strategy for use with goal programs is to utilize both Archimedian and preemptive features but in an interactive context. In this way, it is hoped that the difficulties of Archimedian and preemptive goal programming as discussed in Sections 10.2.2, 10.4, and 10.7 can be overcome.

In the first iteration of the interactive approach, the decision maker

1. Specifies the type of each goal ($\leqq$, $\geqq$, $=$, or range).
2. Specifies target levels for the goals.
3. Groups the goals into priority levels.
4. Specifies *within-priority-level* penalty weights for those priority levels associated with more than one goal.

Then, the GP is solved according to usual lexicographic principles. After the solution is obtained and studied, the decision maker is asked to refine his or her formulation and make any goal type, target value, priority level, and penalty weight changes that may be appropriate. Then, the new problem is solved to generate the second iteration solution, and so forth. Note that this approach subsumes Archimedian GP when only one priority level is specified.

The interactive approach is supported by the fact that when S is bounded only goal-efficient solutions are produced. However, because the formulation can be reshaped in so many different ways, it is not always clear what should be changed in an attempt to move the solution in a desired direction. Hence, the method is essentially an unstructured approach for probing the feasible region. Using goal programming concepts as "dials," interactive GP relies heavily on the skill and intuition of the user.

10.9 MINIMIZING MAXIMUM DEVIATION

If a priority level consists of deviational variables from several goals, one possible construction is to minimize the maximum deviation. This requires the creation of a *minimax* variable $\alpha \in R$. The objective at the priority level becomes

$$\min\{\alpha\}$$

Also, as many additional constraints must be created as there are deviational variables at the priority level.

Example 8. Consider the GP

$$
\begin{array}{ll}
\text{goal}\{\mathbf{c}^1\mathbf{x} = z_1\} & P_1(z_1 = t_1) \\
\text{goal}\{\mathbf{c}^2\mathbf{x} = z_2\} & P_2(z_2 \geqq t_2) \\
\text{goal}\{\mathbf{c}^3\mathbf{x} = z_3\} & P_2(z_3 \geqq t_3) \\
\text{goal}\{\mathbf{c}^4\mathbf{x} = z_4\} & P_2(z_4 \geqq t_4) \\
\text{s.t.} \quad \mathbf{x} \in S &
\end{array}
$$

With the intent of minimizing the maximum deviation at the second-priority level, the lexicographic formulation for the GP is

$$
\begin{aligned}
\text{lex min } & \{(d_1^- + d_1^+), \alpha\} \\
\text{s.t. } & \mathbf{c}^1\mathbf{x} - d_1^+ + d_1^- = t_1 \\
& \mathbf{c}^2\mathbf{x} + d_2^- \geqq t_2 \\
& \mathbf{c}^3\mathbf{x} + d_3^- \geqq t_3 \\
& \mathbf{c}^4\mathbf{x} + d_4^- \geqq t_4 \\
& d_2^- \leqq \alpha \\
& d_3^- \leqq \alpha \\
& d_4^- \leqq \alpha \\
& \mathbf{x} \in S \\
& \text{all } d\text{'s} \geqq 0
\end{aligned}
$$

In this formulation, α is a minimax variable that, when minimized at the second lexicographic level, minimizes the largest of the d_2^-, d_3^-, and d_4^- deviational variables.

Another sophistication that can be used with deviational variables is to rescale them in the goal constraints so that they measure deviations from goal in percentage terms. Consider the following goal constraint

$$
\begin{aligned}
\mathbf{c}^i\mathbf{x} + \quad & d_i^- \geqq t_i \\
20{,}000 + 5{,}000 & \geqq 25{,}000
\end{aligned}
$$

If $\mathbf{c}^i\mathbf{x} = 20{,}000$ and $t_i = 25{,}000$, we have $d_i^- = 5{,}000$ as indicated. In this instance, d_i^- is 20% below goal. Now, if we multiply d_i^- by the coefficient $t_i/100$, we get $d_i^- = 20$ directly as below

$$
\begin{aligned}
\mathbf{c}^i\mathbf{x} + \left(\frac{t_i}{100}\right) & d_i^- \geqq t_i \\
20{,}000 + (250)20 & \geqq 25{,}000
\end{aligned}
$$

10.10 MULTIPLE CRITERION FUNCTION GP

In this variation of goal programming we specify goal types, assign target values, form priority levels, and apply within-priority-level Archimedian weights as usual. However, instead of lexicographically solving the resulting formulation, we solve it as an MOLP. That is, each priority level is treated as an objective function and the endeavor is to search the space of tradeoffs among the priority levels to find the best solution all goals considered. To solve a multiple criterion function GP, any of the multiple objective programming techniques of Chapters 9, 13, or 14 can be employed.

Example 9. Consider the *mixed* GP

$$\begin{array}{ll}
\text{goal}\,\{\mathbf{c}^1\mathbf{x} = z_1\} & P_1(z_1 \geqq t_1) \\
\text{goal}\,\{\mathbf{c}^2\mathbf{x} = z_2\} & P_1(z_2 \geqq t_2) \\
\max\,\{\mathbf{c}^3\mathbf{x} = z_3\} & P_2 \\
\text{goal}\,\{\mathbf{c}^4\mathbf{x} = z_4\} & P_3(z_4 \leqq t_4) \\
\text{goal}\,\{\mathbf{c}^5\mathbf{x} = z_5\} & P_3(z_5 \leqq t_5) \\
\text{goal}\,\{\mathbf{c}^6\mathbf{x} = z_6\} & P_3(z_6 \leqq t_6) \\
\text{s.t.} \quad \mathbf{x} \in S &
\end{array}$$

in which we wish to minimize the average deviation as well as the maximum deviation in P_1. For P_3, assume that we wish to minimize the average percentage deviation. For this problem, the multiple criterion function GP formulation is

$$\begin{array}{llllllll}
\min\,\{\tfrac{1}{2}(d_1^- + d_2^-)\} \\
\min\,\{\alpha\} \\
\max\,\{\mathbf{c}^3\mathbf{x}\} \\
\min\,\{\tfrac{1}{3}(d_4^+ + d_5^+ + d_6^+)\} \\
\text{s.t.} & \mathbf{c}^1\mathbf{x} & +d_1^- & & & & & \geqq t_1 \\
& \mathbf{c}^2\mathbf{x} & & +d_2^- & & & & \geqq t_2 \\
& \mathbf{c}^4\mathbf{x} & & & -\dfrac{t_4}{100}d_4^+ & & & \leqq t_4 \\
& \mathbf{c}^5\mathbf{x} & & & & -\dfrac{t_5}{100}d_5^+ & & \leqq t_5 \\
& \mathbf{c}^6\mathbf{x} & & & & & -\dfrac{t_6}{100}d_6^+ & \leqq t_6 \\
& & d_1^- & & & & & \leqq \alpha \\
& & & d_2^- & & & & \leqq \alpha \\
& & & & & & & \mathbf{x} \in S \\
& & & & & & & \text{all } d\text{'s} \geqq 0
\end{array}$$

Note that in this formulation the criterion value of each objective function has been designed to have meaning.

10.11 GP SOFTWARE

Conventional LP software can be used to solve Archimedian GPs. It can also be used to solve a preemptive GP if the problem is unraveled into a sequence of LPs as described in Section 10.4.

Specialized software is required to implement the lexicographic simplex method for preemptive problems. One such code for small to moderate sized problems is **PAGP** by Arthur and Ravindran (1980). Other preemptive GP codes have been published by Lee (1972, pp. 140–155) and Ignizio (1976, pp. 234–247). For a commercial grade preemptive GP capability, Ignizio and Perlis (1979) suggest a driver program that calls a commercial LP package such as **MPSX** as a subroutine.

Another way to solve a preemptive GP is to use **ADBASE** (described in Section 9.11). With **IFASE2** = 1, **ADBASE** employs lexicographic maximization in Phase II. Thus, by loading the ith row of the criterion matrix with the ith lexicographic objective function, **ADBASE** can be used to find a lex min solution of a preemptive GP.

Example 10. Consider the preemptive GP

$$\begin{array}{llll} \text{goal}\{ & x_1 & = z_1\} & P_1(z_1 \in [2,4]) \\ \text{goal}\{ & 3x_1 + 3x_2 + 3x_3 & = z_2\} & P_2(z_2 \leqq 30) \\ \text{goal}\{ & x_2 & = z_3\} & P_3(z_3 = 5) \\ \text{s.t.} & x_2 & \leqq 4 & \\ & x_3 & \leqq 3 & \\ & x_1, x_2, x_3 & \geqq 0 & \end{array}$$

whose set of lex min solutions is indicated by the shaded area in Fig. 10.12. The vector-maximum formulation inputted to **ADBASE** is as depicted in Table 10.4.

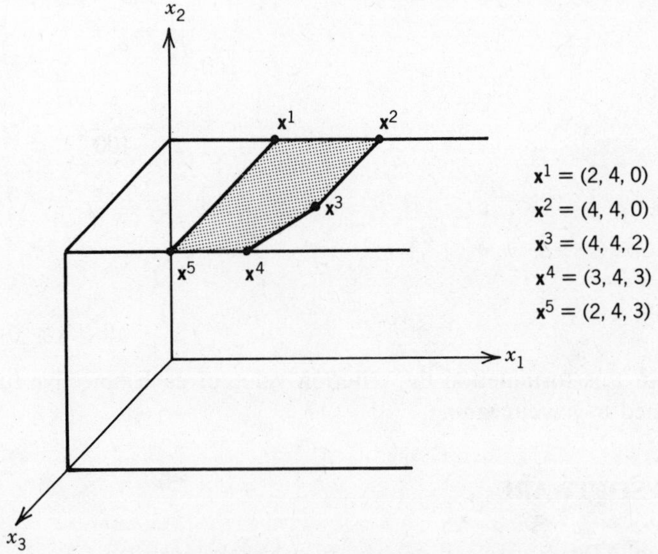

Figure 10.12. Graph of Example 10.

TABLE 10.4
Vector-Maximum Formulation of Example 10

	x_1	x_2	x_3	d_1^-	d_1^+	d_2^+	d_3^+	d_3^-		
Objs				−1					max	
					−1				max	
						−1			max	
							−1		max	
								−1	max	
s.t.	1			1					≥	2
	1				−1				≤	4
	3	3	3			−1			≤	30
		1					−1	1	=	5
		1							≤	4
			1						≤	3

Running **ADBASE** with

$$\texttt{IFASE0} = 0$$
$$\texttt{IFASE2} = 1$$
$$\texttt{IFASE3} = 0$$

we obtain $\mathbf{x}^1$ as a lex min solution of the GP (i.e., as the first efficient point of the vector-maximum problem). At $\mathbf{x}^1$, $d_1^- = d_1^+ = d_2^+ = d_3^+ = 0$, and $d_3^- = 1$. All vertices of the lex min set can be computed by making another **ADBASE** run. In this run, we use a null objective function and delete the deviational variable columns. Then, along with goal constraint RHSs of 2, 4, 30, and 4, respectively, and settings of

$$\texttt{IFASE0} = 0$$
$$\texttt{IFASE2} = 1, 2, 3, 4, \text{ or } 5$$
$$\texttt{IFASE3} = 1$$

we obtain $\mathbf{x}^1$, $\mathbf{x}^2$, $\mathbf{x}^3$, $\mathbf{x}^4$, and $\mathbf{x}^5$.

Readings

1. Armstrong, R. D. and W. D. Cook (1979). "Goal Programming Models for Assigning Search and Rescue Aircraft to Bases," *Journal of the Operational Research Society*, Vol. 30, No. 6, pp. 555–561.

2. Arthur, J. L. and K. D. Lawrence (1980). "A Multiple Goal Blending Problem," *Computers and Operations Research*, Vol. 7, No. 3, pp. 215–224.

3. Arthur, J. L. and K. D. Lawrence (1982). "Multiple Goal Production and Logistics Planning in a Chemical and Pharmaceutical Company," *Computers and Operations Research*, Vol. 9, No. 2, pp. 127–137.

4. Arthur, J. L. and A. Ravindran (1978). "An Efficient Goal Programming Algorithm Using Constraint Partitioning and Variable Elimination," *Management Science*, Vol. 24, No. 8, pp. 867–868.

5. Arthur, J. L. and A. Ravindran (1980). "PAGP: A Partitioning Algorithm for (Linear) Goal Programming Problems," *ACM Transactions on Mathematical Software*, Vol. 6, No. 3, pp. 378–386.

6. Arthur, J. L. and A. Ravindran (1980). "A Branch and Bound Algorithm with Constraint Partitioning for Integer Goal Programming Problems," *European Journal of Operational Research*, Vol. 4, No. 6, pp. 421–425.

7. Arthur, J. L. and A. Ravindran (1981). "A Multiple Objective Nurse Scheduling Model," *AIIE Transactions*, Vol. 13, No. 1, pp. 55–60.

8. Bishop, A. B., R. Narayanan, W. J. Grenney, and P. E. Pugner (1977). "Goal Programming Model for Water Quality Planning," *Journal of the Environmental Engineering Division*, ASCE, EE2, pp. 293–305.

9. Bres, E. S., D. Burns, A. Charnes, and W. W. Cooper (1980). "A Goal Programming Model for Planning Officer Accessions," *Management Science*, Vol. 26, No. 8, pp. 773–783.

10. Charnes, A. and W. W. Cooper (1961). *Management Models and Industrial Applications of Linear Programming* (Appendix B), Vol. I, New York: John Wiley & Sons.

11. Charnes, A. and W. W. Cooper (1977). "Goal Programming and Multiple Objective Optimization—Part 1," *European Journal of Operational Research*, Vol. 1, No. 1, pp. 39–54.

12. Charnes, A., W. W. Cooper, J. K. DeVoe, D. B. Learner, and W. Reinecke (1968). "A Goal Programming Model for Media Planning," *Management Science*, Vol. 14, No. 8, pp. B423–B430.

13. Charnes, A., W. W. Cooper, and R. O. Ferguson (1955). "Optimal Estimation of Executive Compensation by Linear Programming," *Management Science*, Vol. 1, No. 2, pp. 138–151.

14. Charnes, A., W. W. Cooper, K. R. Karwan, and W. A. Wallace (1979). "A Chance-Constrained Goal Programming Model to Evaluate Response Resources for Marine Pollution Disasters," *Journal of Environmental Economics and Management*, Vol. 6, No. 3, pp. 244–274.

15. Charnes, A., W. W. Cooper, and R. J. Niehaus (1972). *Studies in Manpower Planning*, Office of Civilian Manpower Management, Department of the Navy, Washington, D.C.

16. Charnes, A., W. W. Cooper, and R. J. Niehaus (1975). "Dynamic Multi-attribute Models for Mixed Manpower Systems," *Naval Research Logistics Quarterly*, Vol. 22, No. 2, pp. 205–220.

17. Contini, B. (1968). "A Stochastic Approach to Goal Programming," *Operations Research*, Vol. 16, No. 3, pp. 576–586.

18. Cook, W. D. (1984). "Goal Programming and Financial Planning Models for Highway Rehabilitation," *Journal of the Operational Research Society*, Vol. 35, No. 3, pp. 217–223.

19. Dauer, J. P. and R. J. Krueger (1977). "An Iterative Approach to Goal Programming," *Operational Research Quarterly*, Vol. 28, No. 3, pp. 671–681.

20. Deckro, R. F., J. E. Hebert, and E. P. Winkofsky (1982). "Multiple Criteria Job-Shop Scheduling," *Computers and Operations Research*, Vol. 9, No. 4, pp. 279–285.

21. Dyer, J. S. (1977). "On the Relationship Between Goal Programming and Multiattribute Utility Theory." Presented at the International Symposium on Extremal Methods and Systems Analysis, Austin, Texas.

22. Field, R. C., P. E. Dress, and J. C. Fortson (1980). "Complementary Linear and Goal Programming Procedures for Timber Harvest Scheduling," *Forest Science*, Vol. 26, No. 1, pp. 121–133.

23. Fortson, J. C. and R. R. Dince (1977). "An Application of Goal Programming to Management of a Country Bank," *Journal of Bank Research*, Vol. 7, No. 4, pp. 311–319.

24. Freed, N. and F. Glover (1981). "Simple but Powerful Goal Programming Models for Discriminant Problems," *European Journal of Operational Research*, Vol. 7, No. 1, pp. 44–60.

25. Goodman, D. A. (1974). "A Goal Programming Approach to Aggregate Planning of Production and Work Force," *Management Science*, Vol. 20, No. 12, pp. 1569–1575.

26. Hannan, E. L. (1978). "Allocation of Library Funds for Books and Standing Orders—A Multiple Objective Formulation," *Computers and Operations Research*, Vol. 6, No. 2, pp. 109–114.

27. Hannan, E. L. (1978). "The Application of Goal Programming Techniques to the CPM Problem," *Socio-Economic Planning Sciences*, Vol. 12, No. 5, pp. 267–270.

28. Hannan, E. L. (1978). "Nondominance in Goal Programming," *INFOR*, Vol. 18, No. 4, pp. 300–309.

29. Ignizio, J. P. (1976). *Goal Programming and Extensions*, Lexington, Massachusetts: D. C. Heath.

30. Ignizio, J. P. (1978). "A Review of Goal Programming: A Tool for Multiobjective Analysis," *Journal of the Operational Research Society*, Vol. 29, No. 11, pp. 1109–1119.

31. Ignizio, J. P. (1981). "Antenna Array Beam Pattern Synthesis via Goal Programming," *European Journal of Operational Research*, Vol. 6, No. 3, pp. 286–290.

32. Ignizio, J. P. (1981). "The Determination of a Subset of Efficient Solutions via Goal Programming," *Computers and Operations Research*, Vol. 8, No. 1, pp. 9–16.

33. Ignizio, J. P. (1982). *Linear Programming in Single- & Multiple-Objective Systems*, Englewood Cliffs, New Jersey: Prentice-Hall.

34. Ignizio, J. P. (ed.) (1983). *Computers and Operations Research* (Special Issue on Generalized Goal Programming), Vol. 10, No. 4.

35. Ignizio, J. P. (1983). "A Note on Computational Methods in Lexicographic Linear Goal Programming," *Journal of the Operational Research Society*, Vol. 34, No. 6, pp. 539–542.

36. Ignizio, J. P. and J. H. Perlis (1979). "Sequential Linear Goal Programming: Implementation via MPSX," *Computers and Operations Research*, Vol. 6, No. 3, pp. 141–145.

37. Ijiri, Y. (1965). *Management Goals and Accounting for Control*, Chicago: Rand McNally.

38. Isermann, H. (1982). "Lexicographic Goal Programming: The Linear Case." In M. Grauer, A. Lewandowski, and A. P. Wierzbicki (eds.), *Multiobjective and Stochastic Optimization*, International Institute for Applied Systems Analysis, Laxenburg, Austria, pp. 65–78.

39. Jääskeläinen, V. (1969). "A Goal Programming Model of Aggregate Production Planning," *Swedish Journal of Economics*, Vol. 71, No. 1, pp. 14–29.

40. Kendall, K. E. and S. M. Lee (1980). "Formulating Blood Rotation Policies with Multiple Objectives," *Management Science*, Vol. 26, No. 11, pp. 1145–1157.

41. Keown, A. J. (1978). "A Chance-Constrained Goal Programming Model for Bank Liquidity Management," *Decision Sciences*, Vol. 9, No. 1, pp. 93–106.

42. Keown, A. J. and C. P. Duncan (1979). "Integer Goal Programming in Advertising Media Selection," *Decision Sciences*, Vol. 10, No. 4, pp. 577–592.

43. Keown, A. J. and B. W. Taylor (1980). "A Chance-Constrained Integer Goal Programming Model for Capital Budgeting in the Production Area," *Journal of the Operational Research Society*, Vol. 31, No. 7, pp. 579–589.

44. Killough, L. N. and T. L. Souders (1973). "A Goal Programming Model for Public Accounting Firms," *The Accounting Review*, Vol. XLVIII, No. 2, pp. 268–279.

45. Korhonen, P. and J. Laakso (1985). "On Solving Generalized Goal Programming Problems Using a Visual Interactive Approach," *European Journal of Operational Research*, forthcoming.

46. Kornbluth, J. S. H. (1973). "A Survey of Goal Programming," *Omega*, Vol. 1, No. 2, pp. 193–205.

47. Knutson, D. L., L. M. Marquis, D. N. Ricchiute, and G. J. Saunders (1980). "A Goal Programming Model for Achieving Racial Balance in Public Schools," *Socio-Economic Planning Sciences*, Vol. 14, No. 3, pp. 109–116.

48. Lawrence, K. D. and G. R. Reeves (1982). "A Zero-One Goal Programming Model for Capital Budgeting in a Property and Liability Insurance Company," *Computers and Operations Research*, Vol. 9, No. 4, pp. 303–309.

49. Lawrence, S. M., K. D. Lawrence, and G. R. Reeves (1983). "Allocation of Teaching Personnel: A Goal Programming Model," *Socio-Economic Planning Sciences*, Vol. 17, No. 4, pp. 211–216.

50. Lee, S. M. (1972). *Goal Programming for Decision Analysis*, Philadelphia: Auerbach Publishers.

51. Lee, S. M. (1978). "Interactive Integer Goal Programming: Methods and Application," *Lecture Notes in Economics and Mathematical Systems*, No. 155, Berlin: Springer-Verlag, pp. 362–383.

52. Lee, S. M. and D. Chesser (1980). "Goal Programming for Portfolio Management," *The Journal of Portfolio Management*, Vol. 6, No. 3, pp. 22–26.

53. Lee, S. M., E. R. Clayton, and B. W. Taylor (1978). "A Goal Programming Approach to Multi-Period Production Line Scheduling," *Computers and Operations Research*, Vol. 5, No. 3, pp. 205–211.

54. Lee, S. M., G. I. Green, and C. I. Kim (1981). "A Multiple Criteria Model for the Location-Allocation Problem," *Computers and Operations Research*, Vol. 8, No. 1, pp. 1–8.

55. Lee, S. M. and L. J. Moore (1977). "Multi-Criteria School Busing Models," *Management Science*, Vol. 23, No. 7, pp. 703–715.

56. Lee, S. M. and R. L. Morris (1977). "Integer Goal Programming Methods," *TIMS Studies in the Management Sciences*, Vol. 6, pp. 273–289.

57. Lin, W. T. (1979). "Application of Goal Programming in Accounting," *Journal of Business Finance & Accounting*, Vol. 6, No. 4, pp. 559–577.

58. Muhlemann, A. P., A. G. Lockett, and A. E. Gear (1978). "Portfolio Modeling in Multi-Criteria Situations under Uncertainty," *Decision Sciences*, Vol. 9, No. 4, pp. 612–626.

59. Muhlemann, A. P. and A. G. Lockett (1980). "Portfolio Modeling in Multiple-Criteria Situations under Uncertainty: Rejoinder," *Decision Sciences*, Vol. 11, No. 1, pp. 178–180.

60. Olson, D. L. (1984). "Comparison of Four Goal Programming Algorithms," *Journal of the Operational Research Society*, Vol. 35, No. 4, pp. 347–354.

61. Price, W. L. (1978). "Solving Goal-Programming Manpower Models Using Advanced Network Codes," *Journal of the Operational Research Society*, Vol. 29, No. 12, pp. 1231–1239.

62. Price, W. L. and W. G. Piskor (1972). "The Application of Goal Programming to Manpower Planning," *INFOR*, Vol. 10, No. 3, pp. 221–231.

63. Ruefli, T. (1971). "A Generalized Goal Decomposition Model," *Management Science*, Vol. 17, No. 8, pp. 505–518.

64. Sartoris, W. L. and M. L. Spruill (1974). "Goal Programming and Working Capital Management," *Financial Management*, Vol. 3, No. 1, pp. 67–74.

65. Schniederjans, M. J. (1984). *Linear Goal Programming*, Princeton, New Jersey: Petrocelli Books.

66. Steuer, R. E. (1979). "Goal Programming Sensitivity Analysis Using Interval Penalty Weights," *Mathematical Programming*, Vol. 17, No. 1, pp. 16–31.

67. Steuer, R. E. (1983). "Operating Manual for the **ADBASE** Multiple Objective Linear Programming Package," College of Business Administration, University of Georgia, Athens, Georgia.

68. Taylor, B. W. and A. J. Keown (1978). "A Goal Programming Application of Capital Project Selection in the Production Area," *AIIE Transactions*, Vol. 10, No. 1, pp. 52–57.

69. Taylor, B. W., A. J. Keown, and A. G. Greenwood (1983). "An Integer Goal Programming Model for Determining Military Aircraft Expenditures," *Journal of the Operational Research Society*, Vol. 34, No. 5, pp. 379–390.

70. Taylor, B. W., L. J. Moore, and E. R. Clayton (1982). "R & D Project Selection and Manpower Allocation with Integer Nonlinear Programming," *Management Science*, Vol. 28, No. 10, pp. 1149–1158.

71. Wacht, R. F. and D. T. Whitford (1976). "A Goal Programming Model for Capital Investment Analysis in Nonprofit Hospitals," *Financial Management*, Vol. 5, No. 2, pp. 37–47.

72. Welling, P. (1977). "A Goal Programming Model for Human Resource Accounting in a CPA Firm," *Accounting, Organizations and Society*, Vol. 2, No. 4, pp. 307–316.

73. Zanakis, S. H. and M. W. Maret (1981). "A Markovian Goal Programming Approach to Aggregate Manpower Planning," *Journal of the Operational Research Society*, Vol. 32, No. 1, pp. 55–63.

74. Zanakis, S. H. and J. S. Smith (1980). "Chemical Production Planning via Goal Programming," *International Journal of Production Research*, Vol. 18, No. 6, pp. 687–697.

PROBLEM EXERCISES

10-1 Let

$$\bar{\mathbf{x}} = (100, 0)$$

$$\mathbf{x}^1 = (100, 100)$$

$$\mathbf{x}^2 = (95, 50)$$

$$\mathbf{x}^3 = (90, 0)$$

(a) How far is each of $\mathbf{x}^1$, $\mathbf{x}^2$, and $\mathbf{x}^3$ from $\bar{\mathbf{x}}$ according to the L_1-metric?

(b) Lexicographically (with the jth component of priority j), which of the vectors $\mathbf{x}^1$, $\mathbf{x}^2$, and $\mathbf{x}^3$ is closest to $\bar{\mathbf{x}}$? Farthest from $\bar{\mathbf{x}}$?

10-2 Arrange the following vectors, where the jth component is of priority j, in lexicographic order (minimum first).

$$\mathbf{v}^1 = (68, 0, -2, -8)$$

$$\mathbf{v}^2 = (67, 90, 36, 60)$$

$$\mathbf{v}^3 = (67, 90, 35, 98)$$

$$\mathbf{v}^4 = (42, 30, 19, 10)$$

$$\mathbf{v}^5 = (36, 82, 94, 30)$$

$$\mathbf{v}^6 = (67, 90, 36, 42)$$

10-3 For the GP:

$$\begin{aligned}
&\text{goal}\{-x_1 + x_2 = z_1\} && (z_1 = 2)\\
&\text{goal}\{\ \ x_1 \qquad\ = z_2\} && (z_2 \geqq 3)\\
&\text{s.t.} \qquad\quad x_2 \leqq 3 &&\\
&\qquad\quad x_1 + x_2 \leqq 5 &&\\
&\qquad\quad x_1, x_2 \geqq 0 &&
\end{aligned}$$

(a) Specify the utopian set.

(b) Specify the goal-efficient set.

10-4 Consider the GP:

$$\begin{aligned} &\text{goal}\{\ x_1 \qquad\quad = z_1\} \qquad (z_1 \geqq 8) \\ &\text{goal}\{\qquad\quad x_2 = z_2\} \qquad (z_2 \geqq 9) \\ &\text{s.t.}\quad 3x_1 + x_2 \leqq 24 \\ &\qquad\ 2x_1 + 7x_2 \leqq 35 \\ &\qquad\ x_1, x_2 \geqq 0 \end{aligned}$$

(a) Graph S and the utopian set.

(b) Assuming both goals at the same priority level, specify the point in S that minimizes the sum of the undesirable deviations. Specify the point in S that minimizes the maximum deviation.

(c) Specify the point in S that solves the preemptive model with the goals ranked in the order in which they are listed. What is the solution point if we reverse the priorities?

10-5 Using the lexicographic simplex method, solve:

$$\begin{aligned} &\text{goal}\{\ x_1 \qquad\qquad\quad = z_1\} \qquad P_1(z_1 \geqq 4) \\ &\text{goal}\{\qquad x_2 \qquad\quad = z_2\} \qquad P_2(z_2 \geqq 2) \\ &\text{goal}\{\qquad\qquad\quad x_3 = z_3\} \qquad P_3(z_3 \geqq 3) \\ &\text{s.t.}\quad 2x_1 + 3x_2 + 2x_3 \leqq 14 \\ &\qquad\ x_1 + 2x_2 \qquad\quad \leqq 5 \\ &\qquad\ x_1, x_2, x_3 \geqq 0 \end{aligned}$$

10-6 Assume that we wish to locate the optimal point of the MOLP

$$\begin{aligned} &\max\{\ x_1 \qquad = z_1\} \\ &\max\{\qquad x_2 = z_2\} \\ &\text{s.t.}\quad 2x_1 + x_2 \leqq 4 \\ &\qquad\ x_1, x_2 \geqq 0 \end{aligned}$$

where $U = z_1 z_2$.

(a) Specify the optimal point.

(b) Make an assignment of priorities and goal values to the objectives so that the resulting lexicographic GP computes the optimal point.

10-7 Specify the vector-maximum formulation for computing the goal-efficient set of:

$$\begin{aligned} &\text{goal}\{\mathbf{c}^1\mathbf{x} = z_1\} \qquad (z_1 \leqq t_1) \\ &\text{goal}\{\mathbf{c}^2\mathbf{x} = z_2\} \qquad (z_2 \geqq t_2) \\ &\text{goal}\{\mathbf{c}^3\mathbf{x} = z_3\} \qquad (z_3 = t_3) \\ &\text{goal}\{\mathbf{c}^4\mathbf{x} = z_4\} \qquad (z_4 \in [t_4^\ell, t_4^u]) \\ &\text{s.t.}\quad \mathbf{Ax} \leqq \mathbf{b} \\ &\qquad\ \mathbf{x} \geqq \mathbf{0} \end{aligned}$$

10-8 Consider:

$$\begin{array}{ll} \text{goal}\{\mathbf{c}^1\mathbf{x} = z_1\} & P_1(z_1 = t_1) \\ \text{goal}\{\mathbf{c}^2\mathbf{x} = z_2\} & P_1(z_2 = t_2) \\ \text{goal}\{\mathbf{c}^3\mathbf{x} = z_3\} & P_2(z_3 \geqq t_3) \\ \text{goal}\{\mathbf{c}^4\mathbf{x} = z_4\} & P_2(z_4 \geqq t_4) \\ \text{goal}\{\mathbf{c}^5\mathbf{x} = z_5\} & P_2(z_5 \geqq t_5) \\ \text{goal}\{\mathbf{c}^6\mathbf{x} = z_6\} & P_3(z_6 \in [t_6^\ell, t_6^\mu]) \\ \text{s.t.} \quad \mathbf{Ax} \leqq \mathbf{b} & \\ \qquad \mathbf{x} \geqq \mathbf{0} & \end{array}$$

Formulate (in tabular form) the multiple objective function GP when we wish to minimize:

(a) The average as well as the maximum deviation in P_1.

(b) The average percentage as well as the maximum percentage deviation in P_2.

(c) The deviation in P_3.

10-9C Consider the GP:

$$\begin{array}{ll} \text{goal}\{\ -x_1 + 2x_2 = z_1\} & (z_1 \leqq 0) \\ \text{goal}\{-3x_1 + \ x_2 = z_2\} & (z_2 \leqq -3) \\ \text{goal}\{\ \ x_1 + 2x_2 = z_3\} & (z_3 = 8) \\ \text{s.t.} \quad 2x_1 + x_2 \leqq 12 & \\ \quad -3x_1 + 4x_2 \leqq 4 & \\ \quad x_1, x_2 \geqq 0 & \end{array}$$

(a) Graph the GP and indicate the goal-efficient set.

(b) Specify the vector-maximum formulation (in tabular form) for computing the goal-efficient set.

(c) Using a vector-maximum algorithm, compute all goal-efficient vertices.

10-10C Consider:

$$\begin{array}{ll} \text{goal}\{x_1 + x_2 + x_3 = z_1\} & P_1(z_1 \in [9, 11]) \\ \text{goal}\{\qquad\qquad x_3 = z_2\} & P_2(z_2 \geqq 3) \\ \text{s.t.} \quad 0 \leqq x_1, x_2, x_3 \leqq 4 & \end{array}$$

(a) Graph the GP and list the vertices of the set of lex min solutions.

(b) Write the lex min formulation.

(c) Write the vector-maximum formulation in tabular form.

(d) Using a computer, solve for a lex min solution.

(e) Using a computer, solve for all vertices of the set of lex min solutions.

CHAPTER 11

Filtering and Set Discretization

Filtering refers to the process of selecting subsets of points from a larger finite set of points. *Set discretization* refers to the process of characterizing a continuous set by selecting a finite number of points from it. Filtering is a tool that enables us to manage large amounts of finite information. Set discretization enables us to manage continuous information in a finite fashion.

11.1 FORWARD AND REVERSE FILTERING

There are two types of filtering: *forward filtering* and *reverse filtering*. Let V be a finite set of vectors. With regard to forward filtering, let P denote the *forward set size* (the size of the subset to be computed). In forward filtering, we strive to obtain the P vectors in V that are the most different from one another. This is accomplished by finding the P vectors in V that are *furthest* apart from one another according to a given metric.

With regard to reverse filtering, let $\bar{\mathbf{v}} \in V$ and P denote the *reverse set size*. Then, in reverse filtering we compute the $P - 1$ vectors in V that are the most similar to $\bar{\mathbf{v}}$. This is accomplished by finding the $P - 1$ vectors closest to $\bar{\mathbf{v}}$ according to a given metric.

The metrics used in this book come from the family of weighted L_p-metrics (see Section 2.5). Note the difference between p and P: p is the *metric parameter* and P is the set size. Example 1 illustrates the use of forward and reverse filtering in multiple objective linear programming.

Example 1. In Fig. 11.1, let $\mathbf{x}^0$ be the decision maker's optimal point and $\mathbf{x}^5$ be his or her efficient extreme point of greatest utility. Assume that we have just generated all efficient extreme points

$$\{\mathbf{x}^1, \mathbf{x}^2, \ldots, \mathbf{x}^9\}$$

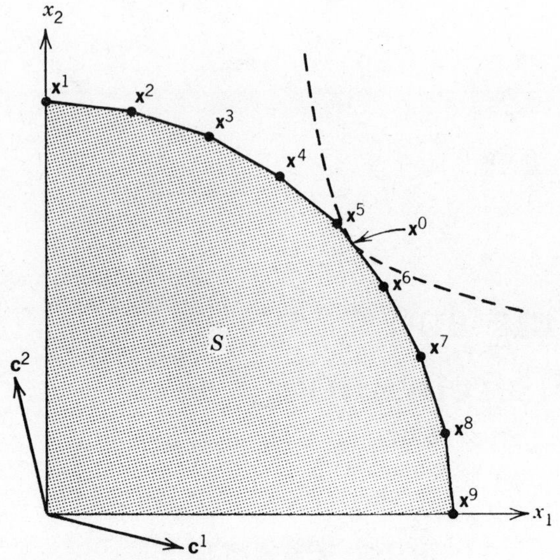

Figure 11.1. Graph of Example 1.

along with their criterion vectors

$$\{\mathbf{z}^1, \mathbf{z}^2, \ldots, \mathbf{z}^9\}$$

Suppose our goal is to locate the efficient extreme point of greatest utility without presenting the DM more than four candidates at a time. Let us assume that by forward filtering the set of criterion vectors, we obtain the dispersed set

$$\{\mathbf{z}^1, \mathbf{z}^4, \mathbf{z}^6, \mathbf{z}^9\}$$

If we present these criterion vectors to the DM, it would be reasonable for him or her to select $\mathbf{z}^6$ as the most preferred. If we reverse filter about $\mathbf{z}^6$ as the reference point, it would be reasonable for us to obtain

$$\{\mathbf{z}^4, \mathbf{z}^5, \mathbf{z}^6, \mathbf{z}^7\}$$

From this set the DM would select $\mathbf{z}^5$. Thus, we are able to identify $\mathbf{x}^5$ as the extreme point of greatest utility. Note that the DM did not have to examine all extreme point criterion vectors to determine which one was best.

11.2 WEIGHTED L_p DISTANCE MEASURE

Let $V \subset R^q$ be a finite set of vectors. The forward and reverse filtering processes operate by comparing distances between pairs of vectors $\mathbf{v}^t, \mathbf{v}^h \in V$ according to a weighted L_p-metric. In accordance with a weighted L_p-metric, the distance between $\mathbf{v}^t$ and $\mathbf{v}^h$ is given by

$$\|\mathbf{v}^t - \mathbf{v}^h\|_p^{\boldsymbol{\pi}} = \left[\sum_{i=1}^{q} \left(\pi_i |v_i^t - v_i^h| \right)^p \right]^{1/p} \tag{11.1}$$

where

q is the length of the vectors being filtered.
π_i is the *range equalization weight* (explained in Section 11.3) associated with the ith component of the vectors being filtered.
p is the *metric parameter*, $p \in (\{1, 2, \ldots\} \cup \{\infty\})$.

Eq. (11.1) is called the weighted L_p *distance measure*.

11.3 RANGE EQUALIZATION WEIGHTS

The π_i range equalization weights are not to be confused with relative importance weights. The π_i weights have a different purpose. The purpose of the π_i weights is to equalize the ranges of the components of the vectors in V. This is done so that the filtering processes will not be biased in favor of the components with the largest ranges.

To compute the range equalization weights, let R_i denote the *range* of the ith component of the vectors in V where

$$R_i = \max_{\mathbf{v} \in V} \{ v_i \} - \min_{\mathbf{v} \in V} \{ v_i \}$$

From the R_i, we obtain the range equalization weights

$$\pi_i = \frac{1}{R_i} \left[\sum_{j=1}^{q} \frac{1}{R_j} \right]^{-1}$$

Example 2. Let $V = \{\mathbf{v}^1, \mathbf{v}^2, \mathbf{v}^3\}$ where

$$\mathbf{v}^1 = (10, 20, 70)$$
$$\mathbf{v}^2 = (8, 140, 60)$$
$$\mathbf{v}^3 = (0, 60, 80)$$

With $R_1 = 10$, $R_2 = 120$, and $R_3 = 20$, we have $\pi_1 = \frac{12}{19}$, $\pi_2 = \frac{1}{19}$, and $\pi_3 = \frac{6}{19}$. Then, according to the weighted L_p distance measure [Eq. (11.1)] with $p = 2$, the distance between $\mathbf{v}^1$ and $\mathbf{v}^2$ is given by

$$\left[\sum_{i=1}^{3} \left(\pi_i |v_i^1 - v_i^2| \right)^2 \right]^{\frac{1}{2}} = \sqrt{\left(\frac{24}{19}\right)^2 + \left(\frac{120}{19}\right)^2 + \left(\frac{60}{19}\right)^2}$$
$$= 7.1734$$

The distance between $\mathbf{v}^1$ and $\mathbf{v}^3$ is given by

$$\left[\sum_{i=1}^{3} \left(\pi_i |v_i^1 - v_i^3| \right)^2 \right]^{\frac{1}{2}} = \sqrt{\left(\frac{120}{19}\right)^2 + \left(\frac{40}{19}\right)^2 + \left(\frac{60}{19}\right)^2}$$
$$= 7.3684$$

Vector $\mathbf{v}^2$ is closer to $\mathbf{v}^1$ than $\mathbf{v}^3$ because the ranges were equalized. Had the ranges not been equalized, $\mathbf{v}^3$ would have been closer to $\mathbf{v}^1$.

11.4 MECHANICS OF FORWARD FILTERING

We now discuss three forward filtering methods and some related topics. The first method is called the method of *first point outside the neighborhoods*. This method is rapid and produces first approximation results. The second method is called the method of *closest point outside the neighborhoods*. The third method is called the method of *furthest point outside the neighborhoods*. The second and third methods require more computation but produce superior results.

In all three methods, we use a *filtering relationship* to compare the weighted L_p distances between points not currently retained by the filter and those retained by the filter. The filtering relationship is given by

$$\left[\sum_{i=1}^{q}\left(\pi_i|v_i^t - v_i^h|\right)^p\right]^{1/p} < d$$

where

t is the identification superscript of a vector not currently retained by the filter.
h is the identification superscript of a vector retained by the filter.
d is the *test-distance* parameter.

The test-distance parameter regulates the forward filtering process in the following way. If $\mathbf{v}^t$ is greater than or equal to d away from $\mathbf{v}^h$, the two points are assumed to be *significantly* different from one another. If $\mathbf{v}^t$ is less than d away from $\mathbf{v}^h$, the two points are assumed to be *insignificantly* different from one another.

11.4.1 Method of First Point Outside the Neighborhoods

This method is described by means of an example. Suppose we have 15 vectors and we wish to obtain a dispersed set of size eight using the method of first point outside the neighborhoods. To begin, we start with some value for the test-distance parameter d. Then, the 15 vectors are placed in a list to be read in by the filtering routine. The vector at the top of the list is the *forward seed point*. The forward seed point "primes" the filter and is the first point retained by the filter. The points retained by the filter after all have been processed constitute the dispersed set computed by the filter.

To determine the second point retained by the filter, we, one by one, begin processing the other 14 vectors. Since the seed point is the only vector so far retained by the filter, the vectors are each compared with the seed point using the filtering relationship. Let us assume that vectors #2 and #3 are insignificantly different from the seed point (#1), but vector #4 is not. Then, we pause so that #4 can be retained by the filter.

Now we resume processing with vector #5. However, beginning with vector #5, the next vector to be retained by the filter must be significantly different from *both* the seed point and vector #4. Let us assume that vectors #5, #6, #7, and #8 do not meet this condition but #9 does. Then, we pause so that vector #9 can be retained by the filter. Now, if we resume processing with vector #10, any further vector to be retained by the filter must be significantly different from *each* of the vectors #1, #4, and #9.

Let us assume that none of the remaining vectors meet this condition. Although we finished processing the list, note that we did not achieve our objective of retaining exactly eight out of the 15. The reason we wound up with three instead of the desired eight was that the value used for d was too large. Observe that when d is large, it is harder to establish significance than when d is small. Therefore, we must reduce the value of d in hopes of increasing the number of vectors retained by the filter.

Unfortunately, there is no way to determine an appropriate value for d other than by experimentation. Suppose we forward filter a list of vectors with a value of d and find that the reduced list produced is too long. Then d would have to be increased and the vectors filtered once again. Suppose the new reduced list turns out too short. Then, d would have to be decreased and the vectors filtered yet another time, and so forth. Thus, in forward filtering, it is necessary to conduct *multiple runs* with different values of d until we converge to a reduced set of desired size.

Example 3. Suppose the seven points of Fig. 11.2 are in a list, in ascending order according to their superscripts. Forward filtering the list using the method of first point outside the neighborhoods with the L_2-metric and d as indicated,

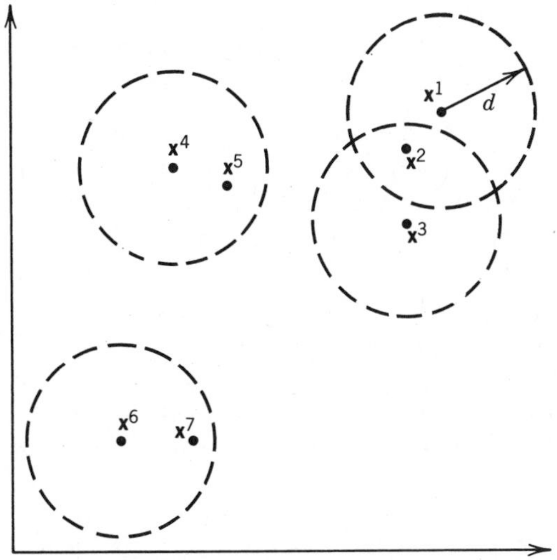

Figure 11.2. Method of first point outside the neighborhoods.

four points are retained by the filter. They are $\mathbf{x}^1$, $\mathbf{x}^3$, $\mathbf{x}^4$, and $\mathbf{x}^6$. Point $\mathbf{x}^1$ is retained because it is the seed point, and $\mathbf{x}^3$ is retained because it was the first point processed by the filtering relationship that was outside the *d-neighborhood* of $\mathbf{x}^1$. Point $\mathbf{x}^4$ is retained because it was the first point encountered outside of the d-neighborhoods of $\mathbf{x}^1$ and $\mathbf{x}^3$. Point $\mathbf{x}^6$ is retained because it was the first point encountered outside the d-neighborhoods of $\mathbf{x}^1$, $\mathbf{x}^3$, and $\mathbf{x}^4$.

In Example 3, we see that the set of points retained by the filter is not only *seed point dependent*, but it is dependent on the order in which the points are processed.

Forward filtering can be defined with *multiple seed points*. With these points placed at the top of the list and unconditionally retained by the filter, the other points to be retained by the filter are determined in the usual way.

11.4.2 Halving and Doubling Strategy

The rule when forward filtering is

1. Reduced set too small: decrease d.
2. Reduced set too large: increase d.

We note that the value of d that produces a given reduced set size comes from an interval. To incrementally adjust d so that the multiple run process of forward filtering converges to a dispersed set of desired size, we utilize a *halving and doubling* strategy. The forward filtering halving and doubling scheme is flowcharted in Fig. 11.3. In Fig. 11.3,

1. Δd is the amount by which we increment (or decrement) d.
2. `ILARGE = 1` means we have already filtered the list of vectors with a d-value that produced a reduced list that was too large.
3. `ISMALL = 1` means we have already filtered the list of vectors with a d-value that produced a reduced list that was too small.

When we start the multiple run forward filtering process, we are in "doubling mode" because we are doubling Δd each iteration. But as soon as `ILARGE` and `ISMALL` both equal one, we switch to "halving mode" because we halve Δd each iteration. We remain in halving mode until termination.

Example 4. Let the interval of d-values that produces a reduced set of the desired size P be $(4.1, 4.9]$. Let the initial test-distance value $d = 9$ and the initial $\Delta d = 1$. Using the halving and doubling algorithm of Fig. 11.3, we converge to a d-value that produces a reduced set of size P as tabulated in Table 11.1. In Fig. 11.4, the circled numbers and arrows indicate the iteration values of d.

11.4.3 Initial Values of d and Δd

Since it is not possible to compute a priori a d-value guaranteed to be in the interval that produces a reduced set of desired size, it is necessary to initialize the halving and doubling algorithm with values for d and Δd. To determine initial

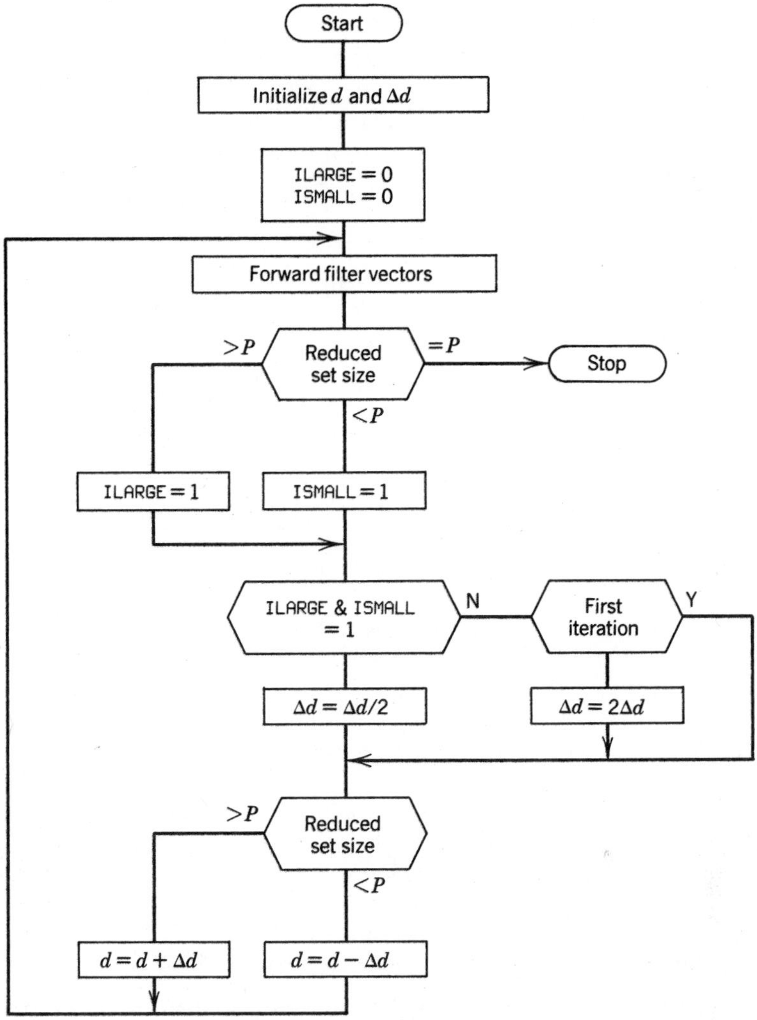

Figure 11.3. Halving and doubling routine.

values for d and Δd, intelligence can be gathered from the range information that is compiled for the computation of the range equalization weights of Section 11.3. The idea is to obtain "good" initial values for d and Δd to give the forward filtering process an advanced start. A method that has worked well is to allow

$$\text{initial } d = \frac{\left[\sum_{i=1}^{q} (\pi_i R_i)^p\right]^{1/p}}{4}$$

and

$$\text{initial } \Delta d = \frac{\text{initial } d}{2}$$

TABLE 11.1
Tabulated Data of Example 4

Iteration	d	Δd	*Set Size*
1	9	1	$< P$
2	8	2	$< P$
3	6	4	$< P$
4	2	2	$> P$
5	4	1	$> P$
6	5	$\frac{1}{2}$	$< P$
7	$4\frac{1}{2}$	–	$= P$

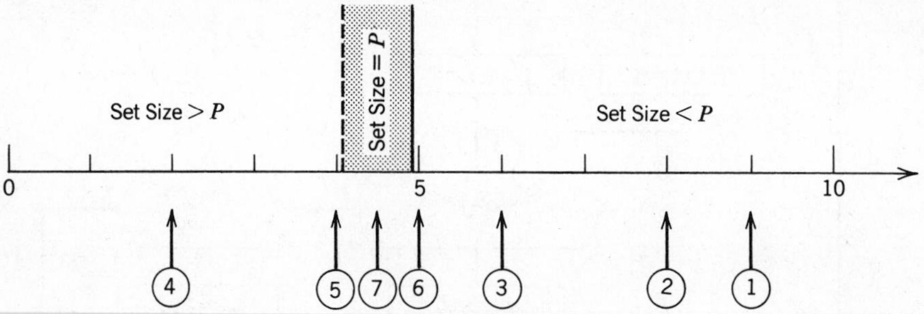

Figure 11.4. Graph of Example 4.

where q is the length of the vectors being filtered, the R_i are the component ranges, the π_i are the range equalization weights, and p is the metric parameter.

11.4.4 Method of Closest Point Outside the Neighborhoods

To explain this method, suppose we wish to forward filter a list of 20 points with a given value for d. As in the method of first point outside the neighborhoods, the first point in the list (the seed point) is retained by the filter. To determine the second point retained by the filter, all of the remaining 19 points are processed by the filtering relationship to determine their distances from the seed point. Let us assume that five of the points are less than d away from the seed point. Hence, they are discarded. Of the remaining 14, the one closest to the seed point is retained as the second point held by the filter.

To determine the third point retained by the filter, the remaining 13 points are processed using the filtering relationship to determine their distances from (a) the seed point, and (b) the second point held by the filter. Let us assume that this results in four of the points being discarded. Of the remaining nine, the point with the smallest sum of the two distances is retained as the third point held by the filter.

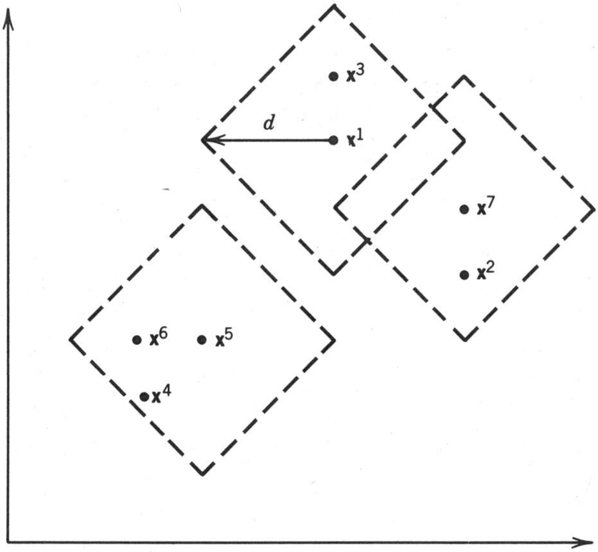

Figure 11.5. Method of closest point outside the neighborhoods.

To determine the fourth point retained by the filter, the remaining eight points are processed using the filtering relationship to determine their distances from (a) the seed point, (b) the second point held by the filter, and (c) the third point held by filter. Let us assume that five of the points do not meet the requirement of being outside of the *d*-neighborhoods of all three. Of the three survivors, the one with the smallest sum of the three distances is retained as the fourth point held by the filter.

To determine if there is a fifth point to be retained by the filter, the two remaining points are processed to determine their distances from the four points held by the filter. If we assume that both points are within the *d*-neighborhood of the last point retained by the filter, the two points are discarded. Thus, for the given value of *d*, the method of closest point outside the neighborhoods yields a forwardly reduced set of size four.

Example 5. Suppose the seven points of Fig. 11.5 are in a list in ascending order according to their superscripts. If we forward filter the list using the method of closest point outside the neighborhoods with the L_1-metric and d as indicated, three points are retained by the filter. They are $\mathbf{x}^1$, $\mathbf{x}^7$, and $\mathbf{x}^5$. Point $\mathbf{x}^1$ is retained because it is the seed point. Point $\mathbf{x}^7$ is the next point retained because it is the closest point outside the *d*-neighborhood of $\mathbf{x}^1$. Point $\mathbf{x}^5$ is the final point retained because it is outside the *d*-neighborhoods of, but closest to, $\mathbf{x}^1$ and $\mathbf{x}^7$. Since no other points are outside of the three *d*-neighborhoods, we are done. Note that if we had used the method of first point outside the neighborhoods, the filter would have retained $\mathbf{x}^1$, $\mathbf{x}^2$, and $\mathbf{x}^4$ in that order.

As seen, the method of closest point outside the neighborhoods requires more processing than the method of first point outside the neighborhoods. Although the method of closest point outside the neighborhoods is seed point dependent, it does not depend on the order in which the rest of the vectors are processed.

11.4.5 Method of Furthest Point Outside the Neighborhoods

This method is identical to the method of closest point outside the neighborhoods, except for the way we pick the next point to be retained by the filter. Instead of picking the significantly different point that has the *smallest* sum of the distances from the points already retained by the filter, we pick the significantly different point with the *largest* sum. For instance, with this method in Example 5, we would obtain $\mathbf{x}^1$, $\mathbf{x}^4$, and $\mathbf{x}^2$ in that order. Although the method of furthest point outside the neighborhoods may well be the best to use in most situations, there are instances, such as in Problem 11-5, when the method is unable to produce desired results and another method might have to be selected.

11.4.6 Maximally Dispersed Subsets

The real goal of forward filtering is to produce *maximally dispersed* subsets. Let V be a finite set of points. By a maximally dispersed subset of V of size P, we mean the set of P points whose members are further apart from one another than any other set of P points in V. That is, a maximally dispersed subset of size P is the set of points that is able to survive the filtering relationship for a larger value of d than any other set of P points.

To obtain a maximally dispersed subset requires considerable computation. With the methods of closest point or furthest point outside the neighborhoods, this would involve filtering the set of points down to a reduced set of size P once for each point as the seed point. Then, the subset with the largest final value of d is maximally dispersed. Because of the amount of computation involved in obtaining a maximally dispersed subset, we typically accept *approximately* maximally dispersed subsets. Such subsets are produced by the methods of first point, closest point, and furthest point outside the neighborhoods for an arbitrary seed point.

Example 6. Consider the set of points $\{\mathbf{x}^1, \mathbf{x}^2, \ldots, \mathbf{x}^9\}$ along the real line in Fig. 11.6. In this set is a maximally dispersed subset of size five, $\{\mathbf{x}^1, \mathbf{x}^3, \mathbf{x}^5, \mathbf{x}^7, \mathbf{x}^9\}$. However, whether we would obtain this set by forward filtering would depend on

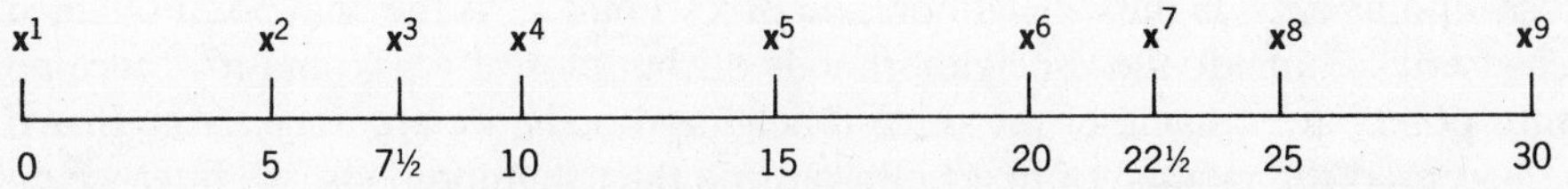

Figure 11.6. Graph of Example 6.

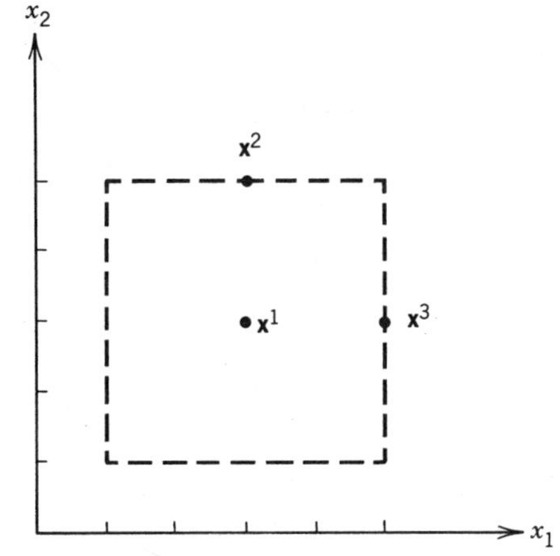

Figure 11.7. Graph of Example 7.

the choice of seed point. If we were to choose $\mathbf{x}^2$, $\mathbf{x}^4$, $\mathbf{x}^6$, or $\mathbf{x}^8$ as the seed point, we would not obtain the maximally dispersed subset.

11.4.7 Unattained Set Sizes

It may not be possible to produce a reduced set of a given size using the methods of first point, closest point, or furthest point outside the neighborhoods because of "ties."

Example 7. Let $\mathbf{x}^1 = (3, 3)$, $\mathbf{x}^2 = (3, 5)$, and $\mathbf{x}^3 = (5, 3)$. With these points and $p = \infty$, there does not exist a maximally dispersed subset of size two. If d is greater than the value indicated in Fig. 11.7, the reduced set size is one. If d is equal to or less than that indicated, the reduced set size is three.

11.5 REVERSE FILTERING

Let V be a finite set of vectors that is to be reverse filtered. Let $\bar{\mathbf{v}} \in V$ be the *reverse seed point* (placed at the top of the list of points of V when inputted to the filter). In reverse filtering, we rank order (closest to farthest) the points in V in accordance with their distances from the reverse seed point using the weighted L_p distance measure of Section 11.2. If P is the reverse set size, the reverse filtering process outputs $\bar{\mathbf{v}}$ and its $P - 1$ closest neighbors. We call this the *reverse list*. The closer a point is to the top of the reverse list, the closer the point is to the seed point.

We use the term "reverse" to describe the above process because in reverse filtering we are concerned about the points closest to the seed point, whereas in

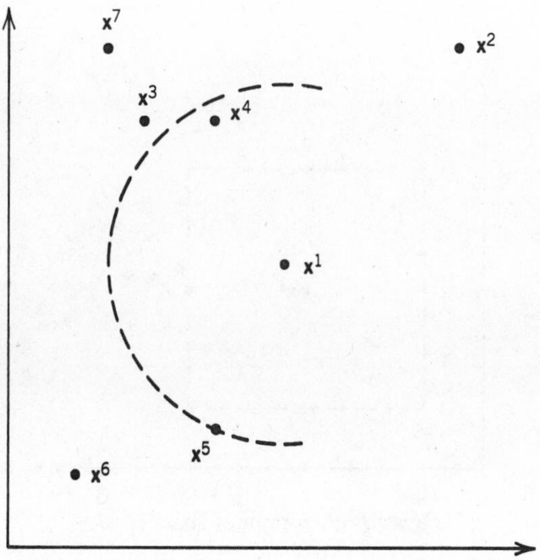

Figure 11.8. Graph of Example 8.

forward filtering, we are concerned about the points furthest away from other points. Thus, forward filtering is used to "thin out" a set of points, whereas reverse filtering is used to reintroduce points in a given locality. Also, in contrast to forward filtering that typically requires multiple runs, a list of points is reverse filtered on one pass.

Example 8. Consider the set of points $\{\mathbf{x}^1, \mathbf{x}^2, \ldots, \mathbf{x}^7\}$ in Fig. 11.8. When we reverse filter the list with $p = 2$ and $P = 4$, the reverse list outputted is $\{\mathbf{x}^1, \mathbf{x}^4, \mathbf{x}^5, \mathbf{x}^3\}$. (The dashed half-circle is drawn for visualizing distances from $\mathbf{x}^1$.)

Reverse filtering can be defined with multiple seed points. With these points placed at the top of the list, the other points in the list are then ranked in terms of their closeness to the multiple seed points. A point's closeness to the multiple seed points is determined by the total of its weighted L_p distances from each of the seed points.

11.6 FILTER PROGRAM

FILTER is one of the three codes (besides **LAMBDA** and **ADBASE**) that constitutes the **ADBASE** multiple objective linear programming package [Steuer (1983)]. **FILTER** is a Fortran program that provides the forward and reverse filtering capabilities explained in this chapter. Described in terms of the parameters that must be set prior to an execution, we have

1. **METRIC**: Value of p. To communicate the L_∞-metric, **METRIC** is set to 9999. *Caution:* If **METRIC** is set to a number greater than about 10 other than 9999, we are likely to encounter overflow because of the computations involved in the left-hand side of the filtering relationship.
2. **WGT(*)**: Enables us to override the (default) internally generated range equalization weights.
3. **D** *and* **DELTD**: Enables us to override the (default) internally computed initial d and Δd values.
4. **IFR**: Type of filter

   = 1 Forward filter.
   = 2 Reverse filter.

5. **NFSPTS**: Number of forward seed points.
6. **IFSIZE**: Value of P (size of the forwardly reduced set that is to be computed).
7. **MX1ST**: Maximum number of iterations for the method of first point outside the neighborhoods.
8. **MXCLOS**: Number of iterations for the method of closest point outside the neighborhoods.
9. **MXFUR**: Number of iterations for the method of furthest point outside the neighborhoods. **MXFUR** cannot be positive if **MXCLOS** is positive, and vice versa.
10. **NRSPTS**: Number of reverse seed points.
11. **IRSIZE**: Value of P (length of the reverse list to be outputted).
12. **IPRT**: Output switch

    = 0 Write to a file.
    = 1 Print hardcopy output.

    FILTER reads in the vectors to be filtered in **ADBASE** six-field format (see Section 9.11.3). When **IPRT** = 0, **FILTER** writes the forwardly reduced or reverse list to a designated file in **ADBASE** six-field format.
13. **IPUFMT**: Format in which the component values of the vectors that are written to a file are expressed:

    = 1 F10.1
    = 2 F10.2
    = 3 F10.3
    = 4 F10.4
    = 5 F10.5
    = 6 F10.6

Example 9. To illustrate **FILTER**, let us forwardly filter the criterion vectors (shown in Fig. 9.18) that were written to a file by the **ADBASE** example of Section

```
7400     4  8
7400     1  1    32.222      6.667   -20.000    58.889
7400     4  1   -80.000    100.000    60.000    40.000
7400     8  1   -19.344     24.590    63.279    35.082
7400    15  1   -26.032    -18.095    20.000    75.873
7400    17  1   -26.414    -17.489   116.806     3.498
7400    18  1   -52.973     60.811   105.946     8.919
7400    21  1   -87.027      8.108    68.649    58.378
7400    25  1   -71.592     11.783   153.376   -16.752
```

Figure 11.9. Output file produced by `FILTER` for Example 9.

9.11.4. If we run `FILTER` with

`IFR`	= 1
`NFSPTS`	= 1
`IFSIZE`	= 8
`MX1ST`	= 30
`MXCLOS`	= 30
`IPRT`	= 0
`IPUFMT`	= 3

the dispersed set written to a file by `FILTER` is given in Fig. 11.9.

Comment is required about `MX1ST` (that can be associated with a Phase I of the forward filtering process) and `MXCLOS` and `MXFUR` (that can be associated with a Phase II). For the sake of discussion, let `MX1ST` = 8, `MXCLOS` = 25, and `MXFUR` = 0. In Phase I, we use the method of first point outside the neighborhoods to compute a dispersed set of size `IFSIZE`. If we have been successful within eight iterations, or if we have not after the eighth iteration, we then exit Phase I and enter Phase II. Utilizing the final Phase I d and Δd values as the starting d and Δd values for Phase II, we employ the method of closest point outside the neighborhoods. We continue with 25 more forward filtering iterations, trying to increase d even after having found a dispersed set of size `IFSIZE`. Our goal in Phase II is to find the dispersed set of size `IFSIZE` that corresponds to the largest value of d. In other words, we are trying to converge to the upper limit of the interval of d-values that produces reduced sets of size `IFSIZE`. The purpose of Phase I is to obtain an advanced start for Phase II.

11.7 INTERACTIVE FORWARD AND REVERSE FILTERING

Suppose we have a relatively large finite set of criterion vectors and our goal is to identify the DM's most preferred. If we use forward and reverse filtering, an interactive procedure can be applied to sample a sequence of progressively smaller neighborhoods in the set of criterion vectors until the criterion vector of greatest preference is obtained.

To illustrate, suppose we have a file that contains $\nu = 28$ criterion vectors. In an effort to locate his or her most preferred, assume that the DM does not want to iterate more than $t = 4$ times and that he or she does not want to be presented with samples that contain more than $P = 5$ criterion vectors each. To calibrate the algorithm so that the neighborhoods being sampled are steadily reduced until, on the tth iteration, the final neighborhood contains exactly P criterion vectors, we utilize an *r-reduction factor*. Since the tth neighborhood is to contain exactly P criterion vectors

$$r^{t-1}(\nu) = P$$

Thus,

$$r = \sqrt[t-1]{P/\nu}$$

With $\nu = 28$, $P = 5$ and $t = 4$, we then have $r = .563$.

Consider Fig. 11.10, in which the numbers denote the criterion vectors. Let #27 be optimal. Since we are starting, all 28 criterion vectors constitute the first neighborhood. To determine the first iteration sample of five criterion vectors, we forwardly filter the 28 criterion vectors. Assume that this yields the dispersed set $\{12, 14, 21, 22, 26\}$ as indicated in Fig. 11.10*a*. Suppose that the DM (for whatever reason) selects #26 as the one he or she likes best.

To develop the second iteration sample, we reverse filter the 28 criterion vectors with #26 as the seed point. Since $r = .563$, we select the top $r(28) \cong 16$ criterion vectors from the reverse list as the second iteration neighborhood. Forward filtering the 16 criterion vectors, assume that we obtain the set $\{1, 5, 14, 22, 26\}$ of Fig. 11.10*b* as the second iteration sample. Let #14 be the DM's second iteration selection.

For the third iteration sample, we reverse filter the 28 criterion vectors, but this time with #14 as the seed point. Taking the topmost $r^2(28) \cong 9$ criterion vectors from the reverse list, we have the third iteration neighborhood. Forward filtering the nine criterion vectors, assume that we obtain the set $\{4, 6, 9, 11, 14\}$ of Fig. 11.10*c* as the third iteration sample. Let #9 be the DM's third iteration selection.

The fourth iteration sample is determined directly by reverse filtering the 28 criterion vectors with #9 as the seed point, and then taking the topmost $r^3(28) \cong 5$ criterion vectors from the reverse list. This yields the set $\{7, 9, 22, 27, 28\}$ of Fig. 11.10*d*. From this set, the DM can select his or her optimal solution #27.

It is very possible that a decision maker will alter his or her expectations as he or she searches for the most preferred solution. Thus, it is desirable to have P and t as large as possible so that r can be as large as possible. This is because the greater r, the greater the procedure's ability to go back into regions discarded in previous iterations. Note that in our problem with 28 criterion vectors, the optimal criterion vector was in the region discarded after the first iteration. However, in the third and fourth iterations, we were able to go back into previously discarded territory and locate #27.

A variation of the procedure outlined in this section is to never present the DM with a "duplicate." That is, never include in any sample a candidate that the DM has seen before. This means that at each iteration, the DM would be asked whether there are any points in the new sample that are better than the best of

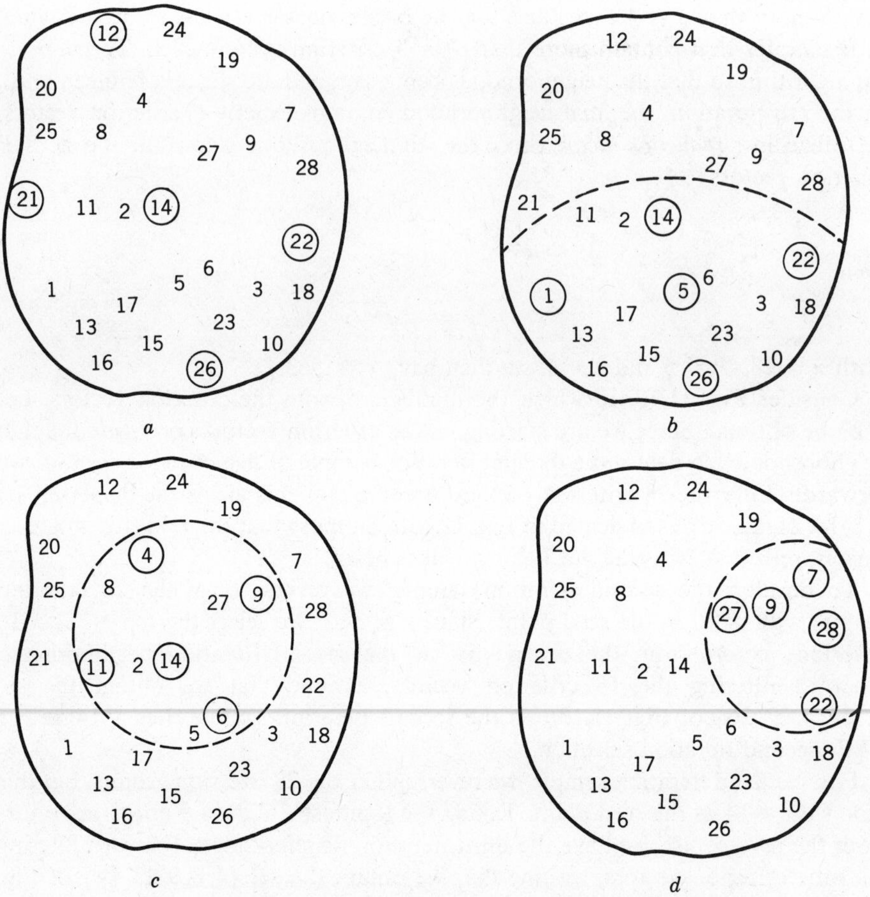

Figure 11.10. Sequence of progressively smaller neighborhoods.

what he or she has seen previously. If the answer is "No," we would then use the best previous reverse seed point two iterations in a row.

11.8 SET DISCRETIZATION

To motivate the need for set discretization techniques, consider the problem in Fig. 11.11, in which it is necessary to represent the shaded set

$$M = \overset{4}{\underset{i=1}{\gamma}} (\mathbf{x}^i)$$

with eight discrete points. To best represent the range of choice available in M with eight points, it is important that the eight points provide a dispersed covering of M. Since there are not enough extreme points, additional points will have to be generated that are not extreme. In fact, we may not even want to use

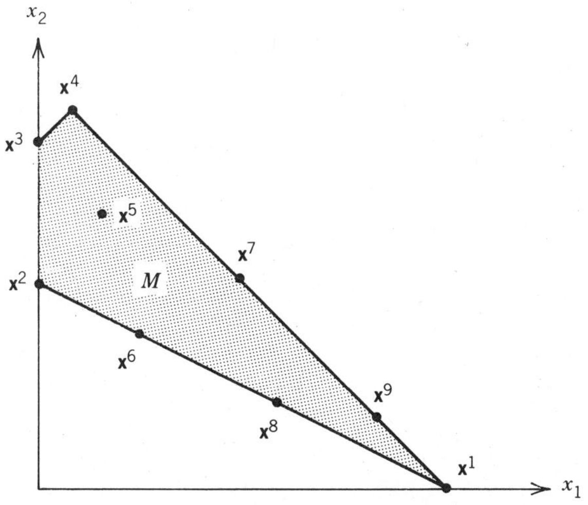

Figure 11.11. Set discretization example.

all extreme points. For instance, because of the similarity between $\mathbf{x}^3$ and $\mathbf{x}^4$, we would probably not want to use both. To represent M with eight points, we would want a group of dispersed points such as $\{\mathbf{x}^1, \mathbf{x}^2, \mathbf{x}^4, \mathbf{x}^5, \mathbf{x}^6, \mathbf{x}^7, \mathbf{x}^8, \mathbf{x}^9\}$.

We now discuss techniques for generating dispersed representations of continuous sets.

11.8.1 Predetermined Convex Combinations

Suppose we wish to characterize a polyhedron with a small finite number of representatives. In the method of this section, we manufacture weighting vectors in advance by formula so as to be as different from one another as possible. Then, the weighting vectors are applied to the extreme points of the polyhedron to obtain a group of convex combinations. Because the weighting vectors have been constructed to be well dispersed, it is hoped that the resulting convex combinations will also be well dispersed. This is not always the case.

To demonstrate, consider the triangle defined by $\mathbf{x}^1$, $\mathbf{x}^2$, and $\mathbf{x}^3$ in Fig. 11.12. To obtain seven dispersed representatives, let us apply the seven weighting vectors

$$\begin{aligned}
\boldsymbol{\lambda}^1 &= (1, 0, 0) \\
\boldsymbol{\lambda}^2 &= (0, 1, 0) \\
\boldsymbol{\lambda}^3 &= (0, 0, 1) \\
\boldsymbol{\lambda}^4 &= \left(\tfrac{1}{9}, \tfrac{4}{9}, \tfrac{4}{9}\right) \\
\boldsymbol{\lambda}^5 &= \left(\tfrac{4}{9}, \tfrac{1}{9}, \tfrac{4}{9}\right) \\
\boldsymbol{\lambda}^6 &= \left(\tfrac{4}{9}, \tfrac{4}{9}, \tfrac{1}{9}\right) \\
\boldsymbol{\lambda}^7 &= \left(\tfrac{1}{3}, \tfrac{1}{3}, \tfrac{1}{3}\right)
\end{aligned}$$

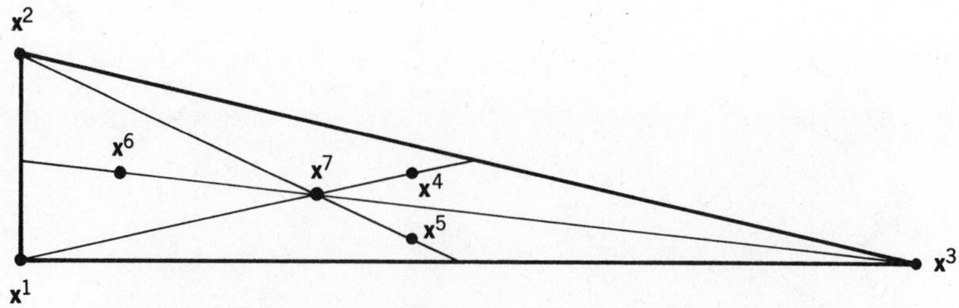

Figure 11.12. Predetermined convex combinations.

It can be argued that these weighting vectors are well dispersed because they correspond to the three extreme points ($\mathbf{x}^1, \mathbf{x}^2, \mathbf{x}^3$) of the triangle, the center point ($\mathbf{x}^7$) of the triangle, and the center points ($\mathbf{x}^4, \mathbf{x}^5, \mathbf{x}^6$) of the three subtriangles that have $\mathbf{x}^7$ and two of the three extreme points as vertices. In Fig. 11.12, the resulting convex combinations $\mathbf{x}^1, \dots, \mathbf{x}^7$ are seen not to be very well distributed.

In this illustration, it was easy to manufacture dispersed weighting vectors because the number chosen was two times the number of extreme points plus one. If some other number were chosen, it might not have been so easy to manufacture the weighting vectors by formula. Thus, along with distortions caused by eccentricities of the polyhedra, the method of predetermining convex combinations by means of manufactured weighting vectors has its limitations.

11.8.2 Convex Combinations Drawn from the Uniform Distribution

A way to overcome the shortcomings of the previous method is as follows. Randomly generate an over-abundance of weighting vectors and then, from the set of convex combinations thus created, filter away the excess by removing those that are the most redundant. The efficiency of this approach, however, is dependent on the distribution that we use to compute the weights. To demonstrate, consider the two-dimensional facet defined by $\mathbf{x}^1$, $\mathbf{x}^2$, $\mathbf{x}^3$, and $\mathbf{x}^4$ in Fig. 11.13. To construct the convex combinations

$$\sum_{i=1}^{4} \lambda_i \mathbf{x}^i$$

let the λ_i convex combination weights be determined by

$$\lambda_i = \frac{v_i}{\sum_{j=1}^{4} v_j} \tag{11.2}$$

in which v_i is drawn from the uniform distribution 0 to 1. As shown in Fig. 11.13, where each dot represents a resultant convex combination, there is a multinormal

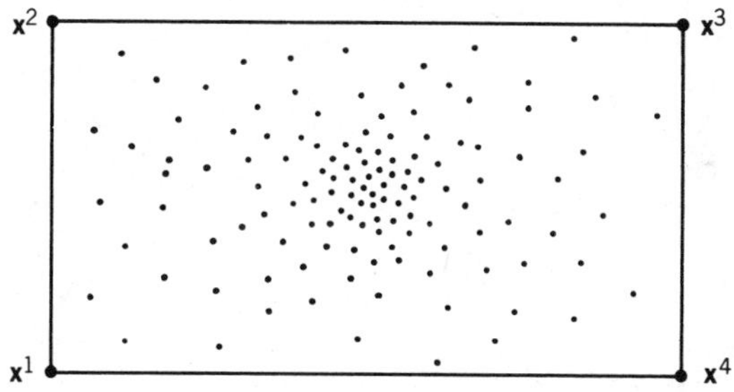

Figure 11.13. Convex combinations from the uniform distribution.

type of buildup about the geometric center of the facet when the λ_i convex combination weights are chosen in this way.

11.8.3 Convex Combinations Drawn from the Weibull Distribution

In an effort to counteract a multinormal type of buildup and place more convex combinations into the corners of a facet, consider making draws for the v_i in Eq. (11.2) from the Weibull distribution. A draw v_i from the Weibull distribution with parameters b and c can be obtained from the expression

$$b[-\ln(R: 0,1)]^{1/c}$$

where R: 0, 1 is a draw from the uniform distribution 0 to 1. By keeping both of the Weibull parameters at low values such as $b = .1$ and $c = .3$, the pattern of resultant convex combinations would take on the form indicated in Fig. 11.14.

11.8.4 50–50 Strategy

Sections 11.8.2 and 11.8.3 tend to suggest that a good "rule of thumb" is to generate 50% of the convex combination weighting vectors using draws from the uniform distribution, and 50% of the convex combination weighting vectors using

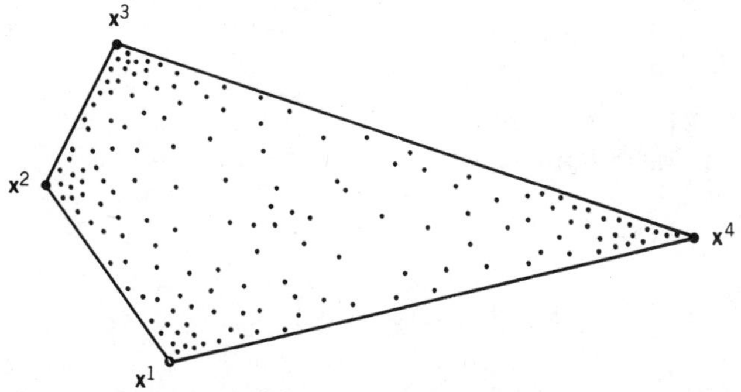

Figure 11.14. Convex combinations from the Weibull distribution.

draws from the Weibull distribution. In this way, we not only create convex combinations in the center of the polyhedron, but in its corners as well. Then, when we forward filter the over-abundance of representatives down to our set of desired size, the representatives surviving the filter should provide a well-dispersed covering of the polyhedron. The 50–50 strategy is implemented in the `LAMBDA` code whose description follows.

11.9 LAMBDA PROGRAM

`LAMBDA` is one of the three codes (besides `FILTER` and `ADBASE`) that comprise the `ADBASE` multiple objective linear programming package. `LAMBDA` is a Fortran program that provides a set discretization capability for gathering representatives from the (continuous) set of weighting vectors

$$\tilde{\Lambda} = \left\{ \lambda \in R^k \mid \lambda_i \in [\ell_i, \mu_i], \quad \sum_{i=1}^{k} \lambda_i = 1 \right\}$$

`LAMBDA` has been written for use with certain of the interactive procedures discussed in Chapters 13, 14, 15, and 16. Described in terms of the parameters that must be set prior to an execution, we have

1. `NOMB`: Iteration number. This helps keep weighting vector data from one iteration from being confused with weighting vector data from another.
2. `NPTS`: Number of weighting vectors from $\tilde{\Lambda}$ to be generated.
3. `IPORIG`: Number of components per weighting vector.
4. `BNDLO(i)`: Specifies the ℓ_i lower bounds on the weighting vector components.
5. `BNDHI(i)`: Specifies the μ_i upper bounds on the weighting vector components.
6. `ISEED`: Initial seed to the random number generator.
7. `BWEIB`: Value of the b parameter of the Weibull distribution.
8. `CWEIB`: Value of the c parameter of the Weibull distribution.
9. `IPRT`: Output switch

   $= 0$ Write to a file.

   $= 1$ Print hardcopy output.

   When `IPRT` $= 0$, `LAMBDA` writes the generated weighting vectors to a designated file in `ADBASE` six-field format (with a 3I5 header card).

Example 10. Let

$$\tilde{\Lambda} = \left\{ \lambda \in R^5 \mid \lambda_i \in [\ell_i, \mu_i], \quad \sum_{i=1}^{5} \lambda_i = 1 \right\}$$

in which $\ell_1 = \ell_2 = \ell_3 = \ell_5 = .1$, $\ell_4 = .3$, $\mu_1 = \mu_3 = \mu_4 = .4$, and $\mu_2 = \mu_5 = .2$. To illustrate how `LAMBDA` can be used with `FILTER` to produce a dispersed set of

```
 2    5   250
2   1  1    0.1545    0.1179    0.2055    0.3497    0.1724
2   2  1    0.1641    0.2000    0.1901    0.3213    0.1245
2   3  1    0.1636    0.1385    0.2205    0.3417    0.1357
2   4  1    0.1950    0.1330    0.1949    0.3387    0.1384
2   5  1    0.2551    0.1149    0.1380    0.3532    0.1388
2   6  1    0.1745    0.2000    0.1738    0.3265    0.1252
2   7  1    0.2095    0.1351    0.1994    0.3344    0.1216
2   8  1    0.2391    0.1312    0.1680    0.3373    0.1244
2   9  1    0.2296    0.1092    0.2000    0.3254    0.1358
2  10  1    0.1982    0.1337    0.1983    0.3336    0.1362
2  11  1    0.2015    0.1289    0.2366    0.3129    0.1201
2  12  1    0.2229    0.1131    0.2230    0.3000    0.1410
2  13  1    0.2166    0.1474    0.2022    0.3273    0.1065
2  14  1    0.1001    0.1000    0.3448    0.3000    0.1551
2  15  1    0.1901    0.1213    0.2124    0.3333    0.1429
2  16  1    0.3158    0.1000    0.1110    0.3720    0.1012
2  17  1    0.1985    0.1172    0.2303    0.3468    0.1072
2  18  1    0.1427    0.1192    0.3046    0.3203    0.1132
2  19  1    0.1054    0.1252    0.3000    0.3374    0.1320
2  20  1    0.1000    0.1681    0.2638    0.3000    0.1681
```

Figure 11.15. Top portion of file generated by LAMBDA.

10 representatives of $\tilde{\Lambda}$, let us run LAMBDA with

```
NOMB      = 2
NPTS      = 250
IPORIG    = 5
BNDLO(1) = .1        BNDHI(1) = .4
BNDLO(2) = .1        BNDHI(2) = .2
BNDLO(3) = .1        BNDHI(3) = .4
BNDLO(4) = .3        BNDHI(4) = .4
BNDLO(5) = .1        BNDHI(5) = .2
ISEED     = 6537
BWEIB     = .1
CWEIB     = .3
IPRT      = 0
```

The top portion of the file containing the first 20 weighting vectors generated by LAMBDA is shown in Fig. 11.15. Using the output of LAMBDA as input to FILTER with

```
IFR    = 1
NFSPTS = 1
IFSIZE = 10
MX1ST  = 30
MXCLOS = 30
IPRT   = 0
IPUFMT = 4
```

```
2     5  10
2     1   1    0.1545  0.1179  0.2055  0.3497  0.1724
2    36   1    0.3414  0.1585  0.1000  0.3000  0.1001
2    39   1    0.2675  0.1083  0.1945  0.3171  0.1126
2    94   1    0.1007  0.1012  0.3217  0.3025  0.1739
2   115   1    0.2253  0.1142  0.1477  0.3865  0.1263
2   144   1    0.1780  0.1824  0.1000  0.3572  0.1824
2   162   1    0.1738  0.1661  0.1994  0.3290  0.1317
2   172   1    0.3215  0.1016  0.1029  0.3002  0.1738
2   206   1    0.2186  0.1943  0.1125  0.3746  0.1000
2   226   1    0.1311  0.1113  0.3388  0.3103  0.1085
```

Figure 11.16. Output file produced by **FILTER** for Example 10.

we obtain the file containing the 10 dispersed representatives of $\tilde{\Lambda}$ shown in Fig. 11.16.

Readings

1. Hastings, N. A. J. and J. B. Peacock (1974). *Statistical Distributions*, London: Butterworth & Company.
2. Miller, G. (1956). "The Magical Number Seven Plus or Minus Two: Some Limits on our Capacity for Processing Information," *Psychological Review*, Vol. 3, No. 2, pp. 81–97.
3. Morse, J. N. (1980). "Reducing the Size of the Nondominated Set: Pruning by Clustering," *Computers and Operations Research*, Vol. 7, No.'s 1–2, pp. 55–66.
4. Steuer, R. E. and F. W. Harris (1980). "Intra-Set Point Generation and Filtering in Decision and Criterion Space," *Computers and Operations Research*, Vol. 7, No.'s 1–2, pp. 41–53.
5. Steuer, R. E. (1983). "Operating Manual for the **ADBASE** Multiple Objective Linear Programming Package," College of Business Administration, University of Georgia, Athens, Georgia.
6. Törn, A. A. (1980). "A Sampling-Search-Clustering Approach for Exploring the Feasible/Efficient Solutions of MCDM Problems," *Computers and Operations Research*, Vol. 7, No.'s 1–2, pp. 67–78.
7. Winkels, H.-M. (1981). "Visualization and Discretization for Convex Polyhedral Sets," Working Paper on Economathematics, No. 8104, Ruhr-Universität, Bochum, West Germany.

PROBLEM EXERCISES

11-1 Consider the weighted L_p-metric distance measure of Section 11.2 with $\pi_1 = \pi_2 = \frac{1}{2}$. Let $\mathbf{x}^1 = (3,2)$, $\mathbf{x}^2 = (10,1)$, and $\mathbf{x}^3 = (9,3)$.

(a) With $p = 1$, how far is $\mathbf{x}^2$ from $\mathbf{x}^1$? How far is $\mathbf{x}^3$ from $\mathbf{x}^1$? Which point is closest to $\mathbf{x}^1$?

(b) Follow the same instructions as in (a) but with $p = \infty$.

11-2 **(a)** Compute the π_i range equalization weights for the following set of vectors:

$$\mathbf{v}^1 = (40, 2, 350) \qquad \mathbf{v}^4 = (30, -2, 400)$$
$$\mathbf{v}^2 = (40, -2, 375) \qquad \mathbf{v}^5 = (30, 0, 275)$$
$$\mathbf{v}^3 = (25, 0, 300) \qquad \mathbf{v}^6 = (20, 1, 500)$$

(b) Let the metric parameter $p = 2$. What is the distance between $\mathbf{v}^1$ and $\mathbf{v}^2$ according to the weighted L_p distance measure? What is the distance between $\mathbf{v}^5$ and $\mathbf{v}^6$?

11-3 With regard to the method of first point outside the neighborhoods and the following set containing:

$$\mathbf{v}^1 = (2, 2) \qquad \mathbf{v}^5 = (6, 6)$$
$$\mathbf{v}^2 = (0, 4) \qquad \mathbf{v}^6 = (8, 4)$$
$$\mathbf{v}^3 = (2, 6) \qquad \mathbf{v}^7 = (6, 2)$$
$$\mathbf{v}^4 = (4, 8) \qquad \mathbf{v}^8 = (4, 0)$$

obtain a forwardly reduced dispersed subset of size $P = 4$. Assume that the metric parameter $p = 2$ and that the points are read into the filtering routine in ascending order according to their superscripts.

(a) Specify the points comprising the dispersed set in the order in which they are retained by the filter.

(b) Specify the maximum value of d that produces a dispersed set of size $P = 4$.

11-4 The same as Problem 11-3 except using the method of closest point outside the neighborhoods.

11-5 The same as Problem 11-3 except using the method of furthest point outside the neighborhoods.

11-6 With regard to the method of closest point outside the neighborhoods and the following points:

$$\mathbf{z}^1 = (0, 4) \qquad \mathbf{z}^3 = (6, 6)$$
$$\mathbf{z}^2 = (4, 0) \qquad \mathbf{z}^4 = (8, 8)$$

obtain a forwardly reduced dispersed set of size $P = 3$. Assume that the metric parameter $p = 2$ and that the points are read into the filtering routine in ascending order according to their superscripts.

(a) Specify the points comprising the dispersed set in the order in which they are retained by the filter.

(b) Specify the maximum value of d that produces a dispersed set of size $P = 3$.

11-7 The same as Problem 11-6 except using the method of furthest point outside the neighborhoods.

11-8 With reference to Problem 11-6, what is the range of d values for which the method of furthest point outside the neighborhoods produces a forwardly reduced set of size $P = 2$.

11-9 Consider the method of furthest point outside the neighborhoods and the following data:

$$\mathbf{z}^1 = (0, -2) \qquad \mathbf{z}^3 = (5, 3)$$
$$\mathbf{z}^2 = (4, 1) \qquad \mathbf{z}^4 = (10, 8)$$

Assume that the points are read into the forward filtering routine in ascending order according to their superscripts.

(a) What are the range equalization weights?

(b) What is the range of d values corresponding to a forwardly reduced dispersed subset of size $P = 3$?

(c) Specify the points comprising the dispersed set in the order in which they are retained by the filter when $d = 2$.

(d) Follow the same instructions as in (c) but with $d = 3$.

11-10 Let the interval of d-values that produces a forwardly filtered dispersed subset of size P be $(14.8, 14.9]$. Let the initial test-distance $d = 12$ and the initial increment $\Delta d = 2$. Using the halving and doubling algorithm of Section 11.4.2, show the audit trail of d and Δd values that leads to the desired set size, as in Table 11.1.

11-11 Consider the following points:

$$\mathbf{x}^1 = (4, 4) \qquad \mathbf{x}^4 = (7, 7)$$
$$\mathbf{x}^2 = (15, 16) \qquad \mathbf{x}^5 = (4, 12)$$
$$\mathbf{x}^3 = (16, 14)$$

that are to be filtered using the method of closest point outside the neighborhoods. Assume that the points are read into the filtering routine in ascending order according to their superscripts. List the points that constitute the resulting reduced set in the order in which they are retained by the filter when:

(a) $p = 1$ and $d = 2$.

(b) $p = \infty$ and $d = 2$.

11-12C Randomly generate 150 $\boldsymbol{\lambda}$-vectors from $\Lambda = \{\boldsymbol{\lambda} \in R^3 \mid \lambda_i > 0, \ \Sigma_{i=1}^{3} \lambda_i = 1\}$ and then filter down the 150 $\boldsymbol{\lambda}$-vectors to a dispersed set of size 15. Using a parametric diagram, plot the 15 points.

11-13C **(a)** Using the last four digits of your social security number (add one if even) as the seed to **LAMBDA**, generate 120 $\boldsymbol{\lambda}$-vectors of length eight.

(b) Using vector 1 as the seed point, use **FILTER** to reduce the 120 $\boldsymbol{\lambda}$-vectors to the 10 most different with **METRIC** = 2, **MX1ST** = 30, and **MXCLOS** = 0.

(c) Same as (b) except with **MXCLOS** = 30.

(d) Same as (b) except with **METRIC** = 1.

(e) Same as (b) except with vector 20 as the seed point.

(f) Using **FILTER**, compute the 19 closest λ-vectors to vector 100.

11-14C **(a)** Using the first three digits of your social security number (add one if even) as the seed to **LAMBDA**, generate 200 λ-vectors of length four from the intervals

$$\begin{bmatrix} .1000, .4000 \\ .2000, .3000 \\ .0000, .5000 \\ .1500, .2500 \end{bmatrix}$$

(b) Using vector 1 as the seed point, use **FILTER** to reduce the 200 λ-vectors to the 16 most different with **METRIC** = 2, **MX1ST** = 30, and **MXCLOS** = 30.

(c) Same instructions as in (b), except with **MXFUR** = 30 instead of **MXCLOS** = 30.

CHAPTER 12

Multiple Objective Linear Fractional Programming

In an otherwise all linear multiple objective program, one or more of the objectives may be *linear fractional*

$$\frac{\mathbf{c}^T\mathbf{x} + \alpha}{\mathbf{d}^T\mathbf{x} + \beta}$$

where α and β are scalar constants. That is, we have a ratio objective with a linear numerator and a linear denominator. This leads to the study of *multiple objective linear fractional* programming in which we have the (MOLFP)

$$\max\left\{\frac{\mathbf{c}^1\mathbf{x} + \alpha_1}{\mathbf{d}^1\mathbf{x} + \beta_1} = z_1\right\}$$

$$\max\left\{\frac{\mathbf{c}^2\mathbf{x} + \alpha_2}{\mathbf{d}^2\mathbf{x} + \beta_2} = z_2\right\}$$

$$\vdots$$

$$\max\left\{\frac{\mathbf{c}^k\mathbf{x} + \alpha_k}{\mathbf{d}^k\mathbf{x} + \beta_k} = z_k\right\}$$

$$\text{s.t. } \mathbf{x} \in S = \{\mathbf{x} \in R^n \,|\, \mathbf{A}\mathbf{x} = \mathbf{b}, \quad \mathbf{x} \geqq \mathbf{0}, \quad \mathbf{b} \in R^m\}$$

It is customary to assume that the MOLFP is well posed in the sense that all denominators are positive everywhere in S. In other words, nowhere in S do any of the denominators vanish.

We note the generalized nature of an MOLFP because it reduces to an MOLP when all **d**'s in the denominators are null vectors. This means that if we could develop an algorithm for solving an MOLFP, we would then have a "super-algorithm" for all of single and multiple objective linear and linear fractional programming.

Linear fractional (ratio) criteria are frequently encountered in finance as illustrated by the following situations.

Corporate Planning

min {debt-to-equity ratio}

max {return on investment}

max {output per employee}

min {actual cost to standard cost}

Bank Balance Sheet Management

min {risk-assets to capital}

max {actual capital to required capital}

min {foreign loans to total loans}

min {residential mortgages to total mortgages}

Also, fractional objectives occur in other areas. Consider, for instance, marine transportation. Instead of maximizing profit from a given unit voyage, a more relevant measure is profit divided by the duration of the unit voyage. In water resources, we may wish to minimize water-temperature elevations in a river due to the cooling of power generation plants in the basin. The objective would then be to minimize the BTUs to be dissipated, divided by the volume of flow. In health care, we may have cost-to-bed, nurse-to-doctor, and doctor-to-patient ratios. In university planning, we may have student-teacher ratios, tenured-to-nontenured faculty ratios, and so forth.

In this chapter, we first review single objective linear fractional programming. Then, the notion of weak efficiency is explored, followed by a discussion of the difficulties involved in solving an MOLFP. Then, an algorithm is discussed for computing all weakly efficient vertices for a certain class of MOLFPs.

12.1 SINGLE OBJECTIVE LINEAR FRACTIONAL PROGRAMMING

Let us consider the *single objective* linear fractional program

$$\max \left\{ \frac{\mathbf{c}^T\mathbf{x} + \alpha}{\mathbf{d}^T\mathbf{x} + \beta} = z \right\}$$

$$\text{s.t. } \mathbf{x} \in S = \{\mathbf{x} \in R^n \mid \mathbf{A}\mathbf{x} = \mathbf{b}, \quad \mathbf{x} \geqq \mathbf{0}, \quad \mathbf{b} \in R^m\}$$

in which the objective function denominator is positive everywhere in S. What is interesting about linear fractional programs is the linearity of their objective function level curves. To demonstrate, consider an arbitrary $\bar{z}$-level curve of the objective function

$$\frac{\mathbf{c}^T\mathbf{x} + \alpha}{\mathbf{d}^T\mathbf{x} + \beta} = \bar{z}$$

Upon rearrangement, we have

$$\mathbf{c}^T\mathbf{x} + \alpha = \bar{z}(\mathbf{d}^T\mathbf{x} + \beta)$$

that yields

$$(\mathbf{c} - \bar{z}\mathbf{d})^T\mathbf{x} = \bar{z}\beta - \alpha$$

as the (linear) expression for the $\bar{z}$-level curve of the linear fractional objective function. Since $\bar{z}$ was arbitrary, it is seen that each level curve of a linear fractional objective is linear over S, provided that nowhere in S is the denominator zero. Thus, if a single objective linear fractional program has an optimal solution, at least one extreme point of S will be optimal.

Despite the linearity of the level curves of the objective functions, the level curves are not parallel (when $\mathbf{c} \neq \mathbf{0}$, $\mathbf{d} \neq \mathbf{0}$, and $\mathbf{c} \neq \omega\mathbf{d}$ for all $\omega \in R$) to one another as they are in linear programming. Instead, they "radiate" from an $n - 2$ dimensional *rotation set*. The rotation set is the set of all points of intersection between the 0-level curve of the numerator and the 0-level curve of the denominator. In R^2 the rotation set is called a *rotation point*, and in R^3 it is called a *rotation axis*. Mathematically, the rotation set is the set of points that simultaneously solve the two linear equations

$$\mathbf{c}^T\mathbf{x} = -\alpha$$

$$\mathbf{d}^T\mathbf{x} = -\beta$$

To introduce the geometry of linear fractional programming, we have the following examples.

Example 1. Consider the single objective linear fractional program

$$\max\left\{\frac{x_1 + x_2 - 1}{5x_1 + x_2 - 1} = z\right\}$$

$$\begin{aligned} \text{s.t.} \quad 3x_1 + 2x_2 &\geq 6 \\ x_1 \qquad &\leq 3 \\ x_2 &\leq 3 \\ x_1, x_2 &\geq 0 \end{aligned}$$

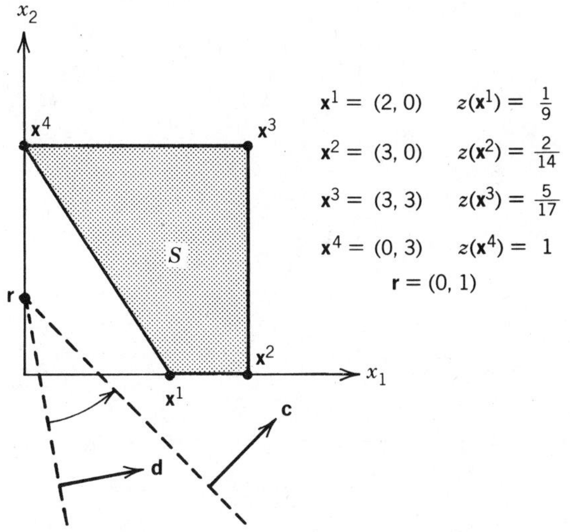

Figure 12.1. Graph of Example 1.

whose graph is in Fig. 12.1. As seen, the problem has four extreme points. Also indicated are the criterion values of the four extreme points. Note that the dashed lines denote the 0-level curves of the numerator and denominator and that the rotation point is $\mathbf{r} = (0, 1)$. The circular arrow denotes the gradient of the linear fractional objective function. Thus, by visualizing a counter-clockwise rotation, we see that $\mathbf{x}^4$ is optimal.

Example 2. Consider the linear fractional program

$$\max \left\{ \frac{-x_1 - 3}{x_2 + 1} = z \right\}$$

$$\begin{aligned} \text{s.t.} \quad 2x_1 + x_2 &\geqq 2 \\ -4x_1 + x_2 &\leqq 2 \\ x_1, x_2 &\geqq 0 \end{aligned}$$

whose graph is in Fig. 12.2. Visualizing a counter-clockwise rotation about $\mathbf{r}$, we observe continual z-value improvement as we move along the unbounded edge emanating from $\mathbf{x}^1$. However, the optimal (maximal) z-value of the objective function is bounded. Its least upper bound (lub), which is never achieved, is $-\frac{1}{4}$.

Now, let us assume that we are to minimize the objective function rather than to maximize it. Visualizing a clockwise rotation about $\mathbf{r}$, we observe continual z-value deterioration as we move along the unbounded edge emanating from $\mathbf{x}^2$. However, in this case, the optimal (minimal) z-value of the objective function is unbounded.

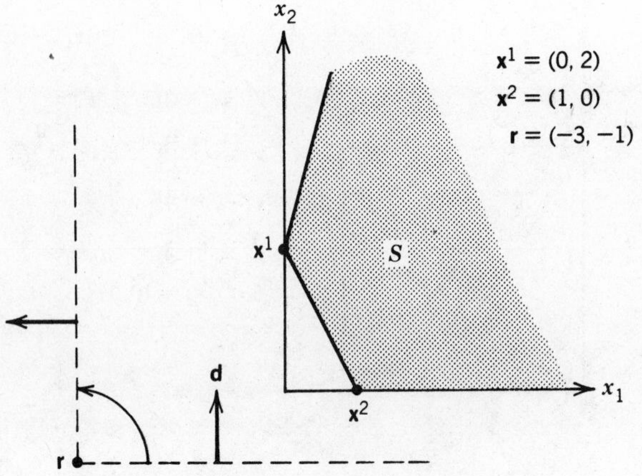

Figure 12.2. Graph of Example 2.

Example 3. Consider the three-dimensional linear fractional program

$$\max \left\{ \frac{x_1 - x_2}{x_2 - 4} = z \right\}$$

$$\begin{aligned} \text{s.t.} \quad x_1 \qquad & \leqq 8 \\ x_2 \quad & \leqq 2 \\ x_3 & \leqq 2 \\ x_1, x_2, x_3 & \geqq 0 \end{aligned}$$

So that the denominator is positive over S, we multiply both the numerator and denominator by -1. Now, our objective function is written

$$\max \left\{ \frac{-x_1 + x_2}{-x_2 + 4} = z \right\}$$

Portraying this problem with $\mathbf{c} = (-1, 1, 0)$ and $\mathbf{d} = (0, -1, 0)$, we have Fig. 12.3. In this problem, the rotation axis is the set $\{\mathbf{x} \in R^3 \mid x_1 = x_2 = 4\}$ and the optimal set is $\gamma(\mathbf{x}^1, \mathbf{x}^2)$ with $z^* = 1$.

Two methods are now discussed for solving single objective linear fractional programs: the *variable transformation* method (credited to Charnes and Cooper (1962)) and the *updated objective function* method (derived from Bitran and Novaes (1973)).

12.1.1 Variable Transformation Method

In this method for solving a single objective linear fractional program, we, as usual, assume that the denominator is positive everywhere in S and make the

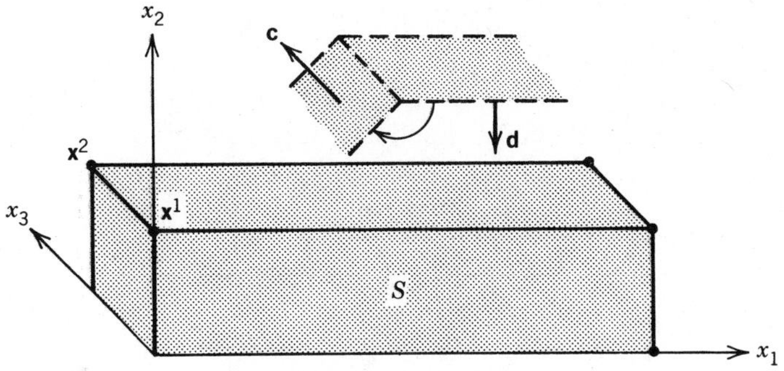

Figure 12.3. Graph of Example 3.

variable change

$$\rho = \frac{1}{\mathbf{d}^T\mathbf{x} + \beta}$$

With this change, the objective function becomes

$$\sum_{i=1}^{n} (c_i x_i \rho) + \alpha\rho$$

If we make the additional variable changes

$$y_i = x_i\rho \quad \text{for all } i$$

the linear fractional program becomes

$$\begin{aligned} &\max \{ \mathbf{c}^T\mathbf{y} + \alpha\rho \} \\ &\text{s.t.} \qquad \frac{\mathbf{Ay}}{\rho} = \mathbf{b} \\ &\qquad (\mathbf{d}^T\mathbf{x} + \beta)\rho = 1 \\ &\qquad \mathbf{0} \leqq \mathbf{y} \in R^n, \quad 0 \leqq \rho \in R \end{aligned}$$

Rearranging, we have the *variable change formulation*

$$\begin{aligned} &\max \{ \mathbf{c}^T\mathbf{y} + \alpha\rho \} \\ &\text{s.t.} \qquad \mathbf{Ay} - \mathbf{b}\rho = \mathbf{0} \\ &\qquad \mathbf{d}^T\mathbf{y} + \beta\rho = 1 \\ &\qquad \mathbf{0} \leqq \mathbf{y} \in R^n, \quad 0 \leqq \rho \in R \end{aligned}$$

with $m + 1$ constraints and $n + 1$ variables that can be solved by the simplex method. (Had the original constraints been in $\mathbf{Ax} \gtreqless \mathbf{b}$ form, the first batch of constraints in the above formulation would have been written $\mathbf{Ay} - \mathbf{b}\rho \gtreqless \mathbf{0}$.)

Example 4. Consider the linear fractional program

$$\max \left\{ \frac{x_2 - 5}{-x_1 - x_2 + 9} = z \right\}$$

$$\begin{aligned} \text{s.t.} \quad 2x_1 + 5x_2 &\geqq 10 \\ 4x_1 + 3x_2 &\leqq 20 \\ -x_1 + x_2 &\leqq 2 \\ x_1, x_2 &\geqq 0 \end{aligned}$$

whose graph is in Fig. 12.4. The variable change formulation is

$$\max \{ y_2 - 5\rho \}$$

$$\begin{aligned} \text{s.t.} \quad 2y_1 + 5y_2 - 10\rho &\geqq 0 \\ 4y_1 + 3y_2 - 20\rho &\leqq 0 \\ -y_1 + y_2 - 2\rho &\leqq 0 \\ -y_1 - y_2 + 9\rho &= 1 \\ y_1, y_2, \rho &\geqq 0 \end{aligned}$$

Solving using the simplex method, we obtain $y_1 = \frac{2}{3}$, $y_2 = \frac{4}{3}$, and $\rho = \frac{1}{3}$. This yields $\mathbf{x}^1 = (2, 4)$ as the optimal solution of the linear fractional program.

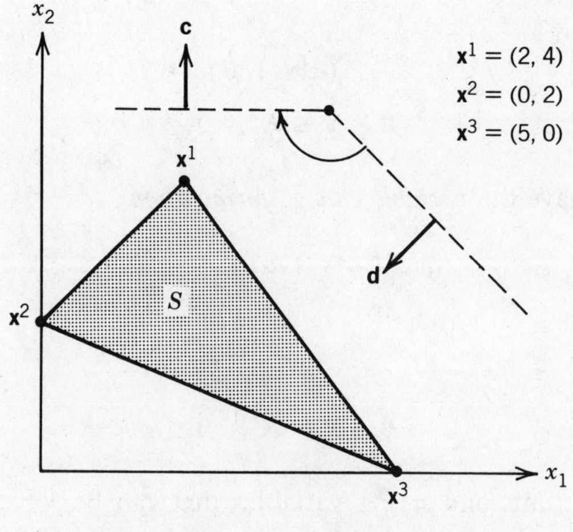

Figure 12.4. Graph of Example 4.

12.1.2 Updated Objective Function Method

In this method, we periodically recompute the local gradient of the fractional objective function at $\bar{\mathbf{x}}$

$$\frac{(\mathbf{d}^T\bar{\mathbf{x}} + \beta)\mathbf{c} - (\mathbf{c}^T\bar{\mathbf{x}} + \alpha)\mathbf{d}}{(\mathbf{d}^T\bar{\mathbf{x}} + \beta)^2}$$

In this method, we solve the linear fractional program by solving a sequence of LPs only recomputing the local gradient of the objective function after each optimization stage. The algorithmic specifications of the method are as follows.

Step 1: Set $i = 0$.

Step 2: Set $i = i + 1$.

Step 3: Determine the local gradient of the objective function at $\mathbf{x}^{(i)}$.

Step 4: Optimize the resulting LP to obtain extreme point $\mathbf{x}^{(i+1)}$.

Step 5: If $\mathbf{x}^{(i+1)} \neq \mathbf{x}^{(i)}$, go to Step 2. Otherwise, go to Step 6.

Step 6: $\mathbf{x}^{(i)}$ is an optimal solution of the linear fractional program.

Example 5. Consider the linear fractional program

$$\max\left\{\frac{-x_1 + x_2 + 2}{x_2 + 2} = z\right\}$$

$$\begin{aligned} \text{s.t.} \quad 4x_1 - 3x_2 &\geqq 2 \\ x_1 \quad\quad &\leqq 5 \\ x_1 \quad\quad &\geqq 2 \\ x_1, x_2 &\geqq 0 \end{aligned}$$

whose graph is in Fig. 12.5. At $\mathbf{x}^{(1)}$, the local gradient is

$$\frac{(2)\mathbf{c} - (2)\mathbf{d}}{(2)^2} = (-\tfrac{1}{2}, 0)$$

Solving the LP with $(-\frac{1}{2}, 0)$ as the objective row, we obtain $\mathbf{x}^{(2)}$. At $\mathbf{x}^{(2)}$, the local gradient is $(-\frac{1}{2}, \frac{1}{2})$. Solving the LP with $(-\frac{1}{2}, \frac{1}{2})$ as the objective row, we obtain $\mathbf{x}^{(3)}$. At $\mathbf{x}^{(3)}$, the local gradient is $(-\frac{8}{64}, \frac{5}{64})$. Solving the LP with $(-\frac{8}{64}, \frac{5}{64})$ as the objective row, we obtain $\mathbf{x}^{(4)}$. At $\mathbf{x}^{(4)}$, the local gradient is $(-\frac{1}{4}, \frac{1}{8})$. Solving the LP with $(-\frac{1}{4}, \frac{1}{8})$ as the objective row, we obtain $\mathbf{x}^{(5)} = \mathbf{x}^{(4)}$. Hence, $(2, 2)$ is the optimal solution with $z^* = \frac{1}{2}$.

Another mode of implementation would be to recompute the local gradient of the objective function at each new extreme point encountered. With the fractional program of Example 5, the approach would have led to the optimal solution in fewer pivots $(\mathbf{x}^{(1)} \to \mathbf{x}^{(2)} \to \mathbf{x}^{(4)})$ instead of $(\mathbf{x}^{(1)} \to \mathbf{x}^{(2)} \to \mathbf{x}^{(4)} \to \mathbf{x}^{(3)} \to \mathbf{x}^{(4)})$. With unbounded feasible regions, however, the updated objective function method is not fail-safe.

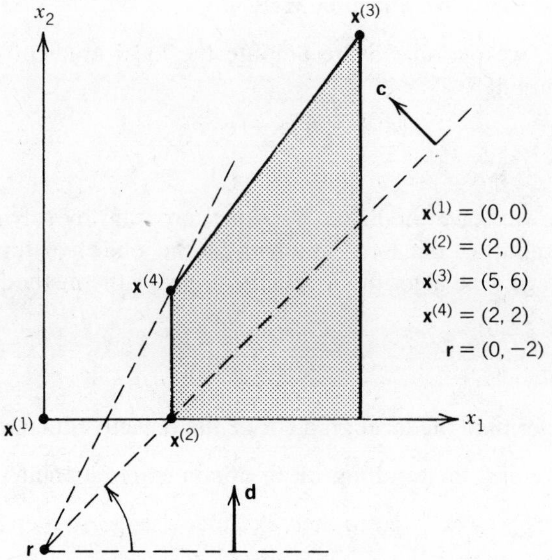

Figure 12.5. Graph of Example 5.

Example 6. Consider the linear fractional program

$$\max \left\{ \frac{-2x_1 + x_2}{x_1 + 5x_2} = z \right\}$$

$$\begin{aligned} \text{s.t.} \quad -x_1 + x_2 &\leqq 1 \\ x_1 - x_2 &\leqq 5 \\ x_1 + 2x_2 &\geqq 5 \\ x_1, x_2 &\geqq 0 \end{aligned}$$

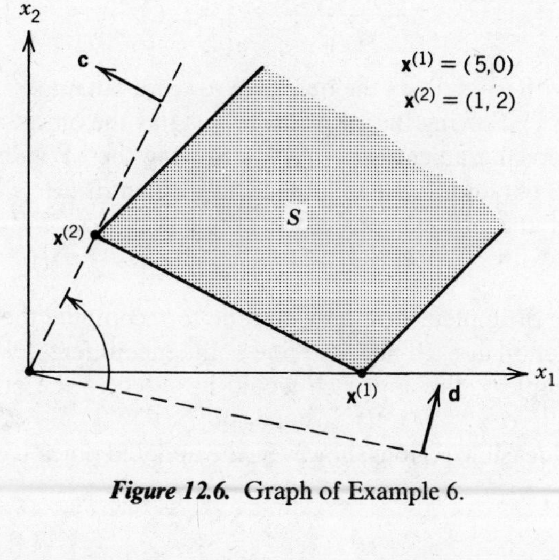

Figure 12.6. Graph of Example 6.

whose graph is presented in Fig. 12.6. With either mode of implementation, after Phase I we would be at $\mathbf{x}^{(1)}$, at which point the local gradient of the objective function is $(0, \frac{11}{5})$. Since the unbounded edge emanating from $\mathbf{x}^{(1)}$ would have the largest positive $c_j - z_j$ value, the updated objective function approach would not find the optimal extreme point $\mathbf{x}^{(2)} = (1, 2)$.

12.2 WHEN THE DENOMINATOR VANISHES

With regard to the functional

$$\frac{\mathbf{c}^T\mathbf{x} + \alpha}{\mathbf{d}^T\mathbf{x} + \beta} = z$$

z is undefined when $\mathbf{d}^T\mathbf{x} + \beta = 0$. Thus, only qualified statements can be made about whether or not the functional possesses finite optima. For problems in which the denominator vanishes, we have the following classifications.

1. *Denominator is both positive and negative in S* (Fig. 12.7*a*). Allowing $\mathbf{x}^+ \in S$ to be a point at which the denominator is positive and $\mathbf{x}^- \in S$ to be a point at which the denominator is negative, there exists an $\mathbf{x}^0 \in S$ on the line segment $\gamma(\mathbf{x}^+, \mathbf{x}^-)$ where the denominator is zero. Now, consider the two sequences of points in $\gamma(\mathbf{x}^+, \mathbf{x}^-)$

$$\mathbf{x}^0 + \Delta\mathbf{x}, \quad \mathbf{x}^0 + \frac{\Delta\mathbf{x}}{2}, \quad \mathbf{x}^0 + \frac{\Delta\mathbf{x}}{4}, \ldots$$

$$\mathbf{x}^0 - \Delta\mathbf{x}, \quad \mathbf{x}^0 - \frac{\Delta\mathbf{x}}{2}, \quad \mathbf{x}^0 - \frac{\Delta\mathbf{x}}{4}, \ldots$$

   With one of the sequences $z \to +\infty$, and with the other $z \to -\infty$. This gives us two different limits at $\mathbf{x}^0$. In such problems, z has neither a finite minimum nor maximum.
2. *Denominator = 0 everywhere* (Fig. 12.7*b*). In this situation, all points in S have undefined z-values.
3. **c** *is co-linear with* **d**. This situation consists of two cases:
   a. *Rotation set is empty* (Fig. 12.7*c*). In this case, there exists a nonzero scalar π such that $\mathbf{c} = \pi\mathbf{d}$ but $\alpha \neq \pi\beta$. In other words, the 0-level curves of the numerator and denominator are parallel, but do not lie on top of one another. In Fig. 12.7*c*, z is unbounded from above and z is undefined at $\mathbf{x}^0$.
   b. *Rotation set is nonempty* (Fig. 12.7*d*). In this case, there exists a nonzero scalar π such that $\mathbf{c}^T\mathbf{x} + \alpha = \pi(\mathbf{d}^T\mathbf{x} + \beta)$. In other words, the numerator and denominator have identical 0-level curves. In Fig. 12.7*d*, z is constant over S except for points in $\gamma(\mathbf{x}^1, \mathbf{x}^2)$, at which z "takes on" $0/0$.
4. **c** *not co-linear with* **d**. This situation consists of three cases:
   a. *S is a subset of the rotation set*. In this case, all points in S "take on" z-values of $0/0$.

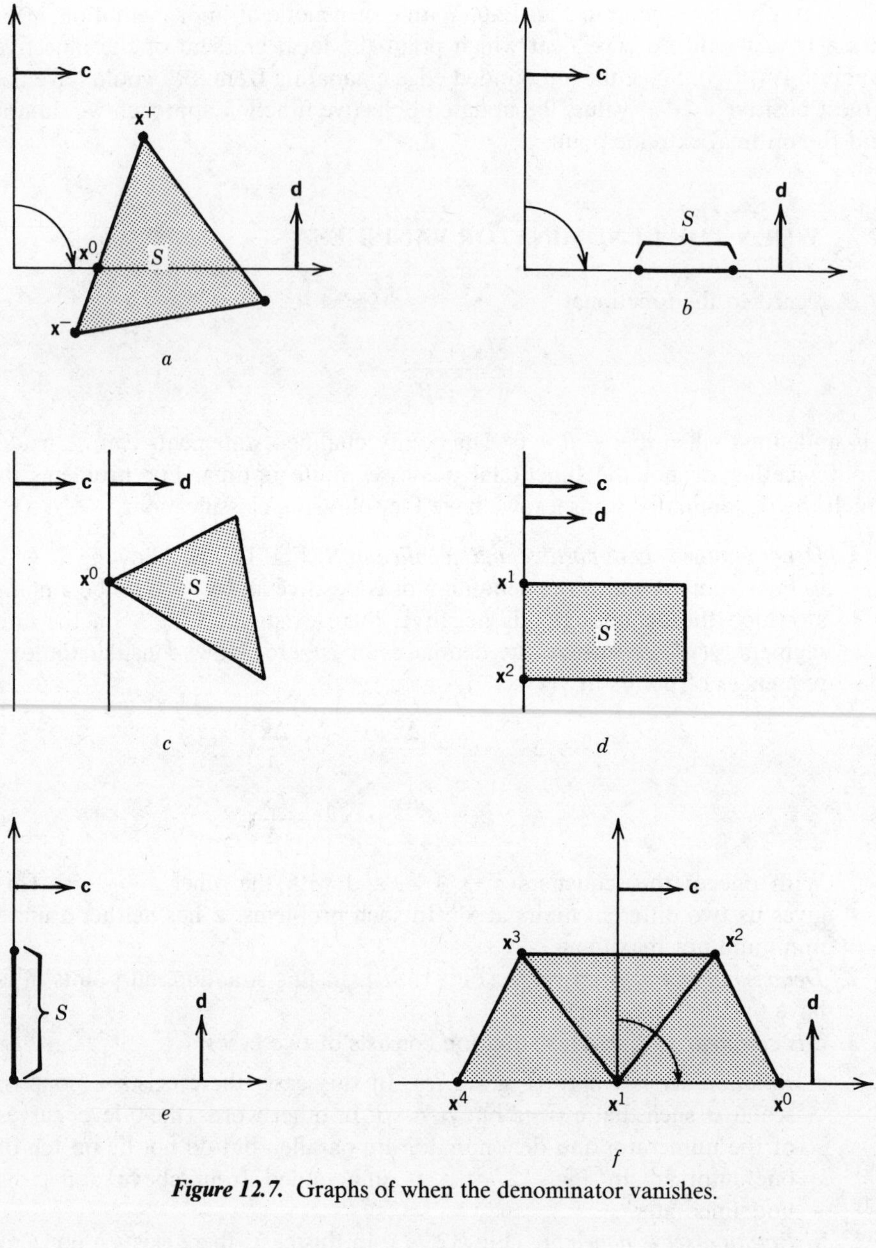

Figure 12.7. Graphs of when the denominator vanishes.

b. *Numerator = 0 everywhere* (Fig. 12.7*e*). In this case, $z = 0$ everywhere except for those points in S in the rotation set, at which z "takes on" $0/0$.

c. *There exist points in S at which numerator ≠ 0* (Fig. 12.7*f*). This case consists of three subcases:

i. *Finite minima and finite maxima.* In Fig. 12.7*f*, consider the feasible region defined by

$$S = \underset{i=1}{\overset{3}{\gamma}} (\mathbf{x}^i)$$

Apart from $\mathbf{x}^1$ at which z "takes on" $0/0$, z achieves a finite maximum at all points in $\gamma(\mathbf{x}^1, \mathbf{x}^2) - \{\mathbf{x}^1\}$ and a finite minimum at all points in $\gamma(\mathbf{x}^1, \mathbf{x}^3) - \{\mathbf{x}^1\}$.

ii. *Finite minima but unbounded maxima* (or finite maxima and unbounded minima). For example, in Fig. 12.7*f*, consider the feasible region defined by

$$S = \underset{i=0}{\overset{2}{\gamma}} (\mathbf{x}^i)$$

Apart from $\mathbf{x}^1$, z achieves a finite minimum at all points along $\gamma(\mathbf{x}^1, \mathbf{x}^2) - \{\mathbf{x}^1\}$, but is undefined at all points along $\gamma(\mathbf{x}^1, \mathbf{x}^0) - \{\mathbf{x}^1\}$.

iii. *Unbounded minima and unbounded maxima.* In Fig. 12.7*f*, consider the feasible region defined by

$$S = \underset{i=0}{\overset{4}{\gamma}} (\mathbf{x}^i)$$

Apart from $\mathbf{x}^1$, z is undefined at all points along $\gamma(\mathbf{x}^1, \mathbf{x}^0) - \{\mathbf{x}^1\}$ and at all points along $\gamma(\mathbf{x}^1, \mathbf{x}^4) - \{\mathbf{x}^1\}$.

12.3 WEAK AND STRONG EFFICIENCY

In multiple objective linear fractional programming, we will often deal with *weakly* efficient (w-efficient) points. Recall from Section 9.4 that $\bar{\mathbf{x}} \in S$ is *w-efficient* if and only if there does not exist another $\mathbf{x} \in S$ such that $z_i(\mathbf{x}) > z_i(\bar{\mathbf{x}})$ for all i. The notion of weak efficiency is to be distinguished from the usual definition of efficiency, called *strong* efficiency (s-efficiency) in this chapter, that states $\bar{\mathbf{x}} \in S$ is *s-efficient* if and only if there does not exist another $\mathbf{x} \in S$ such that $z_i(\mathbf{x}) \geqq z_i(\bar{\mathbf{x}})$ for all i and $z_i(\mathbf{x}) > z_i(\bar{\mathbf{x}})$ for at least one i. Note that the set of all s-efficient points is a subset of the set of all w-efficient points.

A difficulty exists with s-efficiency in MOLFP. It is that the set of all s-efficient points may not be entirely closed. This creates serious difficulties in trying to compute directly the set of all s-efficient points. Because the set of all w-efficient points is closed, it is easier to compute this set.

Normally, the difference between the s-efficient set and the w-efficient set is minor—usually, the w-efficient set is the closure of the s-efficient set. However, in some problems, the difference between the s-efficient set and the w-efficient set can be substantial (see Example 9 in Section 12.6).

12.4 AN MOLFP EXAMPLE AND TERMINOLOGY

To introduce the nature of multiple objective linear fractional programming, consider the MOLFP

$$\max\{-2x_1 + x_2 = z_1\}$$
$$\max\{-3x_1 - x_2 = z_2\}$$
$$\max\left\{\frac{x_1 + x_2 - 2}{-x_1 + 2x_2 + 5} = z_3\right\}$$
$$\text{s.t.} \quad x_1 \leqq 4$$
$$x_2 \leqq 4$$
$$x_1, x_2 \geqq 0$$

that is graphed in Fig. 12.8, in which $S = \gamma(\mathbf{x}^1, \mathbf{x}^2, \mathbf{x}^4, \mathbf{x}^8)$. Using E^w to denote

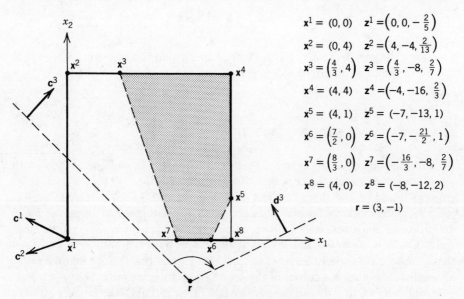

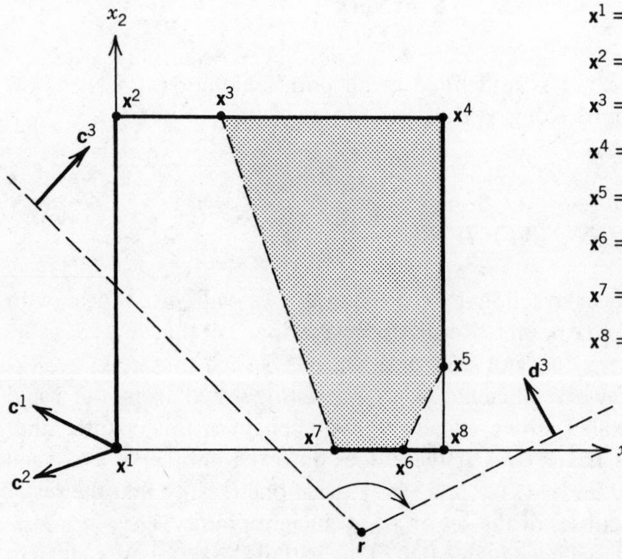

$\mathbf{x}^1 = (0, 0) \quad \mathbf{z}^1 = \left(0, 0, -\frac{2}{5}\right)$

$\mathbf{x}^2 = (0, 4) \quad \mathbf{z}^2 = \left(4, -4, \frac{2}{13}\right)$

$\mathbf{x}^3 = \left(\frac{4}{3}, 4\right) \quad \mathbf{z}^3 = \left(\frac{4}{3}, -8, \frac{2}{7}\right)$

$\mathbf{x}^4 = (4, 4) \quad \mathbf{z}^4 = \left(-4, -16, \frac{2}{3}\right)$

$\mathbf{x}^5 = (4, 1) \quad \mathbf{z}^5 = (-7, -13, 1)$

$\mathbf{x}^6 = \left(\frac{7}{2}, 0\right) \quad \mathbf{z}^6 = \left(-7, -\frac{21}{2}, 1\right)$

$\mathbf{x}^7 = \left(\frac{8}{3}, 0\right) \quad \mathbf{z}^7 = \left(-\frac{16}{3}, -8, \frac{2}{7}\right)$

$\mathbf{x}^8 = (4, 0) \quad \mathbf{z}^8 = (-8, -12, 2)$

$\mathbf{r} = (3, -1)$

Figure 12.8. MOLFP example of Section 12.4.

the w-efficient set and E^s the s-efficient set, we have

$$E^w = \gamma(\mathbf{x}^1, \mathbf{x}^2) \cup \gamma(\mathbf{x}^2, \mathbf{x}^3) \cup \bigcup_{i=3}^{7} \gamma(\mathbf{x}^i) \cup \gamma(\mathbf{x}^6, \mathbf{x}^8)$$

$$E^s = E^w - [\gamma(\mathbf{x}^3, \mathbf{x}^7) - \{\mathbf{x}^3\}] - [\gamma(\mathbf{x}^5, \mathbf{x}^6) - \{\mathbf{x}^6\}]$$

We note several things about E^s in an MOLFP that are not seen in an MOLP.

1. E^s need not be closed. For instance, although w-efficient, the E^s boundary line segments $[\gamma(\mathbf{x}^3, \mathbf{x}^7) - \{\mathbf{x}^3\}]$ and $[\gamma(\mathbf{x}^5, \mathbf{x}^6) - \{\mathbf{x}^6\}]$ are not s-efficient.
2. Some interior points of S may be s-efficient, while others may not. (In an MOLP, if an interior point is s-efficient, all of S is s-efficient.)
3. Two s-efficient extreme points of S need not be connected by a path of s-efficient edges of S. For instance, $\mathbf{x}^1$ and $\mathbf{x}^8$ are not connected by a path of s-efficient edges of S.
4. A given edge may start out as s-efficient but become inefficient. For instance, the $\gamma(\mathbf{x}^4, \mathbf{x}^8)$ edge is s-efficient from $\mathbf{x}^4$ to $\mathbf{x}^5$ but is inefficient from $\mathbf{x}^5$ to $\mathbf{x}^8$.

Edges such as $\gamma(\mathbf{x}^4, \mathbf{x}^8)$ that have both s-efficient and inefficient points in their relative interiors are called *broken edges*. A point at which such an edge ceases to be efficient is called a *break point*. Along $\gamma(\mathbf{x}^4, \mathbf{x}^8)$, $\mathbf{x}^5$ is a break point. Emanating from $\mathbf{x}^4$, edge $\gamma(\mathbf{x}^4, \mathbf{x}^8)$ is *initially* s-efficient.

In this chapter, we will refer to points such as $\mathbf{x}^1, \mathbf{x}^2, \ldots, \mathbf{x}^8$ in Fig. 12.8 as *vertices* and only use the term *extreme point* when referring to points such as $\mathbf{x}^1$, $\mathbf{x}^2$, $\mathbf{x}^4$, $\mathbf{x}^8$ in Fig. 12.8 that are extreme in S. Applying the notation E_v^w, E_x^w, E_v^s, and E_x^s to denote the sets of w-efficient and s-efficient vertices and extreme points, for the MOLFP of Fig. 12.8 we have

$$E_v^w = \{\mathbf{x}^i \mid 1 \leqq i \leqq 8\} \qquad E_v^s = E_v^w - \{\mathbf{x}^5, \mathbf{x}^7\}$$

$$E_x^w = \{\mathbf{x}^1, \mathbf{x}^2, \mathbf{x}^4, \mathbf{x}^8\} \qquad E_x^s = E_x^w$$

Note the boundary line segments $\gamma(\mathbf{x}^3, \mathbf{x}^7)$ and $\gamma(\mathbf{x}^5, \mathbf{x}^6)$ of E^w. They fall along *coincidence planes*. Line segment $\gamma(\mathbf{x}^3, \mathbf{x}^7)$ lies along the plane defined by the $z_2 = -8$ and $z_3 = \frac{2}{7}$ objective function level curves. Line segment $\gamma(\mathbf{x}^5, \mathbf{x}^6)$ lies along the plane defined by the $z_1 = -7$ and $z_3 = 1$ objective function level curves.

12.5 GRAPHICAL DETECTION OF EFFICIENCY

In multiple objective programs that are *not* MOLPs, it is best to use the domination set approach (see Section 6.8) for the graphical detection of efficiency. For s-efficiency, the test is this. Let $D_{\bar{x}}^s = \{\bar{\mathbf{x}}\} \oplus C^{\geq}$ where $C^{\geq}$ is the *semi-positive* polar cone determined by the local objective function gradients at $\bar{\mathbf{x}}$. Then, $\bar{\mathbf{x}} \in S$ is s-efficient if and only if $D_{\bar{x}}^s \cap S = \{\bar{\mathbf{x}}\}$. Since $C^{\geq}$ is likely to vary with $\bar{\mathbf{x}}$, $D_{\bar{x}}^s$ is likely to vary with $\bar{\mathbf{x}}$.

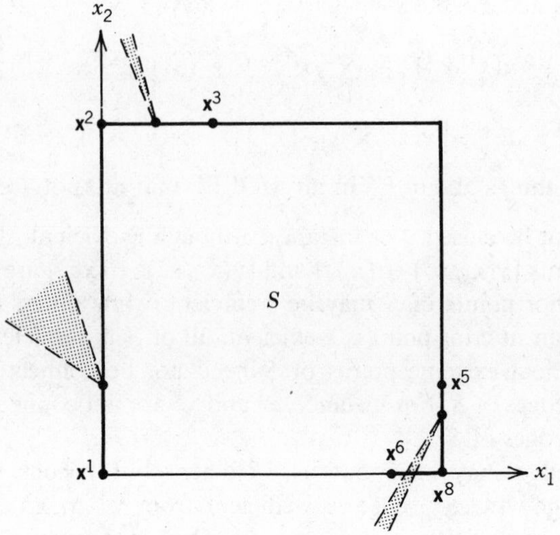

Figure 12.9. D^w domination sets for detecting *w*-efficiency.

For *w*-efficiency, the test is this. Let $D_{\bar{x}}^w = \{\bar{\mathbf{x}}\} \oplus C^>$ where $C^>$ is the *strictly positive* polar cone determined by the local objective function gradients at $\bar{\mathbf{x}}$. Then, $\bar{\mathbf{x}} \in S$ is *w*-efficient if and only if $D_{\bar{x}}^w \cap S = \{\bar{\mathbf{x}}\}$. Since $C^>$ is likely to vary with $\bar{\mathbf{x}}$, $D_{\bar{x}}^w$ is likely to vary with $\bar{\mathbf{x}}$.

Let us now apply D^w and D^s domination sets to the MOLFP of Fig. 12.8 for the graphical detection of *w*- and *s*-efficiency. Figure 12.9 shows why the

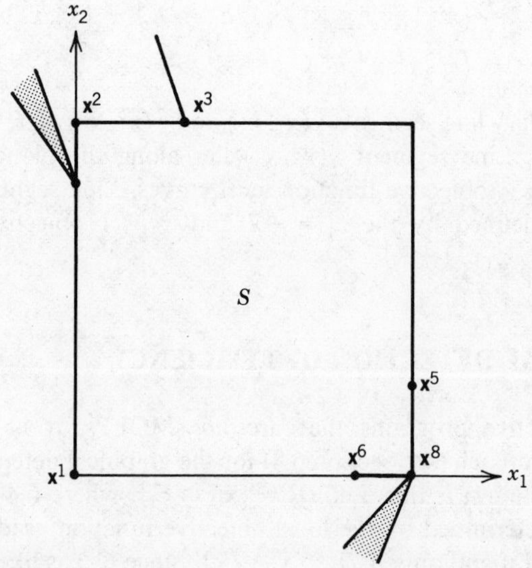

Figure 12.10. D^s domination sets for detecting *s*-efficiency.

$\gamma(\mathbf{x}^1, \mathbf{x}^2)$ and $\gamma(\mathbf{x}^2, \mathbf{x}^3)$ line segments are w-efficient and why the $\gamma(\mathbf{x}^5, \mathbf{x}^8)$ line segment is w-inefficient. Figure 12.10 shows, for instance, why (a) edge $\gamma(\mathbf{x}^1, \mathbf{x}^2)$, (b) point $\mathbf{x}^3$, and (c) the line segment $\gamma(\mathbf{x}^6, \mathbf{x}^8)$ are s-efficient. For all $\bar{\mathbf{x}}$ such that

$$\bar{\mathbf{x}} \in \underset{i=3}{\overset{7}{\gamma}} (\mathbf{x}^i)$$

we note that $D_{\bar{x}}^w = \{\bar{\mathbf{x}}\}$. However, $D_{\bar{x}}^s = \{\bar{\mathbf{x}}\}$ only for

$$\bar{\mathbf{x}} \in \underset{i=3}{\overset{7}{\gamma}} (\mathbf{x}^i) - [\gamma(\mathbf{x}^3, \mathbf{x}^7) - \{\mathbf{x}^3\}] - [\gamma(\mathbf{x}^5, \mathbf{x}^6) - \{\mathbf{x}^6\}]$$

12.6 ADDITIONAL MOLFP EXAMPLES

To gain additional familiarity with the geometry of an MOLFP and the various types of efficiency situations that can occur, consider the following three examples.

Example 7. Consider the MOLFP

$$\begin{aligned} \max \{ -x_1 &= z_1 \} \\ \max \left\{ \frac{x_1 - 2}{x_2 + 2} \right. &= \left. z_2 \right\} \\ \text{s.t.} \quad x_1 &\leqq 4 \\ x_1, x_2 &\geqq 0 \end{aligned}$$

that is graphed in Fig. 12.11, in which

$$E^w = \mu(\mathbf{x}^1, \mathbf{v}) \cup \mu(\mathbf{x}^2, \mathbf{v}) \cup \gamma(\mathbf{x}^2, \mathbf{x}^3) \quad \text{where } \mathbf{v} = (0, 1)$$

$$E_v^w = \{\mathbf{x}^1, \mathbf{x}^2, \mathbf{x}^3\}$$

$$E_x^w = \{\mathbf{x}^1, \mathbf{x}^3\}$$

$$E^s = E^w - \mu(\mathbf{x}^1, \mathbf{v})$$

$$E_v^s = \{\mathbf{x}^2, \mathbf{x}^3\}$$

$$E_x^s = \{\mathbf{x}^3\}$$

In this example, $\mu(\mathbf{x}^2, \mathbf{v})$ lies along a coincidence plane, $\gamma(\mathbf{x}^3, \mathbf{x}^1)$ is a broken edge, and $\mathbf{x}^2$ is a break point. This example is interesting because E^w is disconnected. The disconnectedness of E^w, however, is only possible because of the unboundedness of S. Although E^w is disconnected, E^s is connected. E^s consists of the unbounded vertical line segment emanating from $\mathbf{x}^2$ and the line segment $\gamma(\mathbf{x}^2, \mathbf{x}^3)$.

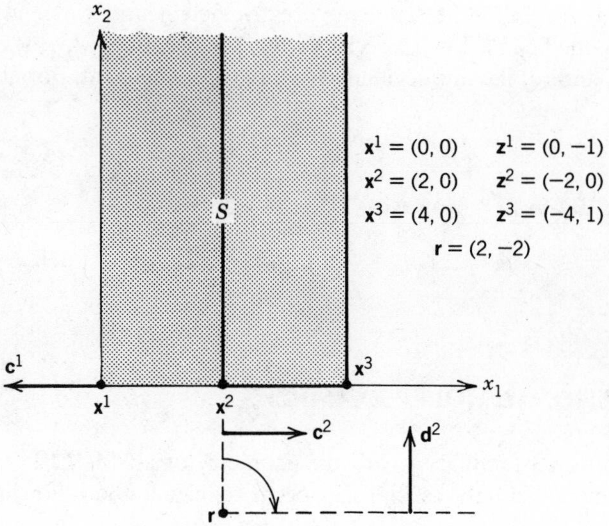

Figure 12.11. Graph of Example 7.

Example 8. Consider the MOLFP

$$\begin{aligned}
&\max \left\{ \frac{x_1 - 4}{-x_2 + 4} = z_1 \right\} \\
&\max \left\{ \frac{-x_1 + 4}{x_2 + 2} = z_2 \right\} \\
&\text{s.t.} \quad x_2 - 2x_3 \leqq 1 \\
&\qquad x_1 \leqq 8 \\
&\qquad x_2 \leqq 2 \\
&\qquad x_3 \leqq 2 \\
&\qquad x_1, x_2, x_3 \geqq 0
\end{aligned}$$

that is graphed in Fig. 12.12. In this MOLFP, the rotation sets of the two fractional objectives are axes. The rotation axis for the first objective is parallel to the x_3-axis and goes through the point $\mathbf{x}^{12} = (4, 4, 0)$. The rotation axis for the second objective is also parallel to the x_3-axis but goes through the point $\mathbf{x}^{13} = (4, -2, 0)$. In this example,

$$E^w = \gamma_{i=1}^{4} (\mathbf{x}^i) \cup \gamma_{i=3}^{7} (\mathbf{x}^i) \cup \gamma_{i=6}^{9} (\mathbf{x}^i) \qquad E^s = E^w$$

$$E_v^w = \{\mathbf{x}^i | 1 \leqq i \leqq 9\} \qquad E_v^s = E_v^w$$

$$E_x^w = \{\mathbf{x}^1, \mathbf{x}^2, \mathbf{x}^8, \mathbf{x}^9\} \qquad E_x^s = E_x^w$$

Only one coincidence plane exists in this MOLFP. It is formed by the $z_1 = 0$ or $z_2 = 0$ objective function level curves. This MOLFP is interesting because of

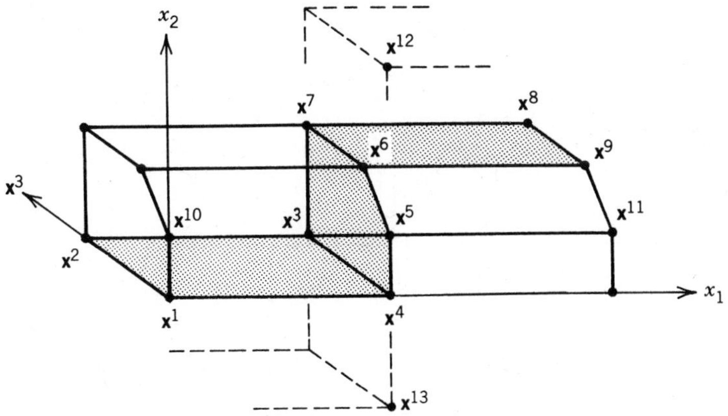

$\mathbf{x}^1 = (0, 0, 0)$	$\mathbf{x}^4 = (4, 0, 0)$	$\mathbf{x}^7 = (4, 2, 2)$	$\mathbf{x}^{10} = (0, 1, 0)$
$\mathbf{x}^2 = (0, 0, 2)$	$\mathbf{x}^5 = (4, 1, 0)$	$\mathbf{x}^8 = (8, 2, 2)$	$\mathbf{x}^{11} = (8, 1, 0)$
$\mathbf{x}^3 = (4, 0, 2)$	$\mathbf{x}^6 = (4, 2, \frac{1}{2})$	$\mathbf{x}^9 = (8, 2, \frac{1}{2})$	$\mathbf{x}^{12} = (4, 4, 0)$
			$\mathbf{x}^{13} = (4, -2, 0)$

Figure 12.12. Graph of Example 8.

the $\gamma(\mathbf{x}^{10}, \mathbf{x}^{11})$ edge of S. Although the edge is s-inefficient as it emanates from each of its extreme points $\mathbf{x}^{10}$ and $\mathbf{x}^{11}$, point $\mathbf{x}^5$, which is in the middle of the edge, is s-efficient.

Example 9. Consider the MOLFP that is graphed in Fig. 12.13.

$$\max \quad \{x_1 = z_1\}$$

$$\max \quad \left\{\frac{1}{x_1 + 1} = z_2\right\}$$

$$\max \quad \left\{\frac{-x_2}{x_1 - 8} = z_3\right\}$$

$$\begin{aligned} \text{s.t.} \quad -x_1 + 3x_2 &\leqq 6 \\ x_1 + 3x_2 &\leqq 12 \\ x_1 \qquad &\leqq 6 \\ x_1, x_2 &\geqq 0 \end{aligned}$$

In this example, the gradient of the second objective has been drawn as $\mathbf{c}^2 = (-1, 0)$ because

$$\max\left\{\frac{1}{x_1 + 1}\right\} \equiv \min\{x_1 + 1\} \equiv \max\{-x_1 - 1\}$$

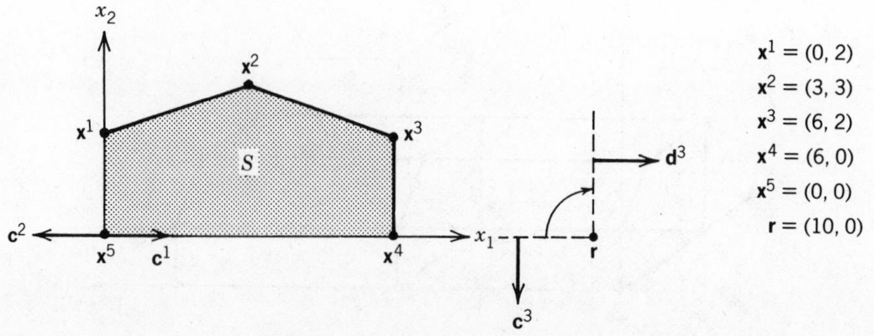

Figure 12.13. Graph of Example 9.

In this problem,

$$E^w = S \qquad E^s = \gamma(\mathbf{x}^1, \mathbf{x}^2) \cup \gamma(\mathbf{x}^2, \mathbf{x}^3)$$

$$E_v^w = \{\mathbf{x}^1, \mathbf{x}^2, \mathbf{x}^3, \mathbf{x}^4, \mathbf{x}^5\} \qquad E_v^s = \{\mathbf{x}^1, \mathbf{x}^2, \mathbf{x}^3\}$$

$$E_x^w = E_v^w \qquad E_x^s = E_v^s$$

In most MOLFPs, E^w seldom amounts to more than the closure of E^s. However, this example shows that the difference between E^w and E^s can be dramatic.

12.7 AN MOLFP EXAMPLE WITH A NONLINEAR E^w BOUNDARY

From Choo and Atkins (1983), we have the following two theorems.

Theorem 12.1. The set of all weakly efficient points E^w is closed.

Theorem 12.2. Let S be bounded and let $\bar{\mathbf{x}}, \hat{\mathbf{x}} \in E^w$. Then, $\bar{\mathbf{x}}$ and $\hat{\mathbf{x}}$ are connected by a finite number of w-efficient line segments.

From the above two theorems and our graphical experience, one might be tempted to conclude that the w-efficient set can be defined in terms of the union of a finite number of w-efficient polyhedra. This is *not* true as shown in the following example developed by Choo.

Example 10. Consider Choo's MOLFP

$$\max \quad \left\{ \frac{x_1}{x_2} \qquad = z_1 \right\}$$

$$\max \quad \{ x_3 \qquad = z_2 \}$$

$$\max \quad \left\{ \frac{-x_1 - x_3}{x_2 + 1} = z_3 \right\}$$

$$\text{s.t.} \quad 1 \leqq x_1, x_2, x_3 \leqq 4$$

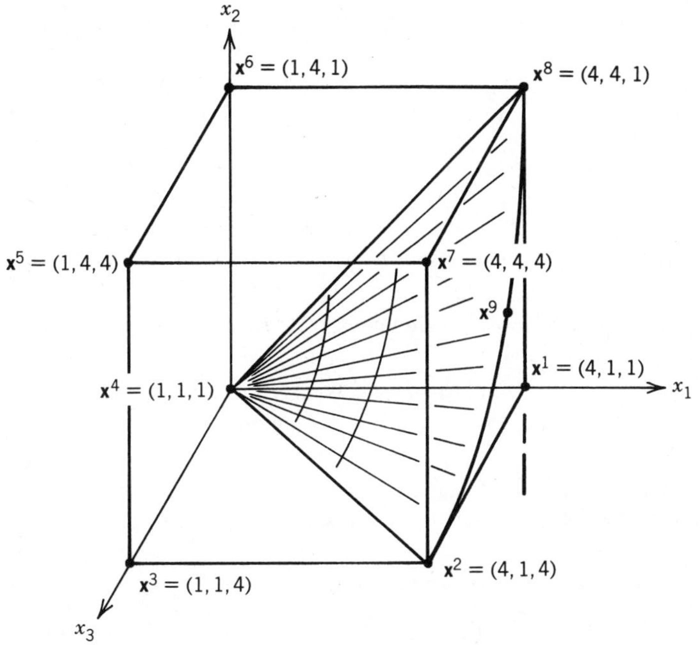

Figure 12.14. Choo's MOLFP with a nonlinear E^w boundary.

that is graphed in Fig. 12.14. In this problem,

$$E^w = \{\mathbf{x} \in S \mid \mathbf{x} = (a, b, c), a = bc\} \cup \{\mathbf{x} \in S \mid \mathbf{x} = (4, b, c), bc \geqq 4\}$$

$$\cup \gamma(\mathbf{x}^5, \mathbf{x}^6, \mathbf{x}^7, \mathbf{x}^8) \cup \gamma(\mathbf{x}^2, \mathbf{x}^3, \mathbf{x}^5, \mathbf{x}^7) \cup \gamma(\mathbf{x}^1, \mathbf{x}^2, \mathbf{x}^4)$$

In the above expression, the first term represents the arced surface given by $\mathbf{x}^2$, $\mathbf{x}^4$, $\mathbf{x}^8$ and the second expression represents the arced portion of the boundary of S given by $\mathbf{x}^2$, $\mathbf{x}^7$, $\mathbf{x}^8$. Note that E^w is not the union of a finite number of w-efficient polyhedra. Consider $\mathbf{x}^1$ and $\mathbf{x}^9 \in \{\mathbf{x} \in S \mid \mathbf{x} = (a, b, c), a = bc\}$. Are $\mathbf{x}^1$ and $\mathbf{x}^9$ connected by means of a finite number of w-efficient line segments? Yes, by $\gamma(\mathbf{x}^1, \mathbf{x}^4) \cup \gamma(\mathbf{x}^4, \mathbf{x}^9)$.

12.8 MOLFP ALGORITHMS

Only modest progress has been made on algorithms for the MOLFP problem. The only algorithm that has been reported is by Kornbluth and Steuer (1981). It computes all w-efficient vertices provided that S is bounded and E^w is the union of a finite number of polyhedral sets. The algorithm searches efficient edges for break points and then inserts into the problem cutting planes that intersect the break points. Thus, the break points become vertices in the modified problem. However, more work needs to be done to determine if the algorithm can be extended to deal with MOLFPs, such as in Example 10, that have nonlinear portions of the E^w boundary.

Readings

1. Ashton, D. J. and D. R. Atkins (1979). "Multicriteria Programming for Financial Planning," *Journal of the Operational Research Society*, Vol. 30, No. 3, pp. 259–270.
2. Awerbuch, S., J. G. Ecker, and W. A. Wallace (1976). "A Note: Hidden Nonlinearities in the Application of Goal Programming," *Management Science*, Vol. 22, No. 8, pp. 918–920.
3. Benson, H. P. (1985). "Finding Certain Weakly-Efficient Vertices in Multiple Objective Linear Fractional Programming," *Management Science*, Vol. 31, No. 2, pp. 240–245.
4. Bitran, G. R. and T. L. Magnanti (1976). "Duality and Sensitivity Analysis for Fractional Programs," *Operations Research*, Vol. 24, No. 4, pp. 675–699.
5. Bitran, G. R. and A. G. Novaes (1973). "Linear Programming with a Fractional Objective Function," *Operations Research*, Vol. 21, No. 4, pp. 22–29.
6. Charnes, A. and W. W. Cooper (1962). "Programming with Linear Fractional Functionals," *Naval Research Logistics Quarterly*, Vol. 9, No.'s 3–4, pp. 181–186.
7. Choo, E.-U. (1984). "Proper Efficiency and the Linear Fractional Vector Maximum Problem," *Operations Research*, Vol. 32, No. 1, pp. 216–220.
8. Choo, E.-U. and D. R. Atkins (1982). "Bicriteria Linear Fractional Programming," *Journal of Optimization Theory and Applications*, Vol. 36, No. 2, pp. 203–220.
9. Choo, E.-U. and D. R. Atkins (1983). "Connectedness in Multiple Linear Fractional Programming," *Management Science*, Vol. 29, No. 2, pp. 250–255.
10. Dinkelbach, W. (1967). "On Nonlinear Fractional Programming," *Management Science*, Vol. 13, No. 7, pp. 492–498.
11. Hannan, E. L. (1977). "Effects of Substituting a Linear Goal for a Fractional Goal in the Goal Programming Problem," *Management Science*, Vol. 24, No. 1, pp. 105–107.
12. Kornbluth, J. S. H. and R. E. Steuer (1980). "On Computing the Set of All Weakly Efficient Vertices in Multiple Objective Linear Fractional Programming," *Lecture Notes in Economics and Mathematical Systems*, No. 177, Berlin: Springer-Verlag, pp. 189–202.
13. Kornbluth, J. S. H. and R. E. Steuer (1981). "Goal Programming With Linear Fractional Criteria," *European Journal of Operational Research*, Vol. 8, No. 1, pp. 58–65.
14. Kornbluth, J. S. H. and R. E. Steuer (1981). "Multiple Objective Linear Fractional Programming," *Management Science*, Vol. 27, No. 9, pp. 1024–1039.
15. Kornbluth, J. S. H. and J. D. Vinso (1982). "Capital Structure and the Financing of the Multinational Corporation: A Fractional Multiobjective Approach," *Journal of Financial and Quantitative Analysis*, Vol. 17, No. 2, pp. 147–178.
16. Martos, B. (1964). "Hyperbolic Programming," *Naval Research Logistics Quarterly*, Vol. 11, No.'s 2–3, pp. 135–155.

17. Nykowski, I. and Z. Zolkiewski (1985). "A Compromise Procedure for the Multiple Objective Linear Fractional Programming Problem," *European Journal of Operational Research*, Vol. 19, No. 1, pp. 91–97.

18. Sakawa, M. and T. Yumine (1983). "Interactive Fuzzy Decisionmaking of Multiobjective Linear Fractional Programming Problems," *Large Scale Systems*, Vol. 5, No. 2, pp. 105–113.

19. Schaible, S. (1981). "Fractional Programming: Applications and Algorithms," *European Journal of Operational Research*, Vol. 7, No. 2, pp. 111–120.

20. Schaible, S. (1983). "Fractional Programming," *Zeitschrift für Operations Research*, Vol. 27, No. 1, pp. 39–54.

21. Soyster, A. L. and B. Lev (1978). "An Interpretation of Fractional Objectives in Goal Programming as Related to Papers by Awerbuch et al. and Hannan," *Management Science*, Vol. 24, No. 14, pp. 1546–1549.

22. Stancu-Minasian, I. M. (1977). "Bibliography of Fractional Programming 1960–1976," Preprint No. 3, Academy of Economic Studies, Department of Economic Cybernetics, Bucarest, Romania.

23. Tigan, S. T. (1975). "Sur le Probleme de la Programmation Vectorielle Fractionnaire," *Mathematica-Revue D'Analyse Numerique et de Theorie de L'Approximation*, Vol. 4, No. 1, pp. 99–103.

24. Zionts, S. (1968). "Programming With Linear Fractional Functionals," *Naval Research Logistics Quarterly*, Vol. 15, No. 3, pp. 449–451.

PROBLEM EXERCISES

12-1 Write the local gradient of:

(a) $\dfrac{x_1 + 3x_2 - 4}{-2x_1 + 2x_2 + 7} = z$ evaluated at $\bar{\mathbf{x}} = (2, 3)$

(b) $\dfrac{x_1 - x_3 + 6}{x_2 + 1} = z$ evaluated at $\bar{\mathbf{x}} = (2, 1, 3)$

12-2 Graphically solve the following linear fractional programs. In doing so:

(a) Indicate the feasible region by shading.

(b) Specify the rotation point.

(c) Draw the 0-level curves of the numerator and denominator.

(d) Indicate the circular direction of the gradient over the feasible region.

(e) Specify the optimal set (if it exists).

(f) Specify the optimal z-value, its lub (least upper bound), or whether the optimal z-value is unbounded, whichever is appropriate.

(i) $$\max \left\{ \frac{2x_1 + x_2}{-x_1 - 2x_2} = z \right\}$$

s.t.
$$x_1 + x_2 \leqq 6$$
$$2x_1 + x_2 \geqq 4$$
$$x_1, x_2 \geqq 0$$

(ii) $$\max \left\{ \frac{x_1 + x_2}{3x_1 - x_2 + 16} = z \right\}$$

s.t.
$$-x_1 + 2x_2 \leqq 4$$
$$x_1, x_2 \geqq 0$$

(iii) $\max\left\{\dfrac{-x_1 + x_2 - 1}{x_1 + 2} = z\right\}$

$$\begin{aligned} \text{s.t.} \quad & -x_1 + 2x_2 \leqq 0 \\ & x_2 \geqq 1 \\ & x_1, x_2 \geqq 0 \end{aligned}$$

(iv) $\max\left\{\dfrac{x_2 + 1}{x_1 + 1} = z\right\}$

$$\begin{aligned} \text{s.t.} \quad & x_1 + x_2 \geqq 2 \\ & x_1, x_2 \geqq 0 \end{aligned}$$

12-3C Using the variable transformation method, solve:

$$\max\left\{\frac{3x_1 - x_2 + 1}{x_1 + 3x_2 - 3} = z\right\}$$

$$\begin{aligned} \text{s.t.} \quad & -x_1 + 3x_2 \geqq 6 \\ & x_2 \leqq 3 \\ & x_1, x_2 \geqq 0 \end{aligned}$$

(a) Specify the variable change formulation in tabular form.

(b) Specify the optimal point.

(c) Specify the optimal z-value.

12-4C **(a)** Graph the following problem:

$$\max\left\{\frac{x_1 + 3x_2 + 2}{x_2 + 1} = z\right\}$$

$$\begin{aligned} \text{s.t.} \quad & 2x_1 + x_2 \geqq 2 \\ & x_1, x_2 \geqq 0 \end{aligned}$$

(b) Solve the above problem using the variable transformation method.

12-5 **(a)** Graph the MOLP:

$$\begin{aligned} \max\{\ & x_1 & & = z_1\} \\ \max\{ & -x_1 & & = z_2\} \\ \max\{\ & x_1 + x_2 & & = z_3\} \\ \text{s.t.} \quad & x_1 & & \leqq 6 \\ & x_2 & & \leqq 2 \\ & & x_3 & \leqq 3 \\ & x_1, x_2, x_3 & & \geqq 0 \end{aligned}$$

(b) Specify E = the set of all efficient points.

(c) Specify E^w = the set of all w-efficient points.

12-6 Graph the following MOLFPs and specify E^w, E_x^w, E_v^w, E^s, E_x^s, and E_v^s.

(a)
$$\begin{aligned} \max \ & \left\{\frac{-x_1 + x_2 + 2}{3x_1 + 2x_2 + 4} = z_1\right\} \\ \max \ & \left\{\frac{-3x_1 - x_2 + 14}{x_2 + 1} = z_2\right\} \\ \text{s.t.} \ & x_1 \leqq 3 \\ & x_2 \leqq 4 \\ & x_1, x_2 \geqq 0 \end{aligned}$$

(b)
$$\begin{aligned} \max \ & \left\{\frac{x_1 - 1}{-x_2 + 3} = z_1\right\} \\ \max \ & \left\{\frac{-x_1 + 1}{x_2 + 1} = z_2\right\} \\ \text{s.t.} \ & x_1 + x_2 \leqq 6 \\ & x_1 + x_2 \geqq 2 \\ & -x_1 + x_2 \leqq 2 \\ & -x_1 + x_2 \geqq -2 \\ & x_1, x_2 \geqq 0 \end{aligned}$$

(c)
$$\begin{aligned} \max \ & \{-x_1 = z_1\} \\ \max \ & \left\{\frac{-x_1 - 1}{-x_2 - 1} = z_2\right\} \\ \text{s.t.} \ & -2x_1 + x_2 \leqq 3 \\ & 3x_1 + x_2 \leqq 3 \\ & x_1, x_2 \geqq 0 \end{aligned}$$

(d)
$$\begin{aligned} \max \ & \{-x_1 + x_3 = z_1\} \\ \max \ & \{-x_1 - x_3 = z_2\} \\ \max \ & \left\{\frac{x_1 + x_3 - 2}{x_2 + 2} = z_3\right\} \\ \text{s.t.} \ & x_1 \leqq 8 \\ & x_2 \leqq 2 \\ & x_3 \leqq 2 \\ & x_1, x_2, x_3 \geqq 0 \end{aligned}$$

12-7 Consider the MOLFP:

$$\begin{aligned} \max \ & \{-x_1 + 4 = z_1\} \\ \max \ & \left\{\frac{-x_1 - 1}{-x_2 - 1} = z_2\right\} \\ \text{s.t.} \ & x_1 \leqq 3 \\ & x_2 \leqq 5 \\ & 5x_1 + 2x_2 \geqq 10 \\ & x_1, x_2 \geqq 0 \end{aligned}$$

(a) Graph the MOLFP in decision space.

(b) Specify E^s.

(c) Graph the set of nondominated criterion vectors in criterion space.

12-8 Specify E^w and E^s for:

$$\max\left\{\frac{-x_1 + x_2 - 3}{x_1} = z_1\right\}$$

$$\max\left\{\frac{-x_1 - x_2 + 2}{x_1} = z_2\right\}$$

$$\begin{aligned} \text{s.t.} \quad x_1 \quad &\geqq 2 \\ x_2 &\geqq 1 \\ x_2 &\leqq 4 \\ x_1, x_2 &\geqq 0 \end{aligned}$$

CHAPTER 13

Interactive Procedures

The future of multiple objective programming is in its interactive application. In interactive procedures, we conduct an exploration over the region of feasible alternatives for an optimal, or satisfactorily near optimal, solution. Interactive procedures are characterized by phases of decision making alternating with phases of computation. We generally establish a pattern and keep repeating it until termination. At each iteration, a solution, or group of solutions, is generated for examination. As a result of the examination, the decision maker inputs information to the solution procedure.

The feedback between man and model enables the decision maker to learn more about his or her problem. It enables the decision maker to appreciate more fully the range of possibilities in his or her feasible region and how the objectives trade off against one another. This, then, should enable the decision maker to know better where to look for improved solutions and how to recognize a final solution upon encountering it. In essence, the strategy of the interactive approach is to facilitate human intervention in order to allow midcourse corrections to the search process as appropriate.

Interactive procedures permit an effective division of labor. They allow the computer to do what it does best (process data and execute algorithms), and they allow the decision maker to do what he or she does best (make improved judgments in the face of new information). This chapter discusses the following six interactive procedures:

1. STEM [Benayoun et al. (1971)].
2. Geoffrion-Dyer-Feinberg Procedure (1972).
3. Zionts-Wallenius Method (1976 and 1983).
4. Interval Criterion Weights/Vector-Maximum Approach [Steuer (1977)].
5. Interactive Weighted-Sums/Filtering Approach [Steuer (1980)].
6. Visual Interactive Approach of Korhonen and Laakso (1985).

In the sections to follow, these algorithms are classified as *feasible region reduction*, *weighting vector space reduction*, *criterion cone contraction*, or *line search* methods. While these algorithms are quite different from one another, they all tend to converge surprisingly rapidly—in k (where k is the number of objectives) iterations more or less.

To compare and contrast interactive procedures, or to help identify the most appropriate to apply, one might ask the following questions.

1. Does the algorithm require the decision maker to be mathematically or computer-sophisticated?
2. Does the algorithm require a skilled analyst-technician to operate it?
3. Is it enjoyable for the decision maker to interface with the algorithm?
4. Does the algorithm present one solution, or a group of solutions, at each iteration?
5. Does the algorithm ask the same pattern of questions each iteration or are different types of questions asked on different iterations?
6. Is the decision maker asked to assess weights, specify relaxation quantities, evaluate tradeoffs, or make pairwise comparisons?
7. Is the algorithm an *ad hoc* procedure? (A procedure is ad hoc if the user would not know how to respond deterministically to the questions posed by the algorithm even if the decision maker's utility function were known mathematically. In other words, with an ad hoc procedure, we proceed primarily on instinct).
8. Can the decision maker make rapid iteration judgments or will he or she want several days to ponder each of his or her decisions?
9. Does the algorithm converge within a prespecified number of iterations?
10. Does the algorithm allow the decision maker to change his or her aspirations during the solution process without necessitating an algorithm restart?
11. Does the algorithm only generate efficient points (nondominated criterion vectors)?
12. Does the algorithm make specific assumptions about the shape of the decision maker's utility function?
13. Does the algorithm have convergence proofs or does it have only heuristic or intuitive convergence properties?
14. Does the algorithm require specialized, as opposed to conventionally available, software?
15. Does the algorithm require a large amount of CPU time?
16. Is the algorithm only applicable to MOLP, or can it be applied to multiple objective integer and nonlinear programming problems as well?

13.1 STEM

Proposed by Benayoun, de Montgolfier, Tergny, and Laritchev (1971), **STEM** is a *reduced feasible region* method for solving the MOLP

$$\begin{aligned}
&\max\,\{\mathbf{c}^1\mathbf{x} = z_1\}\\
&\max\,\{\mathbf{c}^2\mathbf{x} = z_2\}\\
&\quad\vdots\\
&\max\,\{\mathbf{c}^k\mathbf{x} = z_k\}\\
&\text{s.t.} \qquad \mathbf{x} \in S
\end{aligned}$$

where all objectives are bounded over S. Each iteration **STEM** makes a single probe of the efficient set. This is done by computing the point in that iteration's reduced feasible region whose criterion vector is closest in a weighted L_∞ (Tchebycheff) sense to a $\mathbf{z}^* \in R^k$ *ideal criterion vector*. **STEM** was the first interactive procedure to have impact on the field of multiple objective programming. Originally designed for MOLPs, the method can be applied to integer and nonlinear multiple objective programs. In a linear context, the **STEM** algorithm is as follows.

13.1.1 STEM Algorithm

Step 1: By individually optimizing each objective function, construct a payoff table to obtain the ideal criterion vector $\mathbf{z}^* \in R^k$.

As discussed in Section 9.12, a *payoff table* is of the form

	z_1	z_2		z_k
$\mathbf{z}^1$	z_1^*	z_{12}		z_{1k}
$\mathbf{z}^2$	z_{21}	z_2^*		z_{2k}
			$\ddots$	
$\mathbf{z}^k$	z_{k1}	z_{k2}		z_k^*

where the rows are the criterion vectors resulting from individually optimizing each of the objectives. The z_i^* entries along the main diagonal form the $\mathbf{z}^*$ ideal criterion vector.

Step 2: Let iteration counter $h = 0$. Let m_i be the minimum value in the ith column of the payoff table. Calculate π_i values where

$$\pi_i = \begin{cases} \dfrac{z_i^* - m_i}{z_i^*}\left[\sum_{j=1}^{n} (c_{ij})^2\right]^{-1/2} & \text{when} \quad z_i^* > 0 \\ \dfrac{m_i - z_i^*}{m_i}\left[\sum_{j=1}^{n} (c_{ij})^2\right]^{-1/2} & \text{when} \quad z_i^* \leqq 0 \end{cases}$$

Term 1 Term 2

The purpose of Term 1 is to place the most weight on the objectives with the greatest relative ranges. Term 2 normalizes the gradients of the objective functions according to the L_2-norm.

Step 3: Let $S^{(1)} = S$ and index set $J^* = \varnothing$.

$S^{(1)} = S$ means that we start the algorithm with the original (not yet reduced) feasible region. Index set J^* designates the criterion values that are to be relaxed on the next iteration to allow greater achievement of the others. At this point, J^* is empty because no solutions have yet been generated to relax.

Step 4: Let $h = h + 1$. Calculate $\lambda_i^{(h)}$ minimax (Tchebycheff) weights where

$$\lambda_i^{(h)} = \begin{cases} 0 & i \in J^* \\ \dfrac{\pi_i}{\sum_{j=1}^{k} \pi_j} & i \notin J^* \end{cases}$$

These weights define the weighted Tchebycheff metric

$$\|\mathbf{z}^* - \mathbf{z}\|_\infty^{\lambda^{(h)}} = \max_{i=1,\ldots,k} \{\lambda_i^{(h)} |z_i^* - z_i|\}$$

that is used in Step 5 to generate the solution of iteration h. Note that on the first iteration, the $\lambda_i^{(h)}$ sum to one. On all subsequent iterations, the $\lambda_i^{(h)}$ sum to less than one because J^* is not null.

Step 5: Solve the weighted minimax program

$$\begin{aligned} &\min\{\alpha\} \\ \text{s.t.} \quad &\alpha \geqq \lambda_i^{(h)}(z_i^* - \mathbf{c}^i\mathbf{x}) \qquad 1 \leqq i \leqq k \\ &\mathbf{x} \in S^{(h)} \\ &0 \leqq \alpha \in R \end{aligned}$$

for decision space solution $\mathbf{x}^{(h)}$.

In this step, we solve for the point in the reduced feasible region $S^{(h)}$ whose criterion vector is closest to $\mathbf{z}^*$ according to the weighted Tchebycheff metric defined by $\lambda^{(h)} \in R^k$.

Step 6: Let $\mathbf{z}^{(h)} = \mathbf{z}(\mathbf{x}^{(h)})$. Compare $\mathbf{z}^{(h)}$ with $\mathbf{z}^*$.

The idea behind the comparison is that $\mathbf{z}^*$ is a good reference point for assessing the quality of a candidate criterion vector.

Step 7: If all components of $\mathbf{z}^{(h)}$ are satisfactory, stop with $(\mathbf{z}^{(h)}, \mathbf{x}^{(h)})$ as the final solution. Otherwise, go to Step 8.

As long as some criterion vector components are more satisfactory than others, we keep iterating because our situation can be improved by making tradeoffs. If the decision maker is unwilling to trade off any components of $\mathbf{z}^{(h)}$ to acquire more of others, we are done and exit from the algorithm.

Step 8: Specify the index set J^* of criterion values to be relaxed and specify the amounts $(\Delta_j,\ j \in J^*)$ by which they are to be relaxed.

We only arrive at this step if components of $\mathbf{z}^{(h)}$ exist for which the decision maker is willing to sacrifice achievement in order to improve other components. For each component j that the decision maker is willing to trade off achievement, we specify a relaxation quantity Δ_j (i.e., the maximum amount of z_j we are willing to sacrifice).

Step 9: Form reduced feasible region

$$S^{(h+1)} = \left\{ \mathbf{x} \in S \;\middle|\; \begin{array}{ll} \mathbf{c}^j\mathbf{x} \geqq z_j(\mathbf{x}^{(h)}) - \Delta_j & j \in J^* \\ \mathbf{c}^j\mathbf{x} \geqq z_j(\mathbf{x}^{(h)}) & j \notin J^* \end{array} \right\}$$

Then, go to Step 4.

By adding additional constraints as above, we iterate through a series of progressively smaller subsets of S. With regard to the number of iterations, Benayoun et al. mention that if a j is no longer eligible for further relaxation once it has been relaxed, the algorithm will terminate in at most k iterations. In practice, however, we would probably want the freedom to relax several criterion values at once and given criterion values more than once. A graphical example of the relaxation process creating a reduced feasible region is given in Example 1. `STEM` is flowcharted in Fig. 13.1.

Example 1. Consider Fig. 13.2 in which $\mathbf{x}^*$ is the inverse image of $\mathbf{z}^*$. Suppose $\mathbf{x}^{(1)}$ is the first iteration solution (generated in Step 5). Letting $J^* = \{1\}$ with a relaxation of Δ_1, reduced feasible region $S^{(2)}$ is formed in Step 9. Calculating $\lambda_1^{(2)} = 0$ and $\lambda_2^{(2)} > 0$ in Step 4, in Step 5 we obtain $\mathbf{x}^{(2)}$, and so forth.

13.1.2 `STEM` Sample Output

Sample `STEM` output as scrolled at the terminal might appear as in Fig. 13.3. The user's responses follow the @ signs.

13.1.3 `STEM` Comments

1. `STEM` is an ad hoc procedure because even if one knew the decision maker's utility function, it would not be clear which objectives are to be relaxed and by how much. Hence, we might consider `STEM` a semistructured approach for generating a sequence of improved solutions. There is no guarantee that we will converge to an optimal solution.
2. `STEM` was designed for use with conventional single criterion software. After the construction of the payoff table, only one optimization per iteration (in Step 5) is required. In addition to the linear case, `STEM` can also be applied to integer and nonlinear multiple objective programs of any size that can be accommodated by the single criterion software employed.
3. Because of payoff table difficulties (see Section 9.12.1), the ranges of the criterion values over the efficient set may be either understated or overstated. If the ranges are incorrect, this would affect the $\lambda_i^{(h)}$ weights.
4. Rather than being restricted to only extreme points of S, `STEM` can converge to nonextreme final solutions.
5. It is possible for a $\mathbf{z}^{(h)}$ criterion vector presented for comparison in Step 6 to be dominated as in Example 2.

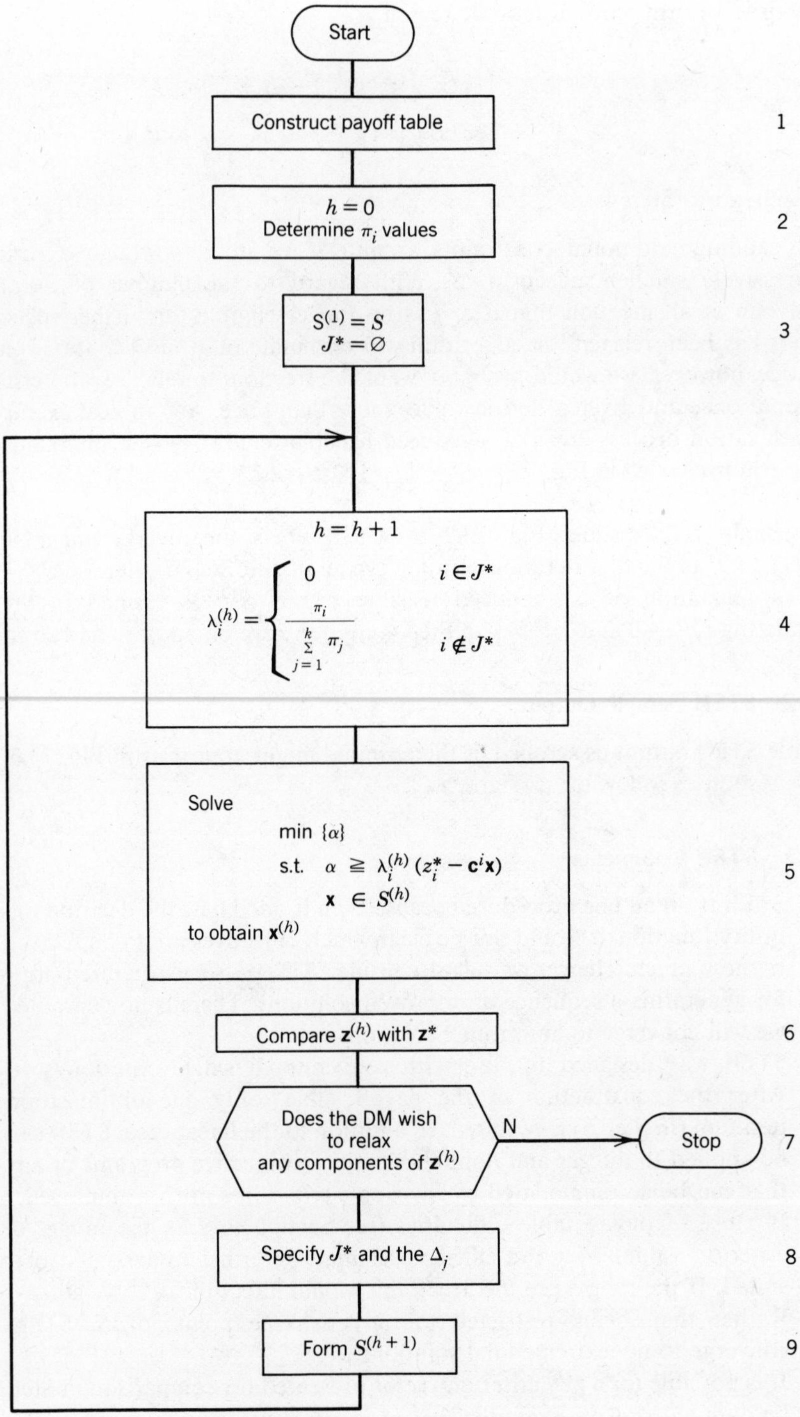

Figure 13.1. Flowchart of **STEM**.

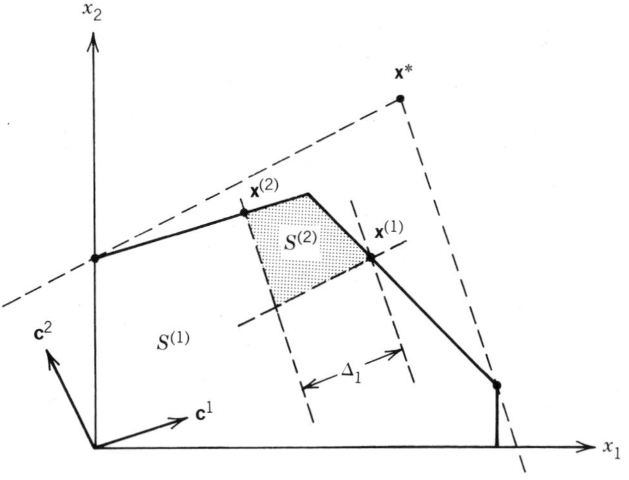

Figure 13.2. Graph of Example 1.

Example 2. Let $\mathbf{x}^{(1)}$ be the solution of the first iteration and let the first objective be relaxed by Δ_1 as in Figure 13.4. Then, $\mathbf{x}^{(2)}$ is a legitimate second-iteration solution. However, the criterion vector of $\mathbf{x}^{(2)}$ is dominated by the criterion vector of $\bar{\mathbf{x}}$.

13.2 GEOFFRION-DYER-FEINBERG (GDF) PROCEDURE

Proposed by Geoffrion, Dyer, and Feinberg (1972), the GDF algorithm is a *line search* method for solving the multiple objective program

$$\begin{aligned} &\max \{f_1(\mathbf{x}) = z_1\} \\ &\max \{f_2(\mathbf{x}) = z_2\} \\ &\quad\vdots \\ &\max \{f_k(\mathbf{x}) = z_k\} \\ &\text{s.t.} \qquad \mathbf{x} \in S \end{aligned}$$

where S is defined by linear constraints and is bounded. Internally, the GDF procedure uses the Frank-Wolfe gradient ascent algorithm. The authors chose the Frank-Wolfe algorithm because of its simplicity and rapid initial convergence properties in nonlinear programming. Statements about convergence to a multiple criteria optimal solution can be made when the decision maker's utility function is differentiable and concave on S.

We begin our study of the GDF procedure with preliminaries about the Frank–Wolfe algorithm in Sections 13.2.1 and 13.2.2. This is followed by discussions in Section 13.2.3 about a pairwise comparison routine that is used in the

```
PAYOFF TABLE
           48.7    9.8   15.1
           21.6   43.2   27.6
           18.0   36.0   63.6

                             IDEAL CRITERION
OBJECTIVE       Z(1)             VECTOR
    1           37.8              48.7
    2           16.5              43.2
    3           36.2              63.6

WISH TO TERMINATE (YES / NO)?

@ NO

WHICH COMPONENTS DO YOU WISH TO RELAX (IF
MORE THAN ONE, SEPARATE BY COMMAS)?

@ 1, 2

SPECIFY AMOUNTS OF RELAXATION

@ 6.0, 4.0

                             IDEAL CRITERION
OBJECTIVE       Z(2)             VECTOR
    1           34.3              48.7
    2           12.5              43.2
    3           44.3              63.6

WISH TO TERMINATE (YES / NO)?

@ NO

WHICH COMPONENTS DO YOU WISH TO RELAX (IF
MORE THAN ONE, SEPARATE BY COMMAS)?

@ 3

SPECIFY AMOUNTS OF RELAXATION
:
:
:
```

Figure 13.3. Sample **STEM** terminal output.

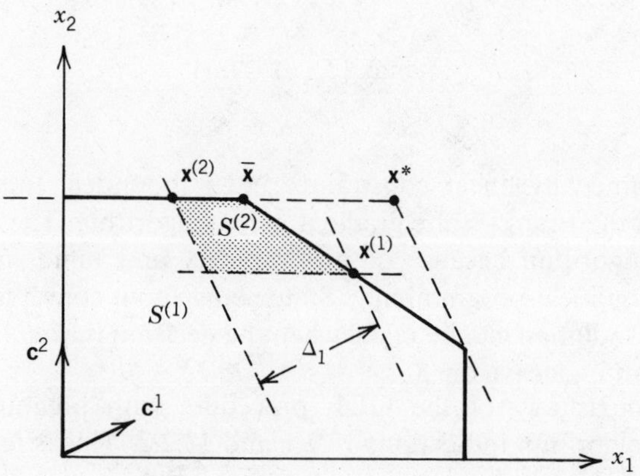

Figure 13.4. Graph of Example 2.

GDF procedure. Then the GDF procedure is specified and illustrated in Sections 13.2.4, 13.2.5 and 13.2.6.

13.2.1 A Steepest Ascent Algorithm

Before introducing the Frank-Wolfe algorithm, let us observe the search behavior of the following steepest ascent algorithm where $F\colon R^n \to R$ is a mathematically known differentiable function to be optimized over S.

1. Let $h = 0$. Obtain an initial point $\mathbf{x}^{(0)} \in S$.
2. Let $h = h + 1$. Define the direction $\mathbf{d}^{(h)} = \nabla_x F(\mathbf{x}^{(h-1)})$ where $\nabla_x F(\mathbf{x}^{(h-1)})$ is the gradient of F at $\mathbf{x}^{(h-1)}$.
3. Conducting a *linear search* in the direction $\mathbf{d}^{(h)}$, solve the *step-size* program

$$\begin{aligned} &\max \{F(\mathbf{x}^{(h-1)} + t\mathbf{d}^{(h)})\} \\ &\text{s.t.} \quad (\mathbf{x}^{(h-1)} + t\mathbf{d}^{(h)}) \in S \\ &\qquad 0 \leqq t \in R \end{aligned}$$

With step size t^* optimal, let $\mathbf{x}^{(h)} = \mathbf{x}^{(h-1)} + t^*\mathbf{d}^{(h)}$. If an appropriate *stopping rule* is satisfied, stop. Otherwise, go the Step 2.

Possible stopping rules are

1. Belief that an optimal solution has been obtained.
2. A $t^* = 0$ value results from the step-size program.
3. Lack of significant progress over the last few iterations.
4. The decision maker has reached the limits of his or her patience.

In the steepest ascent algorithm outlined above, the search direction is that of the local gradient of the function being optimized at $\mathbf{x}^{(h-1)}$. Despite the fact that we search in the direction of greatest increase in F, convergence can be disappointingly slow as illustrated in Example 3.

Example 3. Consider Fig. 13.5 in which the dashed lines are contours of F. Letting $\mathbf{x}^{(0)}$ be the starting point for the steepest ascent algorithm of this section, we observe that it takes a large number of zig-zag linear searches to converge to the optimum. As illustrated in the next section, there are other linear search directions different from the exact direction of the local gradient of F that often yield more rapid convergence.

13.2.2 Frank–Wolfe Algorithm

With the known function $F\colon R^n \to R$ to be optimized over S, the Frank-Wolfe algorithm iterates as follows.

1. Let $h = 0$. Obtain an initial point $\mathbf{x}^{(0)} \in S$.
2. Let $h = h + 1$. Solve the *direction finding* program

$$\max \left\{ \left[\nabla_x F(\mathbf{x}^{(h-1)})\right]^T \mathbf{x} \mid \mathbf{x} \in S \right\}$$

With $\mathbf{y}^{(h)} \in S$ an optimal solution, define the direction $\mathbf{d}^{(h)} = \mathbf{y}^{(h)} - \mathbf{x}^{(h-1)}$.

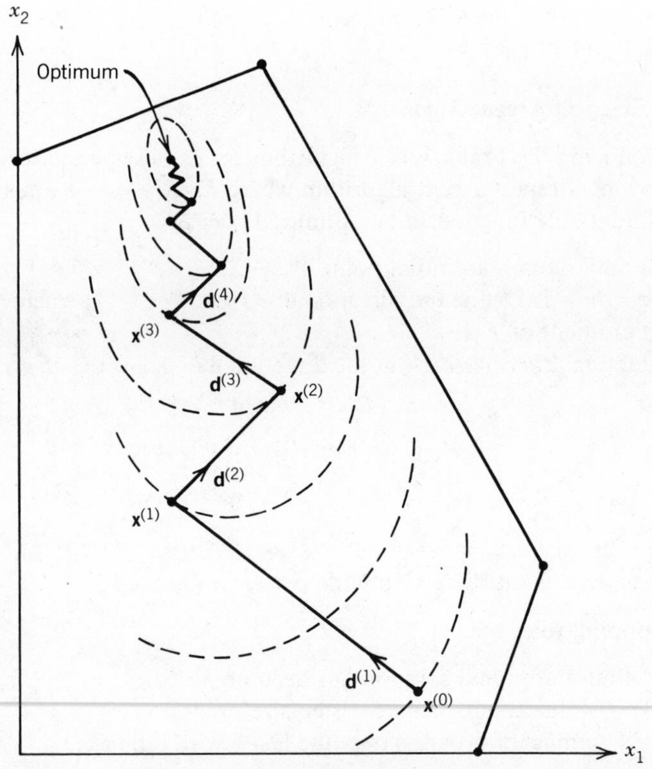

Figure 13.5. Graph of Example 3.

3. Conducting a linear search in the direction $\mathbf{d}^{(h)}$, solve the step-size program

$$\max\left\{F(\mathbf{x}^{(h-1)} + t\mathbf{d}^{(h)}) \mid 0 \leqq t \leqq 1\right\}$$

With step size t^* optimal, let $\mathbf{x}^{(h)} = \mathbf{x}^{(h-1)} + t^*\mathbf{d}^{(h)}$. If an appropriate stopping rule is satisfied, stop. Otherwise, go to Step 2.

The local gradient of F is used in both the Frank–Wolfe algorithm and in the steepest ascent algorithm of Section 13.2.1. However, in the Frank–Wolfe algorithm, the local gradient is used as the gradient of the direction finding program, whereas in the steepest ascent algorithm the local gradient is used as the search direction. We also observe in the Frank–Wolfe algorithm that although $\mathbf{d}^{(h)}$ may point in a direction *similar* to that of the gradient of F at $\mathbf{x}^{(h-1)}$, it is unlikely that it will point precisely in the same direction. This can be an advantage of the Frank–Wolfe algorithm over the steepest ascent algorithm, as shown in Example 4.

Example 4. Figure 13.6 portrays the solution of the problem of Example 3 using the Frank–Wolfe algorithm. If we start at $\mathbf{x}^{(0)}$, the solution to the direction

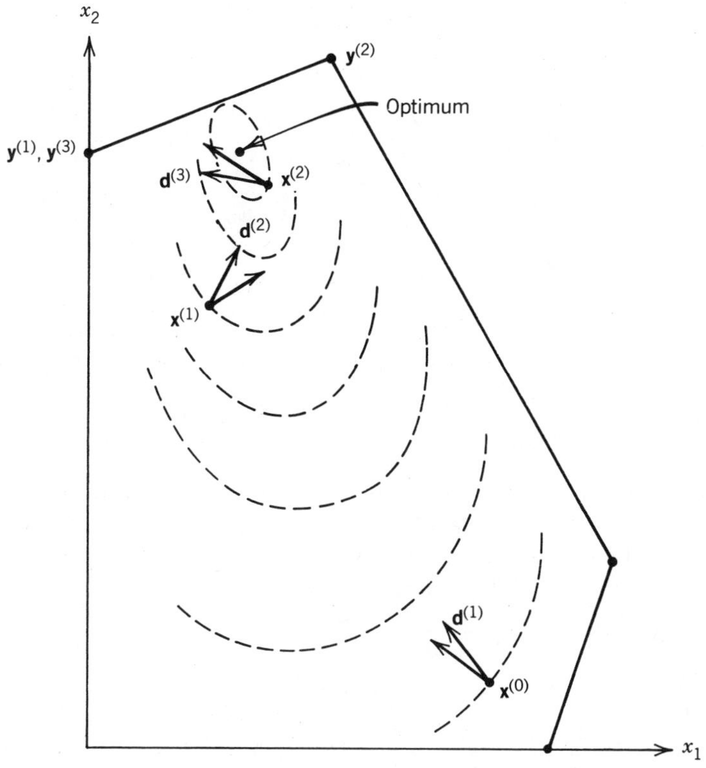

Figure 13.6. Graph of Example 4.

finding program is $\mathbf{y}^{(1)}$. Using $\mathbf{y}^{(1)}$ to define direction $\mathbf{d}^{(1)}$ (which is different from the direction of $\nabla_x F(\mathbf{x}^{(0)})$, we find that the step-size program results in $\mathbf{x}^{(1)}$. Solving the second direction finding program results in $\mathbf{y}^{(2)}$. Using $\mathbf{y}^{(2)}$ to define $\mathbf{d}^{(2)}$, we solve the second step-size program to obtain $\mathbf{x}^{(2)}$. Continuing in this fashion, we make more rapid progress (particularly in the early iterations) than when we used the local gradient directions in Example 3.

When considering the Frank–Wolfe approach in the context of problems with multiple criteria, F is replaced by U. Despite the fact that U is not known mathematically, the GDF algorithm iteratively elicits information about the decision maker's preferences to estimate the local gradient directions of U for the direction finding program. The likelihood that the estimates are not perfectly accurate may actually be a blessing in disguise.

13.2.3 Routine for Inducing Locally Relevant Weights

To estimate the local gradients of U, the GDF algorithm is equipped with a pairwise comparison routine for obtaining the decision maker's preference weights. Without loss of generality, let the first objective be designated as the *reference* criterion. Recall from Section 7.4 that the ith criterion weight is calculated using the ratio

$$\frac{\Delta_1}{\Delta_i}$$

where Δ_i is the amount by which the ith criterion is to be increased to compensate for a Δ_1 decrease in the first (reference) criterion, while holding all other criteria at their unperturbed values. For a given Δ_1, the purpose of the pairwise comparison process is to determine the Δ_i, $i \neq 1$, and hence the preference weights $w_i = \Delta_1/\Delta_i$. To determine the unknown Δ_i, we have the *halving and doubling* routine as follows.

1. Let $\mathbf{z}^a = (z_1, \ldots, z_i, \ldots, z_k)$ be the current criterion vector. Specify Δ_1 and let $i = 2$. Let $w_1 = 1$.
2. Let $\ell = 0$ and specify an initial $\Delta_i^{(1)}$.
3. Let $\ell = \ell + 1$. Compare $\mathbf{z}^a$ and $\mathbf{z}^b = (z_1 - \Delta_1, \ldots, z_i + \Delta_i^{(\ell)}, \ldots, z_k)$.
4. If $\mathbf{z}^a$ is preferred, increase the desirability of $\mathbf{z}^b$ by letting $\Delta_i^{(\ell+1)} = 2\Delta_i^{(\ell)}$. Go to Step 3. If $\mathbf{z}^b$ is preferred, let $\theta^{(\ell)} = \Delta_i^{(\ell)}/2$ and go to Step 6. If the decision maker is indifferent between $\mathbf{z}^a$ and $\mathbf{z}^b$, go to Step 8.
5. Let $\theta^{(\ell+1)} = \theta^{(\ell)}/2$ and let $\ell = \ell + 1$. Compare $\mathbf{z}^a$ and $\mathbf{z}^b$. If $\mathbf{z}^b$ is preferred, go to Step 6. If $\mathbf{z}^a$ is preferred, go to Step 7. If the decision maker is indifferent between $\mathbf{z}^a$ and $\mathbf{z}^b$, go to Step 8.
6. Decrease the desirability of $\mathbf{z}^b$ by letting $\Delta_i^{(\ell+1)} = \Delta_i^{(\ell)} - \theta^{(\ell)}/2$ and go to Step 5.
7. Increase the desirability of $\mathbf{z}^b$ by letting $\Delta_i^{(\ell+1)} = \Delta_i^{(\ell)} + \theta^{(\ell)}/2$ and go to Step 5.
8. Compute $w_i = \Delta_1/\Delta_i^{(\ell)}$. If $i \neq k$, let $i = i + 1$ and go to Step 2. Otherwise, stop.

Example 5. To estimate w_i weights corresponding to the decision maker's tradeoffs at $\mathbf{z}^a = (100, 50)$, let $\Delta_1 = 10$ and $\Delta_2^{(1)} = 4$ in the halving and doubling routine. In Fig. 13.7*a*, $\mathbf{z}^a$ is preferred to $\mathbf{z}^b = (90, 54)$. To increase the desirability of $\mathbf{z}^b$, we increase $\Delta_2^{(1)}$ by four so that $\Delta_2^{(2)} = 8$. In Fig. 13.7*b*, $\mathbf{z}^b = (90, 58)$ is preferred to $\mathbf{z}^a$. To decrease the desirability of $\mathbf{z}^b$, we decrease $\Delta_2^{(2)}$ by two so that $\Delta_2^{(3)} = 6$. In Fig. 13.7*c*, $\mathbf{z}^a$ is preferred to $\mathbf{z}^b = (90, 56)$. To increase the desirability of $\mathbf{z}^b$, we increase $\Delta_2^{(3)}$ by one so $\Delta_2^{(4)} = 7$. In Fig. 13.7*d*, we are indifferent between $\mathbf{z}^a$ and $\mathbf{z}^b = (90, 57)$. Thus, $w_1 = 1$ and $w_2 = 10/7$. Note that they are not perfect weights because they correspond to the dashed line in Fig. 13.7*d* rather than the tangent to the utility contour at $\mathbf{z}^a$.

13.2.4 GDF Algorithm

Step 1: Let iteration counter $h = 0$. Obtain a starting point $\mathbf{x}^{(0)} \in S$.

With S defined by linear constraints, an initial solution can be obtained by performing an ordinary Phase I from linear programming. When the objectives are also linear, another method is to ask the decision maker to supply a criterion

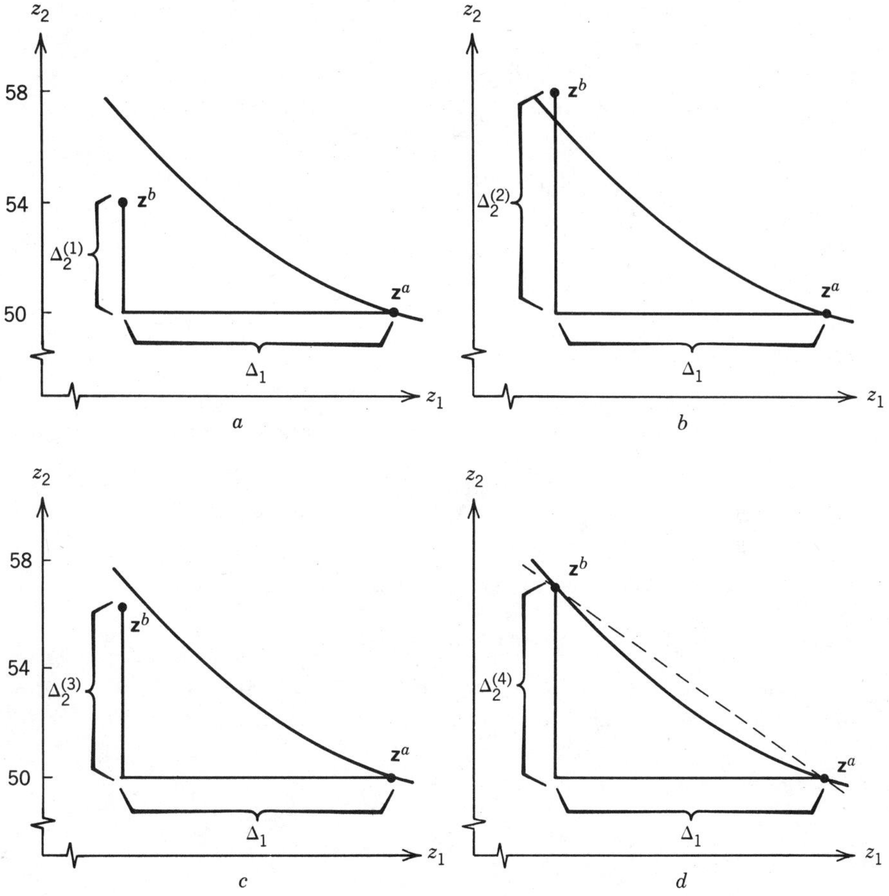

Figure 13.7. Graphs of Example 5.

vector $\bar{\mathbf{z}}$. Then, $\bar{\mathbf{z}}$ can be evaluated for feasibility by solving LP

$$\begin{aligned} &\max \{\mathbf{0}^T\mathbf{x}\} \\ &\text{s.t.} \quad \mathbf{c}^i\mathbf{x} = \bar{z}_i \qquad 1 \leqq i \leqq k \\ &\qquad\quad \mathbf{x} \in S \end{aligned}$$

If the LP is consistent, the solution generated by the program can be used as $\mathbf{x}^{(0)}$.

Step 2: Let $h = h + 1$.

Step 3: To estimate the direction of the gradient of U at $\mathbf{x}^{(h-1)}$, the decision maker must either input a locally relevant weighting vector or respond to a series of pairwise comparison questions so that the computer can induce a set of locally relevant weights.

On most applications, it is expected that the decision maker would be unable (or unwilling) to directly specify weights. Consequently, the decision maker would have to submit to a series of pairwise comparison questions, as in Section 13.2.3, so that the procedure could calculate weights for him or her.

Step 4: Let $\nabla_x f_i(\mathbf{x}^{(h-1)})$ denote the gradient of the ith objective function at $\mathbf{x}^{(h-1)}$. Solve the direction finding program

$$\max\{\mathbf{g}^T\mathbf{x} \mid \mathbf{x} \in S\} \quad \text{where} \quad \mathbf{g} = \sum_{i=1}^{k} w_i \nabla_x f_i(\mathbf{x}^{(h-1)})$$

to obtain $\mathbf{y}^{(h)} \in S$. Let $\mathbf{d}^{(h)} = \mathbf{y}^{(h)} - \mathbf{x}^{(h-1)}$.

Provided that reasonably good weights were obtained in Step 3, $\mathbf{g}$ should point in roughly the same direction as the gradient of U at $\mathbf{x}^{(h-1)}$. This should suffice for generating useful $\mathbf{d}^{(h)}$ linear search directions.

Step 5: If $\mathbf{y}^{(h)} = \mathbf{x}^{(h-1)}$, set $\mathbf{x}^{(h)}$ to $\mathbf{x}^{(h-1)}$ and go to Step 10. Otherwise, go to Step 6.

Step 6: Specify the number of stepwise points P, where $P > 1$.

Step 7: Display the P stepwise criterion vectors.

To conduct the linear search in the direction $\mathbf{d}^{(h)}$, we print out the criterion vectors at equally spaced *grid points* along the line segment $\gamma(\mathbf{x}^{(h-1)}, \mathbf{y}^{(h)})$. With $\mathbf{z}(\mathbf{x}^{(h-1)})$ and $\mathbf{z}(\mathbf{y}^{(h)})$ being the first and last such criterion vectors, the $P - 2$ *intermediate* criterion vectors are given by

$$\mathbf{z}\left(\mathbf{x}^{(h-1)} + \left(\frac{j-1}{P-1}\right)\mathbf{d}^{(h)}\right) \qquad j = 2, \ldots, P-1$$

Step 8: The decision maker selects his or her most preferred of the stepwise criterion vectors, designating it and its inverse image by the pair $(\mathbf{z}^{(h)}, \mathbf{x}^{(h)})$.

By specifying P in Step 6, the user controls the amount of criterion vector information displayed at each iteration.

Step 9: If the decision maker wishes to keep iterating, go to Step 2. Otherwise, go to Step 10.

Step 10: Stop with $(\mathbf{z}^{(h)}, \mathbf{x}^{(h)})$ as the final solution.

The GDF algorithm is flowcharted in Fig. 13.8.

13.2.5 GDF Sample Output

Sample GDF output as scrolled at the terminal might appear as in Fig. 13.9. The user's responses follow the @ signs.

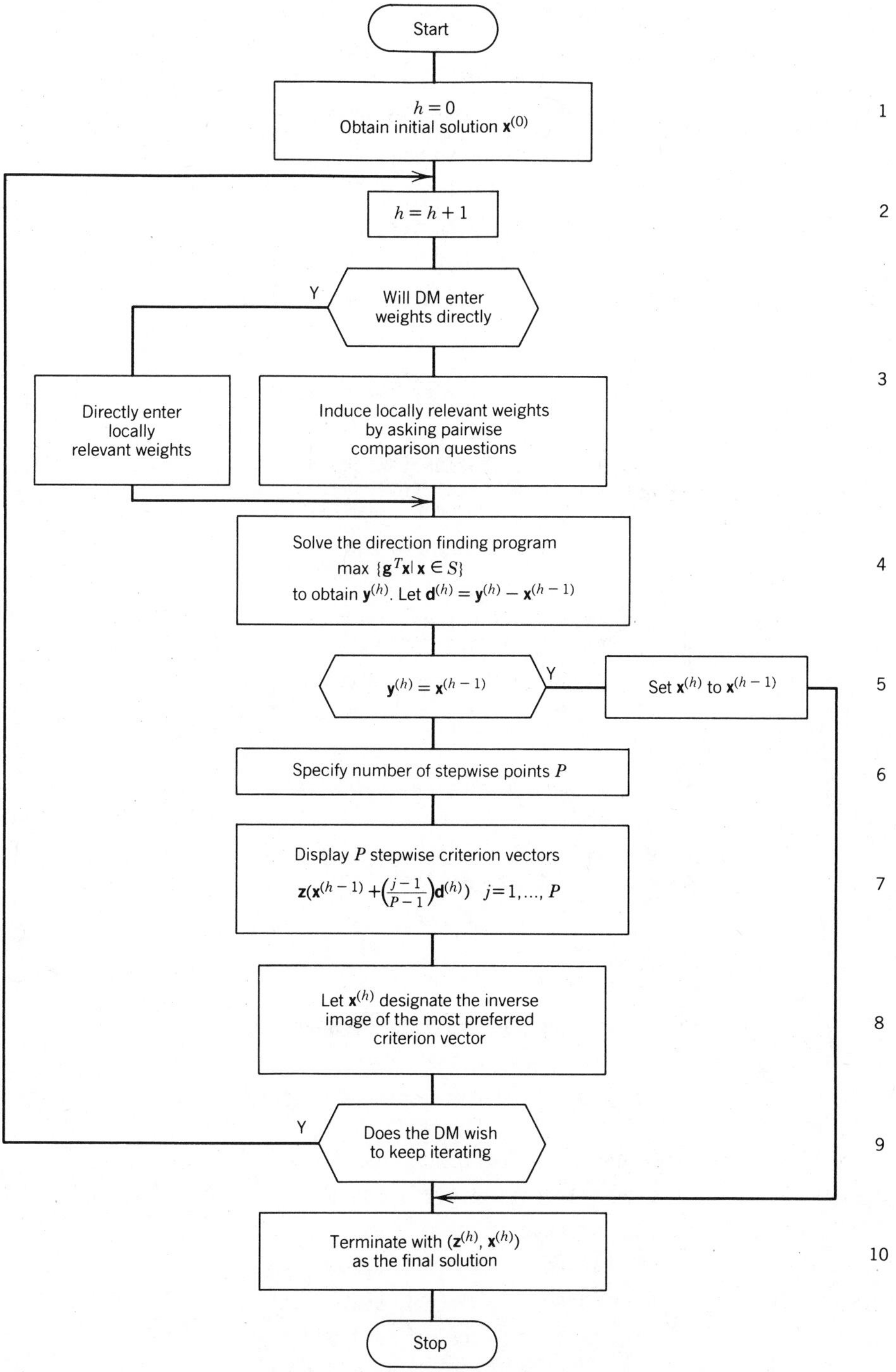

Figure 13.8. Flowchart of GDF algorithm.

```
SPECIFY COMPONENTS OF A FEASIBLE CRITERION
VECTOR

@ 14300.0, 100.0, 650.0

DO YOU WISH TO UTILIZE THE PAIRWISE
COMPARISON ROUTINE IN LIEU OF ENTERING
WEIGHTS DIRECTLY (YES / NO)?

@ YES

YOUR CURRENT CRITERION VECTOR IS

        14300.0    100.0    650.0

SPECIFY CHANGE IN 1ST (REFERENCE) CRITERION
VALUE

@ -1000.

SPECIFY CHANGE IN 2ND CRITERION VALUE

@ 5.

DO YOU PREFER CRITERION VECTOR (1), (2),
OR ARE YOU INDIFFERENT (0)?
   (1)  14300.0    100.0    650.0

   (2)  13300.0    105.0    650.0

@ 2

DO YOU PREFER CRITERION VECTOR (1), (2),
OR ARE YOU INDIFFERENT (0)?

   (1)  14300.0    100.0    650.0

   (2)  13300.0    102.5    650.0

@ 0

SPECIFY CHANGE IN 3RD CRITERION VALUE

@ 20.

DO YOU PREFER CRITERION VECTOR (1), (2),
OR ARE YOU INDIFFERENT (0)?

   (1)  14300.0    100.0    650.0

   (2)  13300.0    100.0    670.0

@ 0

YOUR DIRECTION FINDING CRITERION VECTOR IS

        12700.0    108.0    674.0

SPECIFY NUMBER OF STEPWISE POINTS (2 OR MORE)
@ 5

WHICH OF THE FOLLOWING DO YOU PREFER?

   (1)  14300.0    100.0    650.0

   (2)  13900.0    102.0    656.0

   (3)  13500.0    104.0    662.0

   (4)  13100.0    106.0    668.0

   (5)  12700.0    108.0    674.0
```

Figure 13.9. Sample GDF terminal output.

```
@ 4

DO YOU WISH TO TERMINATE (YES / NO)?

@ NO

DO YOU WISH TO UTILIZE THE PAIRWISE
COMPARISON ROUTINE IN LIEU OF ENTERING
WEIGHTS DIRECTLY (YES / NO)?

@ YES

YOUR CURRENT CRITERION VECTOR IS

          13100.0    106.0    668.0

SPECIFY CHANGE IN 1ST (REFERENCE) CRITERION
VALUE

@ -400.

SPECIFY CHANGE IN 2ND CRITERION VALUE

@ 6.
:
:
```

Figure 13.9. *(Continued)*

13.2.6 GDF Numerical Example

To illustrate the GDF algorithm, consider the MOLP

$$\begin{aligned}
\max \{x_1 &&= z_1\} \\
\max \{ && x_2 = z_2\} \\
\text{s.t.} \quad x_1 & + x_2 & \leqq 10 \\
2x_1 & + x_2 & \leqq 18 \\
5x_1 & + 9x_2 & \geqq 45 \\
& x_1, x_2 & \geqq 0
\end{aligned}$$

that is graphed in Fig. 13.10. To answer the questions posed by the algorithm, let $U = (z_1 + 1)^2(z_2 + 1)^2$. Hence, (5, 5) is the optimal solution and 1,296 is the optimal U value. Let us run the GDF algorithm for three iterations and let $P = 6$ on all iterations. Instead of obtaining weights as in Step 3 and using them to form $\mathbf{g}$ as in Step 4, in this illustration let $\mathbf{g}$ be specified by the gradient of U evaluated at $\mathbf{x}^{(h-1)}$, i.e., by $\nabla_x U(\mathbf{x}^{(h-1)})$.

Let $\mathbf{x}^{(0)} = (0, 5)$ be the starting point. With $\mathbf{g} = \nabla_x U(\mathbf{x}^{(0)}) = (72, 12)$, the solution of the direction finding program is $\mathbf{y}^{(1)} = (9, 0)$. Hence, the linear search direction is $\mathbf{d}^{(1)} = (9, -5)$ and the six stepwise criterion vectors are (0, 5), (1.8, 4), (3.6, 3), (5.4, 2), (7.2, 1) and (9, 0). Of these, $\mathbf{z}^{(1)} = (5.4, 2)$ because its U of 368.6 is the highest.

With $\mathbf{x}^{(1)} = (5.4, 2)$ as the reference point for the second iteration, $\mathbf{g} = \nabla_x U(\mathbf{x}^{(1)}) = (115.2, 245.8)$ and the solution to the direction finding program is $\mathbf{y}^{(2)} = (0, 10)$. Hence, the linear search direction is $\mathbf{d}^{(2)} = (-5.4, 8)$, and so forth. The data for the first three GDF iterations are summarized in Table 13.1. Observe

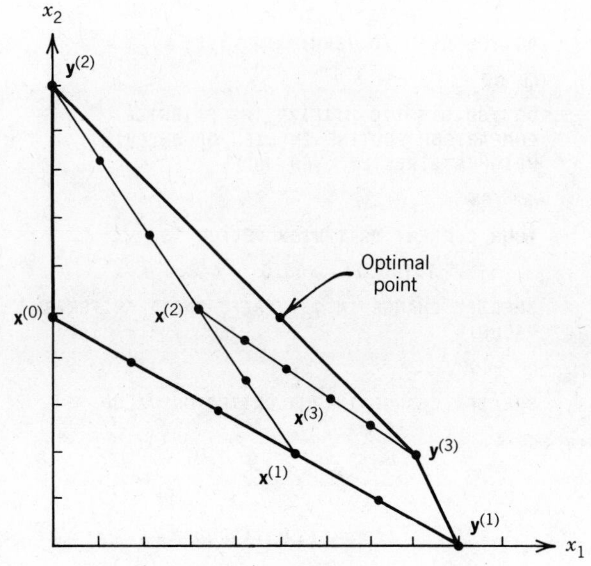

Figure 13.10. Graph of GDF example.

TABLE 13.1
Iteration Data for the GDF Example

	$h = 1$		$h = 2$		$h = 3$	
$\mathbf{x}^{(h-1)}$	(0, 5)		(5.4, 2)		(3.24, 5.2)	
$\mathbf{g}$	(72, 12)		(115.2, 245.8)		(326, 223)	
$\mathbf{y}^{(h)}$	(9, 0)		(0, 10)		(8, 2)	
$\mathbf{d}^{(h)}$	(9, −5)		(−5.4, 8)		(4.76, −3.2)	
	Stepwise Criterion Vectors	*U*	*Stepwise Criterion Vectors*	*U*	*Stepwise Criterion Vectors*	*U*
	(0, 5)	36.0	(5.4, 2)	368.6	(3.24, 5.2)	691.1
	(1.8, 4)	196.0	(4.32, 3.6)	598.9	(4.192, 4.56)	833.3
	(3.6, 3)	338.6	(3.24, 5.2)	691.1	(5.144, 3.92)	913.8
	(5.4, 2)	368.6	(2.16, 6.8)	607.5	(6.096, 3.28)	922.4
	(7.2, 1)	269.0	(1.08, 8.4)	382.3	(7.048, 2.64)	858.2
	(9, 0)	100.0	(0, 10)	121.0	(8, 2)	729.0
$\mathbf{z}^{(h)}$	(5.4, 2)		(3.24, 5.2)		(6.096, 3.28)	
$U(\mathbf{z}^{(h)})$	368.6		691.1		922.4	

that all but one of the stepwise criterion vectors of each iteration are dominated. Also note that the algorithm converges somewhat slowly to the optimal point.

13.2.7 GDF Comments

1. As opposed to STEM, the GDF algorithm is a non-ad hoc procedure. It is *testable* given a mathematical specification of the decision maker's utility function. That is, the pairwise comparison and stepwise responses can be provided deterministically by "plugging" the criterion vectors in question into the utility function.
2. In practice, the decision maker will emerge from Step 3 with only approximations of his or her true weights, and from Step 8 with only an approximation of the optimal stepsize. However, there are results showing that infinite convergence to the optimum holds, provided the approximations become sufficiently more exact as the iteration count increases [see Geoffrion, Dyer, and Feinberg (1972, p. 361)].
3. Although the Frank-Wolfe method was selected by the authors because of its rapid initial convergence properties in nonlinear programming, the Frank–Wolfe method may not give the interactive procedure the initial convergence speed that is needed in a multiple objective environment.
4. Although there is only one optimization per iteration (in Step 4), considerable dialogue occurs between man and model each iteration when the pairwise comparison process of Step 3 is employed.
5. It is likely that many of the stepwise criterion vectors presented to the decision maker (in Step 7) will be dominated.
6. The GDF procedure can be used when S is defined by nonlinear constraints, provided that a nonlinear code is used for solving the direction finding program of Step 4.

13.3 ZIONTS-WALLENIUS (Z–W) METHOD

Introduced by Zionts and Wallenius (1976) and updated in (1983), the Z–W procedure is a *reduced weighting vector space* method for solving the MOLP

$$\begin{aligned}
&\max \{ \mathbf{c}^1\mathbf{x} = z_1 \} \\
&\max \{ \mathbf{c}^2\mathbf{x} = z_2 \} \\
&\quad \vdots \\
&\max \{ \mathbf{c}^k\mathbf{x} = z_k \} \\
&\text{s.t.} \quad \mathbf{x} \in S
\end{aligned}$$

where the objective functions have already been normalized (for instance, using the L_1-norm). The method converges to the efficient extreme point of greatest utility when the decision maker's utility function $U: R^k \to R$ is pseudoconcave.

The method operates by iteratively asking the user questions about adjacent extreme points or tradeoff vectors. From the responses, portions of weighting vector space $\Lambda = \{\boldsymbol{\lambda} \in R^k \mid \lambda_i > 0, \sum_{i=1}^k \lambda_i = 1\}$ are eliminated. The process continues until Λ has been reduced to a small enough region for a final solution to be identified.

13.3.1 Z–W Algorithm

Step 1: Let $\Lambda^{(0)} = \Lambda$. Using an arbitrary $\boldsymbol{\lambda} \in \Lambda^{(0)}$, solve the composite LP

$$\max \{ \boldsymbol{\lambda}^T C\mathbf{x} \mid \mathbf{x} \in S \}$$

Let $\mathbf{x}^{(0)}$ denote the resulting extreme point solution and $\mathbf{z}^{(0)}$ its criterion vector. Let iteration counter $h = 0$.

Although any $\boldsymbol{\lambda} \in \Lambda$ would suffice, a $\boldsymbol{\lambda}$-vector with equal weights is recommended as a good *middlemost* representative of $\Lambda^{(0)}$.

Step 2: Partition the set of efficient nonbasic variables into

A = set of nonbasic variables efficient w.r.t. $\Lambda^{(h)}$

B = set of efficient nonbasic variables that are not in A

Let indicator set $I = A$.

With respect to the basis associated with $\mathbf{x}^{(h)}$, a modification of the Zionts-Wallenius routine for detecting nonbasic variable efficiency is used for the construction of sets A and B. An example of the Zionts-Wallenius routine, as modified to meet the requirements of Step 2, is given in Section 13.3.3. Although set A is of primary interest, it is often necessary to examine set B at least once before terminating.

Step 3: With respect to I, generate all adjacent extreme point criterion vectors. Temporarily disregard those criterion vectors that are not distinctly different from $\mathbf{z}^{(h)}$. For each distinctly different criterion vector, ask the decision maker whether it is preferred to $\mathbf{z}^{(h)}$. There are three possible responses:

1. *Yes* (adjacent criterion vector is preferred).
2. *No* ($\mathbf{z}^{(h)}$ is preferred).
3. *I don't know* (unable to express a preference).

If at least one distinctly different adjacent criterion vector is preferred to $\mathbf{z}^{(h)}$, save one such criterion vector, denoting it $\mathbf{z}^a$, and go to Step 7. Otherwise, go to Step 4.

In this step, we perform pairwise comparisons. We compare with $\mathbf{z}^{(h)}$ each *distinctly* different criterion vector of an adjacent extreme point w.r.t. I. For two criterion vectors to be different enough for a pairwise comparison, Zionts and Wallenius suggest the rule that there be at least a 10% difference between at least one set of corresponding criterion values. Step 3 can be implemented in various ways. One way is to conduct pairwise comparisons with all distinctly different adjacent criterion vectors. Another possibility is to terminate questioning as soon as a *yes* response is obtained.

Step 4: With respect to I, generate tradeoff vectors for all nonbasic variables not asked about in Step 3. For each such tradeoff vector, ask the decision maker whether he or she likes the tradeoff. There are three possible responses:

1. *Yes* (likes the tradeoff).
2. *No* (dislikes the tradeoff).
3. *I don't know* (unable to decide).

If the DM likes a tradeoff pertaining to an unbounded edge, terminate with an unbounded solution. If the DM likes at least one tradeoff associated with a bounded edge, go to Step 7. Otherwise, go to Step 5.

The tradeoff vectors are obtained from the columns of the reduced cost matrix. This step enables us to evaluate the edges connecting efficient extreme points that are too close for pairwise comparisons.

Step 5: With respect to I, ask the decision maker whether he or she likes any of the tradeoff vectors characterizing edges leading to adjacent efficient extreme points not preferred in Step 3. For each such tradeoff vector, there are three responses:

1. *Yes* (likes the tradeoff).
2. *No* (dislikes the tradeoff).
3. *I don't know* (unable to decide).

If at least one *yes* response is obtained, go to Step 7. Otherwise, go to Step 6.

This step enables us to identify edges along which initial movements are desirable, but along which further movements lead to points less preferred than the current extreme point.

Example 6. With $\mathbf{z}^{(h)} = \mathbf{z}(\mathbf{x}^{(h)})$ and $\bar{\mathbf{z}} = \mathbf{z}(\bar{\mathbf{x}})$, let the optimal point be as indicated in Fig. 13.11. In the pairwise comparison in Step 3 of $\bar{\mathbf{z}}$ with $\mathbf{z}^{(h)}$, the DM's response would be *no* ($\bar{\mathbf{z}}$ is not preferred to $\mathbf{z}^{(h)}$). However, for the tradeoff vector question of Step 5, the answer would be *yes* (the DM likes the tradeoff) because initial movements along the $\gamma(\mathbf{x}^{(h)}, \bar{\mathbf{x}})$ edge lead to higher utility values.

Step 6: If $I = A$, let $I = B$ and go to Step 3. Otherwise, stop with $(\mathbf{z}^{(h)}, \mathbf{x}^{(h)})$ as the optimal solution.

Set B, together with set A, enable the decision maker to check the tradeoff vectors and/or adjacent extreme points associated with *all* efficient edges emanating from $\mathbf{x}^{(h)}$ (regardless of whether or not they are consistent with the DM's responses) before terminating.

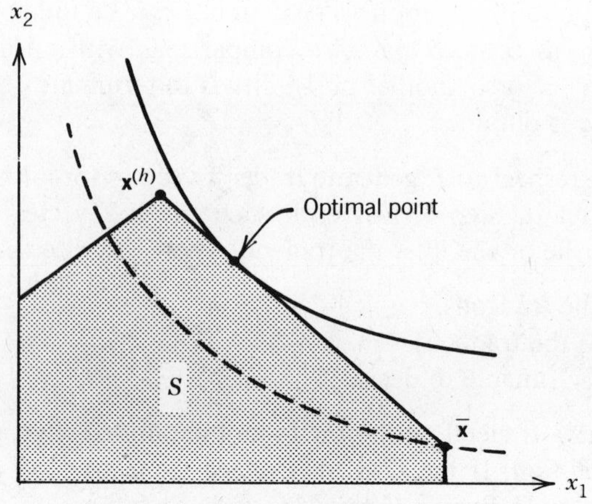

Figure 13.11. Graph of Example 6.

Step 7: Based upon the DM's responses, write $\boldsymbol{\lambda}$-constraints to form $\Lambda^{(h+1)}$.

Depending on whether or not an adjacent criterion vector $\bar{\mathbf{z}}$ is preferred to $\mathbf{z}^{(h)}$ in Step 3, we write $\boldsymbol{\lambda}$-constraints

$\boldsymbol{\lambda}^T(\bar{\mathbf{z}} - \mathbf{z}^{(h)}) \geqq \varepsilon$ For each *yes* response
$\boldsymbol{\lambda}^T(\bar{\mathbf{z}} - \mathbf{z}^{(h)}) \leqq -\varepsilon$ For each *no* response

where ε is a small positive number. Depending on whether or not the DM likes a tradeoff vector $\bar{\mathbf{w}}$ in Steps 4 or 5, we write $\boldsymbol{\lambda}$-constraints

$\boldsymbol{\lambda}^T\bar{\mathbf{w}} \geqq \varepsilon$ For each *yes* response
$\boldsymbol{\lambda}^T\bar{\mathbf{w}} \leqq -\varepsilon$ For each *no* response

Employing the $\boldsymbol{\lambda}$-constraints to reduce $\Lambda^{(h)}$, we form $\Lambda^{(h+1)}$. Zionts and Wallenius do not write $\boldsymbol{\lambda}$-constraints for any of the *I don't know* responses.

Step 8: Find a $\boldsymbol{\lambda}^{(h+1)} \in \Lambda^{(h+1)}$. If no such $\boldsymbol{\lambda}^{(h+1)}$ exists, delete the oldest set of active $\boldsymbol{\lambda}$-constraints, update $\Lambda^{(h+1)}$, and repeat.

Active $\boldsymbol{\lambda}$-constraints are those that have not been deleted. The active $\boldsymbol{\lambda}$-constraints define the reduction of Λ. By mistake or from changes in the DM's aspirations, there may be inconsistencies in the $\boldsymbol{\lambda}$-constraints to the extent that $\Lambda^{(h+1)} = \varnothing$. In such a case, we keep deleting the oldest active $\boldsymbol{\lambda}$-constraints until there exists a $\boldsymbol{\lambda}^{(h+1)} \in \Lambda^{(h+1)}$. To obtain a good representative of $\Lambda^{(h+1)}$, we compute the *middlemost* point (the point that maximizes the minimum slack from the bounding level curves of the $\boldsymbol{\lambda}$-constraints defining $\Lambda^{(h+1)}$).

Example 7. Let $h = 1$ and $2\lambda_1 + 3\lambda_2 - \lambda_3 \geqq \varepsilon$ and $\lambda_1 - 3\lambda_3 \leqq -\varepsilon$ be the active $\boldsymbol{\lambda}$-constraints from the first iteration. With $\bar{\mathbf{z}} = (9, 7, 1)$ preferred to $\mathbf{z}^{(h)} =$

(6, 8, 2), and $\bar{\mathbf{w}} = (0, -1, 2)$ a disliked tradeoff vector, we have the following formulation as the middlemost LP for computing a $\boldsymbol{\lambda}^{(h+1)} \in \Lambda^{(h+1)}$

$$\begin{aligned}
&\max \ \{\varepsilon\} \\
&\text{s.t.} \quad 2\lambda_1 + 3\lambda_2 - \lambda_3 \geqq \varepsilon \\
&\qquad \lambda_1 \qquad\quad - 3\lambda_3 \leqq -\varepsilon \\
&\qquad 3\lambda_1 - \lambda_2 - \lambda_3 \geqq \varepsilon \\
&\qquad\qquad - \lambda_2 + 2\lambda_3 \leqq -\varepsilon \\
&\qquad \lambda_1 + \lambda_2 + \lambda_3 = 1 \\
&\qquad\qquad\qquad \lambda_i \geqq \varepsilon \quad \text{for all } i \\
&\qquad\qquad \text{all vars} \geqq 0
\end{aligned}$$

Step 9: Using $\boldsymbol{\lambda}^{(h+1)}$, solve the composite LP for $\mathbf{z}^b$.

Step 10: If $\mathbf{z}^a$ from Step 3 is null (does not exist), the DM chooses the most preferred of $\mathbf{z}^{(h)}$ and $\mathbf{z}^b$.

1. If $\mathbf{z}^b$ is the most preferred, add a $\boldsymbol{\lambda}$-constraint indicating the preference, let $\mathbf{z}^{(h+1)} = \mathbf{z}^b$, and go to Step 13.
2. If $\mathbf{z}^{(h)}$ is the most preferred, go to Step 11.

If $\mathbf{z}^a$ from Step 3 is not null (i.e., $\mathbf{z}^a$ exists), go to Step 12.

The likelihood that a $\mathbf{z}^a$ in Step 3 does not materialize increases as the number of iterations increases. If $\mathbf{z}^a$ is null, we only cycle for another iteration if $\mathbf{z}^b$ is preferred to $\mathbf{z}^{(h)}$.

Step 11: Terminate with solution $(\mathbf{z}^{(h)}, \mathbf{x}^{(h)})$. Better solutions may exist in the relative interior of an efficient facet. A search procedure (not part of the method) should be used to find the optimum.

Being an extreme point method, the Z–W algorithm cannot find an optimal solution if it lies in the interior of a facet. Perhaps some of the filtering and discretization methods of Chapter 11 could be of help. This, however, would increase the number of iterations.

Step 12: If the DM can choose between $\mathbf{z}^a$ and $\mathbf{z}^b$, add a $\boldsymbol{\lambda}$-constraint indicating the preference, let $\mathbf{z}^{(h+1)}$ designate the most preferred of the two, and go to Step 13. If the DM is unable to choose, let $\mathbf{z}^{(h+1)} = \mathbf{z}^b$ and go to Step 13.

Step 13: Nullify the existence of $(\mathbf{z}^a, \mathbf{x}^a)$, update $\Lambda^{(h+1)}$ if necessary, and let $h = h + 1$. Go to Step 2.

We only need to update $\Lambda^{(h+1)}$ if a $\boldsymbol{\lambda}$-constraint were constructed in Step 10 or Step 12. The Z–W procedure is flowcharted in Fig. 13.12.

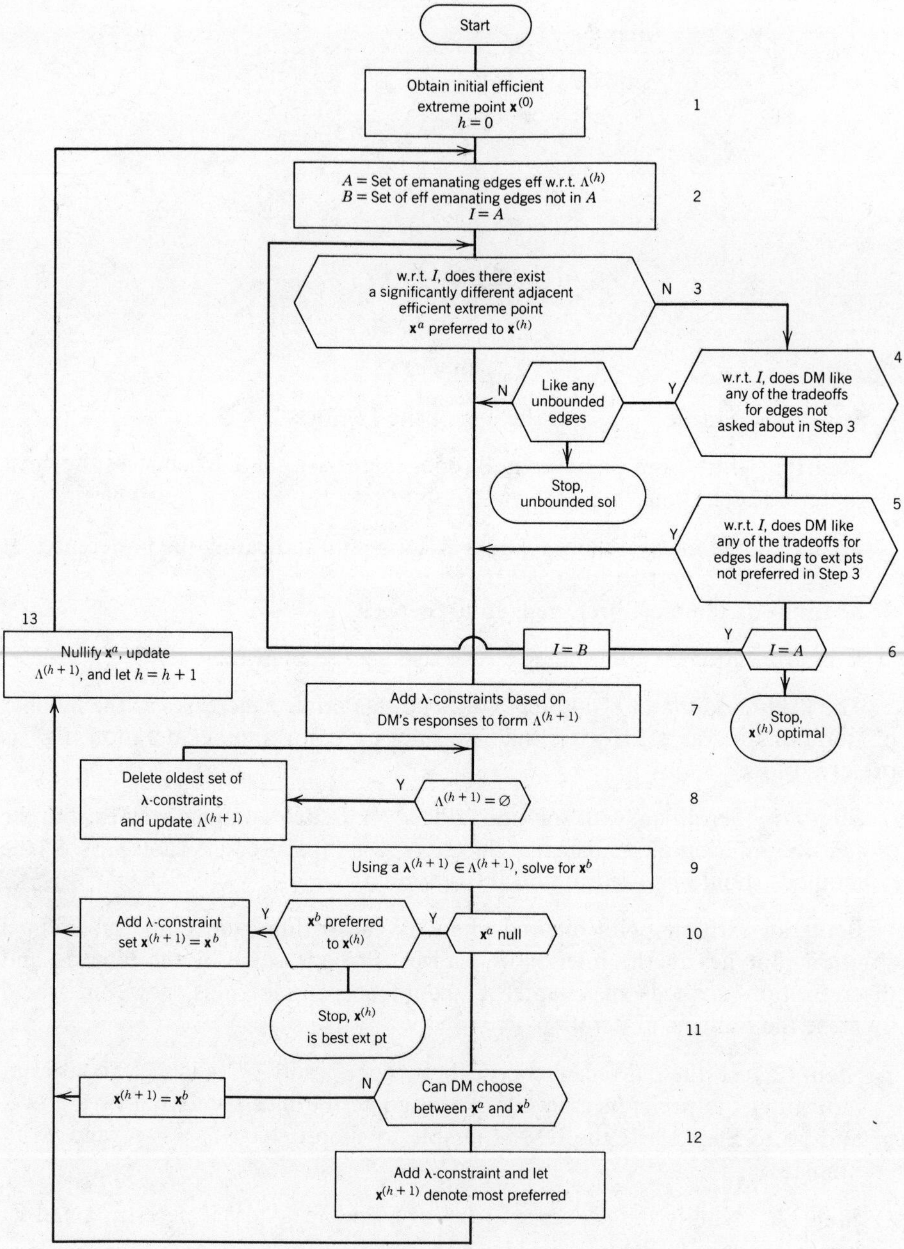

Figure 13.12. Flowchart of Z–W algorithm.

13.3.2 Z–W Numerical Example

To illustrate the Z–W algorithm, consider the MOLP

$$\begin{aligned}
\max\{\, x_1 &&&= z_1\}\\
\max\{ && x_2 &&= z_2\}\\
\max\{ &&& x_3 = z_3\}\\
\text{s.t.}\quad x_1 &+x_2 &+x_3 &\leqq 5\\
x_1 &+3x_2 &+x_3 &\leqq 9\\
3x_1 &+4x_2 &&\leqq 16\\
&& x_1, x_2, x_3 &\geqq 0
\end{aligned}$$

that is graphed in Fig. 13.13, in which the shaded facets constitute the efficient set. To answer questions posed by the algorithm, let us use $U = .45z_1 + .50z_2 + .05z_3$. Hence, $(4,1,0)$ is the optimal solution and 2.30 is the optimal U value.

1. *Step 1*. Using an arbitrary λ-vector, suppose we obtain $\mathbf{x}^{(0)} = (0,2,3)$ whose $U = 1.15$.
2. *Step 3*. With respect to $I = A$, we have three adjacent extreme points. We have extreme points $(0,3,0)$ whose $U = 1.50$, and $(\frac{8}{3}, 2, \frac{1}{3})$ whose $U = 2.22$, preferred to $\mathbf{x}^{(0)}$. But $\mathbf{x}^{(0)}$ is preferred to $(0,0,5)$ whose $U = 0.25$. Designate $\mathbf{x}^a = (0,3,0)$.
3. *Step 7*. Restrict $\Lambda^{(0)}$ with the following λ-constraints

$$\left\langle\begin{array}{r} \lambda_2 - 3\lambda_3 \geqq \varepsilon \\ \frac{8}{3}\lambda_1 \qquad - \frac{8}{3}\lambda_3 \geqq \varepsilon \\ -2\lambda_2 + 2\lambda_3 \leqq -\varepsilon \end{array}\right\rangle \Leftrightarrow \left\langle\begin{array}{r} 3\lambda_1 + 4\lambda_2 \geqq 3 + \varepsilon \\ 2\lambda_1 + \lambda_2 \geqq 1 + \frac{3}{8}\varepsilon \\ \lambda_1 + 2\lambda_2 \geqq 1 + \dfrac{\varepsilon}{2} \end{array}\right\rangle$$

   A parametric diagram for the MOLP is given in Fig. 13.14, in which the shaded area is reduced weighting vector space $\Lambda^{(1)}$ after the above λ-constraints have been applied.
4. *Step 9*. If we use $\lambda^{(1)} = (.25, .65, .10)$ as a representative of $\Lambda^{(1)}$, the weighted-sums LP yields $\mathbf{x}^b = (\frac{12}{5}, \frac{11}{5}, 0)$ whose $U = 2.18$.
5. *Step 12*. Since $\mathbf{x}^b$ is preferred to $\mathbf{x}^a$, we set $\mathbf{x}^{(1)} = (\frac{12}{5}, \frac{11}{5}, 0)$ and create the λ-constraint.

$$\lambda^T(\mathbf{z}^b - \mathbf{z}^a) \geqq \varepsilon \quad \Leftrightarrow \quad 3\lambda_1 - \lambda_2 \geqq \tfrac{5}{4}\varepsilon$$

6. *Step 13*. Updating $\Lambda^{(1)}$, we now have $\Lambda^{(1)} = \gamma(\lambda^1, \lambda^2, \lambda^3)$.
7. *Step 3*. With reference to $\mathbf{x}^{(1)}$ and $I = A$, there is only one adjacent extreme point efficient w.r.t. $\Lambda^{(1)}$. It is $(\frac{8}{3}, 2, \frac{1}{3})$ whose $U = 2.22$ and it is preferred to $\mathbf{x}^{(1)}$. Extreme point $(4,1,0)$ is *not* adjacent to $\mathbf{x}^{(1)}$ according to Step 2 because the edge connecting $\mathbf{x}^{(1)}$ and $(4,1,0)$ is not efficient. Designate $\mathbf{x}^a = (\frac{8}{3}, 2, \frac{1}{3})$.

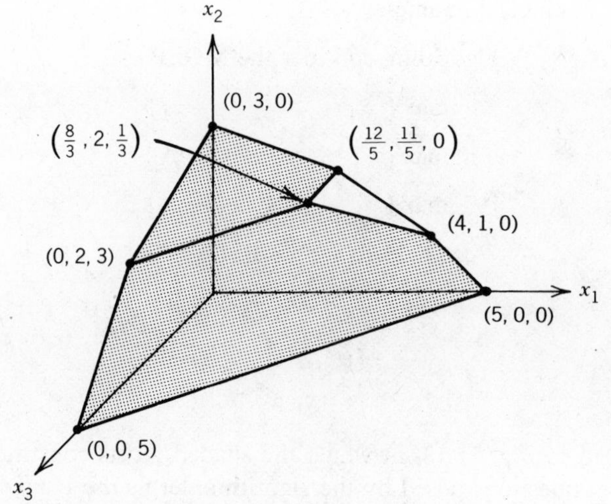

Figure 13.13. Graph of Z–W example.

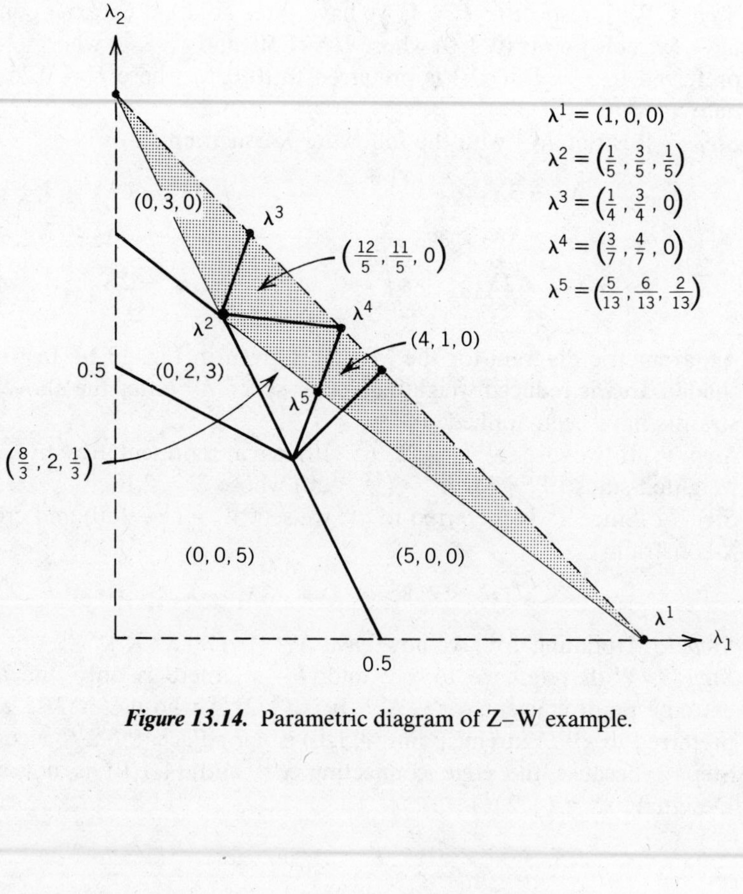

Figure 13.14. Parametric diagram of Z–W example.

8. *Step 7.* Add the λ-constraint

$$\tfrac{4}{15}\lambda_1 - \tfrac{1}{5}\lambda_2 + \tfrac{1}{3}\lambda_3 \geqq \varepsilon \quad \Leftrightarrow \quad \lambda_1 + 8\lambda_2 \leqq 5 - 15\varepsilon$$

Thus, $\Lambda^{(2)} = \gamma(\lambda^1, \lambda^2, \lambda^4)$.

9. *Step 9.* If we use $\lambda^{(2)} = (.45, .50, .05)$ as a representative of $\Lambda^{(2)}$, the weighted-sums LP yields $\mathbf{x}^b = (4, 1, 0)$ whose $U = 2.30$.

10. *Step 12.* Since $\mathbf{x}^b$ is preferred to $\mathbf{x}^a$, we set $\mathbf{x}^{(2)} = (4, 1, 0)$ and create the λ-constraint

$$\tfrac{4}{3}\lambda_1 - \lambda_2 - \tfrac{1}{3}\lambda_3 \geqq \varepsilon \quad \Leftrightarrow \quad 5\lambda_1 - 2\lambda_2 \geqq 1 + 3\varepsilon$$

11. *Step 13.* Updating $\Lambda^{(2)}$, we now have $\Lambda^{(2)} = \gamma(\lambda^1, \lambda^4, \lambda^5)$.

12. *Step 2.* With reference to $\mathbf{x}^{(2)}$, the only adjacent extreme point efficient w.r.t. $\Lambda^{(2)}$ is $(5, 0, 0)$ whose $U = 2.25$. An $\mathbf{x}^a$ does not exist.

13. *Step 6.* Set $I = B$.

14. *Step 3.* No efficient extreme points adjacent to $\mathbf{x}^{(2)}$ are preferred to $\mathbf{x}^{(2)}$.

15. *Step 6.* Stop with $\mathbf{x}^{(2)} = (4, 1, 0)$ as the final solution.

13.3.3 Determining Nonbasic Variables Efficient w.r.t. $\Lambda^{(h)}$

With reference to $\mathbf{x}^{(h)}$ in Step 3, we can determine which nonbasic variables are efficient w.r.t. $\Lambda^{(h)}$ by employing the Zionts-Wallenius routine of Section 9.7.4. The equations of the linear system are of two types. The equations of the first type are formed in the usual way from the nonbasic columns of the reduced cost matrix $\mathbf{W}$. These equations are of the form

$$\lambda^T w_j + v_j = 0$$

The equations of the second type represent the active λ-constraints. They are of the form

$$\lambda^T(\hat{\mathbf{z}} - \bar{\mathbf{z}}) + v_j = 0 \qquad \text{If } \bar{\mathbf{z}} \text{ is preferred to } \hat{\mathbf{z}}$$

$$-\lambda^T\bar{\mathbf{w}} + v_j = 0 \qquad \text{If } \bar{\mathbf{w}} \text{ is a desirable tradeoff vector}$$

$$\lambda\bar{\mathbf{w}} + v_j = 0 \qquad \text{If } \bar{\mathbf{w}} \text{ is an undesirable tradeoff vector}$$

Subject to the equations of the second type, the goal is to determine the efficiency status of each v_j in the equations of the first type. Hence, in Step 2 of the Zionts-Wallenius routine, only the v_j variables pertaining to the first group of equations are designated *status unknown* in the initial tableau.

Let us illustrate the application of the Zionts-Wallenius routine by computing the nonbasic variables that are efficient w.r.t. $\Lambda^{(1)}$ in the Z–W example of the previous section. The columns of the reduced cost matrix at $\mathbf{x}^{(1)} = (\tfrac{12}{5}, \tfrac{11}{5}, 0)$ pertaining to the three adjacent extreme points $(0, 3, 0)$, $(\tfrac{8}{3}, 2, \tfrac{1}{3})$ and $(4, 1, 0)$ yield the first group of equations

$$\begin{aligned} -3\lambda_1 + \lambda_2 \qquad\quad + v_1 &= 0 \\ 4\lambda_1 - 3\lambda_2 + 5\lambda_3 + v_2 &= 0 \\ 4\lambda_1 - 3\lambda_2 \qquad\quad + v_3 &= 0 \end{aligned}$$

The DM's previous responses yield the second group of equations

$$\begin{aligned} -\lambda_2 + 3\lambda_3 + v_4 &= 0 \\ -\lambda_1 \qquad + \lambda_3 + v_5 &= 0 \\ -\lambda_2 + \lambda_3 + v_6 &= 0 \\ -3\lambda_1 + \lambda_2 \qquad + v_7 &= 0 \end{aligned}$$

Only designating v_1, v_2, and v_3 status unknown, we have the following.

1st Iteration (Step 9)

	λ_1	λ_2	λ_3
v_1	-3	1	
v_2	4	-3	5
v_3	4	-3	
v_4		-1	3
v_5	-1		1
v_6		-1	1
v_7	-3	$\boxed{1}$	

2nd Iteration (Steps 3, 4, and 9)

	λ_1	v_7	λ_3	
Ineff v_1		-1		X
v_2	-5	3	5	
v_3	-5	$\boxed{3}$		
v_4	-3	1	3	
v_5	-1		1	
v_6	-3	1	1	
λ_2	-3	1		

3rd Iteration (Steps 5 and 9)

	λ_1	Ineff v_3	λ_3
v_2		-1	5
v_7	$-\frac{5}{3}$	$\frac{1}{3}$	
v_4	$-\frac{4}{3}$	$-\frac{1}{3}$	$\boxed{3}$
v_5	-1		1
v_6	$-\frac{4}{3}$	$-\frac{1}{3}$	1
λ_2	$-\frac{4}{3}$	$-\frac{1}{3}$	

4th Iteration (Step 7)

	λ_1	Ineff v_3	v_4
Eff v_2	$\frac{20}{9}$	$-\frac{4}{9}$	$-\frac{5}{3}$
v_7	$-\frac{5}{3}$	$\frac{1}{3}$	
λ_3	$-\frac{4}{9}$	$-\frac{1}{9}$	$\frac{1}{3}$
v_5	$-\frac{5}{9}$	$\frac{1}{9}$	$-\frac{1}{3}$
v_6	$-\frac{8}{9}$	$-\frac{2}{9}$	$-\frac{1}{3}$
λ_2	$-\frac{4}{9}$	$-\frac{1}{3}$	

13.3.4 Z–W Comments

1. The Z–W algorithm is not an ad hoc procedure because the questions posed by the algorithm can be answered deterministically when one knows the decision maker's utility function.
2. The Z–W method can be very effective in rapidly reducing weighting vector space in the early iterations. However, the method cannot converge to nonextreme final solutions.
3. Some decision makers may feel uncomfortable responding to the tradeoff vectors of Steps 4 and 5.
4. The Z–W algorithm has an error-correcting capability in that it deletes its oldest λ-constraints when inconsistencies occur.
5. As indicated by its flowchart, the Z–W method is more complex than STEM or the GDF procedure.

13.4 INTERVAL CRITERION WEIGHTS / VECTOR-MAXIMUM APPROACH

Introduced by Steuer (1977), the interval criterion weights/vector-maximum approach is a *criterion cone reduction* method for solving the MOLP

$$\begin{aligned} &\max \{ \mathbf{c}^1\mathbf{x} = z_1 \} \\ &\max \{ \mathbf{c}^2\mathbf{x} = z_2 \} \\ &\quad\vdots \\ &\max \{ \mathbf{c}^k\mathbf{x} = z_k \} \\ &\text{s.t.} \quad \mathbf{x} \in S \end{aligned}$$

The purpose of this procedure is to locate the efficient extreme point of greatest utility. The interactive procedure samples the nondominated set and then presents a group of nondominated criterion vectors to the decision maker at each iteration. From each group presented, the decision maker must indicate which criterion vector he or she most prefers.

The criterion vectors presented to the decision maker are generated by maximizing a series of LPs. The objective function gradients of these LPs are formed by taking particular convex combinations of the MOLP objective function gradients. By knowing which criterion vectors the decision maker most prefers, we then know, in an a posteriori sense, which convex combinations of the MOLP objective function gradients most closely correspond to his or her preferences. Then, by successively forming more concentrated convex combinations about the previously selected convex combination, the interactive procedure "focuses in" on the decision maker's final solution. The algorithm is designed to terminate within a fixed number of iterations presenting a fixed number of solutions at each iteration.

13.4.1 Interval Criterion Weights Algorithm

Step 1: To calibrate the algorithm, specify the sample size P and the number of iterations t.

Parameter P is the (prechosen) fixed number of criterion vectors to be presented to the decision maker at each iteration, and t is the (prechosen) fixed number of iterations the procedure is to run. Parameters P and t are negotiated with the decision maker. It is suggested that

$$k \leqq P < 2k + 1$$
$$t \approx k$$

where " $\approx$ " means approximately and k is the number of objectives.

Step 2: Form $\mathbf{C}_1 = \overline{\mathbf{N}}\mathbf{C}$ where $\mathbf{C}$ is the $k \times n$ original criterion matrix whose rows are the $\mathbf{c}^i$, and $\overline{\mathbf{N}}$ is the $k \times k$ diagonal matrix whose diagonal elements are the factors by which the objectives are rescaled (normalized).

Because there may be magnitudes of difference among the z-values generated by the different objectives, the objectives are rescaled (normalized). Rather than use one of the traditional L_p-norms, multiplying each of the objectives by an appropriate power of 10 has been found to be a convenient alternative. With this approach, the only change made to the objective function coefficients is to move the decimal point.

Step 3: Let iteration counter $h = 1$ and form the $2k + 1$ convex combination weighting vectors

$$\begin{aligned}
\lambda^1 &= (1, 0, \ldots, 0) \\
\lambda^2 &= (0, 1, \ldots, 0) \\
&\vdots \\
\lambda^k &= (0, 0, \ldots, 1) \\
\lambda^{k+1} &= (1/k^2, r, \ldots, r) \\
\lambda^{k+2} &= (r, 1/k^2, \ldots, r) \\
&\vdots \\
\lambda^{2k} &= (r, r, \ldots, 1/k^2) \\
\lambda^{2k+1} &= (1/k, 1/k, \ldots, 1/k)
\end{aligned}$$

where $r = (k + 1)/k^2$.

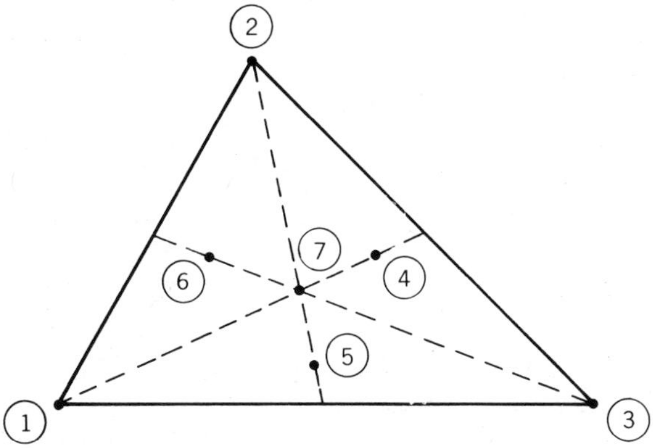

Figure 13.15. Dispersion of convex combination trial gradients.

In practice, a very small but computationally significant value is used in place of the zeros in the weighting vectors to prevent the generation of dominated criterion vectors.

Step 4: Letting $\mathbf{C}_h$ be the $k \times n$ matrix whose rows specify generators for the hth enveloping reduced criterion cone, solve the $2k + 1$ convex combination LPs

$$\max\left\{(\boldsymbol{\lambda}^i)^T \mathbf{C}_h \mathbf{x} \,|\, \mathbf{x} \in S\right\} \quad i = 1, \ldots, 2k + 1$$

where

$$\mathbf{C}_h = \begin{cases} \mathbf{T}_{h-1}\mathbf{T}_{h-2} \cdots \mathbf{T}_1\mathbf{C}_1 & \text{when } h > 1 \\ \mathbf{C}_1 & \text{when } h = 1 \end{cases}$$

On the first iteration, we work with the original criterion cone. On subsequent iterations, we work with *enveloping reduced* subset criterion cones. Details about the *premultiplication* **T**-matrices by which the enveloping reduced criterion cones are created are given in Section 9.10.3.

When applied to the generators of a criterion cone, the weighting vectors create a group of trial gradients radially dispersed over the criterion cone. If we consider the cross-section of a three-dimensional criterion cone in Fig. 13.15, the circled numbers pertain to the convex combination LP gradients resulting from the $2k + 1$ $\boldsymbol{\lambda}$-vectors.

Step 5: Filter the resulting criterion vectors to obtain the P most different. Constituting the sample of iteration h, present the P nondominated criterion vectors to the decision maker.

Because of the possibility of alternative optima and because a given extreme point may maximize more than one of the convex combination LPs, the number of different criterion vectors resulting from Step 4 is not precisely predictable. We do, however, expect the number of different criterion vectors to decrease somewhat as the iterative process proceeds and the trial gradients become more concentrated. Hence, it is advisable for P to be less than $2k + 1$.

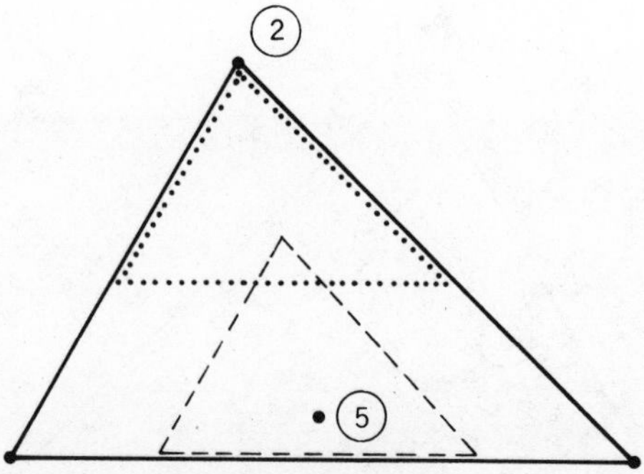

Figure 13.16. Cross-sections of reduced criterion cones.

Step 6: From the sample of P nondominated criterion vectors, the decision maker selects his or her most preferred, designating it $\mathbf{z}^{(h)}$.

Step 7: Contract the hth iteration criterion cone about the trial gradient that produced $\mathbf{z}^{(h)}$ to form the $(h+1)$th iteration reduced criterion cone.

Each iteration the current enveloping reduced criterion cone is reduced to a subset enveloping reduced criterion cone whose cross-section is $(1/k)$th the cross-section of the current cone. The reduced cone is centered (as well as possible) about the convex combination that produced $\mathbf{z}^{(h)}$ in Step 4. To illustrate, consider Figs. 13.15 and 13.16. Suppose trial gradient ⑤ produced $\mathbf{z}^{(h)}$. Then, the cross-section of the newly reduced cone would be as indicated by the dashed lines in Fig. 13.16. Had $\mathbf{z}^{(h)}$ been produced by ②, the cross-section of the new cone would be as indicated by the dotted lines.

Step 8: Let $h = h + 1$.

Step 9: If $h \leqq t - 1$, go to Step 4. Otherwise, go to Step 10.

Step 10: Solve for all extreme points efficient w.r.t. $\mathbf{C}_t$ in the enveloping reduced vector-maximum problem

$$\text{eff}\{\mathbf{C}_t\mathbf{x} \mid \mathbf{x} \in S\}$$

In this step, we compute all efficient extreme points in a neighborhood about the inverse image of $\mathbf{z}^{(t-1)}$. Since the cross-section of the original criterion cone has been reduced $t-1$ times, the reduced cone of the tth iteration is "thin" in relation. Consequently, the number of efficient extreme points generated should only be a small fraction of the total number of efficient extreme points.

Step 11: Filter the criterion vectors of the efficient extreme points generated in Step 10 to obtain the P most different.

Step 12: From the P nondominated criterion vectors, the decision maker selects his or her most preferred, designating it $\mathbf{z}^{(t)}$ and its inverse image $\mathbf{x}^{(t)}$. Stop with $(\mathbf{z}^{(t)}, \mathbf{x}^{(t)})$ as the final solution.

The interval criterion weights/vector-maximum procedure is flowcharted in Fig. 13.17.

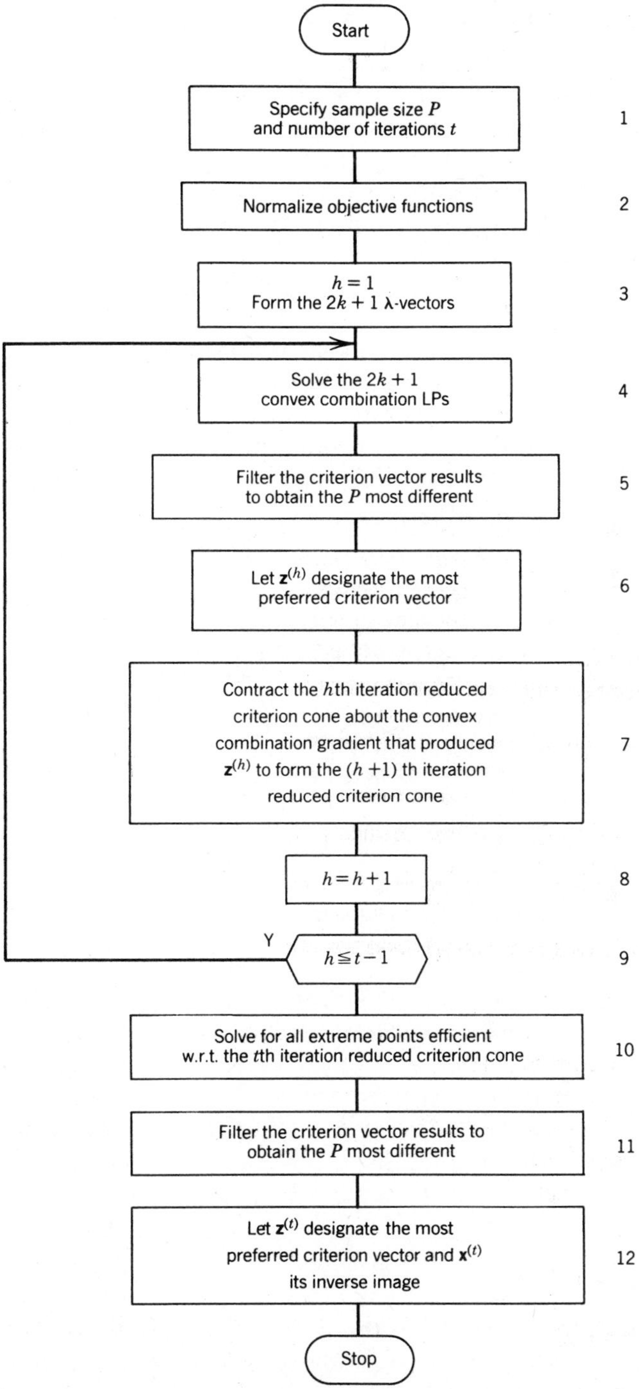

Figure 13.17. Flowchart of interval criterion weights/vector-maximum procedure.

13.4.2 Interval Criterion Weights Comments

1. The algorithm is not an ad hoc procedure.
2. The algorithm possesses definitiveness in the sense that it is predetermined how may iterations the procedure is to run and how many solutions are to be presented to the DM at each iteration.
3. Because of the dispersed sampling of the efficient set, the algorithm possesses a *multiple starting point* property that enables the algorithm to deal with nonconcave utility functions.
4. The algorithm's CPU time requirements are much greater than the three previous algorithms.
5. The algorithm only searches among extreme points to locate the efficient extreme point of greatest utility. However, there is no guarantee that the algorithm will conclude with the efficient extreme point of greatest utility.
6. Because the algorithm uses a vector-maximum capability (on the last iteration), the size of MOLP that can be solved is limited.

13.5 INTERACTIVE WEIGHTED-SUMS / FILTERING APPROACH

Employed in Steuer (1980) and Steuer and Wood (1986), the interactive weighted-sums/filtering approach is a *weighting vector space reduction* method for solving the MOLP

$$\begin{aligned} &\max\,\{\mathbf{c}^1\mathbf{x} = z_1\} \\ &\max\,\{\mathbf{c}^2\mathbf{x} = z_2\} \\ &\quad\vdots \\ &\max\,\{\mathbf{c}^k\mathbf{x} = z_k\} \\ &\text{s.t.} \qquad \mathbf{x} \in S \end{aligned}$$

The approach is similar in philosophy to the interval criterion weights/vector-maximum method. That is, the method strives to locate the efficient extreme point of greatest utility in a fixed number of iterations, presenting a fixed number of solutions per iteration. Likewise, the method makes no assumptions about the decision maker's utility function except that it be coordinatewise increasing in R^k. Instead of reducing the criterion cone, the procedure reduces weighting vector space $\Lambda = \{\boldsymbol{\lambda} \in R^k \mid \lambda_i > 0,\ \Sigma_{i=1}^k \lambda_i = 1\}$. Another difference is that predetermined convex combination $\boldsymbol{\lambda}$-vectors are not used. They are randomly generated instead.

13.5.1 Calibration

The interactive procedure samples the set of nondominated criterion vectors by solving batches of weighted-sums LPs whose weighting vectors come from

$$\Lambda^{(h)} = \left\{\boldsymbol{\lambda} \in R^k \mid \lambda_i \in \text{rel}\left[\ell_i^{(h)}, \mu_i^{(h)}\right],\ \ \sum_{i=1}^{k} \lambda_i = 1\right\}$$

where the $[\ell_i^{(h)}, \mu_i^{(h)}]$ bounds are iteratively tightened as the iteration count h increases. To enable the interactive procedure to focus in with sufficient resolution on a final solution within t iterations, it is also necessary to calibrate the algorithm with values for

$P =$ Iteration sample size.
$w =$ Final iteration $[\ell_i, \mu_i]$ interval width.
$r =$ Λ-reduction factor.

Parameters t and P are negotiated with the decision maker, as in the interval criterion weights/vector-maximum procedure. However, based on experience, the analyst selects a value for the final interval width w (where $w = [\ell_i^{(t)}, \mu_i^{(t)}]$ for all i). At the analyst's discretion, a value for w between $1/(2k)$ and $3/(2k)$ should normally suffice. Then, in connection with w, the analyst fits a value for the Λ-reduction factor r. It is suggested that all this be done in rough accordance with the following *guideline relationships*

$$P \gtrapprox k$$
$$t \approx k$$
$$1/(2k) \lessapprox w \lessapprox 3/(2k)$$
$$\sqrt[k]{1/P} \lessapprox r \lessapprox \sqrt[t-1]{w}$$

where " $\approx$ " means approximately.

Motivation for the fourth relationship is drawn from the k-dimensional hypercube. Suppose we obtain P samples from the unit hypercube and that each sample characterizes a subset hypercube whose volume is greater than or equal to $1/P$. If we allow r to be the subset hypercube's length on a side, $r^k \geqq 1/P$ which yields

$$r \geqq \sqrt[k]{1/P}$$

If we employ r as a reduction factor in order to iterate to a $[\ell_i, \mu_i]$ interval width that is less than or equal to w on the tth iteration, $r^{t-1} \leqq w$ and this yields

$$r \leqq \sqrt[t-1]{w}$$

Hence, we have

$$\sqrt[k]{1/P} \leqq r \leqq \sqrt[t-1]{w}$$

Although Λ is not a hypercube, the above relationship is used as a guide in selecting r in conjunction with the other three guideline relationships. To illustrate, we have Example 9.

Example 9. Let $k = 6$, $P = 8$, and $t = 6$. If we select $w = .20$, we can fit a value for r of .710 because

$$\sqrt[k]{1/P} = .707 \leqq .725 = \sqrt[t-1]{w}$$

From the guideline relationships we have (a) the larger P, the fewer iterations that are required; (b) the smaller r, the faster we reduce weighting vector space;

and (c) the narrower w, the more pressure there is for P and t to be large. Note that the guideline relationships, because they are based on the unit hypercube, are *only* suggestions; they need not be followed strictly.

13.5.2 Interactive Weighted-Sums Algorithm

Step 1: To calibrate the algorithm, specify the sample size P, the number of iterations t, the final $[\ell_i, \mu_i]$ interval width w, and the Λ-reduction factor r.

Step 2: Rescale (normalize) the objective functions.

It is probably most convenient to rescale the objectives using an appropriate power of 10 as suggested in Step 2 of the interval criterion weights/vector-maximum algorithm.

Step 3: Let $h = 0$. Let $[\ell_i^{(1)}, \mu_i^{(1)}] = [0, 1]$ for all i.

Step 4: Let $h = h + 1$. Form

$$\Lambda^{(h)} = \left\{ \lambda \in R^k \mid \lambda_i \in \text{rel}\left[\ell_i^{(h)}, \mu_i^{(h)}\right], \sum_{i=1}^{k} \lambda_i = 1 \right\}$$

On the first iteration, $\Lambda^{(1)} = \Lambda$. On subsequent iterations, $\Lambda^{(h)} \subset \Lambda$ because the $\text{rel}[\ell_i^{(h)}, \mu_i^{(h)}]$ are subintervals of the open unit interval.

Step 5: Randomly generate $50 \times k$ weighting vectors from $\Lambda^{(h)}$.

Step 6: Filter the $50 \times k$ weighting vectors to obtain the $3 \times P$ most different.

The idea of Step 5 is to create a pool of λ-vectors so that after filtering in Step 6, the resulting λ-vectors provide an evenly dispersed finite covering of $\Lambda^{(h)}$. To randomly generate and filter the λ-vectors, we use the `LAMBDA` and `FILTER` programs of Chapter 11. The numbers $50 \times k$ and $3 \times P$ are suggested as "rules of thumb." They can be altered by the analyst if so desired.

Step 7: Using the $3P$ weighting vector representatives of $\Lambda^{(h)}$, solve the $3P$ associated weighted-sums LPs.

Since the components of the λ-vectors come from subintervals of the open unit interval, all maximizing extreme points are efficient. Hence, all associated criterion vectors are nondominated.

Step 8: Filter the criterion vectors resulting from Step 7 to obtain the P most different. Then, present the P nondominated criterion vectors to the decision maker.

Because a given efficient extreme point may, particularly on the later iterations, maximize more than one of the weighted-sums LPs of Step 7, it becomes necessary to solve more than P weighted-sums LPs. Another reason for solving more than P LPs is that the resulting criterion vectors may not be well dispersed even though the λ-vectors are.

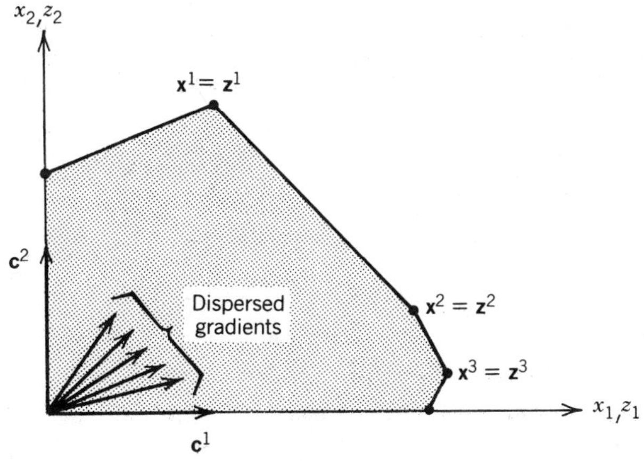

Figure 13.18. Graph of Example 10.

Example 10. Suppose we have five representatives from $\Lambda^{(h)}$ that give the five dispersed weighted-sums objective function gradients in Fig. 13.18. Although the five weighted-sums gradients are dispersed, only three ($\mathbf{z}^1$, $\mathbf{z}^2$, and $\mathbf{z}^3$) distinct criterion vectors are produced and they are not well dispersed.

Step 9: From the sample of P nondominated criterion vectors, the decision maker selects his or her most preferred, designating it $\mathbf{z}^{(h)}$.

Step 10: If $h < t$, go to Step 11. Otherwise, go to Step 12.

Step 11: Let $\lambda^{(h)}$ be the weighting vector that generated $\mathbf{z}^{(h)}$ via the weighted-sums LP in Step 7. Form

$$\Lambda^{(h+1)} = \left\{ \lambda \in R^k \mid \lambda_i \in \text{rel}\left[\ell_i^{(h+1)}, \mu_i^{(h+1)}\right], \quad \sum_{i=1}^{k} \lambda_i = 1 \right\}$$

where

$$\left[\ell_i^{(h+1)}, \mu_i^{(h+1)}\right] = \begin{cases} [0, r^h] \ldots & \text{if } \lambda_i^{(h)} - \dfrac{r^h}{2} \leqq 0 \\ [1 - r^h, 1] \ldots & \text{if } \lambda_i^{(h)} + \dfrac{r^h}{2} \geqq 1 \\ \left[\lambda_i^{(h)} - \dfrac{r^h}{2}, \lambda_i^{(h)} + \dfrac{r^h}{2}\right] \ldots & \text{otherwise} \end{cases}$$

in which r^h is the reduction factor r raised to the hth power. Go to Step 4.

Step 12: With $\mathbf{x}^{(h)}$ the inverse image of $\mathbf{z}^{(h)}$, stop with $(\mathbf{z}^{(h)}, \mathbf{x}^{(h)})$ as the final solution.

The interactive weighted-sums/filtering procedure is flowcharted in Fig. 13.19.

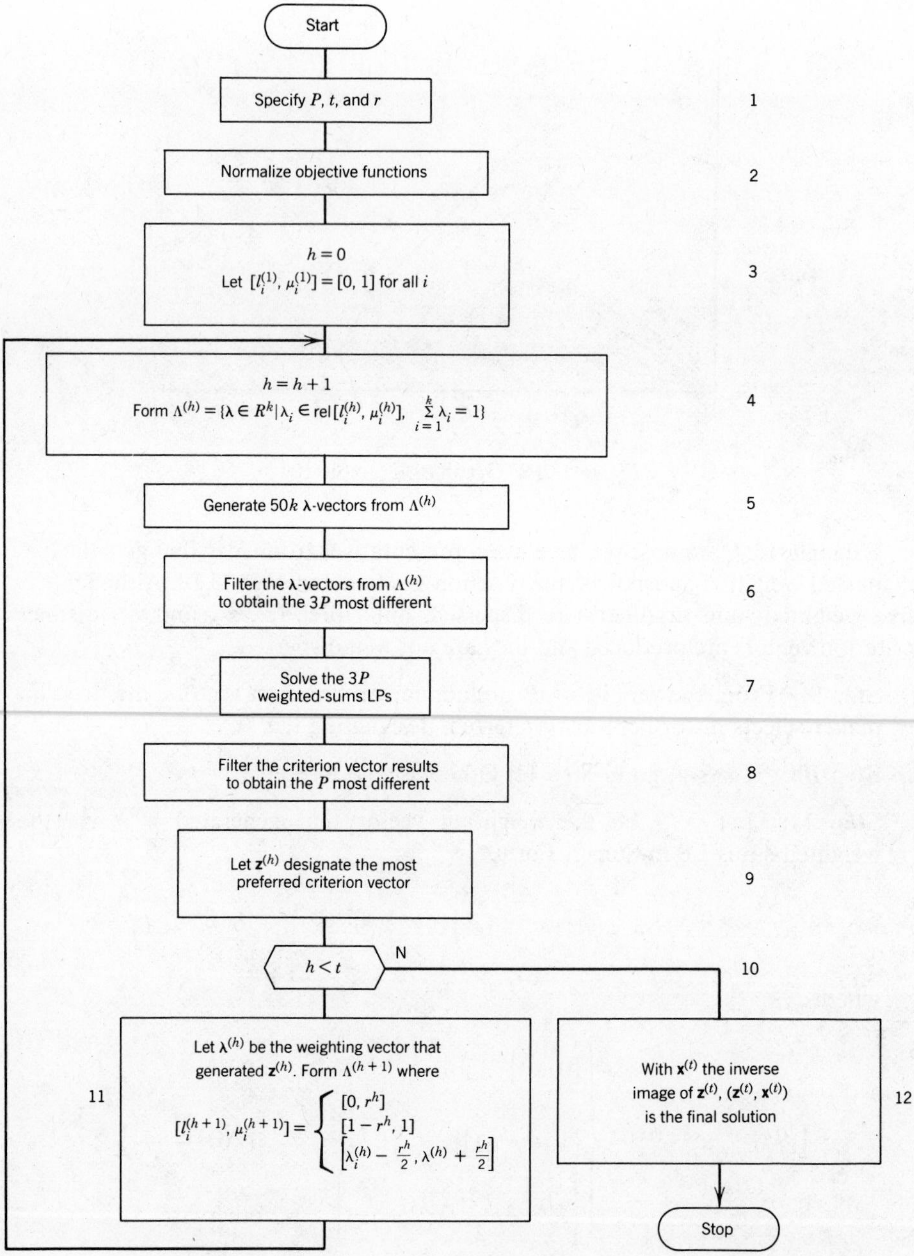

Figure 13.19. Flowchart of interactive weighted-sums/filtering procedure.

13.5.3 Interactive Weighted-Sums Comments

1. The algorithm possesses the same definitiveness as the interval criterion weights procedure in that the number of iterations and the number of solutions to be presented at each iteration are prechosen.
2. Similarly, the algorithm possesses a multiple starting point capability and requires considerable CPU time because of the number of optimizations that must be performed each iteration.
3. Because the algorithm presumes the use of a commercial LP package as its workhorse software, the size of the MOLP that can be solved is not limited by the multiple criteria nature of the problem. It is only limited by the size of problem that can be handled by the commercial package.
4. Since the algorithm only searches among extreme points for the efficient extreme point of greatest utility, the interactive weighted-sums approach is applicable to network problems. On such problems, it would be appropriate to use a network code in place of an LP code to perform the optimizations of each iteration.
5. Because of the way weighting vector space is contracted, the algorithm has an error-correcting capability. That is, the algorithm can adjust to changes in the decision maker's aspirations during the course of the solution procedure as long as they are not too severe. Otherwise, an algorithm restart is advised.

13.6 VISUAL INTERACTIVE APPROACH OF KORHONEN AND LAAKSO

The idea of this method [as described in Korhonen and Laakso (1985)] is to iteratively project an unbounded line segment in criterion space onto the non-dominated surface of Z. The projection onto the nondominated surface is obtained by using an *achievement scalarizing function* (see Section 13.6.1) and RHS parametric programming. The method anticipates the use of computer graphics in the iterative display of solution information. The method is designed for the following linear and nonlinear multiple objective programs

$$
\begin{aligned}
&\max \{f_1(\mathbf{x}) = z_1\} \\
&\max \{f_2(\mathbf{x}) = z_2\} \\
&\quad\vdots \\
&\max \{f_k(\mathbf{x}) = z_k\} \\
&\text{s.t.} \qquad \mathbf{x} \in S
\end{aligned}
$$

If the decision maker's utility function is pseudoconcave, and S is formed by linear constraints and is bounded, a necessary and sufficient condition exists for determining whether a given point is optimal.

13.6.1 Achievement Scalarizing Functions

Let a given $\bar{\mathbf{q}} \in R^k$ be denoted a *reference point* in criterion space. The reference point may or may not be feasible (i.e., be a member of Z). Also, let $\boldsymbol{\lambda}$ be a given vector from Λ. Then, an *achievement scalarizing function* $s(\bar{\mathbf{q}}, \mathbf{z}, \boldsymbol{\lambda})$ is one that projects the reference point $\bar{\mathbf{q}}$ onto the set of nondominated criterion vectors N. The term "achievement scalarizing function" has been taken from the research of Wierzbicki (1982 and 1983).

The projected point in N is defined by the criterion vector that lies on the lowest valued contour of $s(\bar{\mathbf{q}}, \mathbf{z}, \boldsymbol{\lambda})$ that intersects (i.e., is "tangent" to) Z. Such a criterion vector is obtained by solving the *achievement scalarizing program*

$$\min \{ s(\bar{\mathbf{q}}, \mathbf{z}, \boldsymbol{\lambda}) \}$$
$$\text{s.t.} \quad \mathbf{z} \in Z$$

The achievement scalarizing function used in this book [others are discussed in Wierzbicki (1982 and 1983)] is

$$s(\bar{\mathbf{q}}, \mathbf{z}, \boldsymbol{\lambda}) = \max_{i=1,\dots,k} \{ \lambda_i(\bar{q}_i - z_i) \} - \varepsilon \sum_{i=1}^{k} z_i$$

where ε is a very small positive scalar. With this function, the achievement scalarizing program is formulated as

$$\begin{aligned} &\min \{ \alpha - \varepsilon \sum_{i=1}^{k} z_i \} \\ &\text{s.t.} \quad \alpha \geqq \lambda_i(\bar{q}_i - z_i) \qquad 1 \leqq i \leqq k \\ &\qquad\quad f_i(\mathbf{x}) = z_i \qquad 1 \leqq i \leqq k \\ &\qquad\quad \mathbf{x} \in S \end{aligned}$$

in which $\alpha \in R$ and $\mathbf{z} \in R^k$ are, in general, unrestricted in sign.

Example 11. Consider the MOLP

$$\begin{aligned} &\max \{ x_1 \qquad = z_1 \} \\ &\max \{ \qquad x_2 = z_2 \} \\ &\text{s.t.} \quad x_1 + x_2 \leqq 5 \\ &\qquad\quad x_1, x_2 \geqq 0 \end{aligned}$$

along with

$$s(\bar{\mathbf{q}}, \mathbf{z}, \boldsymbol{\lambda}) = \max_{i=1,2} \{ \lambda_i(\bar{q}_i - z_i) \} - .001 \sum_{i=1}^{2} z_i$$

Using the achievement scalarizing program to project $\bar{\mathbf{q}} = (1, 1)$ onto N with $\boldsymbol{\lambda} = (\frac{2}{3}, \frac{1}{3})$, we obtain $\mathbf{z}^3 = (2, 3)$ as shown in Fig. 13.20a. In this case, the value of

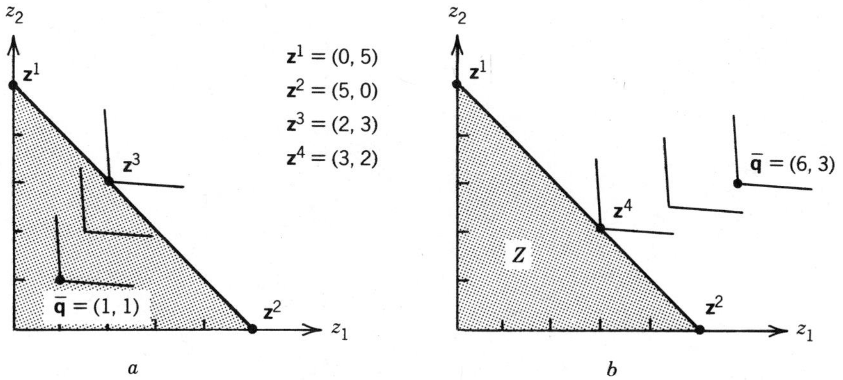

Figure 13.20. Graphs of Example 11.

the contour of s that produces $\mathbf{z}^3$ is $-.672$. Using the achievement scalarizing program to project $\bar{\mathbf{q}} = (6, 3)$ onto N with $\boldsymbol{\lambda} = (\frac{1}{4}, \frac{3}{4})$, we obtain $\mathbf{z}^4 = (3, 2)$ as shown in Fig. 13.20*b*. In this case, the value of the contour of s that produces $\mathbf{z}^4$ is .745.

13.6.2 Projecting a Line Segment onto N

Rearranging the achievement scalarizing program formulation of Section 13.6.1, we have

$$\min \{\alpha - \varepsilon \sum_{i=1}^{k} z_i\}$$

$$\text{s.t.} \quad z_i + \frac{\alpha}{\lambda_i} \geqq \bar{q}_i \qquad 1 \leqq i \leqq k$$

$$f_i(\mathbf{x}) = z_i \qquad 1 \leqq i \leqq k$$

$$\mathbf{x} \in S$$

To project onto N an unbounded line segment $\mu(\bar{\mathbf{q}}, \mathbf{d})$ that emanates from $\bar{\mathbf{q}}$ in the direction $\mathbf{d}$, we solve the *achievement scalarizing parametric program*

$$\min \{\alpha - \varepsilon \sum_{i=1}^{k} z_i\}$$

$$\text{s.t.} \quad z_i + \frac{\alpha}{\lambda_i} \geqq \bar{q}_i + \theta d_i \qquad 1 \leqq i \leqq k$$

$$f_i(\mathbf{x}) = z_i \qquad 1 \leqq i \leqq k$$

$$\mathbf{x} \in S$$

for θ going from $0 \rightarrow \infty$. If the underlying multiple objective program is an

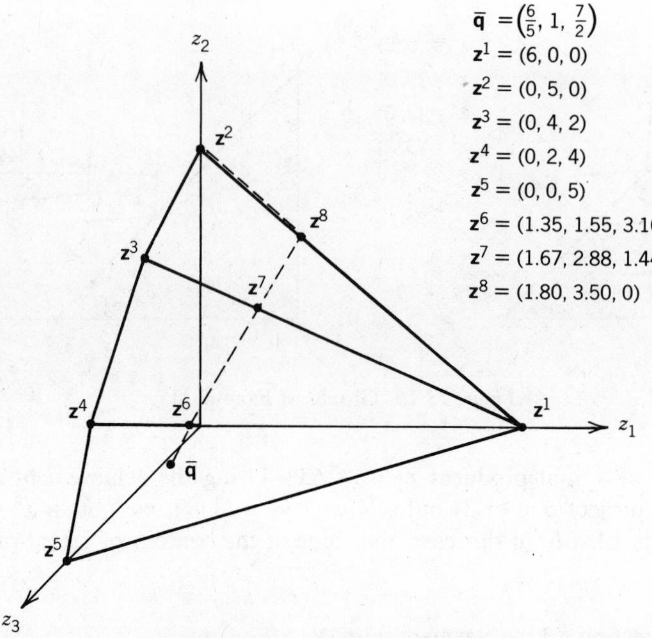

Figure 13.21. Graph of Example 12.

MOLP, the above formulation can be solved using a commercial LP package that possesses RHS parametrics.

Example 12. Consider the MOLP

$$\begin{aligned}
\max\{\ x_1 & = z_1\} \\
\max\{\ x_2 & = z_2\} \\
\max\{\ x_3 & = z_3\} \\
\text{s.t.}\quad 5x_1 + 6x_2 + 3x_3 & \leqq 30 \\
x_1 + x_2 + x_3 & \leqq 6 \\
5x_1 + 3x_2 + 6x_3 & \leqq 30 \\
x_1, x_2, x_3 & \geqq 0
\end{aligned}$$

that is graphed in Fig. 13.21. The MOLP's set of nondominated criterion vectors is

$$N = \gamma(\mathbf{z}^1, \mathbf{z}^2, \mathbf{z}^3) \cup \gamma(\mathbf{z}^1, \mathbf{z}^3, \mathbf{z}^4) \cup \gamma(\mathbf{z}^1, \mathbf{z}^4, \mathbf{z}^5)$$

Assume the achievement scalarizing function of Section 13.6.1 with $\boldsymbol{\lambda} = (\frac{5}{8}, \frac{1}{4}, \frac{1}{8})$

TABLE 13.2
Parametric Program of Example 12

	x_1	x_2	x_3	z_1	z_2	z_3	α		
Obj				$-.001$	$-.001$	$-.001$	1	min	
				1			$\frac{8}{5}$	$\geqq$	$\frac{6}{5}+3\theta$
					1		4	$\geqq$	$1+10\theta$
						1	8	$\geqq$	$\frac{7}{2}-\theta$
	1			-1				$=$	0
s.t.		1			-1			$=$	0
			1			-1		$=$	0
	5	6	3					$\leqq$	30
	1	1	1					$\leqq$	6
	5	3	6					$\leqq$	30

all $x_i \geqq 0$

and $\varepsilon = .001$. To project the unbounded line segment $\mu(\bar{\mathbf{q}}, \mathbf{d})$ where $\bar{\mathbf{q}} = (\frac{6}{5}, 1, \frac{7}{2})$ and $\mathbf{d} = (3, 10, -1)$ onto N, we solve the parametric program of Table 13.2.

The projected path on N is given by the dashed line

$$\gamma(\bar{\mathbf{q}}, \mathbf{z}^6) \cup \gamma(\mathbf{z}^6, \mathbf{z}^7) \cup \gamma(\mathbf{z}^7, \mathbf{z}^8) \cup \gamma(\mathbf{z}^8, \mathbf{z}^2)$$

The *trajectories* of criterion values along the dashed line are plotted in Fig. 13.22.

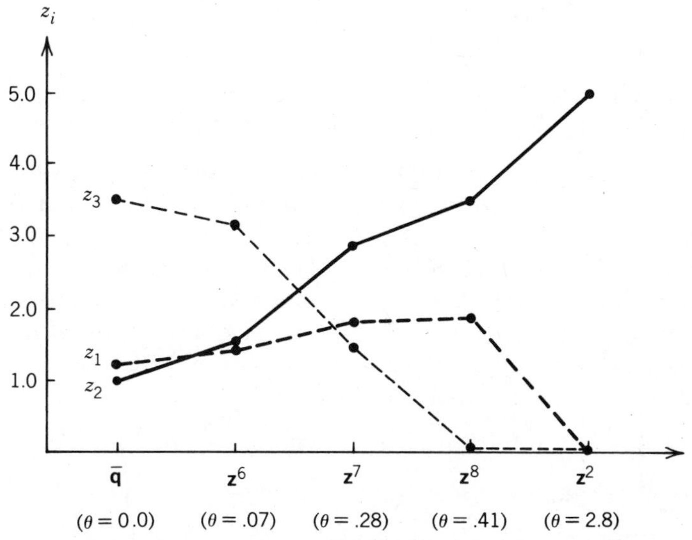

Figure 13.22. Criterion value trajectories of Example 12.

Note that kinks in the trajectories occur at the values of θ at which new bases become optimal.

13.6.3 Visual Interactive Algorithm

Step 1: Select an achievement scalarizing function $s(\bar{\mathbf{q}}, \mathbf{z}, \boldsymbol{\lambda})$ and an initial point $\mathbf{z}^{(0)}$ in criterion space. Let iteration counter $h = 0$.

Step 2: Let $h = h + 1$ and set STALL = 0. Select a reference point $\mathbf{g}^{(h)} \in R^k$. Set direction $\mathbf{d}^{(h)} = \mathbf{g}^{(h)} - \mathbf{z}^{(h-1)}$.

The initial point $\mathbf{z}^{(0)}$ in Step 1 need not be on the surface of Z or even be feasible. With regard to $\boldsymbol{\lambda} \in \Lambda$ in $s(\bar{\mathbf{q}}, \mathbf{z}, \boldsymbol{\lambda})$, Korhonen and Laakso give us no specific guidance as to its selection. STALL is a flag that is set to zero unless the current solution and the previous solution are identical. The reference point $\mathbf{g}^{(h)} \in R^k$ is to be selected according to the DM's aspirations.

Step 3: Solve the achievement scalarizing parametric program to project $\mu(\mathbf{z}^{(h-1)}, \mathbf{d}^{(h)})$ onto the nondominated surface of Z.

If we assume the achievement scalarizing function of Section 13.6.1, the achievement scalarizing parametric program of Step 3 is

$$\min \{ \alpha - \varepsilon \sum_{i=1}^{k} z_i \}$$

$$\begin{aligned} \text{s.t.} \quad z_i + \frac{\alpha}{\lambda_i} &\leqq z_i^{(h-1)} + \theta d_i^{(h)} & 1 \leqq i \leqq k \\ f_i(\mathbf{x}) &= z_i & 1 \leqq i \leqq k \\ \mathbf{x} &\in S \end{aligned}$$

Step 4: Display the criterion value *trajectories* along the projection of $\mu(\mathbf{z}^{(h-1)}, \mathbf{d}^{(h)})$ on N. Designate the most preferred criterion vector $\mathbf{z}^{(h)}$.

The projection of $\mu(\mathbf{z}^{(h-1)}, \mathbf{d}^{(h)})$ will typically result in a piecewise linear path on the surface of Z. To analyze the piecewise linear path of criterion vectors, computer graphics can be employed to plot the trajectories of criterion values, as in the display of Fig. 13.22. Korhonen and Laakso use the following style of implementation on a CRT. As the cursor is moved back and forth at the bottom of the screen, the numerical values of the objectives will appear at the top of the screen above the trajectories. In this way, the user can search the path of criterion vectors for his or her most preferred choice $\mathbf{z}^{(h)}$.

Step 5: If $\mathbf{z}^{(h)} \neq \mathbf{z}^{(h-1)}$, go to Step 2. Otherwise, go to Step 6.

Step 6: If STALL = 1, go to Step 8. Otherwise, go to Step 7.

Step 7: Obtain a set of generators for the *cone of feasible criterion vector directions* at $\mathbf{z}^{(h)}$. Set STALL = 1.

Let S be polyhedral and $\mathbf{z}^{(h)}$ the criterion vector of $\mathbf{x}^{(h)} \in S$. Then, if $\mathbf{x}^{(h)}$ is an extreme point of S, a set of generators for the cone of feasible criterion vector directions can be obtained from the directions of the edges of S emanating from $\mathbf{x}^{(h)}$. If, on the other hand, $\mathbf{x}^{(h)}$ is not extreme, generators for the cone of feasible criterion vector directions must be obtained in some other way (see Section 13.6.4). STALL is set to one so that the set of generators will not have to be recomputed until the current solution changes.

Step 8: If there are any generators that have not been tested as a direction for improvement, go to Step 9. Otherwise, go to Step 10.

Step 9: Let $h = h + 1$. Select an untested generator and designate it $\mathbf{d}^{(h)}$. Go to Step 3.

Step 10: Let $\mathbf{x}^{(h)}$ be an inverse image of $\mathbf{z}^{(h)}$. Terminate with $(\mathbf{x}^{(h)}, \mathbf{z}^{(h)})$ as the final solution.

The algorithm has been designed to exploit the following theoretical result.

Theorem 13.1. Let S be closed, convex, and bounded and let all constraints be continuous and differentiable. Let the decision maker's utility function U: $R^k \to R$ be pseudoconcave. Let $\bar{\mathbf{z}} \in Z$ be a feasible criterion vector and let $V \subset R^k$ be the minimal cone that contains all feasible directions in criterion space at $\bar{\mathbf{z}}$. Let $\{\mathbf{d}^1, \ldots, \mathbf{d}^{\ell}\}$ be a set of generators for V. Then, $\bar{\mathbf{z}}$ is an optimal criterion vector $\Leftrightarrow$ $U(\bar{\mathbf{z}}) \geqq U(\bar{\mathbf{z}} + \alpha_i \mathbf{d}^i)$ for all $\alpha_i \geqq 0$ and $i = 1, \ldots, \ell$.

The theorem is proved in Korhonen and Laakso (1985). The theorem assures us that the algorithm will stop after an optimal solution is computed, provided each generator of the cone of feasible criterion vector directions is tested. The algorithm is flowcharted in Fig. 13.23.

13.6.4 Visual Interactive Comments

1. The method can be applied to nonlinear multiple objective programs with the use of a nonlinear programming code. However, instead of solving the achievement scalarizing parametric program for θ going from $0 \to \infty$, the program can be solved for different specific values of θ.
2. In the Korhonen and Laakso (1985) source article, the $-\varepsilon\sum_{i=1}^{k} z_i$ perturbation term is not included in the achievement scalarizing function. If this causes dominated criterion vectors to be generated, Korhonen and Laakso remove them by inspection at the graphical stage.
3. In nonlinear problems, the minimal number of generators of the cone of feasible criterion vector directions may be infinite. This might cause us to lose the finiteness of our test for detecting an optimal point.

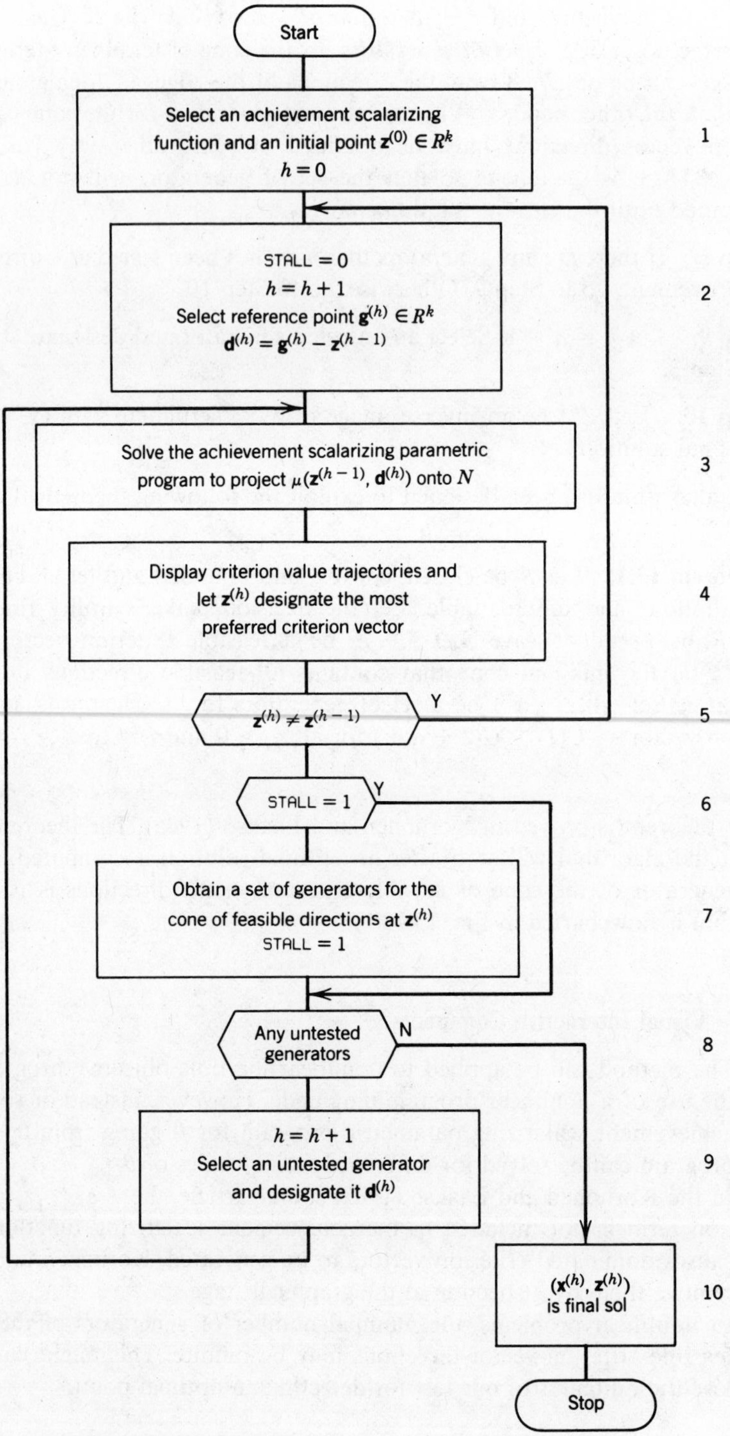

Figure 13.23. Flowchart of visual interactive approach.

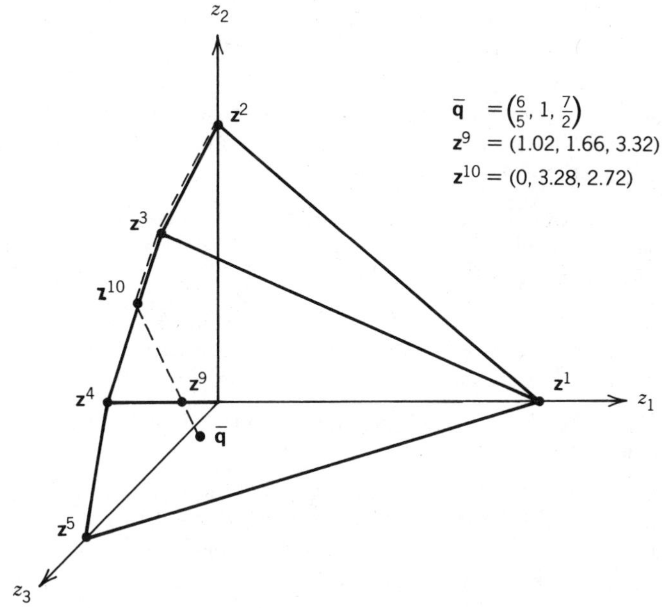

Figure 13.24. Graph of Example 13.

4. If N contains interior points of Z, the projection of $\mu(\mathbf{z}^{(h-1)}, \mathbf{d}^{(h)})$ may not be entirely on the surface of Z.
5. Korhonen and Wallenius (1985) have been working on a method to yield the generators of the cone of feasible directions at $\mathbf{z}^{(h)}$. The hypothesis is that the achievement scalarizing program with $\boldsymbol{\lambda}$-weights such that $\mathbf{z}^{(h)}$ is a solution will provide the generators.
6. The choice of $\boldsymbol{\lambda}$-weights in the achievement scalarizing program can have a strong effect on where the projection onto the surface of N lies.

Example 13. Instead of using $\boldsymbol{\lambda} = (\frac{5}{8}, \frac{1}{4}, \frac{1}{8})$ as in Example 12, suppose we used $\boldsymbol{\lambda} = (\frac{1}{8}, \frac{1}{4}, \frac{5}{8})$. Then, the projection onto N is given by the dashed line $\bar{\mathbf{q}} \to \mathbf{z}^9 \to \mathbf{z}^{10} \to \mathbf{z}^3 \to \mathbf{z}^2$ in Fig. 13.24.

Readings

1. Anderson, M. D., R. G. March, and J. M. Mulvey (1981). "Solving Multi-Objective Problems via Unstructured and Interactive Dialogue," Research Report No. EES–81-8, Department of Civil Engineering, Princeton University, Princeton, New Jersey.
2. Aubin, J. P. and B. Näslund (1972). "An Exterior Branching Algorithm," Working Paper 72–42, European Institute for Advanced Studies in Management, Brussels.

3. Belensen, S. and K. C. Kapur (1973). "An Algorithm for Solving Multicriterion Linear Programming Problems with Examples," *Operational Research Quarterly*, Vol. 24, No. 1, pp. 65–77.

4. Benayoun, R., J. de Montgolfier, J. Tergny, and O. Laritchev (1971). "Linear Programming with Multiple Objective Functions: Step Method (STEM)," *Mathematical Programming*, Vol. 1, No. 3, pp. 366–375.

5. Benayoun, R., J. Tergny, and D. Keuneman (1970). "Mathematical Programming with Multi-Objective Functions: A Solution by P.O.P. (Progressive Orientation Procedure)," *Metra*, Vol. 9, No. 2, pp. 279–299.

6. Dinkelbach, W. and H. Isermann (1980). "Resource Allocation of an Academic Department in the Presence of Multiple Criteria—Some Experience With a Modified STEM-Method," *Computers and Operations Research*, Vol. 7, No.'s 1–2, pp. 99–106.

7. Dyer, J. S. (1972). "Interactive Goal Programming," *Management Science*, Vol. 19, No. 1, pp. 62–70.

8. Dyer, J. S. (1972). "An Empirical Investigation of a Man-Machine Interactive Approach to the Solution of the Multiple Criteria Problem." In J. L. Cochrane and M. Zeleny (eds.), *Multiple Criteria Decision Making*, Columbia, South Carolina: University of South Carolina Press, pp. 202–216.

9. Dyer, J. S. (1973). "A Time-Sharing Computer Program for the Solution of the Multiple Criteria Problem," *Management Science*, Vol. 19, No. 12, pp. 1379–1383.

10. Dyer, J. S. (1974). "The Effects of Errors in the Estimation of the Gradient on the Frank-Wolfe Algorithm, with Implications for Interactive Programming," *Operations Research*, Vol. 22, No. 1, pp. 160–174.

11. Geoffrion, A. M., J. S. Dyer, and A. Feinberg (1972). "An Interactive Approach for Multicriterion Optimization, with an Application to the Operation of an Academic Department," *Management Science*, Vol. 19, No. 4, pp. 357–368.

12. Grauer, M. (1983). "A Dynamic Interactive Decision Analysis and Support System (DIDASS): User's Guide," WP–83-60, International Institute for Applied Systems Analysis, Laxenburg, Austria.

13. Haimes, Y. Y. and W. A. Hall (1974). "Multiobjectives in Water Resources Systems Analysis: The Surrogate Worth Trade-Off Method," *Water Resources Research*, Vol. 10, No. 4, pp. 615–623.

14. Hemming, T. (1976). "A New Method for Interactive Multiobjective Optimization: A Boundary Point Ranking Method," *Lecture Notes In Economics and Mathematical Systems*, No. 130, Berlin: Springer-Verlag, pp. 333–339.

15. Ho, J. K. (1979). "Holistic Preference Evaluation in Multiple Criteria Optimization," Management Science Program, University of Tennessee, Knoxville, Tennessee.

16. Isermann, H. (1979). "Strukturierung von Entscheidungsprozessen bei Mehrfacher Zielsetzung," *OR Spektrum*, Vol. 1, No. 1, pp. 3–26.

17. Kallio, M. and M. Soismaa (1983). "Some Improvements to the Reference Point Approach for Dynamic MCLP." In C. Carlsson and Y. Kochetkov (eds.), *Theory and Practice of Multiple Criteria Decision Making*, Amsterdam: North-Holland, pp. 77–90.

18. Kok, M. and F. A. Lootsma (1985). "Pairwise-Comparison Methods in Multiple Objective Programming, with Applications in a Long-Term Energy-Planning Model," *European Journal of Operational Research*, Vol. 22, No. 1, pp. 44–55.

19. Köksalan, M. M., M. H. Karwan, and S. Zionts (1983). "An Approach for Solving Discrete Alternative Multiple Criteria Problems Involving Ordinal Criteria," Working Paper No. 571, School of Management, State University of New York at Buffalo, Buffalo, New York.

20. Köksalan, M. M., M. H. Karwan, and S. Zionts (1983). "Approaches for Discrete Alternative Multiple Criteria Problems for Different Types of Criteria," Working Paper No. 572, School of Management, State University of New York at Buffalo, Buffalo, New York.

21. Köksalan, M., M. H. Karwan, and S. Zionts (1984). "An Improved Method for Solving Multiple Criteria Problems Involving Discrete Alternatives," *IEEE Transactions on Systems, Man and Cybernetics*, Vol. 13, No. 1, pp. 24–34.

22. Korhonen, P. and J. Laakso (1985). "A Visual Interactive Method for Solving the Multiple Criteria Problem," *European Journal of Operational Research* (forthcoming).

23. Korhonen, P. and M. Siosmaa (1981). "An Interactive Multiple Criteria Approach in Ranking Alternatives," *Journal of the Operational Research Society*, Vol. 32, No. 7, pp. 577–585.

24. Korhonen, P. and J. Wallenius (1985). "A Modification of the Zionts–Wallenius Multiple Criteria Method for Nonlinear Utility Functions," Helsinki School of Economics, Helsinki.

25. Korhonen, P., J. Wallenius and S. Zionts (1984). "Solving the Discrete Multiple Criteria Problem Using Convex Cones," *Management Science*, Vol. 30, No. 11, pp. 1336–1345.

26. Lee, S. M. and L. S. Franz (1983). "An Interactive Decision Support System for Solving the Multiple Objective Problem," College of Business Administration, University of Nebraska, Lincoln, Nebraska.

27. Lotov, A. V. (1980). "On the Concept of Generalized Sets of Accessibility and their Construction of Linear Controlled Systems," *Soviet Physics Doklaty*, Vol. 25, No. 2, pp. 82–84.

28. Nakayama, H., T. Tanino, and Y. Sawaragi (1980). "An Interactive Optimization Method in Multicriteria Decision Making, *IEEE Transactions on Systems, Man and Cybernetics*, Vol. 10, No. 3, pp. 163–169.

29. Oppenheimer, K. R. (1978). "A Proxy Approach to Multi-Attribute Decision Making," *Management Science*, Vol. 24, No. 6, pp. 675–689.

30. Roy, B. (1968). "Classement et Choix en Presence de Criteres Multiples," *RAIRO*, No. 8, pp. 57–75.

31. Roy, B. (1973). "How Outranking Relation Helps Multiple Criteria Decision Making." In J. L. Cochrane and M. Zeleny (eds.), *Multiple Criteria Decision Making*, Columbia, South Carolina: University of South Carolina Press, pp. 179–201.

32. Roy, B. and P. Bertier (1971). "La Methode ELECTRE II: Une Methode de Classement en Presence de Criteres Multiples," *Note de Travail No. 142*, Direction Scientifique, Groupe METRA, Paris.

33. Sakawa, M. (1982). "Interactive Multiobjective Decision Making by the Sequential Proxy Optimization Technique: SPOT," *European Journal of Operational Research*, Vol. 9, No. 4, pp. 386–396.

34. de Samblanckx, S., P. Depraetere, and H. Muller (1982). "Critical Considerations Concerning the Multicriteria Analysis by the Method of Zionts and Wallenius," *European Journal of Operational Research*, Vol. 10, No. 1, pp. 70–76.

35. Soland, R. M. (1979). "Multicriteria Optimization: A General Characterization of Efficient Solutions," *Decision Sciences*, Vol. 10, No. 1, pp. 26–38.

36. Steuer, R. E. (1977). "An Interactive Multiple Objective Linear Programming Procedure," *TIMS Studies in the Management Sciences*, Vol. 6, pp. 225–239.

37. Steuer, R. E. and E.-U. Choo (1983). "An Interactive Weighted Tchebycheff Procedure for Multiple Objective Programming," *Mathematical Programming*, Vol. 26, No. 1, pp. 326–344.

38. Steuer, R. E. and A. T. Schuler (1978). "An Interactive Multiple Objective Linear Programming Approach to a Problem in Forest Management," *Operations Research*, Vol. 25, No. 2, pp. 254–269.

39. Steuer, R. E. and A. T. Schuler (1981). "Interactive Multiple Objective Linear Programming Applied to Multiple Use Forestry Planning." In M. C. Vodak, W. A. Leuschner, and D. I. Navon (eds.), *Symposium on Forest Management Planning: Present Practice and Future Directions*, FWS-1-81, School of Forestry and Wildlife Resources, VPI, Blacksburg, Virginia, pp. 80–93.

40. Steuer, R. E. and E. F. Wood (1986). "A Multiple Objective Markov Reservoir Release Policy Model," College of Business Administration, University of Georgia, Athens, Georgia.

41. Wallenius, J. (1975). "Comparative Evaluation of Some Interactive Approaches to Multicriterion Optimization," *Management Science*, Vol. 21, No. 12, pp. 1387–1396.

42. White, D. J. (1980). "Multi-Objective Interactive Programming," *Journal of the Operational Research Society*, Vol. 31, No. 6, pp. 517–523.

43. Wierzbicki, A. P. (1982). "A Mathematical Basis for Satisficing Decision Making," *Mathematical Modeling*, Vol. 3, No. 5, pp. 391–405.

44. Wierzbicki, A. P. (1983). "Critical Essay on the Methodology of Multiobjective Analysis," *Regional Science and Urban Economics*, Vol. 13, No. 1, pp. 5–29.

45. Zionts, S. (1981). "A Multiple Criteria Method for Choosing Among Discrete Alternatives," *European Journal of Operational Research*, Vol. 7, No. 2, pp. 143–147.

46. Zionts, S. (1982). "Multiple Criteria Decision Making: An Overview and Several Approaches," Working Paper No. 454, School of Management, State University of New York at Buffalo, Buffalo, New York.

47. Zionts, S. and J. Wallenius (1976). "An Interactive Programming Method for Solving the Multiple Criteria Problem," *Management Science*, Vol. 22, No. 6, pp. 652–663.

48. Zionts, S. and J. Wallenius (1980). "Identifying Efficient Vectors: Some Theory and Computational Results," *Operations Research*, Vol. 28, No. 3, pp. 785–793.

49. Zionts, S. and J. Wallenius (1983). "An Interactive Multiple Objective Linear Programming Method for a Class of Underlying Nonlinear Utility Functions," *Management Science*, Vol. 29, No. 5, pp. 519–529.

PROBLEM EXERCISES

13-1 Let $U = K - \left[\sum_{i=1}^{k} \pi_i \left(z_i^{**} - z_i\right)^p\right]^{1/p}$ in which the z_i are variables.

(a) Determine $(\partial U/\partial z_j)$.

(b) Specify $\nabla_z U(\bar{\mathbf{z}})$ at $\bar{\mathbf{z}} = (39, 12, 42)$ when:

$$\begin{array}{lll} p = 4 & \pi_1 = 1 & z_1^{**} = 40 \\ K = 100 & \pi_2 = 3 & z_2^{**} = 20 \\ k = 3 & \pi_3 = 2 & z_3^{**} = 50 \end{array}$$

13-2 Let $U = \prod_{i=1}^{k} \left(\dfrac{z_i}{k_i}\right)^{\beta_i}$ in which all quantities are fixed except the z_i.

(a) Determine $(\partial U/\partial z_j)$.

(b) Specify $\nabla_z U(\bar{\mathbf{z}})$ at $\bar{\mathbf{z}} = (39, 12, 42)$ when:

$$\begin{array}{lll} & k_1 = 1 & \beta_1 = 1.5 \\ k = 3 & k_2 = 2 & \beta_2 = 2.0 \\ & k_3 = 4 & \beta_3 = 2.5 \end{array}$$

13-3 Consider an MOLP whose payoff table is:

	z_1	z_2	z_3
$\mathbf{z}^1$	60.00	−19.25	−66.50
$\mathbf{z}^2$	−12.25	8.75	7.00
$\mathbf{z}^3$	−4.75	8.75	13.00

(a) Allowing

$$\sum_{j=1}^{8} (c_{1j})^2 = 36, \quad \sum_{j=1}^{8} (c_{2j})^2 = 100, \quad \sum_{j=1}^{8} (c_{3j})^2 = 64$$

compute the π_i weights in Step 2 of **STEM**.

(b) Suppose that on the third iteration, $J^* = \{2\}$. Calculate the components of $\boldsymbol{\lambda}^{(3)}$.

13-4C Consider the MOLP (that has 27 efficient extreme points):

	x_1	x_2	x_3	x_4	x_5	x_6	x_7	x_8		
Objs	7	5		4	−2	7	6	6	max	
	−3	−7	8		6	−7	−3		max	
	1	4		3	−1	7	−4	−7	max	
s.t.		8		1	−3	7	−1		≦	5
	7	7	6	4	5				≦	9
	8			1	8		−3	−3	≦	6
	8	3	5		1		8		≦	8
	1	6	3	3	3	3	4		≦	9
		5	−1				3	−3	≦	9
	7	−2		2	8	8	4	−1	≦	7
		5	5	−2	−1			6	≦	8

whose payoff table is:

	z_1	z_2	z_3
$\mathbf{z}^1$	26.02	-3.87	-5.82
$\mathbf{z}^2$	.00	12.00	.00
$\mathbf{z}^3$	11.33	-2.79	9.11

(a) Run **STEM** for three iterations. To simulate decision-maker judgment, let $U = 5z_1 + 10(z_2 + 4) + (z_3 + 12)$.

(b) Using the utility function from (a), solve for the optimal point, the optimal criterion vector, and the optimal U value.

13-5C If feasible, what are the inverse images in decision space of:

(a) $\mathbf{z}^1 = (7, 3, 6)$

(b) $\mathbf{z}^2 = (12, 7, 6)$

in the MOLP:

	x_1	x_2	x_3	x_4		
Objs	2	4	1		max	
	2		4		max	
	3		2	3	max	
s.t.	3		3	4	$\leqq$	8
				3	$\leqq$	10
	5			2	$\leqq$	10
	3	3	5		$\leqq$	10

13-6 Consider the pairwise comparison process of Step 3 of the GDF algorithm. Compute w_1 and w_2 at $\mathbf{z}^a = (5, 2)$ using the halving and doubling algorithm. Let $U = z_1(z_2)^2$ and $\Delta_1 = 3$. Assume indifference if $U(\mathbf{z}^b)$ is within $\pm 5\%$ of $U(\mathbf{z}^a)$. Continue with the step-by-step results as stated below.

Step	
1	$\mathbf{z}^a = (5, 2)$ $\Delta_1 = 3$ $i = 2$
2	$\ell = 0$ $\Delta_2^{(1)} = 1$
3	$\ell = 1$ $U(\mathbf{z}^a) = 20, \quad U(\mathbf{z}^b = (2, 3)) = 18$
4	$\Delta_2^{(2)} = 2$
3	$\ell = 2$ $U(\mathbf{z}^a) = 20, \quad U(\mathbf{z}^b = (2, 4)) = 32$
4	$\theta^{(2)} = 1$

13-7 Using the GDF algorithm, graphically solve the MOLP:

$$\begin{aligned} \max\{\ x_1 \qquad\quad &= z_1\} \\ \max\{\qquad\quad x_2 &= z_2\} \\ \text{s.t.}\quad x_1 \qquad\quad &\leqq 7 \\ 6x_1 + 5x_2 &\leqq 47 \\ x_1 + 4x_2 &\leqq 30 \\ x_1, x_2 &\geqq 0 \end{aligned}$$

Let $\mathbf{x}^{(0)} = (1,5)$, and let $P = 5$ on all iterations. To simulate decision-maker judgment, let $U = 200 - (z_1 - 8)^2 - (z_2 - 8)^2$. Hence, $(242/61, 283/61)$ is the optimal solution and 172.44 is the optimal U value. Proceed for three iterations. Ignore Step 3 and use $\nabla_x U(\mathbf{x}^{(h)})$ for $\mathbf{g}$ in Step 4. Display the iteration data as in Table 13.1.

13-8 Using the GDF algorithm, graphically solve the MOLP:

$$\begin{aligned} \max\{x_1 \qquad &= z_1\} \\ \max\{\qquad x_2 &= z_2\} \\ \text{s.t.} \qquad \mathbf{x} &\in S \end{aligned}$$

where S is the convex hull defined by $(1,5)$, $(1,9)$, $(7,5)$, and $(9,1)$. Let $\mathbf{x}^{(0)} = (1,5)$, and let $P = 5$ on all iterations. To simulate decision-maker judgment, let $U = (z_1)^2(z_2)^2$. Ignore Step 3 and use the gradient of U evaluated at $\mathbf{x}^{(h)}$ for $\mathbf{g}$ in Step 4.

(a) Display the iteration data as in Table 13.1.

(b) Specify the optimal solution and the optimal U value.

13-9C **(a)** Compute the middlemost points of:

$$Q = \{\mathbf{x} \in R^2 \mid \mathbf{x} \geqq \mathbf{0};\quad x_1 + 2x_2 \geqq 6;\quad x_1 + x_2 \leqq 4\}$$

$$R = \{\mathbf{x} \in R^2 \mid \mathbf{x} \geqq \mathbf{0};\quad x_1 + 2x_2 \leqq 6;\quad x_1 + x_2 \leqq 4\}$$

(b) What would the middlemost points of Q and R be if the $x_1 + x_2 \leqq 4$ constraint were written $8x_1 + 8x_2 \leqq 32$?

(c) Graph Q and R and their respective middlemost points from (a) and (b).

13-10C Let $\Lambda^{(1)}$ be defined by:

$$\begin{aligned} \lambda_2 - 3\lambda_3 &\geqq 0 \\ \tfrac{8}{3}\lambda_1 \qquad - \tfrac{8}{3}\lambda_3 &\geqq 0 \\ -2\lambda_2 + 2\lambda_3 &\leqq 0 \\ \lambda_1 + \lambda_2 + \lambda_3 &= 1 \\ \lambda_1, \lambda_2, \lambda_3 &\geqq 0 \end{aligned}$$

(a) Using a parametric diagram, graph $\Lambda^{(1)}$.

(b) Compute the middlemost point of $\Lambda^{(1)}$.

(c) Compute the middlemost point if the third constraint (which is redundant) is deleted.

(d) Compute the middlemost point if the second and third constraints are rescaled and respectively written:

$$\begin{aligned} \lambda_1 \qquad - \lambda_3 &\geqq 0 \\ - \lambda_2 + \lambda_3 &\leqq 0 \end{aligned}$$

13-11 Use the Z–W algorithm to solve manually the MOLP:

$$\begin{aligned} &\max \{x_1 &&= z_1\} \\ &\max \{ \quad x_2 &&= z_2\} \\ &\max \{ \qquad x_3 &&= z_3\} \\ &\text{s.t.} \qquad \mathbf{x} \in S \end{aligned}$$

where S is the convex hull defined by $(0,0,0)$, $(9,0,0)$, $(0,9,0)$, $(0,0,9)$, and $(5,5,0)$. To answer the questions posed by the algorithm, let:

$$U = \prod_{i=1}^{3} (z_i + 1)$$

(a) Graph the MOLP.

(b) Specify the optimal solution and the optimal U value.

(c) Using Section 13.3.2 as a model, describe the course of the solution procedure complete with a parametric diagram.

13-12 Use the Z–W algorithm to solve manually the MOLP:

$$\begin{aligned} &\max \{ \ x_1 &&= z_1\} \\ &\max \{ \qquad x_2 &&= z_2\} \\ &\text{s.t.} \quad 3x_1 + 4x_2 &&\leqq 595 \\ &\qquad 4x_1 + 3x_2 &&\leqq 595 \\ &\qquad 3x_1 + 2x_2 &&\leqq 426 \\ &\qquad x_1, x_2 &&\geqq 0 \end{aligned}$$

To answer questions posed by the algorithm, let:

$$U = 200 + \min\{(z_1 - 158),\ (z_2 - 151)\} + .001(z_1 + z_2)$$

(a) Graph the MOLP.

(b) Specify the optimal solution and the optimal U value.

(c) Using Section 13.3.2 as a model, describe the course of the solution procedure complete with a parametric diagram.

13-13 Consider the MOLP of Fig. 13.25 in which $N = \gamma(\mathbf{z}^1, \mathbf{z}^2, \mathbf{z}^5) \cup \gamma(\mathbf{z}^2, \mathbf{z}^4, \mathbf{z}^5) \cup \gamma(\mathbf{z}^2, \mathbf{z}^3, \mathbf{z}^4)$.

(a) Without pivoting to the extreme points, construct the parametric diagram for the MOLP.

(b) Assume that $\mathbf{z}^5$ is preferred to $\mathbf{z}^1$ is the only active previous response. With this knowledge, use the Z–W routine to determine which edges emanating from $\mathbf{z}^2$ are efficient w.r.t. $\Lambda^{(h)}$.

13-14C Solve the MOLP of this problem using a method that converges to an extreme point. To simulate decision-maker judgment, let U be one of the following:

(a) $U = 3z_1 + 7z_2 + 5z_3 + 2z_4$
(b) $U = 9z_1 + 2z_2 + z_3 + 5z_4$
(c) $U = 2z_1 + z_2 + 4z_3 + 3z_4$
(d) $U = 3z_1 + 4z_2 + 8z_3 + 4z_4$
(e) $U = 2z_1 + z_2 + 3z_3 + 4z_4$

	x_1	x_2	x_3	x_4	x_5	x_6	x_7	x_8	x_9	x_{10}	x_{11}	x_{12}		
	3		2	2	3	1	−1	2	3	−1	1	4	max	
Objs	−1	4			2		−1	4	−1		−1		max	
		3	2	2	−1	2	4	1			−1	3	max	
		2	1			−1		−1		1	5	3	max	
		2	−1					4	2				≤	10
						3						3	≤	10
	−1	4	4			3					−1		≤	10
		4		3					2			4	≤	7
					−1		4	2		5			≤	10
			2	3								5	≤	5
	2		1					2	2	−1	2	4	≤	5
s.t.									2		3		≤	6
				−1			2	1	4				≤	5
	4	−1								4			≤	7
				−1					2	5			≤	6
											2		≤	7
	−1	−1											≤	9
	2			3		5							≤	5
		2	1		3	5							≤	9
		2	5		4		3						≤	5

13-15C Solve the MOLP of this problem using a method that converges to an extreme point. To simulate decision-maker judgment, let U be one of the following:

(a) $U = 3z_1 + z_2 + 3z_3 + 4z_4 + 2z_5$
(b) $U = 9z_1 + 3z_2 + z_3 + 5z_4 + 2z_5$
(c) $U = 5z_1 + 3z_2 + 8z_3 + 2z_4 + 6z_5$
(d) $U = 2z_1 + 7z_2 + 4z_3 + 9z_4 + z_5$
(e) $U = 5z_1 + 6z_2 + 4z_3 + 6z_4 + 6z_5$

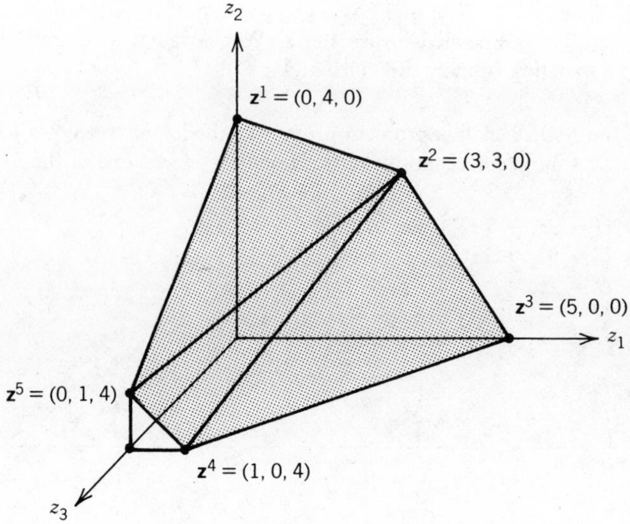

Figure 13.25. Graph of Problem 13-13.

	x_1	x_2	x_3	x_4	x_5	x_6	x_7	x_8	x_9	x_{10}		
	1		−1	1	2	3	−2		−1	3	max	
	1		1	2				2	3		max	
Objs	1		−2	1	2	1				2	max	
	3	2		1	1	2	3	−1	1	3	max	
	2	3		−1		−1	3	−1	−2	3	max	
	1						4			3	≦	10
		−1		1				5			≦	8
	2	5	4		2		−1				≦	7
		−1	2			4			5		≦	10
			1	1			3				≦	8
		2						1	−1	3	≦	8
						2					≦	5
s.t.	1		5		1				−1		≦	9
									5		≦	7
	3	4			4			−1		3	≦	9
		3	4					3			≦	7
				3		2				5	≦	7
	5		1								≦	7
	−1	2		5				2	1		≦	7

13-16C Consider the MOLP:

$$
\begin{aligned}
\max\{\ & x_1 && = z_1\} \\
\max\{\ & && x_2 = z_2\} \\
\text{s.t.}\quad & -x_1 + && x_2 \leqq 5 \\
& x_1 + && 3x_2 \leqq 23 \\
& x_1 + && x_2 \leqq 11 \\
& 3x_1 + && x_2 \leqq 25 \\
& x_1 && \leqq 8 \\
& && x_1, x_2 \geqq 0
\end{aligned}
$$

(a) Using the achievement scalarizing function of Section 13.6.1 with $\boldsymbol{\lambda} = (\frac{4}{5}, \frac{1}{5})$ and $\varepsilon = .001$, project $\mu(\bar{\mathbf{q}}, \mathbf{d})$ onto N where $\bar{\mathbf{q}} = (1, 4)$ and $\mathbf{d} = (5, 1)$.

(i) Show the numerical coefficients of the achievement scalarizing program in tabular form.

(ii) Graph Z and show the projection of $\mu(\bar{\mathbf{q}}, \mathbf{d})$ on N.

(b) Same as (a) but with $\boldsymbol{\lambda} = (\frac{1}{5}, \frac{4}{5})$.

13-17C Solve the MOLP of this problem using a procedure that can converge to a nonextreme final solution. To simulate decision-maker judgment, let U be one of the following:

(a) $U = 100 - [4(71 - z_1)^4 + (69 - z_2)^4 + 2(42 - z_3)^4]^{1/4}$

(b) $U = 100 - [2(71 - z_1)^4 + 4(69 - z_2)^4 + (42 - z_3)^4]^{1/4}$

(c) $U = 100 - [3(71 - z_1)^4 + 2(69 - z_2)^4 + (42 - z_3)^4]^{1/4}$

(d) $U = (z_1/20)^4(z_2/20)^2(z_3/20)^{0.2}$

	x_1	x_2	x_3	x_4	x_5	x_6	x_7		
	3		6	4		5		max	
Objs	6	3	5	6	6	4	6	max	
	6	1	1		5		7	max	
	2			5	4		4	$\leqq$	27
		5			3		6	$\leqq$	40
	5			6	5	3	6	$\leqq$	24
s.t.	1			4		1		$\leqq$	33
				3	2			$\leqq$	41
		2	4			1		$\leqq$	26

13-18C Solve the MOLP of this exercise using a procedure that can converge to a nonextreme final solution. To simulate decision-maker judgment, let U be one of the following:

(a) $U = 100 - [(75 - z_1)^4 + 6(94 - z_2)^4 + 2(56 - z_3)^4]^{1/4}$

(b) $U = 100 - [(75 - z_1)^4 + 3(94 - z_2)^4 + 5(56 - z_3)^4]^{1/4}$

(c) $U = 100 - [2(75 - z_1)^4 + 4(94 - z_2)^4 + (56 - z_3)^4]^{1/4}$

(d) $U = (z_1/20)^{1.4}(z_2/20)^{1.5}(z_3/20)^{1.6}$

(e) $U = (z_1/20)^{2.5}(z_2/20)^{2.5}(z_3/20)^{2.5}$

	x_1	x_2	x_3	x_4	x_5	x_6	x_7		
	4	1	1	2	4	4	5	max	
Objs	3	4	5	6	2			max	
		4			1	1		max	
				2			2	$\leqq$	43
	1	1	5			2		$\leqq$	37
			4	6		5		$\leqq$	31
s.t.		2				2		$\leqq$	26
	2			5	4		4	$\leqq$	41
	1		6		1	4		$\leqq$	42
	6		2		6	6		$\leqq$	21

CHAPTER 14

Interactive Weighted Tchebycheff Procedure

This chapter discusses the interactive weighted Tchebycheff procedure of Steuer and Choo (1983). With two versions (*augmented* and *lexicographic*), the interactive Tchebycheff procedure is a *weighting vector space reduction* method for solving the multiple objective program

$$\begin{aligned} &\max \{f_1(\mathbf{x}) = z_1\} \\ &\max \{f_2(\mathbf{x}) = z_2\} \\ &\quad\vdots \\ &\max \{f_k(\mathbf{x}) = z_k\} \\ &\text{s.t.} \qquad \mathbf{x} \in S \end{aligned}$$

where the f_i need not be linear and S need not be convex. It is assumed that each objective is bounded over S and that there does not exist a point in S at which all objectives are simultaneously maximized.

Externally, the method is similar to the interval criterion weights/vector-maximum and interactive weighted-sums/filtering approaches of Chapter 13. That is, the procedure follows a progressively more concentrated sampling strategy until a final solution is obtained. The procedure operates in a fixed number of iterations, presenting a fixed number of solutions per iteration. However, internally, there is a difference in the way weighting vectors are employed. Instead of using weighting vectors $\boldsymbol{\lambda} \in \Lambda = \{\boldsymbol{\lambda} \in R^k \mid \lambda_i > 0, \Sigma_{i=1}^k \lambda_i = 1\}$ to capture the relative importance of the different objectives, weighting vectors $\boldsymbol{\lambda} \in \bar{\Lambda}$ where

$$\bar{\Lambda} = \left\{ \boldsymbol{\lambda} \in R^k \mid \lambda_i \geqq 0, \quad \sum_{i=1}^{k} \lambda_i = 1 \right\}$$

are used to define different weighted Tchebycheff metrics. As a consequence, the Tchebycheff approach enjoys the following advantages.

1. It can converge to nonextreme final solutions in linear multiple objective programs.
2. The method can compute *unsupported* (Section 14.6) and *improperly* (Section 14.7) nondominated criterion vectors, making the approach generalizable to integer and nonlinear multiple objective programs.
3. The method uses conventional single criterion mathematical programming software, thereby enabling application to large problems. For example, we can employ software such as MPSX [IBM (1979)] with multiple objective linear programs, MPSX - MIP with multiple objective integer programs, or commercial nonlinear codes with nonlinear multiple objective programs.

Before beginning this chapter, readers should note that the preponderance of graphs are drawn in criterion space, not decision space. Also, it is advised that readers review the material on L_∞-metrics in Sections 2.5.2 and 2.5.3.

14.1 THE z** IDEAL CRITERION VECTOR

To utilize the weighted Tchebycheff approach, we first compute a $\mathbf{z}^{**}$ *ideal criterion vector*. The k components of $\mathbf{z}^{**}$ are given by

$$\begin{aligned} z_i^{**} &= z_i^* + \varepsilon_i \\ &= \max\{f_i(\mathbf{x}) \mid \mathbf{x} \in S\} + \varepsilon_i \end{aligned}$$

where $\varepsilon_i \geqq 0$. In general, it suffices for each ε_i to be positive. However, in many, if not most, multiple objective programs, it is permissible for one or more of the ε_i's to be zero. The only time a given ε_i *must* be positive is when

1. There is more than one criterion vector that maximizes the ith objective, or
2. There is only one criterion vector that maximizes the ith objective, but this criterion vector also maximizes one of the other objectives.

If neither of these two conditions occur, the given ε_i can be (but does not have to be) set to zero. Example 1 provides a problem in which a given ε_i must be positive.

Example 1. Consider the MOLP

$$\begin{aligned} &\max\{x_1 &&= z_1\} \\ &\max\{ \quad x_2 &&= z_2\} \\ &\max\{ \quad\quad x_3 &&= z_3\} \\ &\text{s.t.} \quad x_1 + x_2 + x_3 &&\leqq 3 \\ &\quad\quad x_2 &&\leqq 2 \\ &\quad x_1, x_2, x_3 &&\geqq 0 \end{aligned}$$

whose graph in criterion space is given in Fig. 14.1. In this problem, the set of all

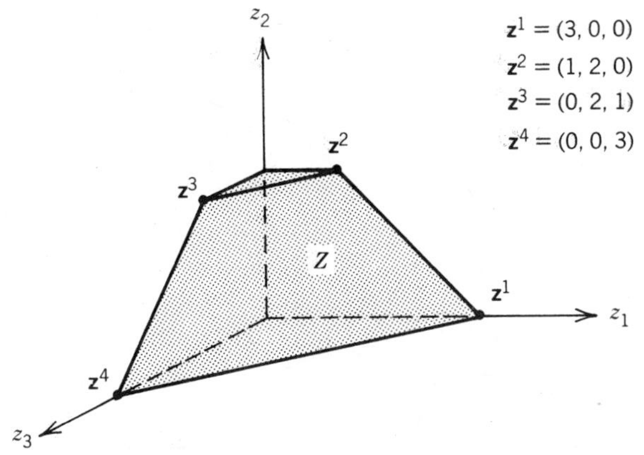

Figure 14.1. Graph of Example 1.

nondominated criterion vectors $N = \gamma(\mathbf{z}^1, \mathbf{z}^2, \mathbf{z}^3, \mathbf{z}^4)$ and $\mathbf{z}^* = (3, 2, 3)$. Since more than one criterion vector maximizes the second objective, ε_2 must be positive. It is, however, permissible for ε_1 and ε_3 to be zero. We might then decide on $(3, 2.1, 3)$ as the $\mathbf{z}^{**}$ ideal criterion vector.

14.2 SELECTING ε_i VALUES

When a given ε_i must be positive, it is set to a moderately small value. Example 2 illustrates how we may wish to let convenience determine the ε_i values.

Example 2. Consider the MOLP

$$
\begin{aligned}
\max\{ \quad x_1 \qquad\quad &= z_1\} \\
\max\{ \ -x_1 + 2x_2 &= z_2\} \\
\text{s.t.} \quad -8x_1 + 6x_2 &\leqq 0 \\
7x_1 - 18x_2 &\leqq 0 \\
11x_1 + 30x_2 &\leqq 102 \\
x_1, x_2 &\geqq 0
\end{aligned}
$$

whose Z is graphed in Fig. 14.2. In this problem, $N = \gamma(\mathbf{z}^1, \mathbf{z}^2)$ and $\mathbf{z}^* = (\frac{9}{2}, \frac{10}{3})$. Although neither ε_i need be positive, it might be convenient to let $\varepsilon_1 = \frac{1}{2}$ and $\varepsilon_2 = \frac{2}{3}$ so that we simply have

$$
\begin{aligned}
\mathbf{z}^{**} &= \mathbf{z}^* + \boldsymbol{\varepsilon} \\
&= \begin{bmatrix} \frac{9}{2} \\ \frac{10}{3} \end{bmatrix} + \begin{bmatrix} \frac{1}{2} \\ \frac{2}{3} \end{bmatrix} \\
&= \begin{bmatrix} 5 \\ 4 \end{bmatrix}
\end{aligned}
$$

as our ideal criterion vector.

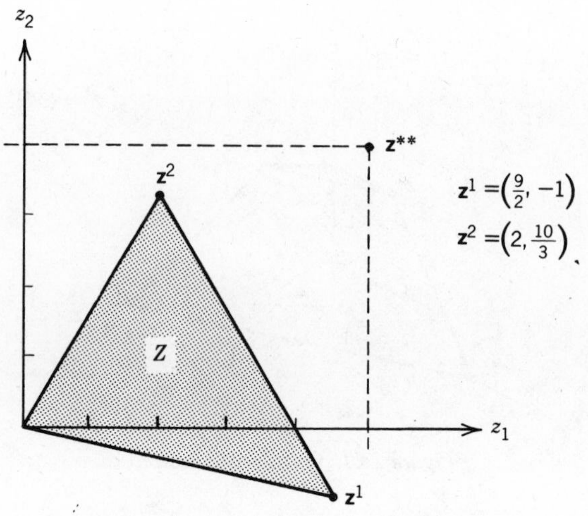

Figure 14.2. Graph of Example 2.

Examples 1 and 2 demonstrate that there is leeway in selecting values for the ε_i. Are there any guidelines? Possible rules of thumb are:

1. ε_i is to be between 1 and 10% of the range of z_i over N.
2. ε_i is to be between 1 and 10% of the range of z_i over Z.
3. ε_i is to be between 1 and 10% of z_i^*.

Care must be taken when selecting a rule of thumb because widely different results can be obtained, as in Example 3.

Example 3. Consider Fig. 14.3. If we let ε_2 be 2% of (a) the range of z_2 over N, (b) the range of z_2 over Z, or (c) z_2^*, we obtain .02, .08, or .12, respectively.

14.3 AUGMENTED WEIGHTED TCHEBYCHEFF METRICS

In this section, we discuss augmented weighted Tchebycheff metrics and their *contours*, *diagonals*, and *definition points*.

With a $\mathbf{z}^{**}$ ideal criterion vector and $\lambda \in \bar{\Lambda} = \{\lambda \in R^k \mid \lambda_i \geqq 0, \sum_{i=1}^k \lambda_i = 1\}$, we recognize

$$\|\mathbf{z}^{**} - \mathbf{z}\|_\infty^\lambda = \max_{i=1,\dots,k} \{\lambda_i |z_i^{**} - z_i|\} \tag{14.1}$$

as a member of the family of weighted Tchebycheff metrics for measuring the distance between $\mathbf{z} \in R^k$ and $\mathbf{z}^{**}$. For also measuring the distance between $\mathbf{z}$ and $\mathbf{z}^{**}$, let us define

$$|||\mathbf{z}^{**} - \mathbf{z}|||_\infty^\lambda = \|\mathbf{z}^{**} - \mathbf{z}\|_\infty^\lambda + \rho \sum_{i=1}^k |z_i^{**} - z_i| \tag{14.2}$$

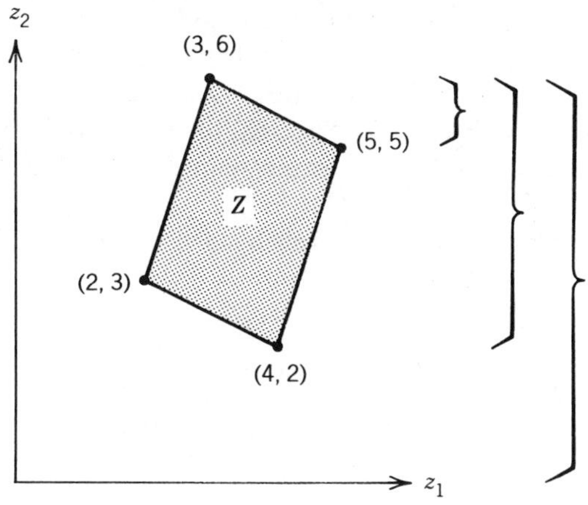

Figure 14.3. Graph of Example 3.

as a member of the family of *augmented* weighted Tchebycheff metrics. We have *families* of $\|\mathbf{z}^{**} - \mathbf{z}\|_\infty^\lambda$ and $|||\mathbf{z}^{**} - \mathbf{z}|||_\infty^\lambda$ metrics because $\lambda \in \overline{\Lambda}$. We use the word "augmented" with reference to the $|||\mathbf{z}^{**} - \mathbf{z}|||_\infty^\lambda$ metric because of the $\rho\sum_{i=1}^k |z_i^{**} - z_i|$ term, in which ρ is a sufficiently small positive number. Restrictions on the size of ρ are discussed in Sections 14.4 (Example 8), 14.8.1, and 14.8.2.

14.3.1 Contours

Suppose that in Fig. 14.4 we wish to determine the point $\bar{\mathbf{z}} \in Z$ that is closest to the $\mathbf{z}^{**}$ ideal criterion vector according to the $\lambda = (\frac{2}{5}, \frac{3}{5})$ augmented weighted Tchebycheff metric. Then, $\bar{\mathbf{z}}$ is graphically determined as shown.

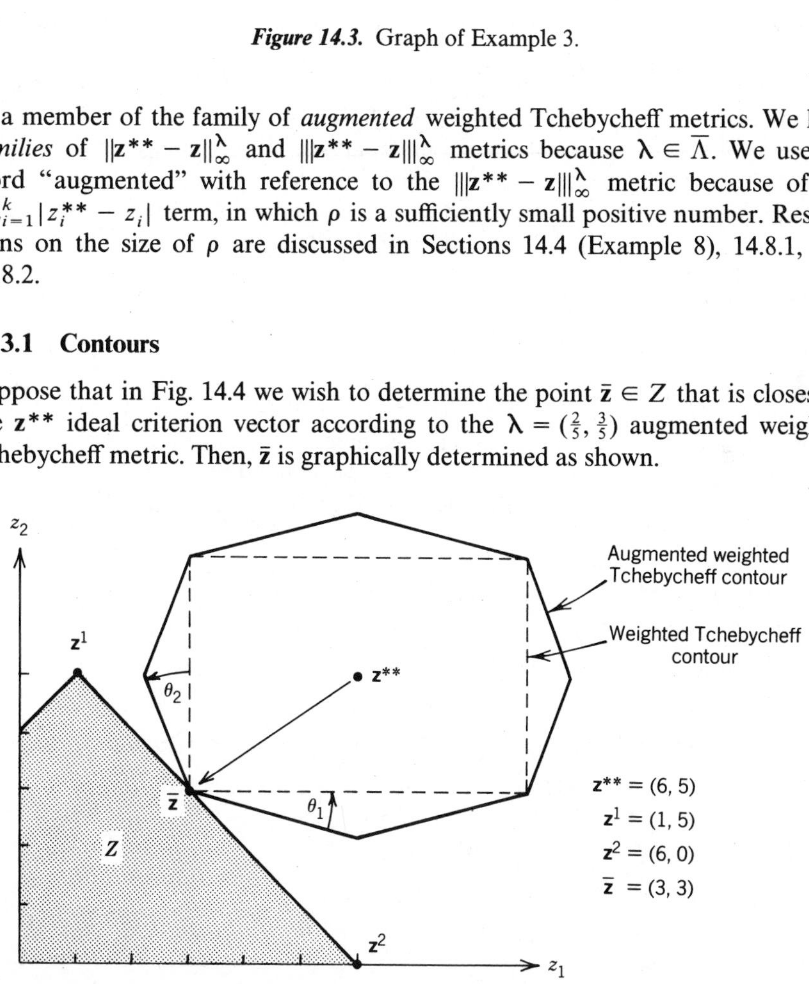

Figure 14.4. Contours of the $\|\mathbf{z}^{**} - \mathbf{z}\|_\infty^\lambda$ and $|||\mathbf{z}^{**} - \mathbf{z}|||_\infty^\lambda$ metrics intersecting Z.

In Fig. 14.4, the dashed rectangular line centered about $\mathbf{z}^{**}$ signifies the lowest valued *contour* of the $\lambda = (\frac{2}{5}, \frac{3}{5})$ weighted Tchebycheff metric that intersects Z. The solid octagonal line signifies the lowest valued contour of the $\lambda = (\frac{2}{5}, \frac{3}{5})$ augmented weighted Tchebycheff metric that intersects Z. (The contours, of course, consist of the loci of points equidistant from $\mathbf{z}^{**}$ according to their respective metrics.)

With regard to augmented weighted Tchebycheff metrics, in Fig. 14.4 we have the angles θ_1 and θ_2 that are given by

$$\theta_1 = \tan^{-1}\left(\frac{\rho}{1 - \lambda_1 + \rho}\right)$$

$$\theta_2 = \tan^{-1}\left(\frac{\rho}{1 - \lambda_2 + \rho}\right)$$

We see that the smaller ρ, the more closely the augmented weighted Tchebycheff metric approximates the weighted Tchebycheff metric (and when $\rho = 0$, the two are identical).

Because of the rules by which $\mathbf{z}^{**}$ is constructed, only the "lower left-hand" portions of the $\|\mathbf{z}^{**} - \mathbf{z}\|_\infty^\lambda$ or $|||\mathbf{z}^{**} - \mathbf{z}|||_\infty^\lambda$ contours can intersect Z (that is, only with the portions of the contours that are in the nonpositive orthant translated to $\mathbf{z}^{**}$). Therefore, since we will only be using these metrics to find points in Z closest to $\mathbf{z}^{**}$, we can drop the absolute value signs in (14.1) and (14.2) because $(z_i^{**} - z_i)$ will never be negative. Thus, we can write

$$|||\mathbf{z}^{**} - \mathbf{z}|||_\infty^\lambda = \max_{i=1,\ldots,k} \left\{\lambda_i\left(z_i^{**} - z_i\right)\right\} + \rho \sum_{i=1}^{k} \left(z_i^{**} - z_i\right)$$

14.3.2 Weighted and Augmented Weighted Tchebycheff Programs

We now formulate programs for finding the points in Z closest to $\mathbf{z}^{**}$ according to the weighted Tchebycheff and augmented weighted Tchebycheff metrics. For the $\|\mathbf{z}^{**} - \mathbf{z}\|_\infty^\lambda$ metric, we have the *weighted Tchebycheff program*

$$\begin{aligned} &\min \{\alpha\} \\ \text{s.t.} \quad &\alpha \geqq \lambda_i(z_i^{**} - z_i) && 1 \leqq i \leqq k \\ &f_i(\mathbf{x}) = z_i && 1 \leqq i \leqq k \\ &\mathbf{x} \in S \end{aligned} \tag{14.3}$$

and for the $|||\mathbf{z}^{**} - \mathbf{z}|||_\infty^\lambda$ metric, we have the *augmented weighted Tchebycheff program*

$$\begin{aligned} &\min \{\alpha + \rho \sum_{i=1}^{k} (z_i^{**} - z_i)\} \\ \text{s.t.} \quad &\alpha \geqq \lambda_i(z_i^{**} - z_i) && 1 \leqq i \leqq k \\ &f_i(\mathbf{x}) = z_i && 1 \leqq i \leqq k \\ &\mathbf{x} \in S \end{aligned} \tag{14.4}$$

A solution to either (14.3) or (14.4) is a vector of the form $(\bar{\mathbf{x}}, \bar{\mathbf{z}}, \bar{\alpha}) \in R^{n+k+1}$ where $\bar{\mathbf{z}}$ is a closest criterion vector and $\bar{\mathbf{x}}$ is its inverse image in decision space.

In the above formulations, α is a "minimax" variable in the sense that it is used to minimize the maximum weighted z_i deviation from z_i^{**}. Variable α will never be negative. However, the z_i variables are, in general, unrestricted.

14.3.3 Definition Points, Vertices, and Diagonals

In Fig. 14.4, $\bar{\mathbf{z}}$ is both the *definition point* and *vertex* of each of the contours. The line segment connecting $\mathbf{z}^{**}$ with $\bar{\mathbf{z}}$ is the *diagonal* of each contour, and the arrow from $\mathbf{z}^{**}$ to $\bar{\mathbf{z}}$ specifies the *diagonal direction* of each contour. As definitions of these concepts, we have the following.

Definition 14.5. Let $\bar{\mathbf{z}} \leqq \mathbf{z}^{**}$ and $\boldsymbol{\lambda} \in \bar{\Lambda}$. Then, $\bar{\mathbf{z}}$ is a *definition point* of the $\|\mathbf{z}^{**} - \bar{\mathbf{z}}\|_\infty^\lambda$ and $|||\mathbf{z}^{**} - \bar{\mathbf{z}}|||_\infty^\lambda$ contours if and only if

$$\lambda_i = \begin{cases} \dfrac{1}{\left(z_i^{**} - \bar{z}_i\right)} \left[\displaystyle\sum_{i=1}^{k} \dfrac{1}{\left(z_i^{**} - \bar{z}_i\right)} \right]^{-1} & \text{if } \bar{z}_i \neq z_i^{**} \text{ for all } i \\ 1 & \text{if } \bar{z}_i = z_i^{**} \\ 0 & \text{if } \bar{z}_i \neq z_i^{**} \text{ but } \exists j \ni \bar{z}_j = z_j^{**} \end{cases} \tag{14.6}$$

Definition 14.7. Let $\bar{\mathbf{z}} < \mathbf{z}^{**}$. Then, $\bar{\mathbf{z}}$ is a *vertex* of a given $\|\mathbf{z}^{**} - \bar{\mathbf{z}}\|_\infty^\lambda$ or $|||\mathbf{z}^{**} - \bar{\mathbf{z}}|||_\infty^\lambda$ contour if and only if $\bar{\mathbf{z}}$ is an extreme point of the closed convex set in R^k whose boundary is the contour.

Definition 14.8. Let $\bar{\mathbf{z}} \in R^k$ be a definition point of a $\|\mathbf{z}^{**} - \bar{\mathbf{z}}\|_\infty^\lambda$ or $|||\mathbf{z}^{**} - \bar{\mathbf{z}}|||_\infty^\lambda$ contour. Then, the line segment connecting $\mathbf{z}^{**}$ with $\bar{\mathbf{z}}$ is the *diagonal* of the contour.

From Definition 14.5, (a) if $\bar{\mathbf{z}}$ is a definition point such that $\bar{\mathbf{z}} < \mathbf{z}^{**}$, the $\boldsymbol{\lambda}$-vector associated with the contour is such that $\boldsymbol{\lambda} > \mathbf{0}$; or (b) if $\bar{\mathbf{z}}$ is a definition point such that $\bar{\mathbf{z}} \not< \mathbf{z}^{**}$, $\boldsymbol{\lambda} \geqq \mathbf{0}$ with $k - 1$ λ_i's equal to 0. Because of the premise in Definition 14.7, $\bar{\mathbf{z}}$ must be strictly less than $\mathbf{z}^{**}$ for $\bar{\mathbf{z}}$ to be a vertex of a contour. If $\bar{\mathbf{z}}$ is a definition point and $\bar{\mathbf{z}} < \mathbf{z}^{**}$, then $\bar{\mathbf{z}}$ is the vertex of the contour. A contour can have at most, one vertex. With regard to the diagonal direction of a contour, we have

Definition 14.9. Let $\boldsymbol{\lambda} \in \bar{\Lambda}$ and $\bar{\mathbf{z}}$ be a definition point of a contour. Then,

(i) When $\bar{\mathbf{z}}$ is a vertex or when $k = 2$, the direction specified by $(\bar{\mathbf{z}} - \mathbf{z}^{**})$ is *the* diagonal direction of the contour.

(ii) When $\bar{\mathbf{z}}$ is not a vertex and $k > 2$, the direction specified by $(\bar{\mathbf{z}} - \mathbf{z}^{**})$ is *a* diagonal direction of the contour.

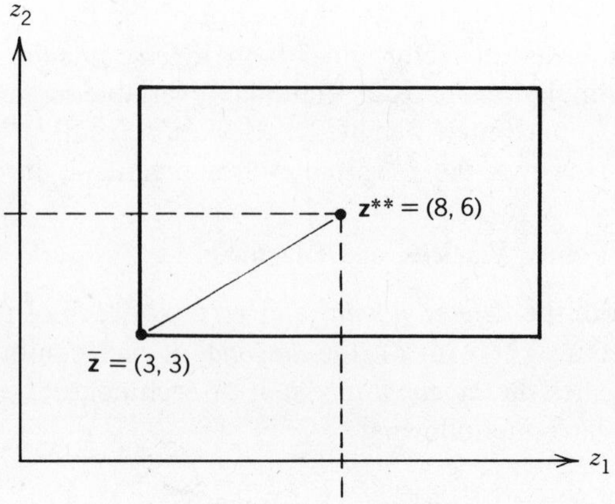

Figure 14.5. Graph of Example 4.

The point of Definition 14.9 is that when $k > 2$ and the contour does not possess a vertex, the contour has an infinite number of diagonal directions. When the contour *has a vertex* (which happens whenever $\boldsymbol{\lambda} > \mathbf{0}$), its unique diagonal direction is given by

$$-\left(\frac{1}{\lambda_1}, \frac{1}{\lambda_2}, \ldots, \frac{1}{\lambda_k}\right)$$

To illustrate the concepts of this section, consider Examples 4 and 5.

Example 4. In Fig. 14.5, $\bar{\mathbf{z}}$ is the vertex of the weighted Tchebycheff contour. None of the other "corner points" of the contour are vertices because they are not strictly less than $\mathbf{z}^{**}$. Because of the equivalence between vertices and definition points when the point in question is strictly less than $\mathbf{z}^{**}$, $\bar{\mathbf{z}}$ is also the definition point of the contour. In applying (14.6), we compute

$$\left[\sum_{i=1}^{k} \frac{1}{(z_i^{**} - \bar{z}_i)}\right]^{-1} = \left[\tfrac{1}{5} + \tfrac{1}{3}\right]^{-1} = \tfrac{15}{8}$$

because $\bar{z}_i \neq z_i^{**}$ for all i. Then, we have

$$\lambda_1 = \left(\tfrac{1}{5}\right)\left(\tfrac{15}{8}\right) = \tfrac{3}{8}$$

$$\lambda_2 = \left(\tfrac{1}{3}\right)\left(\tfrac{15}{8}\right) = \tfrac{5}{8}$$

Thus, Fig. 14.5 portrays the $\boldsymbol{\lambda} = (\tfrac{3}{8}, \tfrac{5}{8})$ $\|\mathbf{z}^{**} - \mathbf{z}\|_\infty^\lambda$ contour that has $\bar{\mathbf{z}}$ as its vertex and definition point. Computing the diagonal direction of the contour, we have

$$-\left(\frac{1}{\lambda_1}, \frac{1}{\lambda_2}\right) = -\left(\tfrac{8}{3}, \tfrac{8}{5}\right)$$

This is proportional to $(\bar{\mathbf{z}} - \mathbf{z}^{**}) = (-5, -3)$.

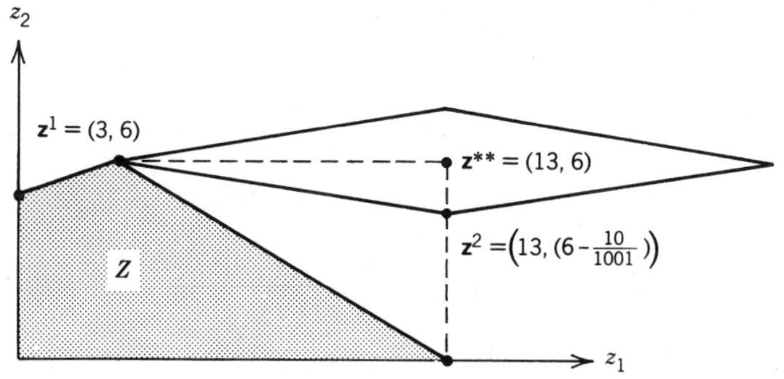

Figure 14.6. Graph of Example 5.

Example 5. Consider Fig. 14.6, in which is drawn the $|||\mathbf{z}^{**} - \mathbf{z}|||_{\infty}^{\lambda} = .01$ contour where $\boldsymbol{\lambda} = (0, 1)$ and $\rho = .001$. From Definition 14.5, it is confirmed that $\mathbf{z}^1$ is a definition point of the $|||\mathbf{z}^{**} - \mathbf{z}^1|||_{\infty}^{\lambda}$ contour. Although $\mathbf{z}^1$ is the only definition point of the contour, $\mathbf{z}^1$ is not a vertex because $\mathbf{z}^1 \nless \mathbf{z}^{**}$. Thus, the contour in Fig. 14.6 is a $k = 2$ example of a contour that does not possess a vertex. (Alternatively, we could have recognized that the contour does not have a vertex because $\boldsymbol{\lambda}$ has a zero component). Note that $\mathbf{z}^2$ is not a definition point because its weighting vector indicated by (14.6) is $(1, 0)$, not the $(0, 1)$ we are assuming. From Definition 14.9, the diagonal direction of the contour is given by $(\mathbf{z}^1 - \mathbf{z}^{**})$ which is proportional to $(-1, 0)$.

14.3.4 Points of Intersection with *Z*

Sometimes, the points of intersection between Z and the lowest valued intersecting contour will include a definition point or vertex, and sometimes not.

Example 6

- **a.** In Figs. 14.7*a* and *b*, let $Z = \gamma(\mathbf{z}^1, \mathbf{z}^2, \mathbf{z}^3)$ where $\mathbf{z}^2 = (5, 6)$ and $\mathbf{z}^3 = (8, 0)$. Hence, $N = \gamma(\mathbf{z}^2, \mathbf{z}^3)$. Let $\boldsymbol{\lambda} = (.8, .2)$. Then, in Fig. 14.7*a* the point of intersection between Z and the lowest valued weighted Tchebycheff contour is $\mathbf{z}^4 = (\frac{22}{3}, \frac{4}{3})$, where $\mathbf{z}^4$ is both a definition point and vertex of the intersecting contour.
- **b.** With $\boldsymbol{\lambda} = (.2, .8)$ in Fig. 14.7*b*, the point of intersection is $\mathbf{z}^2$ which is neither a definition point nor vertex of the lowest valued intersecting contour.

With the solid contours in Example 6, we note that $\mathbf{z}^4$ and $\mathbf{z}^2$ would both have been generated had we used the augmented weighted Tchebycheff metric (with $\rho > 0$ small). So why do we have the $\rho\sum_{i=1}^{k}(z_i^{**} - z_i)$ term? Because otherwise, we might have trouble identifying a nondominated criterion vector as in Example 7.

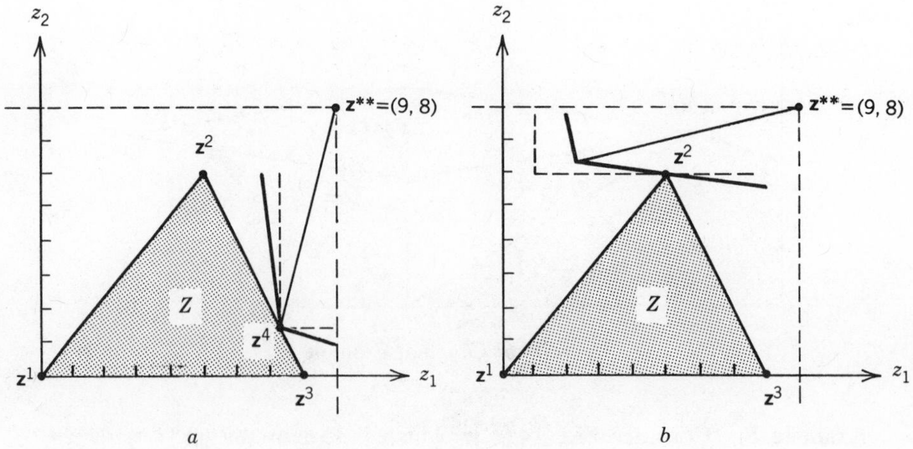

Figure 14.7. Graphs of Example 6.

Example 7. In the example of Fig. 14.8a, $N = \gamma(\mathbf{z}^2, \mathbf{z}^3)$. The set of criterion vectors tied for being closest to $\mathbf{z}^{**}$ according to the drawn weighted Tchebycheff metric (i.e., with $\rho = 0$) is $\gamma(\mathbf{z}^1, \mathbf{z}^2)$, of which only $\mathbf{z}^2$ is nondominated. That is, each criterion vector in $\gamma(\mathbf{z}^1, \mathbf{z}^2)$ minimizes the weighted Tchebycheff program (14.3). Unfortunately, when using conventionally available software to solve the weighted Tchebycheff program, we are not assured that $\mathbf{z}^2$ will be outputted. In Fig. 14.8b, we do not encounter the difficulty of a. For the augmented weighted Tchebycheff metric (with $\rho > 0$), the criterion vector closest to $\mathbf{z}^{**}$ is uniquely $\mathbf{z}^2$.

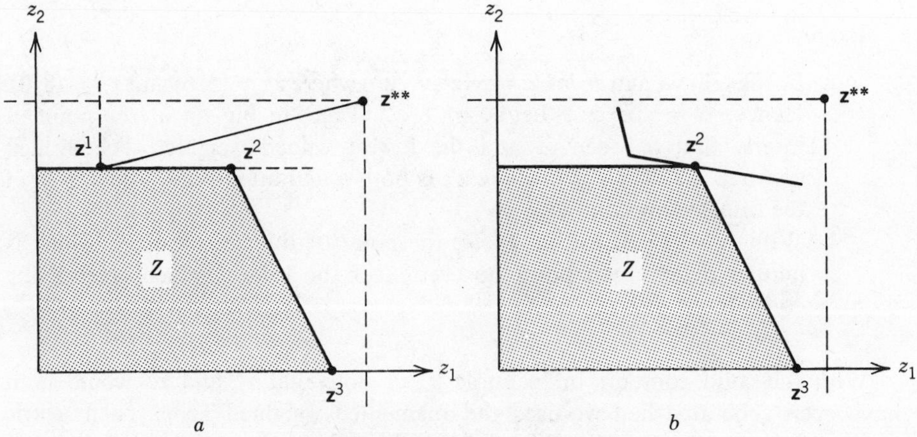

Figure 14.8. Graphs of Example 7.

14.4 SELECTING ρ VALUES

The reason for the $\rho\sum_{i=1}^{k}(z_i^{**} - z_i)$ term is to give the contour a "slight slope." As long as the slope is not too great, the augmented weighted Tchebycheff program will not only be guaranteed to return a nondominated criterion vector, but it will be able to generate *any* particular nondominated criterion vector, given an appropriate λ-vector. To demonstrate the slope of a contour, consider a two-dimensional augmented weighted Tchebycheff contour in which the slope is characterized (see Fig. 14.4) by the angles.

$$\theta_1 = \tan^{-1}\left(\frac{\rho}{1 - \lambda_1 + \rho}\right)$$

$$\theta_2 = \tan^{-1}\left(\frac{\rho}{1 - \lambda_2 + \rho}\right)$$

A difficulty with the angles is that they are not constant. They are a function of ρ, $\mathbf{z}^{**}$, and the location of the vertex of the contour. Example 8 illustrates how the angles are a function of ρ and $\mathbf{z}^{**}$, and Example 9 illustrates how the angles change with the location of the vertex. Example 9 also illustrates what can happen if ρ is too large.

Example 8. With (1, 1) as the vertex, let $\rho = .001$. In Fig. 14.9*a*, $\lambda = (\frac{4}{5}, \frac{1}{5})$, $\theta_1 = .285°$, and $\theta_2 = .072°$. In Fig. 14.9*b*, $\lambda = (\frac{4}{7}, \frac{3}{7})$, $\theta_1 = .133°$, and $\theta_2 = .100°$. Note that θ_i becomes bigger as λ_i becomes bigger. If we were to assume $\rho = .01$, then θ_1 and θ_2 in Fig. 14.9*a* would be 2.726° and .707°, and θ_1 and θ_2 in Fig. 14.9*b* would be 1.306° and .985°.

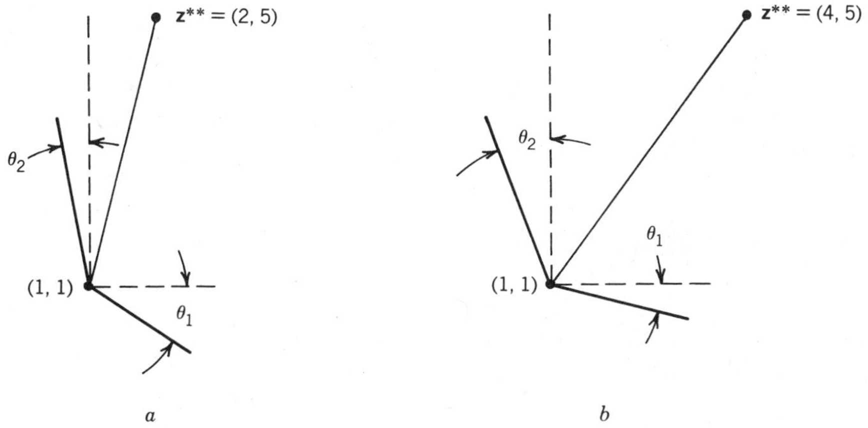

Figure 14.9. Graphs of Example 8.

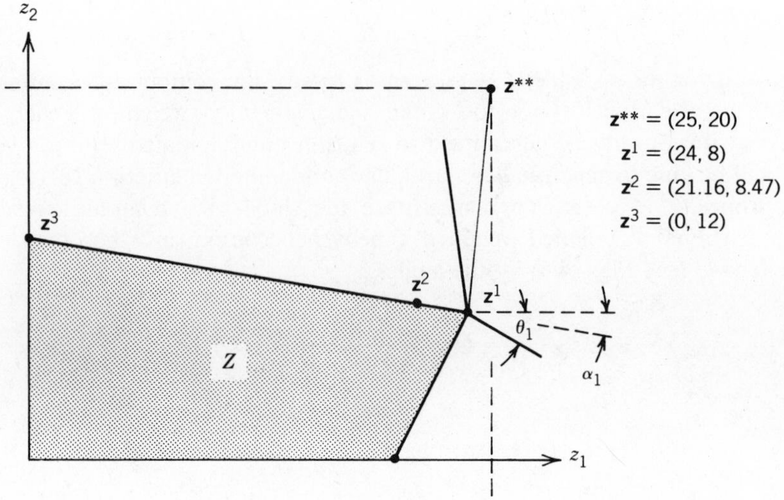

Figure 14.10. Graph of Example 9.

Example 9. Consider Fig. 14.10 in which the slope of the $\gamma(\mathbf{z}^1, \mathbf{z}^3)$ edge is given by $\alpha_1 = 9.46°$. Letting $\rho = .05$ yields

a. $\theta_1 = 21.50°$ at $\mathbf{z}^1$.
b. $\theta_1 = 9.46°$ at $\mathbf{z}^2$.
c. $\theta_1 = 3.54°$ at $\mathbf{z}^3$.

We see that θ_1 decreases as we move along the $\gamma(\mathbf{z}^1, \mathbf{z}^3)$ edge from $\mathbf{z}^1$ to $\mathbf{z}^3$. Because $\theta_1 > \alpha_1$ at $\mathbf{z}^1$, criterion vectors on the portion of the nondominated edge $\gamma(\mathbf{z}^1, \mathbf{z}^3)$ between $\mathbf{z}^1$ and $\mathbf{z}^2$ cannot be generated by the augmented weighted Tchebycheff program.

In theory, the difficulties of Example 9 can be avoided by making $\rho > 0$ *sufficiently* small. In practice, we set ρ small enough to avoid problems such as those in Example 9, but large enough to be numerically significant in a computer. Values between .0001 and .01 should normally suffice.

14.5 DIAGONAL DIRECTION OF A TCHEBYCHEFF METRIC

Consider the family of all positive valued contours of the $|||\mathbf{z}^{**} - \mathbf{z}|||_\infty^{\boldsymbol{\lambda}}$ metric for $\boldsymbol{\lambda} \in \bar{\Lambda}$, $\boldsymbol{\lambda} > \mathbf{0}$, and fixed. Then, the diagonal direction of each positive valued contour in the family is the same and is given by

$$-\left(\frac{1}{\lambda_1}, \frac{1}{\lambda_2}, \ldots, \frac{1}{\lambda_k}\right) \tag{14.10}$$

Thus, all vertices of contours in the family lie along the half-ray emanating from

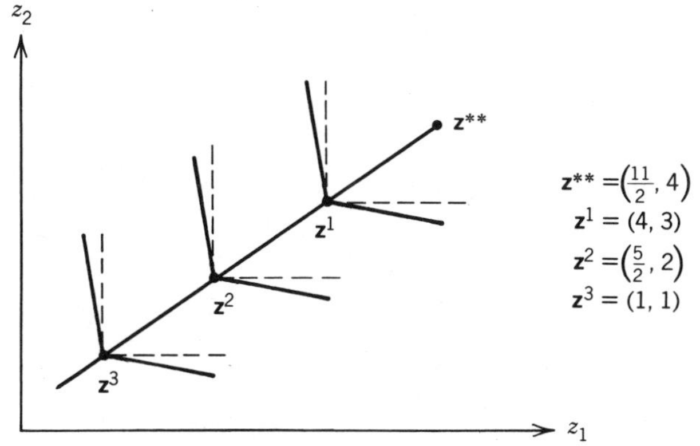

Figure 14.11. Graph of Example 10.

z** in the direction of expression (14.10). In this sense, the half-ray is the diagonal of the *metric* and expression (14.10) specifies the diagonal direction of the *metric*. We note that as long as $\bar{\lambda}$ is the same, the weighted Tchebycheff metric shares the same diagonal and diagonal direction as the augmented weighted Tchebycheff metric.

As with the diagonal and diagonal direction of a contour in Section 14.3.3, the diagonal and diagonal direction of the $\|\mathbf{z}^{**} - \mathbf{z}\|_\infty^\lambda$ and $|||\mathbf{z}^{**} - \mathbf{z}|||_\infty^\lambda$ metrics are nonunique whenever λ has a zero component and $k > 2$.

Example 10. With $\lambda = (.4, .6)$, Fig. 14.11 shows the weighted Tchebycheff (dashed line) and augmented weighted Tchebycheff (solid line) contours defined by $\mathbf{z}^{**}$, $\mathbf{z}^1$, $\mathbf{z}^2$, and $\mathbf{z}^3$. The diagonal direction of both metrics is given by

$$-\left(\frac{1}{\lambda_1}, \frac{1}{\lambda_2}\right) = \left(-\tfrac{10}{4}, -\tfrac{10}{6}\right)$$

The values of the $\|\mathbf{z}^{**} - \mathbf{z}\|_\infty^\lambda$ contours are 0.6, 1.2, and 1.8 at $\mathbf{z}^1$, $\mathbf{z}^2$ and $\mathbf{z}^3$, respectively. With $\rho = .001$, the values of the $|||\mathbf{z}^{**} - \mathbf{z}|||_\infty^\lambda$ contours are 0.6025, 1.2050, and 1.8075 at $\mathbf{z}^1$, $\mathbf{z}^2$, and $\mathbf{z}^3$, respectively.

14.6 UNSUPPORTED NONDOMINATED CRITERION VECTORS

Let

$$Z^{\leqq} = \text{Convex hull of } \left[N \oplus \{\mathbf{z} \in R^k \mid \mathbf{z} \leqq \mathbf{0}\}\right]$$

where $\oplus$ signifies set addition. Then, we have the following definitions.

Definition 14.11. Let $\mathbf{z} \in N$. Then, if $\mathbf{z}$ is on the boundary of $Z^{\leqq}$, $\mathbf{z}$ is a *supported* nondominated criterion vector. Otherwise, $\mathbf{z}$ is an *unsupported* (*convex dominated*) nondominated criterion vector.

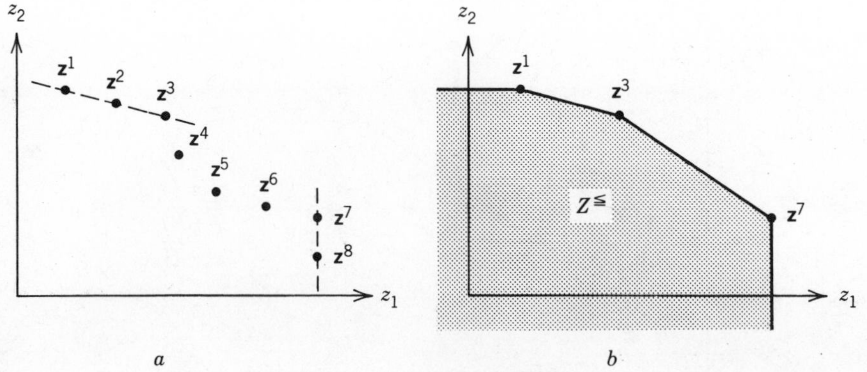

Figure 14.12. Graphs of Example 11.

Unsupported nondominated criterion vectors are dominated by some convex combination of other nondominated criterion vectors.

Definition 14.12. Let $\mathbf{z} \in N$ be supported. Then, $\mathbf{z}$ is *supported-extreme* if it is an extreme point of $Z^{\leqq}$. Otherwise, $\mathbf{z}$ is *supported-nonextreme*.

Example 11. Let $Z = \{\mathbf{z}^i \mid 1 \leqq i \leqq 8\}$ in Fig. 14.12*a*. Note that $\mathbf{z}^2$ is a convex combination of $\mathbf{z}^1$ and $\mathbf{z}^3$. As shown via Fig. 14.12*b*, of the nondominated criterion vectors,

a. $\mathbf{z}^1$, $\mathbf{z}^3$, $\mathbf{z}^7$ are supported-extreme.
b. $\mathbf{z}^2$ is supported-nonextreme.
c. $\mathbf{z}^4$, $\mathbf{z}^5$, $\mathbf{z}^6$ are unsupported.

The terms are not applied to $\mathbf{z}^8$ because $\mathbf{z}^8 \notin N$.

Inverse images of supported nondominated criterion vectors are said to be *supported* efficient points (in decision space) and inverse images of unsupported nondominated criterion vectors are said to be *unsupported* efficient points (in decision space).

14.6.1 Weighted-Sums Approaches and Unsupportedness

Unsupported nondominated criterion vectors do not occur in MOLPs. However, in multiple objective integer and nonlinear programs, unsupported nondominated criterion vectors are a likely occurrence. Because each unsupported member of N is dominated by some convex combination of other nondominated criterion vectors, it is not possible to generate unsupported criterion vectors using the *weighted-sums* program

$$\max\left\{ \sum_{i=1}^{k} \lambda_i f_i(\mathbf{x}) \mid \mathbf{x} \in S \right\} \tag{14.13}$$

regardless of which $\boldsymbol{\lambda} \in \Lambda$ is employed.

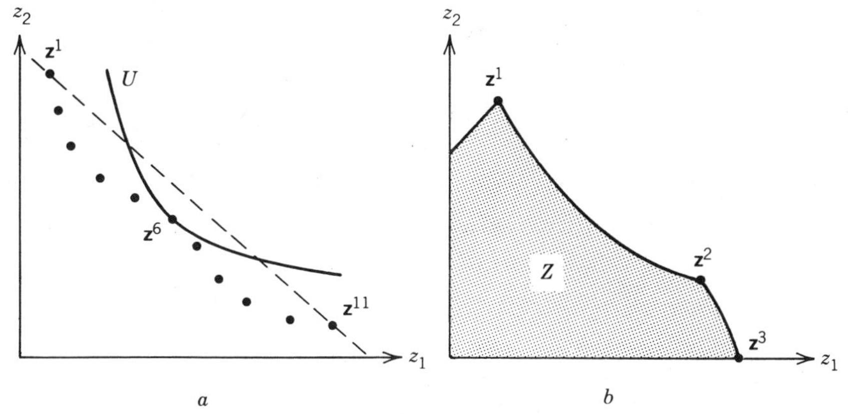

Figure 14.13. Graphs of Example 12.

Example 12. Let $Z = N = \{\mathbf{z}^i \mid 1 \leqq i \leqq 11\}$ in the multiple objective integer program of Fig. 14.13*a*. In this problem, all criterion vectors are unsupported except $\mathbf{z}^1$ and $\mathbf{z}^{11}$. Assume that the DM's utility function makes $\mathbf{z}^6$ optimal. Using the weighted-sums program, we would not be able to generate $\mathbf{z}^6$. We would only be able to generate $\mathbf{z}^1$ or $\mathbf{z}^{11}$, neither of which is a good approximation of $\mathbf{z}^6$. In the multiple objective nonlinear program of Fig. 14.13*b*, N consists of the boundary of Z from $\mathbf{z}^1$ to $\mathbf{z}^2$ to $\mathbf{z}^3$. Only $\mathbf{z}^1$ and the criterion vectors on the boundary portion from $\mathbf{z}^2$ to $\mathbf{z}^3$ are supported members of N. If the optimal criterion vector were elsewhere, the weighted-sums program of (14.13) would not be able to compute it.

As illustrated by the two problems of Example 12, the shortcoming of the weighted-sums approach is that it may not be able to get us close to an optimal point if the optimal solution is unsupported.

14.6.2 Weighted-Sums Approaches and Supported Criterion Vectors

Even if supported, there may be difficulties locating an optimal criterion vector using the weighted-sums approach. Consider Example 13.

Example 13. Supported-nonextreme criterion vector $\mathbf{z}^2$ is optimal in the MOLP of Fig. 14.14*a*. However, the best an LP code can do with the weighted-sums formulation is generate either $\mathbf{z}^1$ or $\mathbf{z}^3$. Supported-nonextreme criterion vector $\mathbf{z}^5$ is optimal in the multiple objective integer program (MOIP) of Fig. 14.14*b*. To generate $\mathbf{z}^5$ with the weighted-sums model, it is necessary to use a $\boldsymbol{\lambda}$-vector such that the weighted-sums objective function gradient is precisely perpendicular to the dashed line connecting $\mathbf{z}^4$ and $\mathbf{z}^6$. Even if we were to use such a gradient, $\mathbf{z}^5$ might not be included among the solutions outputted by an IP code because of the alternatively optimal nature of $\mathbf{z}^4$, $\mathbf{z}^5$, and $\mathbf{z}^6$.

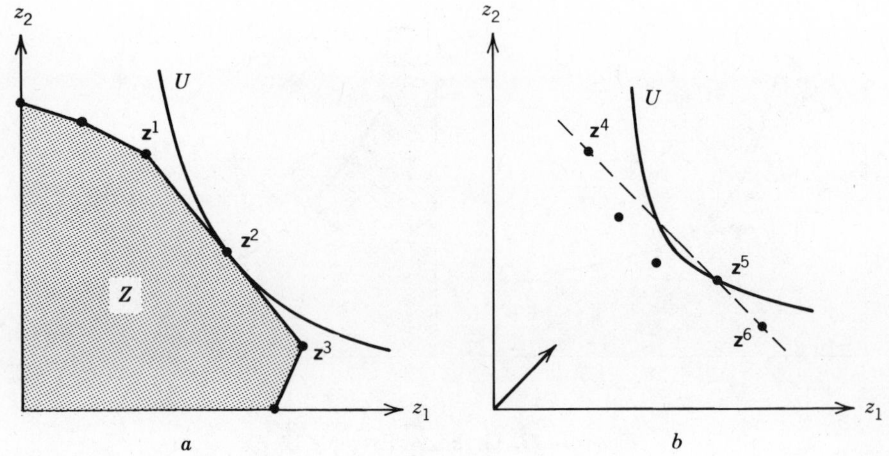

Figure 14.14. Graphs of Example 13.

14.6.3 Subproblem Test for Unsupportedness

Let N be finite. Assuming that $\mathbf{z}^i \neq \mathbf{z}^j$ for all $i \neq j$, let $I_N = \{j \mid \mathbf{z}^j \in N\}$. Because an unsupported nondominated criterion vector is dominated by some convex combination of other nondominated criterion vectors, we have the subproblem LP

$$\max \left\{ \sum_{i=1}^{k} r_i = v \right\}$$

$$\text{s.t.} \quad \sum_{j \in I_N - \{h\}} \lambda_j z_i^j - r_i = z_i^h \qquad 1 \leqq i \leqq k$$

$$\sum_{j \in I_N - \{h\}} \lambda_j = 1$$

$$\text{all } \lambda_j, r_i \geqq 0$$

for determining the supported/unsupported status of a given nondominated criterion vector $\mathbf{z}^h$. If we allow v^* to denote the optimal objective function value, $\mathbf{z}^h$ is

1. Supported-extreme if the subproblem LP is inconsistent.
2. Supported-nonextreme if $v^* = 0$.
3. Unsupported if $v^* > 0$.

Example 14. With the first and second *maximizing* objectives, and the third a *minimizing* objective, let

$$\mathbf{z}^1 = \begin{bmatrix} 1 \\ -2 \\ -7 \end{bmatrix} \quad \mathbf{z}^2 = \begin{bmatrix} 2 \\ 4 \\ -1 \end{bmatrix} \quad \mathbf{z}^3 = \begin{bmatrix} 5 \\ 2 \\ 1 \end{bmatrix} \quad \mathbf{z}^4 = \begin{bmatrix} 2 \\ 1 \\ -2 \end{bmatrix} \quad \mathbf{z}^5 = \begin{bmatrix} 3 \\ 0 \\ -3 \end{bmatrix}$$

be members of N. To determine whether $\mathbf{z}^4$ is unsupported with respect to the other four criterion vectors, we solve the subproblem LP

	λ_1	λ_2	λ_3	λ_5	r_1	r_2	r_3		
Obj					1	1	1	max	
	1	2	5	3	-1			=	2
s.t.	-2	4	2			-1		=	1
	-7	-1	1	-3			1	=	-2
	1	1	1	1				=	1

Since $v^* = 1.75$, $\mathbf{z}^4$ is unsupported. Should we apply the subproblem test to $\mathbf{z}^5$, we would find $v^* = 0$, meaning that with respect to $\{\mathbf{z}^1, \mathbf{z}^2, \mathbf{z}^3, \mathbf{z}^4\}$, $\mathbf{z}^5$ is supported-nonextreme. Although $\mathbf{z}^5$ is supported-nonextreme with respect to the other four criterion vectors, $\mathbf{z}^5$ might be unsupported if N contained other criterion vectors.

14.6.4 Domination Sets and Unsupportedness

For unsupported nondominated criterion vectors to exist, Z, the feasible region in criterion space, must be nonconvex. Therefore, unsupported members of N can only occur in multiple objective discrete alternative and multiple objective nonlinear programs. Since it is not necessary that every such program possess unsupported nondominated criterion vectors, what property is it that leads to unsupportedness? The answer is the degree to which there is conflict among the objectives. The greater the degree to which the gradients of the objective functions are radially dispersed, the smaller the domination set (see Section 6.8). And the smaller the domination set, the greater the likelihood of unsupportedness. To illustrate, consider Examples 15, 16, and 17 that are graphed in decision space.

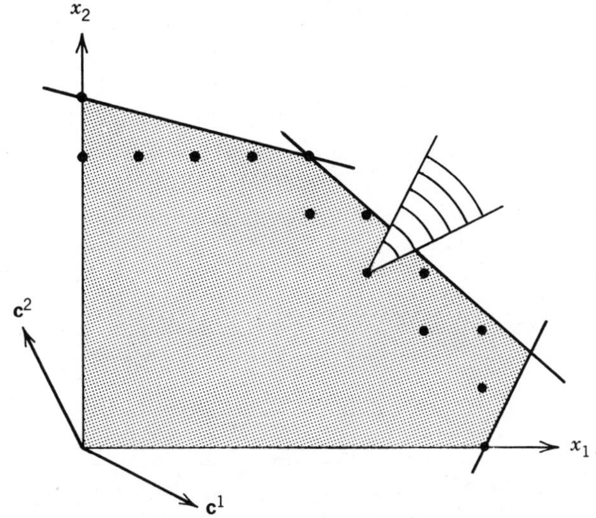

Figure 14.15. Graph of Example 15.

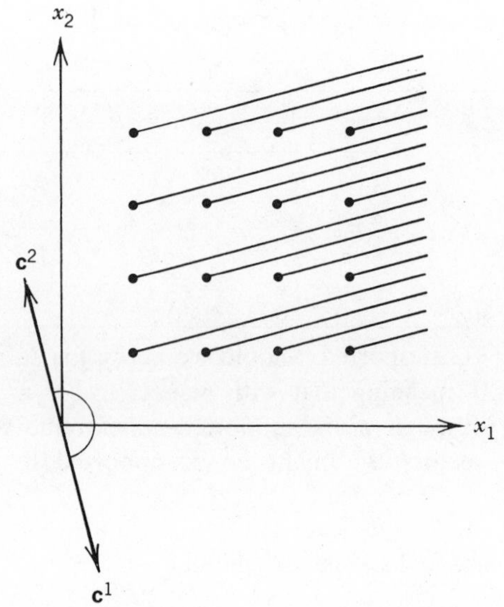

Figure 14.16. Graph of Example 16.

Example 15. Assume the multiple objective integer program

$$\max\{\ 2x_1 - \ x_2 = z_1\}$$
$$\max\{-x_1 + 2x_2 = z_2\}$$
$$\text{s.t.} \quad x_1 + 4x_2 \leqq 24$$
$$8x_1 + 9x_2 \leqq 77$$
$$2x_1 - \ x_2 \leqq 14$$
$$x_1, x_2 \geqq 0 \qquad \text{and integer}$$

As graphed in Fig. 14.15, there is considerable conflict between the objectives. This results in 14 efficient points (in decision space). Now visualize the domination set as indicated. Of the 14 efficient points, seven have unsupported nondominated criterion vectors, four have supported-extreme nondominated criterion vectors, and three have supported-nonextreme nondominated criterion vectors.

Example 16. In the MOIP of Fig. 14.16, the objectives are almost in total conflict. This results in the "very thin" domination sets as indicated by the straight lines. As seen, all 16 feasible points are efficient, seven with supported and nine with unsupported nondominated criterion vectors.

Example 17. Consider three different versions of the mixed 0–1 MOIP

$$\max\left\{\mathbf{c}^1\mathbf{x} = z_1\right\}$$
$$\max\left\{\mathbf{c}^2\mathbf{x} = z_2\right\}$$
$$\text{s.t.} \quad \mathbf{x} \in S = \left\{\mathbf{x} \in R^2 \mid 0 \leqq x_1 \leqq 1, \quad x_2 = 0 \text{ or } 1\right\}$$

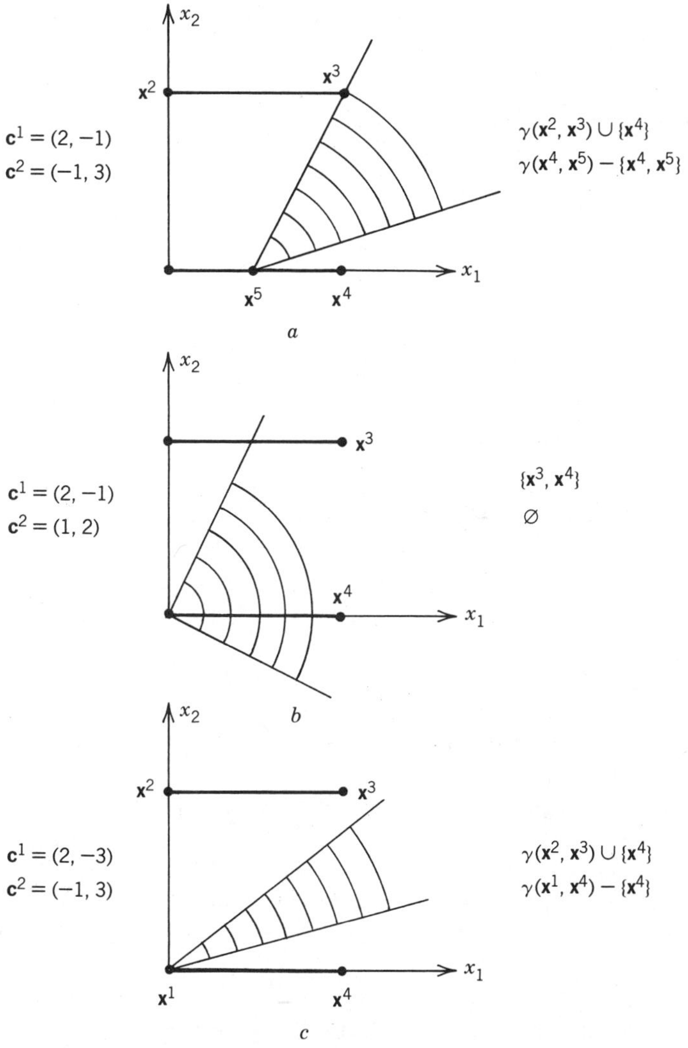

Figure 14.17. Graphs of Example 17.

Let $\mathbf{x}^1 = (0,0)$, $\mathbf{x}^2 = (0,1)$, $\mathbf{x}^3 = (1,1)$, $\mathbf{x}^4 = (1,0)$, and $\mathbf{x}^5 = (\frac{1}{2},0)$. With different sets of objective function coefficient vectors on the left, domination sets drawn in the center, and sets of supported efficient and unsupported efficient points listed on the right, we have Figs. 14.17*a*, *b*, and *c*.

14.7 IMPROPERLY NONDOMINATED CRITERION VECTORS

Let $\bar{\mathbf{x}} \in S$ such that $\mathbf{z}(\bar{\mathbf{x}}) \in N$. Apart from whether $\mathbf{z}(\bar{\mathbf{x}})$ is supported or nonsupported, $\mathbf{z}(\bar{\mathbf{x}})$ is either *properly* or *improperly* nondominated. From Geoffrion

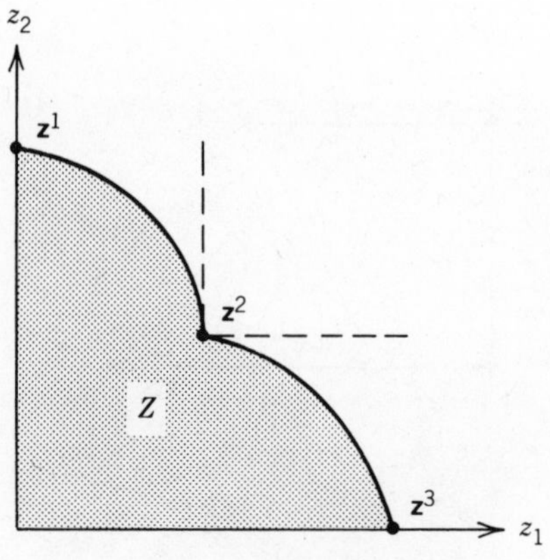

Figure 14.18. Graph of Example 18.

(1968), a criterion vector $\mathbf{z}(\bar{\mathbf{x}}) \in N$ is properly nondominated if there exists a $\pi > 0$ such that, for each i, we have

$$\frac{f_i(\mathbf{x}) - f_i(\bar{\mathbf{x}})}{f_j(\bar{\mathbf{x}}) - f_j(\mathbf{x})} \leqq \pi$$

for some j such that $f_j(\mathbf{x}) < f_j(\bar{\mathbf{x}})$ whenever $\mathbf{x} \in S$ and $f_i(\mathbf{x}) > f_i(\bar{\mathbf{x}})$. Otherwise, the criterion vector is improperly nondominated. In other words, $\mathbf{z}(\bar{\mathbf{x}}) \in N$ is improperly nondominated if there exist criteria for which the marginal gain-to-loss ratio at $\bar{\mathbf{x}}$ can be made arbitrarily large.

Inverse images of properly and improperly nondominated vectors in criterion space are *properly* and *improperly* efficient points in decision space. Improperly nondominated criterion vectors cannot occur in MOLPs and MOIPs that have a finite number of constraints. They can only occur in certain types of discrete problems that have an infinite number of alternatives and in nonlinear multiple objective programs.

Example 18. In the multiple objective nonlinear program of Fig. 14.18, N consists of the boundary portion from $\mathbf{z}^1$ to $\mathbf{z}^2$ to $\mathbf{z}^3$. In this problem, $\mathbf{z}^2$ (in the cusp) is an improperly nondominated criterion vector.

Example 19. Consider the discrete alternative multiple objective program

$$\max \{x = z_1\}$$

$$\max \left\{1 - (x)^2 = z_2\right\}$$

$$\text{s.t.} \quad x \in S = \left\{x^i \in R \,|\, x^i = \frac{1}{2^i}, \quad i = 1, 2, \ldots\right\} \cup \{x^0 = 0\}$$

With N consisting of the sequence of criterion vectors converging to $\mathbf{z}^0$ in Fig. 14.19, $\mathbf{z}^0$ is improperly nondominated. This is because the slope of the line connecting $\mathbf{z}^0$ with $\mathbf{z}^j$ becomes arbitrarily close to being horizontal as $j \to \infty$. Since $\mathbf{z}^0$ is improperly nondominated, its inverse image x^0 is improperly efficient.

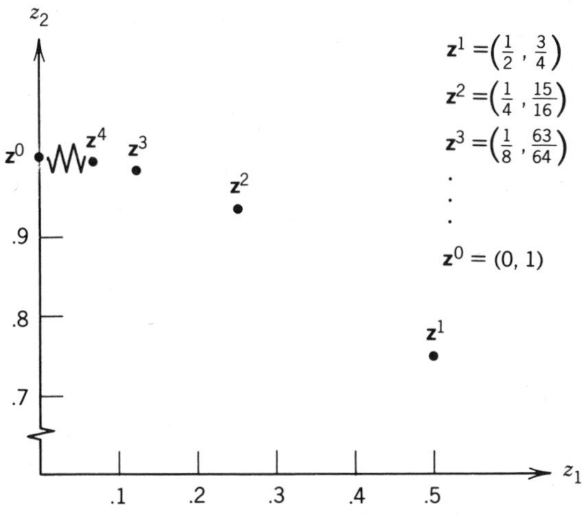

Figure 14.19. Graph of Example 19.

Even when supported, we note that improperly nondominated criterion vectors cannot be generated by the weighted-sums program unless at least one objective is ascribed a zero weight. Further, because of the "slope" given to the Tchebycheff contours when $\rho > 0$, improperly nondominated criterion vectors cannot be generated by the augmented weighted Tchebycheff program.

The question is: Do we have to worry about improperly nondominated criterion vectors? If all we know about the decision maker's utility function is that it is coordinatewise increasing, as shown in Example 20, the answer may be *yes*.

Example 20. In Fig. 14.20, N consists of the boundary segment from $\mathbf{z}^1$ to $\mathbf{z}^2$ (but not including $\mathbf{z}^2$) and the boundary segment from $\mathbf{z}^3$ to $\mathbf{z}^4$. Suppose that, as

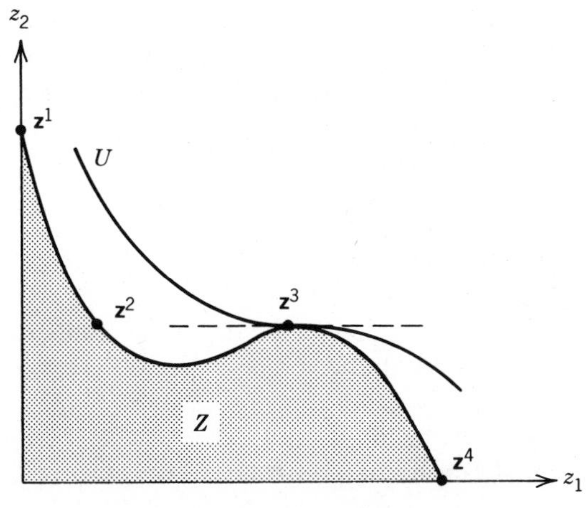

Figure 14.20. Graph of Example 20.

in Fig. 14.20, the decision maker has a utility function contour with an inflection point at $\mathbf{z}^3$. Then, $\mathbf{z}^3$, which is improperly nondominated, would be optimal.

Although unlikely, Example 20 shows that under certain conditions, it is possible for an improperly nondominated criterion vector to be optimal.

14.8 TCHEBYCHEFF THEORY

This section presents the theoretical results upon which the application of the interactive Tchebycheff approach to multiple objective linear, integer, and nonlinear programming problems is based.

14.8.1 Theory for the Finite-Discrete Case

Let S be finite. Then, Z and, hence, N are finite. In this section, we show how each of the finite number of nondominated criterion vectors is *uniquely computable* via the augmented weighted Tchebycheff program

$$
\begin{array}{llr}
\min \left\{ \alpha + \rho \mathbf{e}^T(\mathbf{z}^{**} - \mathbf{z}) \right\} & & \\
\text{s.t.} \quad \alpha \geqq \lambda_i \left(z_i^{**} - z_i \right) & 1 \leqq i \leqq k & \\
\quad f_i(\mathbf{x}) = z_i & 1 \leqq i \leqq k & (14.14) \\
\quad \mathbf{x} \in S & &
\end{array}
$$

where $\rho > 0$ and $\mathbf{e}^T$ is the sum vector of ones. By uniquely computable, we mean that for each $\bar{\mathbf{z}} \in N$ there exists a $\boldsymbol{\lambda} \in \bar{\Lambda}$ such that $\bar{\mathbf{z}}$ is the unique optimal solution of the augmented weighted Tchebycheff program. In this way, none of the finite number of nondominated criterion vectors can remain hidden (even if unsupported).

In the theoretical development that follows, Theorem 14.15 shows that among the optimal solutions of the weighted Tchebycheff program

$$
\begin{array}{ll}
\min \{ \alpha \} & \\
\text{s.t.} \quad \alpha \geqq \lambda_i (z_i^{**} - z_i) & 1 \leqq i \leqq k \\
\quad f_i(\mathbf{x}) = z_i & 1 \leqq i \leqq k \\
\quad \mathbf{x} \in S &
\end{array}
$$

can be found at least one nondominated criterion vector. However, our satisfaction with the weighted Tchebycheff program is diminished because of difficulties in separating nondominated criterion vectors from the dominated in situations of alternative optima (which are not uncommon).

Theorem 14.15. Let Z be finite and $M = \{\mathbf{z} \in Z \,|\, (\mathbf{x}, \mathbf{z}, \alpha)$ is a minimal solution of the weighted Tchebycheff program$\}$. Then, there exists a $\bar{\mathbf{z}} \in M$ such that $\bar{\mathbf{z}} \in N$.

Proof. Since Z is finite, $N \neq \varnothing$. Let $\tilde{\alpha}$ be the minimum value of the weighted Tchebycheff program. Suppose there does not exist a $\mathbf{z} \in M$ such that $\mathbf{z} \in N$. Let $\hat{\mathbf{z}}$ be a member of M that is not dominated by another member of M. If $\hat{\mathbf{z}} \notin N$, then there exists a $\bar{\mathbf{z}} \in N$ such that $\bar{\mathbf{z}} \geqq \hat{\mathbf{z}}$, $\bar{\mathbf{z}} \neq \hat{\mathbf{z}}$. But with $\tilde{\alpha}$ optimal, this implies that $\bar{\mathbf{z}} \in M$, which is a contradiction. Thus, there exists a $\bar{\mathbf{z}} \in M$ such that $\bar{\mathbf{z}} \in N$. ◀

We note that by finding the minimal $\alpha \in R$ of the weighted Tchebycheff program, we find the smallest (in a subset sense) set

$$\Phi(\alpha) = \left\{ \mathbf{z} \in R^k \,|\, z_i \in \left[z_i^{**} - \frac{\alpha}{\lambda_i}, +\infty \right) \quad \text{when} \quad \lambda_i > 0 \right\}$$

in the nested family of sets $\{\Phi(\alpha)\}_{\alpha \geqq 0}$ that intersects Z. Also, let

1. Z be finite, $\mathbf{z}^p \in N$, and $\mathbf{z}^q \in Z$ such that $\mathbf{z}^p \neq \mathbf{z}^q$.

2.

$$\lambda_i^p = \begin{cases} \dfrac{1}{\left(z_i^{**} - z_i^p\right)} \left[\displaystyle\sum_{i=1}^{k} \frac{1}{\left(z_i^{**} - z_i^p\right)} \right]^{-1} & \text{if } z_i^p \neq z_i^{**} \text{ for all } i \\ 1 & \text{if } z_i^p = z_i^{**} \\ 0 & \text{if } z_i^p \neq z_i^{**} \text{ but } \exists j \ni z_j^p = z_j^{**} \end{cases}$$

3. α_{pp} be the minimal objective function value of

$$\min\{\alpha\}$$
$$\text{s.t.} \quad \alpha \geqq \lambda_i^p(z_i^{**} - z_i^p) \qquad 1 \leqq i \leqq k$$

4. α_{pq} be the minimal objective function value of

$$\min\{\alpha\}$$
$$\text{s.t.} \quad \alpha \geqq \lambda_i^p(z_i^{**} - z_i^q) \qquad 1 \leqq i \leqq k$$

Now, we prove Lemmas 14.16 and 14.17 that are used as intermediate results for Theorem 14.18. Theorem 14.18 shows that for a given $\mathbf{z}^p \in N$ with $\boldsymbol{\lambda}^p \in R^k$ and ρ computed according to certain rules, $\mathbf{z}^p$ is uniquely optimal in the program of (14.19). In this sense, since $\mathbf{z}^p \in N$ is arbitrary, each nondominated criterion vector is uniquely computable. Then, in contrast to the weighted Tchebycheff program, Theorem 14.20 proves that *all* criterion vectors returned by the augmented weighted Tchebycheff program are nondominated.

Lemma 14.16. Assume 1, 2, and 3. Then, there does not exist a $\mathbf{z}^q \neq \mathbf{z}^p$ such that $\mathbf{z}^q \in Z$ lies in

$$\Phi(\alpha_{pp}) = \left\{ \mathbf{z} \in R^k \,|\, z_i \in \left[z_i^{**} - \frac{\alpha_{pp}}{\lambda_i^p}, +\infty \right) \quad \text{when} \quad \lambda_i^p > 0 \right\}$$

Proof. *Case 1*: $z_i^p \neq z_i^{**}$ for all i. Substituting for λ_i^p in each of the k constraints

$$\alpha \geqq \lambda_i^p\left(z_i^{**} - z_i^p\right) \qquad 1 \leqq i \leqq k$$

we have

$$\alpha \geqq \left[\sum_{i=1}^{k} \frac{1}{\left(z_i^{**} - z_i^p\right)} \right]^{-1}$$

Thus,

$$\alpha_{pp} = \lambda_i^p \left(z_i^{**} - z_i^p\right) \qquad 1 \leqq i \leqq k$$

or

$$z_i^p = z_i^{**} - \frac{\alpha_{pp}}{\lambda_i^p} \qquad 1 \leqq i \leqq k$$

Since $\mathbf{z}^p$ is nondominated, there does not exist a $q \neq p$ such that $\mathbf{z}^q \in \Phi(\alpha_{pp})$.

Case 2: $\exists j \ni z_j^p = z_j^{**}$. With only one j such that $z_j^p = z_j^{**}$, $\lambda_i^p = 0$ for all $i \neq j$. Thus, $\alpha_{pp} = 0$ and

$$\Phi(\alpha_{pp}) = \left\{ \mathbf{z} \in R^k \,|\, z_j \in \left[z_j^p, +\infty\right) \right\}$$

Since $\mathbf{z}^p$ is the member of Z for which the jth component is greater than or equal to z_j^{**}, there does not exist a $q \neq p$ such that $\mathbf{z}^q \in \Phi(\alpha_{pp})$. ◀

Lemma 14.17. Assume 1, 2, 3, and 4. Then, $\alpha_{pp} < \alpha_{pq}$ for all $\mathbf{z}^q \in Z$, $q \neq p$.

Proof. Since $\mathbf{z}^p$ is nondominated, by Lemma 14.16 no other $\mathbf{z}^q \in Z$ lies in

$$\Phi(\alpha_{pp}) = \left\{ \mathbf{z} \in R^k \,|\, z_i \in \left[z_i^{**} - \frac{\alpha_{pp}}{\lambda_i^p}, +\infty\right) \quad \text{when} \quad \lambda_i^p > 0 \right\}$$

Thus, all $\mathbf{z}^q \in Z$, $q \neq p$ lie in supersets of $\Phi(\alpha_{pp})$. Hence, $\alpha_{pp} < \alpha_{pq}$ for all $\mathbf{z}^q \in Z$, $q \neq p$. ◀

Theorem 14.18. Let Z be finite, $I_Z = \{i \,|\, \mathbf{z}^i \in Z\}$, $I_N = \{i \,|\, \mathbf{z}^i \in N\}$, and $\mathbf{z}^p \in N$. Then, $\mathbf{z}^p$ uniquely minimizes

$$\begin{aligned} &\min \left\{ \alpha + \rho_p \mathbf{e}^T(\mathbf{z}^{**} - \mathbf{z}) \right\} \\ &\text{s.t.} \quad \alpha \geqq \lambda_i^p \left(z_i^{**} - z_i\right) \qquad 1 \leqq i \leqq k \\ &\qquad f_i(\mathbf{x}) = z_i \qquad\qquad 1 \leqq i \leqq k \\ &\qquad \mathbf{x} \in S \end{aligned} \tag{14.19}$$

where

$$0 < \rho_p < \min_{q \in I_Z - \{p\}} \left\{ \frac{\alpha_{pq} - \alpha_{pp}}{\mathbf{e}^T(\mathbf{z}^q - \mathbf{z}^p)} \,|\, \mathbf{e}^T(\mathbf{z}^q - \mathbf{z}^p) > 0 \right\}$$

and the λ_i^p are as specified in 2.

Proof. From Lemma 14.17, it is clear that there exists a $\rho_p > 0$ as above. Suppose $\mathbf{z}^q \in Z$, $\mathbf{z}^q \neq \mathbf{z}^p$ minimizes (14.19). Then, a lower bound for the minimal value of the objective function is $\alpha_{pq} + \rho_p \mathbf{e}^T(\mathbf{z}^{**} - \mathbf{z}^q)$. The minimality of $\mathbf{z}^p$ is preserved, however, if

$$\alpha_{pp} + \rho_p \mathbf{e}^T(\mathbf{z}^{**} - \mathbf{z}^p) < \alpha_{pq} + \rho_p \mathbf{e}^T(\mathbf{z}^{**} - \mathbf{z}^q)$$

Since by Lemma 14.17, $\alpha_{pp} < \alpha_{pq}$, the optimality of $\mathbf{z}^p$ can be thwarted only when $\mathbf{e}^T(\mathbf{z}^{**} - \mathbf{z}^q) < \mathbf{e}^T(\mathbf{z}^{**} - \mathbf{z}^p)$; that is, when $\mathbf{e}^T(\mathbf{z}^q - \mathbf{z}^p) > 0$. To assure the optimality of $\mathbf{z}^p$, we must have

$$\rho_p \mathbf{e}^T(\mathbf{z}^q - \mathbf{z}^p) < \alpha_{pq} - \alpha_{pp}$$

for all $q \in I_Z - \{p\}$. This means that we must have

$$\rho_p < \frac{\alpha_{pq} - \alpha_{pp}}{\mathbf{e}^T(\mathbf{z}^q - \mathbf{z}^p)}$$

whenever $\mathbf{e}^T(\mathbf{z}^q - \mathbf{z}^p) > 0$ for all $q \in I_Z - \{p\}$. Thus, it suffices for ρ_p to be defined as in the theorem for $\mathbf{z}^p$ to uniquely minimize (14.19) with the weights defined as in 2. ◀

Theorem 14.20. Let N be finite and

$$0 < \rho < \min_{i \in I_N} \left[\min_{j \in I_Z - \{i\}} \left\{ \frac{\alpha_{ij} - \alpha_{ii}}{\mathbf{e}^T(\mathbf{z}^j - \mathbf{z}^i)} \middle| \mathbf{e}^T(\mathbf{z}^j - \mathbf{z}^i) > 0 \right\} \right]$$

Then, $\mathbf{z}^p \in N \Leftrightarrow$ there exists a $\boldsymbol{\lambda} \in \bar{\Lambda} = \{\boldsymbol{\lambda} \in R^k \mid \lambda_i \geqq 0, \sum_{i=1}^k \lambda_i = 1\}$ such that $\mathbf{z}^p$ minimizes the augmented weighted Tchebycheff program (14.14).

Proof. $\Rightarrow$ Follows from Theorem 14.18 when the λ_i are defined as the λ_i^p in 2.

$\Leftarrow$ Suppose $\mathbf{z}^p \notin N$ minimizes the augmented weighted Tchebycheff program for some $\boldsymbol{\lambda} \in \bar{\Lambda}$. With $\mathbf{z}^p$ dominated, there exists a $\mathbf{z}^q$ such that $\mathbf{z}^q \geqq \mathbf{z}^p$, $\mathbf{z}^q \neq \mathbf{z}^p$. Since $(\mathbf{z}^{**} - \mathbf{z}^q) \leqq (\mathbf{z}^{**} - \mathbf{z}^p)$ and $\rho\mathbf{e}^T(\mathbf{z}^{**} - \mathbf{z}^q) < \rho\mathbf{e}^T(\mathbf{z}^{**} - \mathbf{z}^p)$, $\mathbf{z}^q$ would have a smaller objective function value than $\mathbf{z}^p$. However, this contradicts the minimality of $\mathbf{z}^p$; thus, $\mathbf{z}^p \in N$. ◀

Thus, we have the result that all criterion vectors returned by the augmented weighted Tchebycheff program are nondominated (Theorem 14.20) and all nondominated criterion vectors are uniquely computable (Theorem 14.18).

14.8.2 Theory for the Polyhedral Case

For the polyhedral case (e.g., when $S = \{\mathbf{x} \in R^n \mid \mathbf{A}\mathbf{x} \lesseqqgtr \mathbf{b}, \mathbf{x} \geqq \mathbf{0}\}$), let $I_{Z_v} = \{i \mid \mathbf{z}^i$ is the image of an extreme point of $S\}$ and $I_{N_v} = \{i \in I_{Z_v} \mid \mathbf{z}^i$ is nondominated$\}$. Then, we have Theorems 14.21 and 14.23.

Theorem 14.21. Let I_{Z_v} and I_{N_v} be defined as above and let

$$0 < \rho < \min_{i \in I_{N_v}} \left[\min_{j \in I_{Z_v} - \{i\}} \left\{ \frac{\alpha_{ij} - \alpha_{ii}}{\mathbf{e}^T(\mathbf{z}^j - \mathbf{z}^i)} \middle| \mathbf{e}^T(\mathbf{z}^j - \mathbf{z}^i) > 0 \right\} \right]$$

Then, $\bar{\mathbf{z}} \in N$ uniquely minimizes the augmented weighted Tchebycheff program

$$\begin{aligned} &\min\{\alpha + \rho\mathbf{e}^T(\mathbf{z}^{**} - \mathbf{z})\} \\ &\text{s.t.} \quad \alpha \geqq \lambda_i(z_i^{**} - z_i) \qquad 1 \leqq i \leqq k \\ &\qquad f_i(\mathbf{x}) = z_i \qquad 1 \leqq i \leqq k \\ &\qquad \mathbf{x} \in S \end{aligned}$$

where

$$\lambda_i = \begin{cases} \dfrac{1}{(z_i^{**} - \bar{z}_i)}\left[\displaystyle\sum_{i=1}^{k} \dfrac{1}{(z_i^{**} - \bar{z}_i)}\right]^{-1} & \text{if } \bar{z}_i \neq z_i^{**} \text{ for all } i \\ 1 & \text{if } \bar{z}_i = z_i^{**} \\ 0 & \text{if } \bar{z}_i \neq z_i^{**} \text{ but } \exists j \in \bar{z}_j = z_j^{**} \end{cases} \tag{14.22}$$

Proof. Follows from the piecewise linearities of the contours of the augmented weighted Tchebycheff metric, Theorem 14.18, and the polyhedral properties of S. ◀

Theorem 14.23. Let ρ be defined as in Theorem 14.21. Then, $\bar{\mathbf{z}} \in N \Leftrightarrow$ there exists a $\boldsymbol{\lambda} \in \bar{\Lambda}$ such that $\bar{\mathbf{z}}$ minimizes the augmented weighted Tchebycheff program

$$\begin{aligned} &\min\{\alpha + \rho\mathbf{e}^T(\mathbf{z}^{**} - \mathbf{z})\} \\ &\text{s.t.} \quad \alpha \geqq \lambda_i(z_i^{**} - z_i) \qquad 1 \leqq i \leqq k \\ &\qquad f_i(\mathbf{x}) = z_i \qquad 1 \leqq i \leqq k \\ &\qquad \mathbf{x} \in S \end{aligned}$$

Proof. Follows from the piecewise linearities of the contours of the augmented weighted Tchebycheff metric, Theorem 14.20, and the polyhedral properties of S. ◀

Thus, with Theorems 14.21 and 14.23 taken together, we obtain the same result in the polyhedral case as we did in the finite-discrete case. That is, all criterion vectors returned by the augmented weighted Tchebycheff program are nondominated and each nondominated criterion vector is uniquely computable.

14.8.3 Theory for the Nonlinear and Infinite-Discrete Case

Because of the possible existence of improperly nondominated criterion vectors in the nonlinear and infinite-discrete feasible region case, we need the power of the *lexicographic weighted Tchebycheff program*

$$\begin{aligned} &\text{lex min}\,\{\alpha, \mathbf{e}^T(\mathbf{z}^{**} - \mathbf{z})\} \\ &\text{s.t.} \quad \alpha \geqq \lambda_i(z_i^{**} - z_i) \qquad 1 \leqq i \leqq k \\ &\qquad f_i(\mathbf{x}) = z_i \qquad 1 \leqq i \leqq k \\ &\qquad \mathbf{x} \in S \end{aligned}$$

to compute any given nondominated criterion vector. Only if the first-stage minimization of α does not yield a unique criterion vector is it necessary for the second-stage minimization of $\mathbf{e}^T(\mathbf{z}^{**} - \mathbf{z})$ to be invoked. Whereas the first-stage is a weighted L_∞-metric optimization, the second-stage is an L_1-metric optimization. Another way of looking at the lexicographic program is that the first-stage is the weighted Tchebycheff program (14.3) and the second-stage is used to break ties in cases of alternative optima. Without ρ, we have Theorems 14.24 and 14.25.

Theorem 14.24. Let $\bar{\mathbf{z}} \in N$. Then, there exists a $\boldsymbol{\lambda} \in \bar{\Lambda}$ such that $\bar{\mathbf{z}}$ uniquely minimizes the lexicographic weighted Tchebycheff program.

Proof. Let $\bar{\alpha}$ minimize the weighted Tchebycheff program with the λ_i as defined in (14.22). Consider the set

$$\Phi(\bar{\alpha}) = \left\{\mathbf{z} \in R^k \,|\, z_i \in \left[z_i^{**} - \frac{\bar{\alpha}}{\lambda_i}, +\infty\right) \quad \text{when} \quad \lambda_i > 0\right\}$$

Clearly, $\bar{\mathbf{z}} \in \Phi(\bar{\alpha})$. From the proof of Lemma 14.16 (which holds for all Z in general), there does not exist another $\mathbf{z} \in N$ such that $\mathbf{z} \in \Phi(\bar{\alpha})$. This means that the associated weighted Tchebycheff program has a unique solution. Since the first stage of the lexicographic weighted Tchebycheff program is the weighted Tchebycheff program, $\bar{\mathbf{z}} \in N$ uniquely minimizes the lexicographic weighted Tchebycheff program for $\boldsymbol{\lambda}$ as defined in (14.22). ◀

Theorem 14.25. $\bar{\mathbf{z}} \in N \Leftrightarrow$ there exists a $\boldsymbol{\lambda} \in \bar{\Lambda}$ such that $\bar{\mathbf{z}}$ minimizes the lexicographic weighted Tchebycheff program.

Proof. $\Rightarrow$ Follows from Theorem 14.24 when the λ_i are defined as in (14.22).

$\Leftarrow$ Suppose $\bar{\mathbf{z}} \notin N$ minimizes the lexicographic weighted Tchebycheff program. Then, there exists a $\hat{\mathbf{z}} \in Z$ such that $\hat{z}_i \geqq \bar{z}_i$ for all i and $\hat{z}_i > \bar{z}_i$ for at least one i. Then, $\mathbf{e}^T(\mathbf{z}^{**} - \hat{\mathbf{z}}) < \mathbf{e}^T(\mathbf{z}^{**} - \bar{\mathbf{z}})$ which contradicts the minimality of $\bar{\mathbf{z}}$ and the theorem is proved. ◀

Theorems 14.24 and 14.25 tell us, regardless of feasible region, that all criterion vectors returned by the lexicographic program are nondominated and that *all* nondominated criterion vectors are uniquely computable. Thus, not only can the lexicographic program be used in the nonlinear and infinite-discrete cases, but it can be used in the finite-discrete and polyhedral cases as well. The only disadvantage of the lexicographic approach is that two stages of optimization are

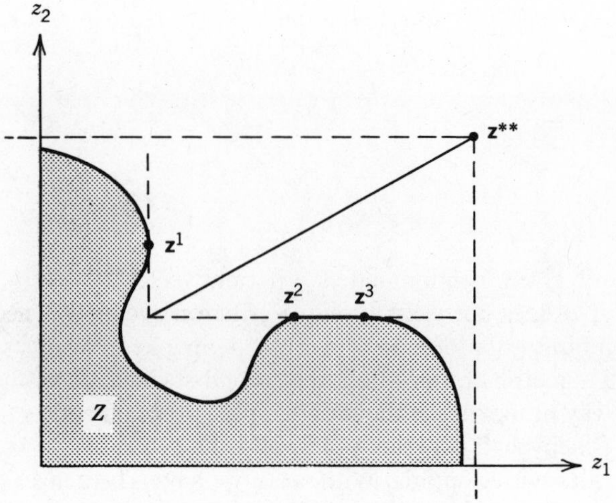

Figure 14.21. Graph of Example 21.

required when the first stage results in alternative optima. However, this may not happen often enough to be important. Probably, the main advantage of the lexicographic approach is that it gives us peace of mind: We do not have to worry about improperly efficient points and estimating ρ. Example 21 demonstrates the lexicographic weighted Tchebycheff program.

Example 21. Let the weighting vector used in the lexicographic weighted Tchebycheff program produce the weighted Tchebycheff contour shown in Fig. 14.21. Then, the set of criterion vectors minimizing the first-stage optimization is

$$\{\mathbf{z}^1\} \cup \gamma(\mathbf{z}^2, \mathbf{z}^3)$$

of which only $\mathbf{z}^1$ and $\mathbf{z}^3$ are nondominated, both being improperly so. After the second-stage optimization, $\mathbf{z}^3$ is uniquely generated (because it is the closest to $\mathbf{z}^{**}$ according to the L_1-metric).

Observe that if we were to use the augmented weighted Tchebycheff approach on problems with improperly efficient points, for any $\rho > 0$ (because of the "slope" of the contour), there would be a neighborhood of properly nondominated criterion vectors that could not be computed about each improperly nondominated criterion vector.

14.9 TCHEBYCHEFF ALGORITHM

Sampling a sequence of progressively smaller subsets of N, the interactive (augmented or lexicographic) weighted Tchebycheff approach is overviewed as follows.

First iteration. After we obtain a $\mathbf{z}^{**}$ ideal criterion vector, the first iteration begins by forming a dispersed group of λ-weighting vectors. The λ-vectors define a group of weighted Tchebycheff metrics whose diagonal directions are radially dispersed. These metrics are used to sample the (entire) set of nondominated criterion vectors N. The nondominated criterion vectors forming the sample are computed by solving the augmented (or lexicographic) weighted Tchebycheff program for each of the λ's. Then, the decision maker's first-iteration selection $\mathbf{z}^{(1)}$ is his most preferred criterion vector from the thus generated sample. Let $\lambda^{(1)}$ be a weighting vector corresponding to $\mathbf{z}^{(1)}$.

Second iteration. To begin the second iteration, another group of λ's is formed, but this time, more concentrated than the first, and centered about $\lambda^{(1)}$. The new λ's form a group of weighted Tchebycheff metrics whose diagonal directions are radially dispersed, but more concentrated than previously for use in sampling a neighborhood in N about $\mathbf{z}^{(1)}$. Developing the second sample of nondominated criterion vectors by repetitively solving the augmented (or lexicographic) weighted Tchebycheff program, the decision maker makes his second iteration selection $\mathbf{z}^{(2)}$ from the new sample. Let $\lambda^{(2)}$ be a weighting vector corresponding to $\mathbf{z}^{(2)}$.

Third iteration. Now, an even more concentrated group of λ's, but this time centered around $\lambda^{(2)}$, is formed. The purpose is to sample an even smaller neighborhood in N but this time centered about $\mathbf{z}^{(2)}$, and so forth.

As with the interval criterion weights/vector-maximum and interactive weighted-sums/filtering approaches, this algorithm is tailored to conclude in a (prechosen) fixed number of iterations with a (prechosen) fixed number of solutions presented per iteration. In detail, we have the following.

Step 1: To calibrate the algorithm, specify the sample size P, the number of iterations t, and the $\overline{\Lambda}$-reduction factor r. Obtain a $\mathbf{z}^{**}$ ideal criterion vector.

Letting w be the final $[\ell_i, \mu_i]$ interval width as in Section 13.5.1, we utilize the same guideline relationships

$$P \gtrapprox k$$

$$t \approx k$$

$$1/(2k) \lessapprox w \lessapprox 3/(2k)$$

$$\sqrt[k]{1/P} \lessapprox r \lessapprox \sqrt[t-1]{w}$$

for establishing values for P, t, and r as in the interactive weighted-sums/filtering algorithm.

Step 2: Normalize (rescale) the objective functions.

Step 3: Let $h = 0$. Let $[\ell_i^{(1)}, \mu_i^{(1)}] = [0, 1]$ for all i.

Step 4: Let $h = h + 1$. Form

$$\overline{\Lambda}^{(h)} = \left\{ \lambda \in R^k \mid \lambda_i \in \left[\ell_i^{(h)}, \mu_i^{(h)}\right], \quad \sum_{i=1}^{k} \lambda_i = 1 \right\}$$

Step 5: Randomly generate $50 \times k$ weighting vectors from $\overline{\Lambda}^{(h)}$.

Step 6: Filter the $50 \times k$ weighting vectors to obtain the $2 \times P$ most different.

Step 7: Using the $2P$ weighting vector representatives of $\overline{\Lambda}^{(h)}$, solve the $2P$ associated augmented (or lexicographic) weighted Tchebycheff programs.

Whereas $3P$ optimizations were recommended with the interactive weighted-sums/filtering method, only $2P$ optimizations are recommended here. The reason is that we can expect a better covering of the neighborhood of N undergoing sampling because it is less likely for different weighting vectors to generate the same solution in the Tchebycheff method than in the interactive weighted-sums approach.

Step 8: Filter the criterion vectors resulting from Step 7 to obtain the P most different. Then, present the P nondominated criterion vectors to the decision maker.

Step 9: From the sample of P nondominated criterion vectors, the decision maker selects his or her most preferred, designating it $\mathbf{z}^{(h)}$.

Step 10: If the decision maker wishes to stop iterating prematurely, go to Step 15. Otherwise, go to Step 11.

Steps 10 and 14 enable the decision maker to override the originally intended number of iterations t. He or she can either stop early via Step 10 or extend the number of iterations via Step 14.

Step 11: Let $\lambda^{(h)}$ be the λ-vector whose components are given by

$$\lambda_i^{(h)} = \begin{cases} \dfrac{1}{\left(z_i^{**} - z_i^{(h)}\right)} \left[\displaystyle\sum_{i=1}^{k} \dfrac{1}{\left(z_i^{**} - z_i^{(h)}\right)} \right]^{-1} & \text{if } z_i^{(h)} \neq z_i^{**} \text{ for all } i \\ 1 & \text{if } z_i^{(h)} = z_i^{**} \\ 0 & \text{if } z_i^{(h)} \neq z_i^{**} \text{ but } \exists j \ni z_j^{(h)} = z_j^{**} \end{cases}$$

This step is included because $\mathbf{z}^{(h)}$ may not have been generated by the vertex of the intersecting contour in Step 7. In such a case, $\lambda^{(h)}$ would be a more appropriate weighting vector to associate with $\mathbf{z}^{(h)}$, as illustrated in Example 22.

Example 22. In Figs. 14.22*a* and *b*, let $\mathbf{z}^{**} = (4, 5)$ and $\bar{\mathbf{z}} = (3, 3)$. If we use $\lambda = (\frac{5}{6}, \frac{1}{6})$ to define the augmented (or lexicographic) weighted Tchebycheff

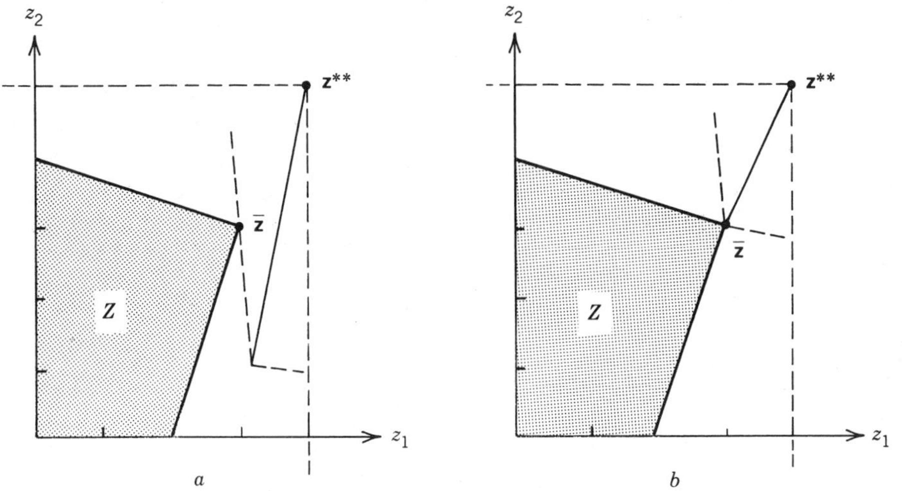

Figure 14.22. Graphs of Example 22.

program in Step 7 to sample N, $\bar{\mathbf{z}}$ would be produced, as in Fig. 14.22*a*. However, a more appropriate weighting vector to associate with $\bar{\mathbf{z}}$ is $(\frac{2}{3}, \frac{1}{3})$ because it causes the augmented (or unaugmented) weighted Tchebycheff metric to generate $\bar{\mathbf{z}}$ at a vertex, as in Fig. 14.22*b*.

Step 12: Let $\lambda^{(h)}$ be the weighting vector computed in Step 11. Form

$$\overline{\Lambda}^{(h+1)} = \left\{ \lambda \in R^k \,|\, \lambda_i \in \left[\ell_i^{(h+1)}, \mu_i^{(h+1)}\right], \quad \sum_{i=1}^{k} \lambda_i = 1 \right\}$$

where

$$\left[\ell_i^{(h+1)}, \mu_i^{(h+1)}\right] = \begin{cases} [0, r^h] & \text{if } \lambda_i^{(h)} - \dfrac{r^h}{2} \leqq 0 \\ [1 - r^h, 1] & \text{if } \lambda_i^{(h)} + \dfrac{r^h}{2} \geqq 1 \\ \left[\lambda_i^{(h)} - \dfrac{r^h}{2}, \lambda_i^{(h)} + \dfrac{r^h}{2}\right] & \text{otherwise} \end{cases}$$

in which r^h is r raised to the hth power.

Step 13: If $h < t$, go to Step 4. If $h \geqq t$, go to Step 14.

Step 14: If the decision maker wishes to keep iterating, go to Step 4. Otherwise, go to Step 15.

Step 15: With $\mathbf{x}^{(h)}$ an inverse image of $\mathbf{z}^{(h)}$, stop with $(\mathbf{z}^{(h)}, \mathbf{x}^{(h)})$ as the final solution.

The Tchebycheff algorithm is flowcharted in Fig. 14.23.

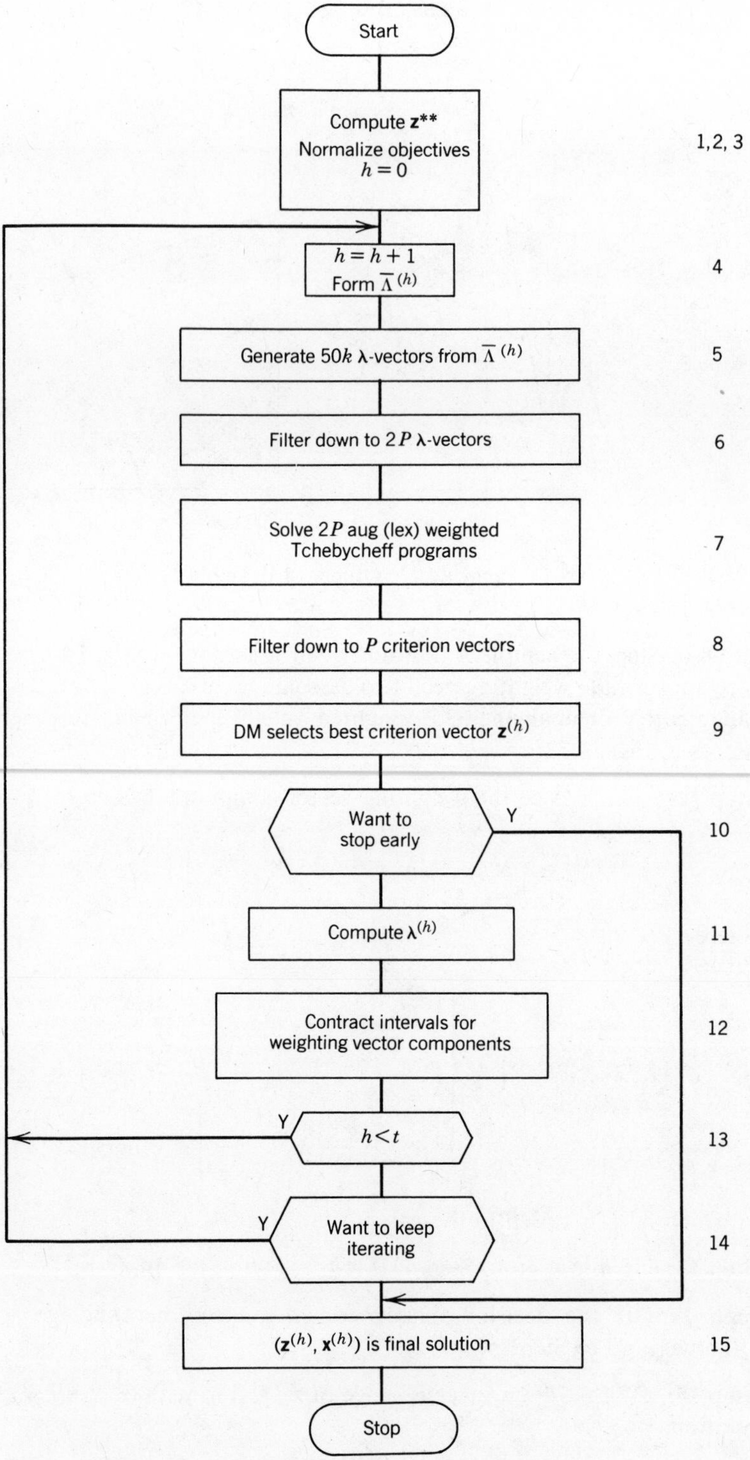

Figure 14.23. Flowchart of the Tchebycheff algorithm.

Readings

1. Benson, H. P. (1979). "An Improved Definition of Proper Efficiency for Vector Maximization with Respect to Cones," *Journal of Mathematical Analysis and Applications*, Vol. 71, No. 1, pp. 232–241.

2. Benson, H. P. and T. L. Morin (1977). "The Vector Maximization Problem: Proper Efficiency and Stability," *SIAM Journal of Applied Mathematics*, Vol. 32, No. 1, pp. 64–72.

3. Borwein, J. (1977). "Proper Efficient Points for Maximizations with Respect to Cones," *SIAM Journal on Control and Optimization*, Vol. 15, No. 1, pp. 57–63.

4. Bowman, V. J. (1976). "On the Relationship of the Tchebycheff Norm and the Efficient Frontier of Multiple-Criteria Objectives," *Lecture Notes in Economics and Mathematical Systems*, No. 135, Berlin: Springer-Verlag, pp. 76–85.

5. Choo, E.-U. and D. R. Atkins (1983). "Proper Efficiency in Nonconvex Multicriteria Programming," *Mathematics of Operations Research*, Vol. 8, No. 2, pp. 467–470.

6. Ecker, J. G. and N. E. Shoemaker (1981). "Selecting Subsets from the Set of Nondominated Vectors in Multiple Objective Linear Programming," *SIAM Journal on Control and Optimization*, Vol. 19, No. 4, pp. 505–515.

7. Geoffrion, A. M. (1968). "Proper Efficiency and the Theory of Vector Maximization," *Journal of Mathematical Analysis and Applications*, Vol. 22, No. 3, pp. 618–630.

8. IBM Document No. GH19–1091-1. (1979). "IBM Mathematical Programming System Extended/370: Primer," IBM Corporation, Data Processing Division, White Plains, New York.

9. Isermann, H. (1974). "Proper Efficiency and the Linear Vector Maximum Problem," *Operations Research*, Vol. 22, No. 1, pp. 189–191.

10. Kaliszewski, I. (1985). "A Modified Weighted Tchebycheff Metric for Multiple Objective Programming," *Mathematical Programming* (forthcoming).

11. Sakawa, M. and N. Mori (1983). "Interactive Multiobjective Decisionmaking for Nonconvex Problems Based on the Weighted Tchebycheff Norm," *Large Scale Systems*, Vol. 5, No. 1, pp. 69–82.

12. Soyster, A. L., B. Lev, and D. I. Toof (1977). "Conservative Linear Programming with Mixed Multiple Objectives," *Omega*, Vol. 5, No. 2, pp. 193–205.

13. Steuer, R. E. and E.-U. Choo (1983). "An Interactive Weighted Tchebycheff Procedure for Multiple Objective Programming," *Mathematical Programming*, Vol. 26, No. 1, pp. 326–344.

14. Steuer, R. E. and F. W. Harris (1980). "Intra-Set Point Generation and Filtering in Decision and Criterion Space," *Computers and Operations Research*, Vol. 7, No.'s 1–2, pp. 41–53.

15. Steuer, R. E. and E. F. Wood (1986). "On the 0–1 Implementation of the Tchebycheff Solution Approach: A Water Quality Illustration," *Large Scale Systems*, forthcoming.

16. Törn, A. A. (1980). "A Sampling-Search-Clustering Approach for Exploring the Feasible/Efficient Solutions of MCDM Problems," *Computers and Operations Research*, Vol. 7, No.'s 1–2, pp. 67–79.

17. Wierzbicki, A. P. (1980). "The Use of Reference Objectives in Multiobjective Optimization," *Lecture Notes in Economics and Mathematical Systems*, No. 177, Berlin: Springer-Verlag, pp. 468–486.

18. Yu, P. L. (1973). "A Class of Solutions for Group Decision Problems," *Management Science*, Vol. 19, No. 8, pp. 936–946.

19. Zeleny, M. (1973). "Compromise Programming." In J. L. Cochrane and M. Zeleny (eds.), *Multiple Criteria Decision Making*, Columbia, South Carolina: University of South Carolina Press, pp. 262–301.

20. Zeleny, M. (1976). "The Theory of the Displaced Ideal," *Lecture Notes in Economics and Mathematical Systems*, No. 123, Berlin: Springer-Verlag, pp. 153–206.

21. Zionts, S. and J. Wallenius (1980). "Identifying Efficient Vectors: Some Theory and Computational Results," *Operations Research*, Vol. 28, No. 3, pp. 785–793.

PROBLEM EXERCISES

14-1 Let:

$$\begin{array}{ll} \mathbf{c}^1 = (4,1,0) & \mathbf{x}^1 = (0,0,0) \\ \mathbf{c}^2 = (-3,1,0) & \mathbf{x}^2 = (0,0,4) \\ & \mathbf{x}^3 = (1,2,2) \\ & \mathbf{x}^4 = (8,0,0) \\ & \mathbf{x}^5 = (5,0,4) \\ & \mathbf{x}^6 = (5,2,2) \end{array}$$

(a) Graph $S = \gamma_{i=1}^{6}(\mathbf{x}^i)$.

(b) Write the MOLP formulation.

(c) Graph Z.

14-2 For the MOLP:

$$\begin{array}{lrcl} \max \{ & x_1 - x_2 & = & z_1 \} \\ \max \{ & 5x_2 & = & z_2 \} \\ \text{s.t.} & x_1 - x_2 & \leqq & 0 \\ & x_1 + x_2 & \leqq & 12 \\ & x_2 & \leqq & 8 \\ & x_2 & \geqq & 3 \\ & x_1, x_2 & \geqq & 0 \end{array}$$

specify 5% of:

(a) z_1^*.

(b) The range of z_1 over S.

(c) The range of z_1 over the efficient set.

(d) z_2^*.

(e) The range of z_2 over S.

(f) The range of z_2 over the efficient set.

14-3 With $\lambda_1 = \frac{1}{4}$ and $\lambda_2 = \frac{3}{4}$, compute θ_1 and θ_2 for:

(a) $\rho = .001$

(b) $\rho = .005$

(c) $\rho = .010$

(d) $\rho = .050$

(e) $\rho = .100$

14-4 With $\rho = .01$, compute θ_1 and θ_2 for:

(a) $\boldsymbol{\lambda} = (\frac{1}{4}, \frac{3}{4})$

(b) $\boldsymbol{\lambda} = (\frac{1}{2}, \frac{1}{2})$

(c) $\boldsymbol{\lambda} = (\frac{7}{8}, \frac{1}{8})$

14-5 Let $\mathbf{z}^{**} = (5, 8)$. What is the $\boldsymbol{\lambda} \in R^2$ of the augmented weighted Tchebycheff contour that has as a definition point:

(a) $\mathbf{z}^1 = (-2, 8)$

(b) $\mathbf{z}^2 = (1, 6)$

(c) $\mathbf{z}^3 = (2, 5)$

(d) $\mathbf{z}^4 = (3, 2)$

14-6 Consider the MOLP:

$$\begin{aligned}
\max\{\ 2x_1 - \ \ x_2 &= z_1\} \\
\max\{-x_1 + 2x_2 &= z_2\} \\
\text{s.t.} \quad x_1 \qquad\quad &\leqq 3 \\
x_1 + 3x_2 &\leqq 9 \\
x_1, x_2 &\geqq 0
\end{aligned}$$

(a) Graph S.

(b) Graph Z.

(c) What are the $\boldsymbol{\lambda}$-vectors that have $(5, -1) \in Z$ as the vertex of a weighted Tchebycheff contour when:

(i) $\mathbf{z}^{**} = (8, 6)$
(ii) $\mathbf{z}^{**} = (8, 8)$
(iii) $\mathbf{z}^{**} = (12, 8)$

14-7 With Z the convex hull of $\mathbf{z}^1 = (0, 0)$, $\mathbf{z}^2 = (2\frac{1}{2}, 5)$, and $\mathbf{z}^3 = (4\frac{1}{2}, 1)$, let $\mathbf{z}^{**} = (5, 6)$.

(a) Specify the set of nondominated criterion vectors.

(b) Graphing the intersecting contours, specify the coordinates of the points in Z closest to $\mathbf{z}^{**}$ according to the weighted L_1-metric for $\boldsymbol{\lambda} = (\frac{1}{2}, \frac{1}{2})$ and $\boldsymbol{\lambda} = (\frac{4}{5}, \frac{1}{5})$.

(c) Follow the same instructions as in (b), but use the weighted L_∞-metric.

14-8 Let $Z = \gamma(\mathbf{z}^1, \mathbf{z}^2, \mathbf{z}^3)$ where $\mathbf{z}^1 = (0, 3, 3)$, $\mathbf{z}^2 = (3, 3, 0)$, and $\mathbf{z}^3 = (3, 1, 0)$.

(a) Specify N.

(b) Which ε_i can be zero?

14-9 Let $Z = \gamma(\mathbf{z}^1, \mathbf{z}^2, \mathbf{z}^3, \mathbf{z}^4)$ where $\mathbf{z}^1 = (4,0,0)$, $\mathbf{z}^2 = (4,4,0)$, $\mathbf{z}^3 = (0,4,4)$, and $\mathbf{z}^4 = (0,0,4)$. Specify:

(a) N.

(b) The set of weakly nondominated criterion vectors.

(c) The set of criterion vectors that optimize the (nonaugmented) weighted Tchebycheff program for $\mathbf{z}^{**} = (5,5,5)$ and $\boldsymbol{\lambda} = (\frac{1}{3}, \frac{1}{3}, \frac{1}{3})$.

(d) Which of the criterion vectors of (c) are members of N.

14-10 For the MOIP:

$$\begin{aligned} \max \{ \quad x_1 - 8x_2 &= z_1 \} \\ \max \{ -4x_1 + 9x_2 &= z_2 \} \\ \text{s.t.} \quad x_1 + 2x_2 &\leqq 8 \\ x_1 - 2x_2 &\leqq 0 \\ x_1, x_2 &\geqq 0 \quad \text{and integer} \end{aligned}$$

(a) Graph the feasible region identifying those points that are efficient.

(b) Graph the set of all nondominated criterion vectors identifying those that are unsupported.

14-11 Graph the criterion vectors:

$\mathbf{z}^1 = (-1,5)$	$\mathbf{z}^6 = (5,2)$
$\mathbf{z}^2 = (1,5)$	$\mathbf{z}^7 = (5,1)$
$\mathbf{z}^3 = (2,4)$	$\mathbf{z}^8 = (6,1)$
$\mathbf{z}^4 = (3,2)$	$\mathbf{z}^9 = (7,-1)$
$\mathbf{z}^5 = (4,3)$	$\mathbf{z}^{10} = (8,-2)$

Which are:

(a) Dominated.

(b) Nondominated.

(c) Weakly nondominated.

(d) Unsupported nondominated.

(e) Supported-extreme nondominated.

(f) Supported-nonextreme nondominated.

14-12 Let the following be the criterion vectors pertaining to the extreme points of S. Which are dominated and which are nondominated?

$\mathbf{z}^1 = (2,1,8)$	$\mathbf{z}^4 = (5,6,3)$
$\mathbf{z}^2 = (4,15,1)$	$\mathbf{z}^5 = (2,0,6)$
$\mathbf{z}^3 = (11,4,2)$	$\mathbf{z}^6 = (1,1,1)$

14-13 Specify the efficient set of the problem of Fig. 14.24 (that has two linear objectives and a nonlinear feasible region).

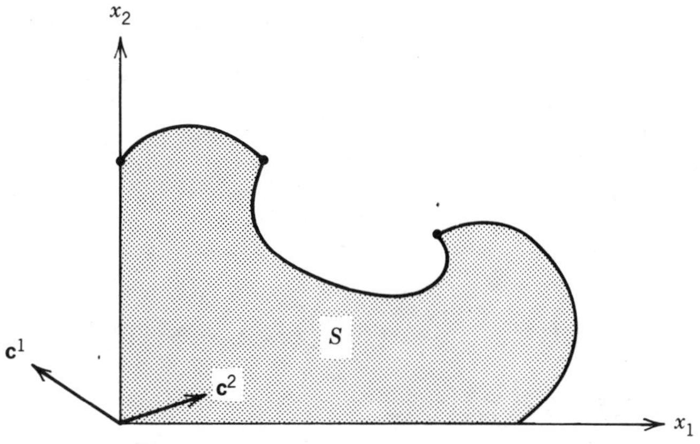

Figure 14.24. Graph of Problem 14–13.

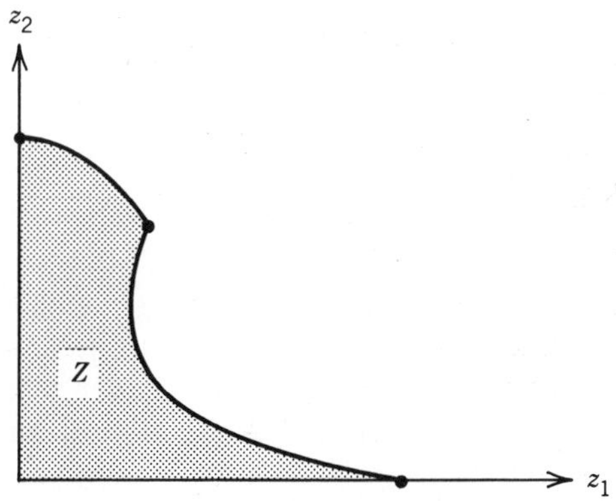

Figure 14.25. Graph of Problem 14–14.

14-14 Specify N (the set of all nondominated criterion vectors) for the problem of Fig. 14.25.

14-15 Let $\mathbf{z}^{**} = (6,5)$ and $N = \{\mathbf{z}^1, \mathbf{z}^2, \mathbf{z}^3\}$ where $\mathbf{z}^1 = (2,4)$, $\mathbf{z}^2 = (4,2)$, and $\mathbf{z}^3 = (5,1)$.

(a) What subset of $\overline{\Lambda}$ yields $\mathbf{z}^2$ as the unique solution of the lexicographic weighted Tchebycheff program?

(b) What $\boldsymbol{\lambda}$-vector causes $\mathbf{z}^1$ and $\mathbf{z}^2$ to be alternative optima of the lexicographic program?

14-16 Let $Z = \{\mathbf{z}^1, \ldots, \mathbf{z}^9\}$ where

$\mathbf{z}^1 = (2,4)$	$\mathbf{z}^4 = (5,8)$	$\mathbf{z}^7 = (12,5)$
$\mathbf{z}^2 = (3,11)$	$\mathbf{z}^5 = (9,7)$	$\mathbf{z}^8 = (13,3)$
$\mathbf{z}^3 = (4,9)$	$\mathbf{z}^6 = (11,7)$	$\mathbf{z}^9 = (13,0)$

Which nondominated criterion vectors possess supporting hyperplanes and which do not?

CHAPTER 15

Tchebycheff/Weighted-Sums Implementation

Because of similarities in the

1. Augmented weighted Tchebycheff;
2. Lexicographic weighted Tchebycheff;
3. Interactive weighted-sums/filtering

procedures, a common computer package can be constructed for implementing all three. The common computer package is referred to as an *automated package* because it automates the repetitive optimizations of each iteration.

This chapter deals with the design and application of an **MPSX**-based automated package for multiple objective linear programming problems. It is **MPSX**-based in the sense that the automated package calls **MPSX** as a subroutine to perform the required LP optimizations. The presentation, however, is sufficiently general for the logic of the chapter to serve as a model for the construction of automated packages that call other commercially available software such as **MINOS** [Murtagh and Saunders (1977, 1980); Preckel (1980); Saunders (1977)] or **SAS/OR** [SAS (1985)]. The chapter can also be used as a guide for the construction of multiple objective integer and nonlinear programming automated packages.

15.1 PROGRAM FORMULATIONS

In many MOLPs, we have *mixed* (minimization and maximization) objectives. Let

$$J^{+} = \{ i \mid \text{objective } \mathbf{c}^{i}\mathbf{x} \text{ is to be maximized} \}$$

$$J^{-} = \{ i \mid \text{objective } \mathbf{c}^{i}\mathbf{x} \text{ is to be minimized} \}$$

Then, with mixed objectives, we compute the $\mathbf{z}^{**}$ *ideal criterion vector* whose components are given by

$$z_i^{**} = \begin{cases} z_i^* + \varepsilon_i & i \in J^+ \\ z_i^* - \varepsilon_i & i \in J^- \end{cases}$$

where

$$z_i^* = \begin{cases} \max \{\mathbf{c}^i\mathbf{x} \mid \mathbf{x} \in S\} & i \in J^+ \\ \min \{\mathbf{c}^i\mathbf{x} \mid \mathbf{x} \in S\} & i \in J^- \end{cases}$$

and the $\varepsilon_i \geqq 0$ are subject to the rules of Sections 14.1 and 14.2.

For purposes of implementation, we rewrite the augmented weighted Tchebycheff program of Section 14.3.2 as

$$\begin{aligned} &\min \left\{ \alpha + \rho \sum_{i=1}^{k} w_i \right\} \\ &\text{s.t.} \quad \alpha \geqq \lambda_i w_i && 1 \leqq i \leqq k \\ &\qquad w_i = z_i^{**} - z_i && i \in J^+ \\ &\qquad w_i = z_i - z_i^{**} && i \in J^- \\ &\qquad \mathbf{Cx} = \mathbf{z} \\ &\qquad \mathbf{x} \in S \\ &0 \leqq \alpha \in R, \quad \mathbf{0} \leqq \mathbf{w} \in R^k, \quad \mathbf{z} \in R^k \text{ unrestricted} \end{aligned}$$

and the lexicographic weighted Tchebycheff program of Section 14.8.3 as

$$\begin{aligned} &\operatorname{lex\,min} \left\{ \alpha, \quad \sum_{i=1}^{k} w_i \right\} \\ &\text{s.t.} \quad \alpha \geqq \lambda_i w_i && 1 \leqq i \leqq k \\ &\qquad w_i = z_i^{**} - z_i && i \in J^+ \\ &\qquad w_i = z_i - z_i^{**} && i \in J^- \\ &\qquad \mathbf{Cx} = \mathbf{z} \\ &\qquad \mathbf{x} \in S \\ &0 \leqq \alpha \in R, \quad \mathbf{0} \leqq \mathbf{w} \in R^k, \quad \mathbf{z} \in R^k \text{ unrestricted} \end{aligned}$$

In the Tchebycheff formulations, the w_i are utilized as *convenience* variables. Although the w_i create additional rows and columns, they make the updating of the formulations with different $\boldsymbol{\lambda}$-vectors easier. By construction, the w_i are nonnegative variables. The z_i variables are, in general, unrestricted because the values of the objective functions can be of any sign over S.

To share a common *layout matrix* (see next section) with the Tchebycheff formulations, we write the weighted-sums program as follows

$$\begin{aligned} &\max \left\{ \boldsymbol{\lambda}^T \mathbf{D}_k \mathbf{z} \right\} \\ &\text{s.t.} \quad \mathbf{Cx} = \mathbf{z} \\ &\qquad\quad \mathbf{x} \in S \\ &\mathbf{z} \in R^k \text{ unrestricted} \end{aligned}$$

where $\mathbf{D}_k$ is a diagonal matrix of order k such that

$$(i, i) \text{ element of } \mathbf{D}_k = \begin{cases} 1 & \text{if } i \in J^+ \\ -1 & \text{if } i \in J^- \end{cases}$$

Example 1. Let $k = 5$ and the first and fourth be minimization objectives. Then,

$$\mathbf{D}_k = \begin{bmatrix} -1 & & & & \\ & 1 & \text{zeros} & & \\ & & 1 & & \\ \text{zeros} & & & -1 & \\ & & & & 1 \end{bmatrix}$$

and

$$\boldsymbol{\lambda}^T \mathbf{D}_k \mathbf{z} = -\lambda_1 z_1 + \lambda_2 z_2 + \lambda_3 z_3 - \lambda_4 z_4 + \lambda_5 z_5$$

TABLE 15.1
MPSX Layout Matrix

	x	z	α	w_1	w_2	$\cdots$	w_k		
AUG			1	ρ	ρ	$\cdots$	ρ	min	
LEX1			1					min	
LEX2				1	1	$\cdots$	1	min	
WGTSUM		$\boldsymbol{\lambda}^T\mathbf{D}_k$						max	
TCHEBYi			-1	λ_1				$\leqq$	0
			-1		λ_2			$\leqq$	0
			$\vdots$			$\ddots$		$\vdots$	$\vdots$
			-1				λ_k	$\leqq$	0
ZDSTARi		$\mathbf{I}_k$			$\mathbf{D}_k$			$=$	$\mathbf{z}^{**}$
OBJi	$\mathbf{C}$	$-\mathbf{I}_k$						$=$	$\mathbf{0}$
AROWi	$\mathbf{A}$							$\gtreqless$	$\mathbf{b}$

$\mathbf{x}, \alpha, \mathbf{w} \geqq \mathbf{0}$
$\mathbf{z}$ unrestricted

15.2 MPSX LAYOUT MATRIX

With regard to the program formulations of Section 15.1, consider the **MPSX** *layout matrix* of Table 15.1.

The **MPSX** layout matrix has $(m + 3k + 4)$ rows and $(n + 2k + 1)$ LHS columns. The first four rows are objective rows. The remaining $m + 3k$ rows are constraint rows. However, depending on the procedure selected, different portions of the **MPSX** layout matrix are ignored.

To solve the augmented weighted Tchebycheff program, we optimize **AUG**, ignoring the **LEX1**, **LEX2**, and **WGTSUM** rows. To solve the first-stage lexicographic weighted Tchebycheff program, we optimize **LEX1**, ignoring the **AUG**, **LEX2**, and **WGTSUM** rows. For the second-stage, **AUG** and **WGTSUM** are ignored and **LEX1** is converted to a constraint. For the weighted-sums program, we optimize **WGTSUM**, ignoring the **AUG**, **LEX1**, **LEX2**, **TCHEBYi**, and **ZDSTARi** rows, and the α and w_i columns.

15.3 SAVE/REVISE/RESTORE SEQUENCE

SAVE, **REVISE**, and **RESTORE** are three **MPSX** control language statements. The **SAVE** command stores in a problem file information that allows for the recreation of the current optimization status of a problem at a later time. The **REVISE** feature allows a user to change problem coefficients, change row-type indicators, change bound-type indicators, add rows, add columns, and so on. Also, the **REVISE** command adapts to a revised problem any basis saved on the problem file. The **RESTORE** command reestablishes the optimization status of a problem in accordance with the information in the problem file.

Example 2. Suppose we have the following **MPSX** control language segment:

```
SAVE('NAME','BASIS6')
REVISE
RESTORE('NAME','BASIS6')
```

The **SAVE** command stores the current basis under the name **BASIS6**. The **REVISE** command allows changes to be made to the model. If the changes are such that the saved basis must be updated, it is automatically done so by **REVISE**. The name **BASIS6** is now assigned to the updated basis. The **RESTORE** command reestablishes the problem with the basis presently stored under the name **BASIS6**.

15.4 COMPUTING THE z* VECTOR

With the Tchebycheff procedures, we must compute the vector of z_i^* values. The $\mathbf{z}^*$ vector is required in order to form the $\mathbf{z}^{**}$ ideal criterion vector that forms the right-hand side of the **ZDSTARi** constraints in the **MPSX** layout matrix. The automated package recomputes $\mathbf{z}^*$ and $\mathbf{z}^{**}$ at each iteration to allow the user to make model changes at each iteration.

To compute the z_i^*, we individually optimize the objectives of the MOLP in question. This is done by reading into the automated package the MOLP in regular `MPSX` format (see Section 3.7), with each of the k objectives indicated by an `N` row-type indicator. Then, the objectives are optimized independently, one after the other. A `SAVE/RESTORE` sequence is used between optimizations to utilize preceding optimal bases. After the z_i^* optimizations, and the construction of the z_i^{**}, the `REVISE` feature is called to

1. Convert the `OBJi` rows to equalities (with `E` row-type indicators).
2. Create the `ZDSTARi` constraints and install the z_i^{**} as their right-hand side values.
3. Complete the rest of the `MPSX` layout matrix.

Note that this section does not apply if we are using the interactive weighted-sums procedure.

15.5 REPETITIVE OPTIMIZATION ECONOMIES

In the repetitive optimizations of any of the three given interactive procedures, we see from the `MPSX` layout matrix that the only thing that changes from optimization to optimization is the λ-vector. Since the programs in a series of optimizations are identical except for their λ-vectors, we utilize the optimal basis of the most recent optimization in order to give its subsequent optimization an advanced start. In this way, economies of scale are achieved because we can *warm start*, as opposed to *cold start*, each subsequent optimization. (See Section 3.11.)

Between optimizations in the weighted-sums case, the `REVISE` feature of `MPSX` is used to replace the λ-vector in the `WGTSUM` row. Then, the optimal basis of the most recently completed program is used as the initial basis of the next program.

Between optimizations in the augmented weighted Tchebycheff case, the `REVISE` feature is called on to replace the old diagonal of λ_i weights with a new diagonal of λ_i weights in the `TCHEBYi` constraints. Then, `REVISE` updates the optimal basis of the previous problem (perhaps performing a few pivots to restore feasibility) so that the next optimization can commence.

In the lexicographic weighted Tchebycheff case, the `REVISE` feature is used as follows. After the first-stage optimization of `LEX1`, `REVISE` converts the `LEX1` row to an equality constraint whose right-hand side is the optimal value of α. Then, `LEX2` is optimized, subject to all rows of the `MPSX` layout matrix as constraints, except `AUG` and `WGTSUM` that are ignored. After this, `REVISE` is again called to convert `LEX1` to its original status as an `N`-type row (with a right-hand side of 0), replace the λ-vector, and update the basis as in the augmented case, to be ready for the next lexicographic program.

TABLE 15.2
Illustrative MOLP

	x_1	x_2	x_3	x_4	x_5	x_6	x_7	x_8		
	−1		−1	1	2	−1	6		max	
Objs		5				2		4	max	
		4		6	2	3	4	−1	max	
			8	6	1	4	−3	7	≦	6
	−1			1		4		3	≦	7
	3		3	5		3	1		≦	7
			−2	7	−2		−2	6	≦	10
s.t.	3	4	−2	1			−1	−1	≦	10
		7	2		7		4		≦	5
	−2	−1	1				−3	4	≦	6
	6	−3		3			6	4	≦	6

all vars ≧ 0

15.6 ILLUSTRATIVE MOLP

To illustrate the automated package, let us solve the MOLP of Table 15.2 using, for instance, the lexicographic weighted Tchebycheff procedure. To simulate decision-maker judgment, let us use the utility function

$$U = (z_1)^3(1 + z_2) + z_3$$

to answer the questions posed by the algorithm. With this utility function, the efficient extreme point of greatest utility has $U = 496.67474$. Its criterion vector is (4.583333, 4.055556, 9.916667), and its decision variable coordinates are (0, .11111, 0, 0, 0, 1.75, 1.05556, 0). This extreme point was found by using **ADBASE** [Steuer (1983)] to compute all of the MOLP's 14 efficient extreme points and then determining the best one with regard to U.

In this problem, however, the optimal point is nonextreme with $U = 547.07745$. Its criterion and decision variable vectors, along with the λ-vector that has the optimal point as a vertex (definition point) of a weighted Tchebycheff contour, are as follows:

$$\mathbf{z}^0 = \begin{bmatrix} 5.32008 \\ 2.58205 \\ 7.70641 \end{bmatrix} \qquad \mathbf{x}^0 = \begin{bmatrix} 0 \\ .11111 \\ 0 \\ 0 \\ 0 \\ 1.01325 \\ 1.05556 \\ 0 \end{bmatrix} \qquad \lambda^0 = \begin{bmatrix} .5878 \\ .1822 \\ .2300 \end{bmatrix}$$

The optimal point was found by solving the (nonlinear) utility function program

$$\max \{ U(\mathbf{z}) | C\mathbf{x} = \mathbf{z}, \quad \mathbf{x} \in S \}$$

using **MINOS**.

To calibrate the interactive procedure, recall the guideline relationships

$$P \gtrapprox k$$
$$t \approx k$$
$$1/(2k) \lessapprox w \lessapprox 3/(2k)$$
$$\sqrt[k]{1/P} \lessapprox r \lessapprox \sqrt[t-1]{w}$$

where " $\approx$ " means approximately and

P is the number of solutions to be presented at each iteration.
t is the intended number of iterations the procedure is to run.
w is the final iteration criterion weight width (of the $[\ell_i^{(t)}, \mu_i^{(t)}]$).
r is the $\bar{\Lambda}$-reduction factor.

Accordingly, let us set $P = 6$, $t = 4$, $w = \frac{1}{6}$, and $r = .5500$ when we solve the problem in Section 15.8.

15.7 STRUCTURE OF AUTOMATED PACKAGE

To illustrate the possible structure of an **MPSX**-based automated package, let us consider the **AUTOPAKX** Tchebycheff/weighted-sums automated package of Steuer and Lambert (1984). The JCL deck for submitting **AUTOPAKX** on the University of Georgia IBM 3081D is given in Fig. 15.1. As seen, the automated package

```
//JCLXXXX1 JOB USER=YYYYYY,PASSWORD=ZZZZZZ,NOTIFY=DEPTBA,TIME=2,
//    MSGCLASS=6
//*FORMAT PR,DDNAME=,DEST=NCT19
//*MAIN LINES=10,CLASS=B
//S1 EXEC MPSECL,REGION.GO=512K,PRM='NA,NAG,NOXREF,SMSG,NSTG',
//       PARM.GO='NOMAP,PRINT,NORES/ISASIZE(10K)',PAPER='*'
//PLI.SYSIN    DD DSN=DEPTBA.GA(ECL),UNIT=SYSDA,DISP=SHR
//GO.SYSLOUT   DD SYSOUT=*
//GO.SYSPRINT  DD SYSOUT=*
//GO.MPIN      DD DSN=DEPTBA.GA(MXXXX),UNIT=SYSDA,DISP=SHR
//GO.SYSIN     DD DSN=DEPTBA.GA(PXXXX),UNIT=SYSDA,DISP=SHR
//GO.FILEL     DD DSN=DEPTBA.GA(LXXXX1),DISP=SHR
//GO.DUMPRNT   DD SYSOUT=*
//GO.PUNCHY    DD DSN=DEPTBA.GA(OXXXX1),DISP=SHR
//
```

Figure 15.1. JCL for automated package.

employs the following five datasets where **xxxx** can be substituted for by the user with a mnemonic for the problem being solved.

1. **ECL**. This file contains the 900 statement PL/1 driver program that uses extended control features to call **MPSX** as a subroutine.
2. **Mxxxx**. This is the input file that contains the MOLP to be solved in **MPSX** format.

```
NAME          XXXX
ROWS
 N  OBJ1
 N  OBJ2
 N  OBJ3
 L  S1
 L  S2
 L  S3
 L  S4
 L  S5
 L  S6
 L  S7
 L  S8
COLUMNS
    X1        OBJ1      -1.       S2        -1.
    X1        S3         3.       S5         3.
    X1        S7        -2.       S8         6.
    X2        OBJ2       5.       OBJ3       4.
    X2        S5         4.       S6         7.
    X2        S7        -1.       S8        -3.
    X3        OBJ1      -1.       S1         8.
    X3        S3         3.       S4        -2.
    X3        S5        -2.       S6         2.
    X3        S7         1.
    X4        OBJ1       1.       OBJ3       6.
    X4        S1         6.       S2         1.
    X4        S3         5.       S4         7.
    X4        S5         1.       S8         3.
    X5        OBJ1       2.       OBJ3       2.
    X5        S1         1.       S4        -2.
    X5        S6         7.
    X6        OBJ1      -1.       OBJ2       2.
    X6        OBJ3       3.       S1         4.
    X6        S2         4.       S3         3.
    X7        OBJ1       6.       OBJ3       4.
    X7        S1        -3.       S3         1.
    X7        S4        -2.       S5        -1.
    X7        S6         4.       S7        -3.
    X7        S8         6.
    X8        OBJ2       4.       OBJ3      -1.
    X8        S1         7.       S2         3.
    X8        S4         6.       S5        -1.
    X8        S7         4.       S8         4.
RHS
    RHS       S1         6.       S2         7.
    RHS       S3         7.       S4        10.
    RHS       S5        10.       S6         5.
    RHS       S7         6.       S8         6.
ENDATA
```

Figure 15.2. **MPSX** deck for the illustrative MOLP.

3. `Pxxxx`. This is the input file that passes parameters to the automated package.
4. `Lxxxxh`. This is the input file that contains the λ-vectors to be used in the repetitive optimizations of iteration *h*.
5. `Oxxxxh`. To this file, we write the criterion vectors resulting from the repetitive optimizations of iteration *h*.

15.7.1 Mxxxx Deck

This file, called the **MPSX** *deck*, contains the MOLP to be solved in **MPSX** input format (Section 3.7). Note that each objective is assigned an **N** row-type indicator in the **ROWS** section. The **Mxxxx** file for the illustrative MOLP of Section 15.6 is shown in Fig. 15.2.

15.7.2 Pxxxx Deck

This file, called the *parameterization deck*, specifies the various options that are available with the Tchebycheff/weighted-sums automated package. The **Pxxxx** deck consists of $k + 3$ cards and is formatted as in Fig. 15.3. The different parameters are explained as follows.

1. *Iteration number*. The iteration number of this particular sequence of optimizations.
2. *Number of objectives*. The number of objectives of the MOLP.
3. *Tchebycheff option switch*. The types of optimizations that are automated:

   = 1 Augmented weighted Tchebycheff.
   = 2 Lexicographic weighted Tchebycheff.
   = 3 Weighted-sums.

4. *Hardcopy output switch*. Controls the amount of printed output:

   = 1 Regular printed output.
   = 2 Regular printed output plus decision variable values.
   = 3 Regular printed output with the resulting criterion vectors written to file **Oxxxxh**.
   = 4 Regular printed output plus decision variable values with the resulting criterion vectors written to file **Oxxxxh**.

5. *Output format switch*. Controls the format of the outputted criterion values that are printed out and/or written to file **Oxxxxh**:

   = 1 F10.1
   = 2 F10.2
   = 3 F10.3
   = 4 F10.4
   = 5 F10.5
   = 6 F10.6

6. *ρ Value (floating point)*. This field is ignored if the Tchebycheff option switch = 2 or 3.

First Card

1	2	3	4	5	6
Iteration Number 1–5	Number of Objectives 6–10	Tcheby Option Switch 15	Hardcopy Output Switch 20	Output Format Switch 25	ρ Value 26–35

Second Card

7	8	9
ε_i Option 5	Scale Factor Option 10	Lex Output Switch 15

Third Card

10	11	12	13
Name on the MPSX NAME Card 1–8	MPSX RHS Vector Name 11–18	MPSX Bounds Vector Name 21–28	PICTURE Option 35

Next k Cards

14	15	16	17	18	19
ith Obj MPSX Name 1–8	MIN or MAX 11–13	z_i MPSX Variable Name 21–28	ith Scale Factor 31–40	ε_i Value 41–50	δ_i Divisibility Factor Value 51–60

Figure 15.3. Format of Pxxxx parameterization deck.

7. ε_i *Option.* This parameter creates the $\mathbf{z}^{**}$ ideal criterion vector from the $\mathbf{z}^*$ vector of maximal criterion values:

$$= 1 \qquad z_i^{**} = \begin{cases} z_i^* + \varepsilon_i & i \in J^+ \\ z_i^* - \varepsilon_i & i \in J^- \end{cases}$$

$$= 2 \qquad z_i^{**} = \begin{cases} (\text{smallest integer divisible by } \delta_i) \geqq z_i^* + \varepsilon_i & i \in J^+ \\ (\text{largest integer divisible by } \delta_i) \leqq z_i^* - \varepsilon_i & i \in J^- \end{cases}$$

When the parameter is set to 1, the δ_i *divisibility factor* fields are ignored.

8. *Scale factor option.* Internally rescales the MOLP objective function rows:

= 1	Assumes a value of 1.0 as the scale factor for each MOLP objective regardless of what is in the scale factor fields.
= 2	Rescales each MOLP objective row according to the value in its scale factor field and "unscales" the criterion vectors upon output.
= 3	Same as = 2 but does not "unscale" the outputted criterion vectors.

9. *Lexicographic output switch.* Controls the hardcopy criterion vector output when the Tchebycheff option switch = 2:

= 1	The criterion vectors from both the first and second lexicographic stages are printed out.
= 2	Only the second-stage lexicographic criterion vectors are printed out.

10. *Name on the* **MPSX NAME** *card.* Columns 1–8 are to contain the name on the **NAME** card in the **MPSX** deck exactly as it appears in columns 15–22.
11. **MPSX** *RHS vector name.* Left justified, this is the name of the RHS vector in the **MPSX** deck that is to be used.
12. **MPSX** *bounds vector name.* Left justified, this is the name of the **BOUNDS** vector in the **MPSX** deck that is to be used. If bounds are not used, write **NONE** left justified in this field.
13. **PICTURE** *Option.*

= 0	Option is ignored.
= 1	Prints a **PICTURE** of the **MPSX** layout matrix.

14. i*th Objective* **MPSX** *name.* Left justified, this must be the same as the name given to the ith objective row in the **MPSX** deck.
15. **MAX** *or* **MIN**. Write **MAX** or **MIN** in this field depending on the objective.
16. z_i **MPSX** *variable name.* This is the name to be associated with the ith criterion value in the printed output. It is also the name given to the z_i column of the **MPSX** layout matrix. This name must not be the same as any other name in the **MPSX** deck.
17. i*th Scale factor (floating point).* This is the value by which the ith row of the criterion matrix **C** in the **MPSX** layout matrix is to be rescaled.
18. ε_i *Value (floating point).* A nonnegative value to be used in the construction of z_i^{**}.
19. δ_i *Divisibility factor (floating point).* Ensures that z_i^{**} is divisible by δ_i. If the ε_i option parameter = 1, this field is ignored.

The **Pxxxx** file does not change from iteration to iteration except for the iteration number in the first field of the first card. A copy of the first-iteration **Pxxxx** deck for the illustrative MOLP of Section 15.6 is given in Fig. 15.4.

```
     1    3    2    4    6
     2    1    1
XXXX      RHS       NONE      1
OBJ1      MAX       Z1                            .001      1.0
OBJ2      MAX       Z2                            .001      1.0
OBJ3      MAX       Z3                            .001      1.0
```

Figure 15.4. Parameterization deck for the illustrative MOLP.

Header Card

Iteration Number 1–5	Number of Vector Components 6–10	Number of Weighting Vectors 11–15

Weighting Vector Cards

Iteration Number 1–4	Weighting Vector Number 5–8	Vector Card Number 9–10	1st Component Value 11–20	. . .	6th Component Value 61–70

Figure 15.5. Format of Lxxxxh weighting vector deck.

15.7.3 Lxxxxh Deck

To prepare the Lxxxxh deck, we use the **LAMBDA** and **FILTER** programs from the **ADBASE** package, as described in Chapter 11. The **LAMBDA** program is used to produce an overabundance of weighting vectors. In the case of the illustrative MOLP of Section 15.6, the number is 150 (i.e., $50 \times k$). The **FILTER** program reduces the 150 weighting vectors to a small representative subset. In the illustrative MOLP, the number of representatives is 12 (i.e., $2 \times P$). The Lxxxxh deck is written in **ADBASE** six-field format (see Section 9.11.3), as depicted in Fig.

```
   1       3   12
   1   1   1     .2779     .2756     .4465
   1   3   1     .0162     .2693     .7145
   1   8   1     .3708     .4691     .1601
   1  14   1     .0139     .0234     .9627
   1  20   1     .9122     .0000     .0878
   1  22   1     .0169     .9829     .0002
   1  23   1     .5175     .0437     .4388
   1  24   1     .7149     .1278     .1573
   1  27   1     .0745     .4643     .4612
   1  38   1     .1672     .6587     .1741
   1  43   1     .2823     .0319     .6858
   1  79   1     .4684     .2485     .2831
```

Figure 15.6. Lxxxx1 weighting vector deck for the illustrative MOLP.

```
1     3   12
1   1 1  3.909616   4.883825   10.076556
1   3 1  3.114864   4.576783   10.709766
1   8 1  3.762010   5.440532    8.476936
1  14 1  4.600000   2.127273   11.490909
1  20 1  6.305050    .579125    4.779801
1  22 1   .000000   7.000000    2.000000
1  23 1  5.069257   1.608953    9.722973
1  24 1  5.800534   1.290311    6.548644
1  27 1  1.271642   5.616984    9.600966
1  38 1   .677458   6.395128    5.928035
1  43 1  4.733267   1.888944   11.066931
1  79 1  4.740545   3.741132    9.445031
```

Figure 15.7. `Oxxxx1` output file for the illustrative MOLP.

15.5. A copy of the first-iteration **Lxxxxh** deck for the illustrative MOLP is given in Fig. 15.6.

15.7.4 Oxxxxh File

The **Oxxxxh** file receives the criterion vectors generated by the automated package. They are stored in the same **ADBASE** six-field format utilized in **Lxxxxh**. A copy of the **Oxxxxh** file from the first iteration of the lexicographic weighted Tchebycheff application for the illustrative MOLP is given in Fig. 15.7.

15.8 NUMERICAL SOLUTION OF THE ILLUSTRATIVE MOLP

Applying the lexicographic weighted Tchebycheff procedure according to the **Pxxxx** deck specifications of Fig. 15.4, we have the following step-by-step solution of the illustrative MOLP.

1. *Step 1.* The automated package maximizes each objective to obtain

$$\mathbf{z}^* = \begin{bmatrix} 6.333333 \\ 7.000000 \\ 11.490909 \end{bmatrix}$$

Let the ε_i option parameter = 2. Then, with all $\varepsilon_i = .001$ and all $\delta_i = 1.0$, we have

$$\mathbf{z}^{**} = \begin{bmatrix} 7.000000 \\ 8.000000 \\ 12.000000 \end{bmatrix}$$

2. *Step 5.* Using **LAMBDA**, 150 ($50 \times k$) λ-vectors are generated from the intervals

$$[\ell^{(1)}, \mu^{(1)}] = \begin{bmatrix} .0000, & 1.0000 \\ .0000, & 1.0000 \\ .0000, & 1.0000 \end{bmatrix}$$

TABLE 15.3
Nondominated Criterion Vectors of the First Iteration

Criterion Vector	z_1	z_2	z_3
1–1	3.909616	4.883825	10.076556
1–14	4.600000	2.127273	11.490909
1–22	.0	7.000000	2.000000
√ 1–24	5.800534	1.290311	6.548644
1–27	1.271642	5.616984	9.600966
1–38	.677458	6.395128	5.928035

3. *Step 6.* Using `FILTER`, the 150 λ-vectors are reduced to 12 dispersed representatives and stored in `Lxxxx1` (shown in Fig. 15.6).
4. *Step 7.* With the automated package, the 12 associated lexicographic weighted Tchebycheff programs are solved and the resulting criterion vectors are stored in `Oxxxx1` (shown in Fig. 15.7).
5. *Step 8.* With `FILTER`, the criterion vectors in `Oxxxx1` are reduced to the six most different. Then, the six criterion vectors are presented to the decision maker in the form of Table 15.3.
6. *Step 9.* Of the criterion vectors in Table 15.3, 1–24 is the best with $U = 453.53924$.
7. *Step 11.* The λ-vector that has criterion vector 1–24 as a definition point of the weighted Tchebycheff (first-stage lexicographic) contour is computed in Table 15.4, in which

$$\text{Term } A = \frac{1}{z_i^{**} - z_i}$$

$$\text{Term } B = \left[\sum_{i=1}^{k} \frac{1}{z_i^{**} - z_i}\right]^{-1}$$

8. *Steps 12 and 5.* Using `LAMBDA`, 150 λ-vectors are generated from the intervals

$$[\ell^{(2)}, \mu^{(2)}] = \begin{bmatrix} .4399, & .9899 \\ .0000, & .5500 \\ .0000, & .5500 \end{bmatrix}$$

TABLE 15.4
Calculation of $\lambda^{(2)}$

$\mathbf{z}^{**}$	$\mathbf{z}^{1-24}$	*Term A*	*Term B*	$\lambda^{(2)}$
7.000000	5.800534	.8337043	.8574984	.7149
8.000000	1.290311	.1490382	.8574984	.1278
12.000000	6.548644	.1834405	.8574984	.1573

Table 15.5
Nondominated Criterion Vectors of the Second Iteration

Criterion Vector	z_1	z_2	z_3
1–24	5.800534	1.290311	6.548644
2–1	5.380533	1.776738	8.111714
2–24	6.241880	.631766	5.032479
2–31	4.996080	2.000217	9.732579
2–73	3.969579	5.090513	8.840792
√ 2–85	4.846459	3.529305	9.127290
2–121	5.788636	1.101629	6.766464

9. *Step 6.* Using `FILTER`, the 150 λ-vectors are reduced to 12 dispersed representatives and stored in `Lxxxx2`.

10. *Step 7.* With the automated package, the 12 associated lexicographic weighted Tchebycheff programs are solved and the resulting criterion vectors are stored in `Oxxxx2`.

11. *Step 8.* With `FILTER`, the criterion vectors in `Oxxxx2` are reduced to the six most different. Then, the six criterion vectors are presented to the decision maker in the form of Table 15.5.

12. *Step 9.* Of the criterion vectors in Table 15.5, 2–85 is the best with $U = 524.71813$.

13. *Step 11.* The λ-vector that has criterion vector 2–85 as the definition point of a weighted Tchebycheff contour is computed in Table 15.6. The weighting vector that generated $\mathbf{z}^{2-85}$ via the lexicographic weighted Tchebycheff program was $\boldsymbol{\lambda}^{2-85} = (.6394, .3080, .0526)$. We note that this is different from $\boldsymbol{\lambda}^{(3)}$. This is the only time in this problem that $\boldsymbol{\lambda}^{(h)}$ differs from the λ-vector in `Lxxxxh` that generated the criterion vector.

14. *Steps 12 and 5.* Using `LAMBDA`, 150 λ-vectors are generated from the intervals

$$[\boldsymbol{\ell}^{(3)}, \boldsymbol{\mu}^{(3)}] = \begin{bmatrix} .2970, & .5995 \\ .0647, & .3672 \\ .1847, & .4872 \end{bmatrix}$$

Table 15.6
Calculation of $\boldsymbol{\lambda}^{(3)}$

$\mathbf{z}^{**}$	$\mathbf{z}^{2-85}$	*Term A*	*Term B*	$\boldsymbol{\lambda}^{(3)}$
7.000000	4.846459	.4643515	.9651263	.4482
8.000000	3.529305	.2236789	.9651263	.2159
12.000000	9.127290	.3481034	.9651263	.3359

TABLE 15.7
Nondominated Criterion Vectors of the Third Iteration

Criterion Vector	z_1	z_2	z_3
2–85	4.846459	3.529305	9.127290
3–1	4.827692	3.334036	9.383137
3–18	4.899960	2.022273	10.166813
3–24	5.173513	2.230331	8.698868
3–27	4.435605	4.304666	9.657707
√ 3–34	5.364758	2.401326	7.650720
3–94	3.985921	4.994288	9.235598

15. *Steps 6, 7, and 8.* Reducing the 150 λ-vectors to 12 dispersed representatives, and then reducing the criterion vectors resulting from the 12 associated lexicographic weighted Tchebycheff optimizations to the six most different, we obtain Table 15.7.
16. *Step 9.* Of the criterion vectors in Table 15.7, 3–34 is the best with $U = 532.81922$.
17. *Step 11.* The λ-vector that has criterion vector 3–34 as the definition point of a weighted Tchebycheff contour is computed in Table 15.8.
18. *Steps 12 and 5.* Using `LAMBDA`, 150 λ-vectors are generated from the intervals

$$[\ell^{(4)}, \mu^{(4)}] = \begin{bmatrix} .5163, & .6827 \\ .0919, & .2583 \\ .1422, & .3086 \end{bmatrix}$$

19. *Steps 6, 7, and 8.* Reducing the 150 λ-vectors to 12 dispersed representatives, and then reducing the criterion vectors resulting from the 12 associated lexicographic weighted Tchebycheff optimizations to the six most different, we have Table 15.9.
20. *Step 9.* Of the criterion vectors in Table 15.9, 4–27 is the best with $U = 544.89867$.

TABLE 15.8
Calculation of $\lambda^{(4)}$

$\mathbf{z}^{**}$	$\mathbf{z}^{3-34}$	*Term A*	*Term B*	$\lambda^{(4)}$
7.000000	5.364758	.6115303	.9803276	.5995
8.000000	2.401326	.1786137	.9803276	.1751
12.000000	7.650720	.2299231	.9803276	.2254

TABLE 15.9
Nondominated Criterion Vectors of the Fourth Iteration

Criterion Vector	z_1	z_2	z_3
3–34	5.364758	2.401326	7.650720
4–1	5.349584	2.174888	7.916341
4–3	5.193693	2.031962	8.773765
4–20	5.437774	1.301855	8.248903
4–24	5.547522	1.713054	7.379064
√ 4–27	5.178493	2.865236	8.131187
4–94	4.940268	3.341687	8.845864

21. *Step 15.* Assuming that the decision maker does not wish to keep iterating, we terminate with the following as our final solution.

$$\mathbf{z}^{4-27} = \begin{bmatrix} 5.178493 \\ 2.865236 \\ 8.131187 \end{bmatrix} \qquad \mathbf{x}^{4-27} = \begin{bmatrix} 0 \\ .11111 \\ 0 \\ 0 \\ 0 \\ 1.15484 \\ 1.05556 \\ 0 \end{bmatrix}$$

In the above problem, the Tchebycheff procedure has located a solution whose $U = 544.89867$. This is better than the best efficient extreme point whose $U = 496.67474$. In utility value terms, this is 95.7% of the distance between the best efficient extreme point and the optimal point whose $U = 547.07745$. Further improvement is possible if we were to continue iterating.

Although we used the lexicographic weighted Tchebycheff procedure, no second-stage optimizations were necessary in this problem because each first-stage optimization resulted in a unique optimum. This is not unusual.

15.9 SOME FINAL TCHEBYCHEFF GRAPHICAL EXAMPLES

To appreciate some final points about the Tchebycheff approach, consider the following four examples.

Example 3. Let $\mathbf{z}^{**} = (10, 10, 10)$. Consider the $|||\mathbf{z}^{**} - \mathbf{z}|||_\infty^\lambda = 1$ contours with $\rho = 0$ for

(a) $\boldsymbol{\lambda} = (\frac{1}{6}, \frac{3}{4}, \frac{1}{12})$
(b) $\boldsymbol{\lambda} = (\frac{1}{2}, 0, \frac{1}{2})$
(c) $\boldsymbol{\lambda} = (0, 1, 0)$

With $\bar{\mathbf{z}}$ its vertex, the contour of (a) is shown in Fig. 15.8*a*. The contour of (b),

$\bar{z} = \left(4, \frac{26}{3}, -2\right)$

a   b

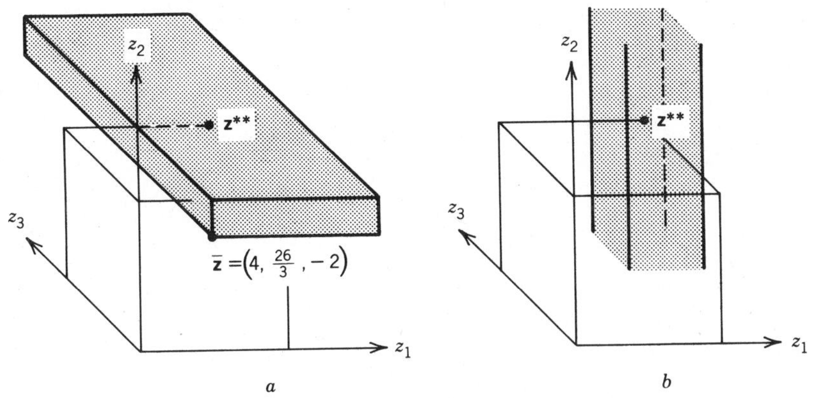

Figure 15.8. Graphs of Example 3.

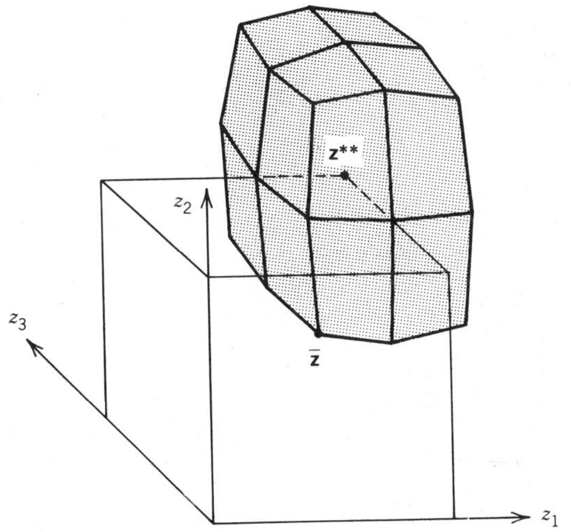

Figure 15.9. Graph of Example 4.

$z^1 = (0, 3, 3)$
$z^2 = (3, 3, 0)$
$z^3 = (3, 0, 3)$
$z^4 = \left(\frac{3}{2}, 3, \frac{3}{2}\right)$

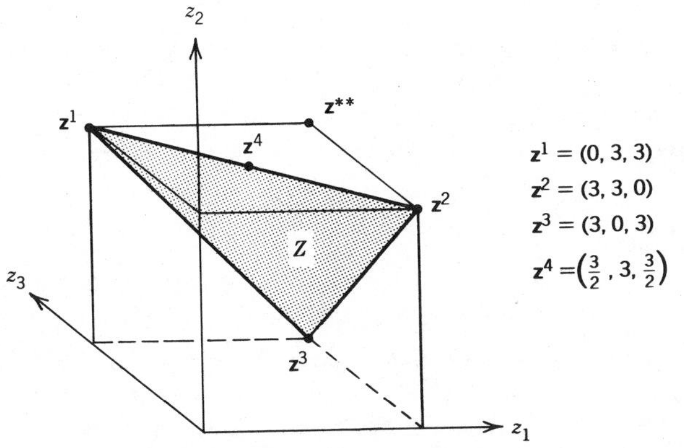

Figure 15.10. Graph of Example 5.

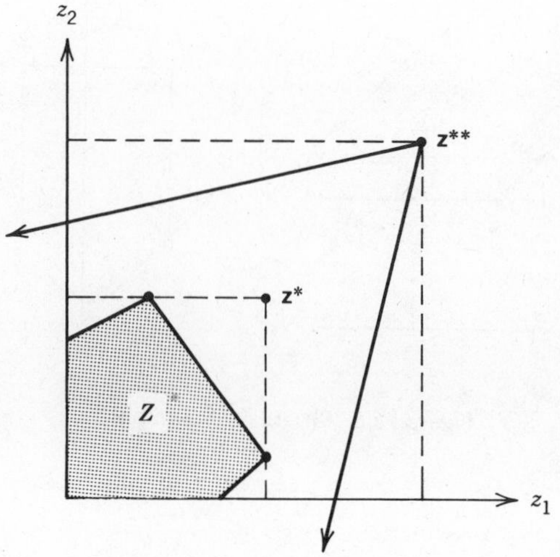

Figure 15.11. Graph of Example 6.

being unbounded from above and below, is without a vertex because $\lambda_2 = 0$. It is shown in Fig. 15.8*b*. The contour of (c) does not have a vertex either. Because $\lambda_1 = \lambda_3 = 0$, the contour of (c) consists of two horizontal planes intersecting the x_2-axis at $(0, 9, 0)$ and $(0, 11, 0)$, respectively.

Example 4. Let $\mathbf{z}^{**} = (10, 10, 10)$. The $|||\mathbf{z}^{**} - \mathbf{z}|||_{\infty}^{\lambda} = 1$ contour with $\rho = .01$ and $\boldsymbol{\lambda} = (.4, .2, .4)$ is shown in Fig. 15.9. In this example, the vertex of the contour occurs at $\bar{\mathbf{z}} = (\frac{85}{11}, \frac{60}{11}, \frac{85}{11})$. This is in contrast to $(7.5, 5, 7.5)$ which is where the vertex would have been had ρ equaled 0.

Example 5. Consider the multiple objective program whose Z is the shaded area in Fig. 15.10. In this problem, $N = \gamma(\mathbf{z}^1, \mathbf{z}^2, \mathbf{z}^3)$. By the conditions of Section 14.1, all ε_i's should be set to positive scalars. Suppose, however, we let $\varepsilon_i = 0$ for all i, in which case $\mathbf{z}^{**} = (3, 3, 3)$. Consider $\mathbf{z}^4 \in N$. The only weighting vector for which $\mathbf{z}^4$ minimizes the augmented (or lexicographic) weighted Tchebycheff program is $\boldsymbol{\lambda} = (0, 1, 0)$. Since all criterion vectors on the $\gamma(\mathbf{z}^1, \mathbf{z}^2)$ edge minimize the augmented (or lexicographic) weighted Tchebycheff program for $\boldsymbol{\lambda} = (0, 1, 0)$, $\mathbf{z}^4$ could never be uniquely computed. This would be unfortunate if $\mathbf{z}^4$ were optimal. However, if the ε_i were set to positive scalars in accordance with Section 14.1, there then would exist a $\boldsymbol{\lambda} \in \overline{\Lambda}$ for which $\mathbf{z}^4$ is uniquely computable.

Example 6. This example demonstrates why it is undesirable to use large ε_i values. In the multiple objective program of Fig. 15.11, let $\mathbf{z}^{**}$ be as indicated even though it is permissible for $\varepsilon_1 = \varepsilon_2 = 0$. From Fig. 15.11, we can see the effect of overly large ε_i values. They cause a sizable proportion of the Tchebycheff diagonal directions to "miss" Z. Thus, overly large ε_i values make ineffective use of unnecessarily large portions of $\overline{\Lambda}$. This is the reason behind the rules of thumb of Section 14.2 for keeping the ε_i values moderately small.

15.10 SOME FINAL IMPLEMENTATION COMMENTS

For some final implementation comments, let us discuss the following topics.

1. Software other than MPSX.
2. Criterion value ranges over the efficient set.
3. Number of solutions presented at each iteration.
4. Most preferred criterion vector as filtering seed point.
5. Insertion of criterion value lower bounds.
6. Selection of ε_i values in integer and nonlinear cases.

15.10.1 Software Other than MPSX

Automated packages for multiple objective linear programming problems can be written that use "workhorse" single criterion software other than MPSX. Also, automated packages can be written for solving multiple objective integer and multiple objective nonlinear programming problems. Candidate single criterion workhorse software systems that utilize MPSX input format are the following.

1. SAS/OR [SAS (1985)]. Automated Tchebycheff/weighted-sums MOLP packages could be written that call the SAS/OR package. The LP procedure of SAS/OR has capabilities similar to the SAVE / REVISE / RESTORE features of MPSX. The NETFLOW procedure of SAS/OR could be used to support automated packages for multiple objective network problems.
2. MPSX–MIP [IBM (1975)]. The automated package of this chapter could be modified to call MPSX–MIP. Then, the package could be used for solving both linear and mixed integer multiple objective programs.
3. MINOS [Murtagh and Saunders (1977, 1980); Preckel (1980); Saunders (1977)]. An automated package could be written to call MINOS. MINOS has traditional warm starting capabilities for linear problems and has "flying start" capabilities with nonlinear problems. Thus, written to access MINOS, we could have automated packages for solving linear and nonlinear multiple objective problems.

15.10.2 Criterion Value Ranges Over the Efficient Set

Prior to the commencement of any interactive procedure, it is advisable to schedule some sort of problem "warm-up" exercise for the decision maker. One goal of a warm-up exercise is to adapt the decision maker's (perhaps overoptimistic) aspirations to the confines of his or her feasible region. One thing that can be done in this regard is to display the ranges of the different criterion values over the efficient set as in Fig. 15.12.

The difficulty with the criterion value range idea is that a special algorithm is necessary to obtain the minimum criterion values over the efficient set (see Section 9.13).

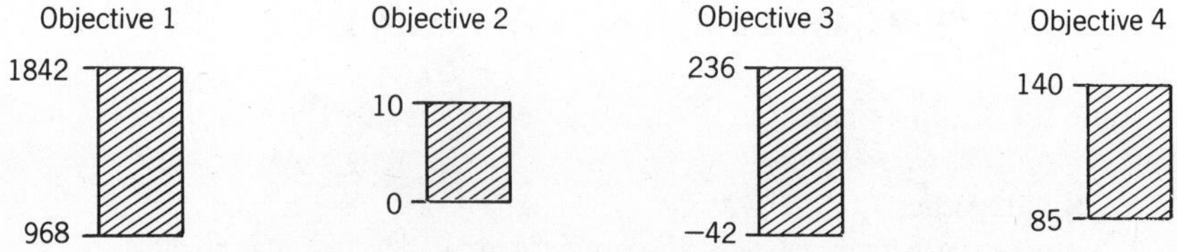

Figure 15.12. Criterion value ranges over the efficient set.

15.10.3 Number of Solutions Presented at Each Iteration

In Miller (1956), we have the well-known article about the number "seven, plus, or minus two." This article has sometimes been interpreted to imply that a decision maker should not be presented with more than about seven solutions at a time. Otherwise, the limits of his or her information processing ability might be surpassed. In other words, the DM might suffer from "information overload" and encounter "decision-maker burnout." This may be true for some decision makers, but it is not true for all. Some decision makers, people who are familiar with the concept of nondominance and know what they are looking for, can process up to 15 or more criterion vectors in one sitting. It is a matter of practice. This is good news because the larger the number of solutions that can be presented to the decision maker at each iteration, the faster algorithms such as the Tchebycheff method can converge.

15.10.4 Most Preferred Criterion Vector as Filtering Seed Point

Consider any iteration of the automated package but the first. If we filter the criterion vectors from the repetitive optimizations without regard to the decision maker's most recent criterion vector selection, one of the new criterion vectors presented to the decision maker might be similar to this criterion vector. To prevent this (in the interest of extracting the most information from each iteration), we should use the most recent criterion vector selection as the filtering seed point. Then, as the decision maker examines the P new criterion vectors versus his or her most recent selection, he or she is assured that none of the $(P + 1)$ vectors is unnecessarily similar to any of the others. In this way, given that we can only present P new criterion vectors at a time, we are presenting the maximum amount of information possible per iteration.

15.10.5 Insertion of Criterion Value Lower Bounds

In the process of examining solutions at each iteration, the decision maker may conclude that values below certain levels for some of the criteria would never be acceptable under any circumstances. In such cases, it would behoove procedures such as the Tchebycheff method to add lower bound constraints on the criterion values of the pertinent objectives before proceeding with subsequent iterations. Of course, it would make no sense to add a lower bound constraint with a RHS less than the objective's minimum criterion value over the efficient set. By giving the decision maker the option to configure any scheme of lower bounds on his or her criterion values, the efficient set is reduced and the convergence of the algorithm can only, if anything, be favorably affected.

15.10.6 Selection of ε_i Values in Integer and Nonlinear Cases

Assume that it is necessary for z_i^{**} to be greater than $z_i^* = \max\{f_i(\mathbf{x}) | \mathbf{x} \in S\}$. If $f_i(\mathbf{x})$ is linear, there is no problem. Any ε_i value greater than zero will do. However, if $f_i(\mathbf{x})$ is nonlinear or S is integer, we may have to work with an approximation of z_i^* because of computational reasons. Then, we must be careful about the value of ε_i. One way of selecting a value for ε_i is as follows. If we are able to obtain an *upper bound* on the value of $f_i(\mathbf{x})$ over S as in branch and bound integer programming, we can select an ε_i value that is greater than the difference between the highest known value and the upper bound. In this way, we ensure that z_i^{**} is greater than z_i^*.

Readings

1. Gabbani, D. and M. J. Magazine (1981). "An Interactive Heuristic Approach for Multi-Objective Integer Programming Problems," Working Paper No. 148, Department of Management Sciences, University of Waterloo, Waterloo, Ontario, Canada.
2. IBM Document No. SH19–1127-0 (1976). "IBM Mathematical Programming System Extended (MPSX/370): Basic Reference Manual," IBM Corporation, Data Processing Division, White Plains, New York.
3. IBM Document No. SH19–147-0 (1978). "IBM Mathematical Programming System Extended/370 (MPSX/370): Introduction to the Extended Control Language," IBM Corporation, Data Processing Division, White Plains, New York.
4. IBM Document No. SH19–1099-1 (1975). "IBM Mathematical Programming System Extended/370 (MPSX/370), Mixed Integer Programming/370 (MIP/370): Program Reference Manual," IBM Corporation, Data Processing Division, White Plains, New York.
5. IBM Document No. GH19–1091-1 (1979). "IBM Mathematical Programming System Extended/370: Primer," IBM Corporation, Data Processing Division, White Plains, New York.
6. Isermann, H. and R. E. Steuer (1985). "Payoff Tables and Minimum Criterion Values Over the Efficient Set," College of Business Administration, University of Georgia, Athens, Georgia.
7. Kallio, M., A. Lewandowski, and W. Orchard-Hays (1980). "An Implementation of the Reference Point Approach for Multiobjective Optimization," WP–80-35, International Institute for Applied Systems Analysis, Laxenburg, Austria.
8. Leitch, R. A., R. E. Steuer, and J. T. Godfrey (1985). "An Interactive Multiple Objective Budget Model," College of Business Administration, University of Georgia, Athens, Georgia.

9. Lewandowski, A. and M. Grauer (1982). "The Reference Point Optimization Approach—Methods of Efficient Implementation." In M. Grauer, A. Lewandowski, and A. P. Wierzbicki (eds.), *Multiobjective and Stochastic Optimization*, Laxenburg, Austria: International Institute for Applied Systems Analysis, pp. 353–376.

10. Miller, G. (1956). "The Magical Number Seven Plus or Minus Two: Some Limits on Our Capacity for Processing Information," *Psychological Review*, Vol. 63, No. 2, pp. 81–97.

11. Murtagh, B. A. and M. A. Saunders (1977). "MINOS User's Guide," Report SOL 77–7, Department of Operations Research, Stanford University, Stanford, California.

12. Murtagh, B. A. and M. A. Saunders (1980). "MINOS/AUGMENTED User's Manual," Report SOL 80–14, Department of Operations Research, Stanford University, Stanford, California.

13. Preckel, P. V. (1980). "Modules for Use with MINOS/AUGMENTED in Solving Sequences of Mathematical Programs," Report SOL 80–15, Department of Operations Research, Stanford University, Stanford, California.

14. SAS Institute (1985). SAS/OR *User's Guide*, Cary, North Carolina: SAS Institute, Inc.

15. Saunders, M. A. (1977). "MINOS System Manual," Report SOL 77-31, Department of Operations Research, Stanford University, Stanford, California.

16. Silverman, J., R. E. Steuer, and A. W. Whisman (1985). "A Multi-Period, Multiple Criteria Optimization System for Manpower Planning," College of Business Administration, University of Georgia, Athens, Georgia.

17. Steuer, R. E. (1983). "Multiple Criterion Function Goal Programming Applied to Managerial Compensation Planning," *Computers and Operations Research*, Vol. 10, No. 4, pp. 299–309.

18. Steuer, R. E. (1983). "Operating Manual for the ADBASE Multiple Objective Linear Programming Package," College of Business Administration, University of Georgia, Athens, Georgia.

19. Steuer, R. E. and E.-U. Choo (1983). "An Interactive Weighted Tchebycheff Procedure for Multiple Objective Programming," *Mathematical Programming*, Vol. 26, No. 1, pp. 326–344.

20. Steuer, R. E., J. M. Dominey, and A. W. Whisman (1985). "A MINOS-Based Tchebycheff/Weighted-Sums Automated Package for Multiple Objective Linear and Nonlinear Programming," College of Business Administration, University of Georgia, Athens, Georgia.

21. Steuer, R. E. and F. W. Harris (1980). "Intra-Set Point Generation and Filtering in Decision and Criterion Space," *Computers and Operations Research*, Vol. 7, No.'s 1–2, pp. 41–53.

22. Steuer, R. E. and T. D. Lambert (1984). "AUTOPAKX: An MPSX-Based Tchebycheff/Weighted-Sums Automated Package for Multiple Objective Linear Programming," College of Business Administration, University of Georgia, Athens, Georgia.

23. Wierzbicki, A. P. (1980). "The Use of Reference Objectives in Multiobjective Optimization," *Lecture Notes in Economics and Mathematical Systems*, No. 177, Berlin: Springer-Verlag, pp. 468–486.

24. Winkels, H.- M. (1983). "A Graphical Subroutine for Multiobjective Linear Programing," *OR Spektrum*, Vol. 5, No. 3, pp. 175–192.

PROBLEM EXERCISES

15-1 With $\varepsilon_1 = \varepsilon_2 = 1$, $\rho = .001$, and $\boldsymbol{\lambda} = (.3, .7)$, create the **MPSX** layout matrix for the MOLP:

$$
\begin{aligned}
\max\{ -x_1 + 2x_2 &= z_1 \} \\
\max\{ 3x_1 + x_2 &= z_2 \} \\
\text{s.t.} \quad x_1 &\leqq 4 \\
2x_1 + 3x_2 &\leqq 11 \\
-x_1 + x_2 &\leqq 2 \\
x_1, x_2 &\geqq 0
\end{aligned}
$$

15-2 Prepare an **MPSX** input deck for the LP:

$$
\begin{aligned}
\max\{ x_1 + x_2 + x_3 &= z \} \\
\text{s.t.} \quad 2x_1 + 2x_2 + 2x_3 &\leqq 10 \\
x_3 &\leqq 4 \\
x_1 \qquad + x_3 &\geqq 1 \\
x_1, x_2, x_3 &\geqq 0
\end{aligned}
$$

Title the problem **MPSXTEST** and denote the objective **OBJ**; the constraints **AROW1**, **AROW2**, and **AROW3**, the variables **X1**, **X2**, and **X3**; and the right-hand side **RVECT**.

15-3 With $\varepsilon_1 = 2$ and $\varepsilon_2 = 1$, consider the MOLP:

$$
\begin{aligned}
\max\{ 5x_1 \qquad &= z_1 \} \\
\max\{ \qquad 5x_2 &= z_2 \} \\
\text{s.t.} \quad x_1 + x_2 &\leqq 6 \\
16x_1 - 3x_2 &\geqq 20 \\
x_1, x_2 &\geqq 0
\end{aligned}
$$

(a) With the ε_i option = 1 (see Section 15.7.2), what is $\mathbf{z}^{**}$?

(b) With the ε_i option = 2, what is $\mathbf{z}^{**}$ with all $\delta_i = 5$?

(c) Same as (a) but assume that the second objective is $\min\{5x_2 = z_2\}$.

(d) Same as (b) but assume that the second objective is $\min\{5x_2 = z_2\}$.

15-4 With all $\varepsilon_i = .001$, all $\delta_i = 2$, $\rho = .005$, and $\lambda = (.3, .1, .2, .4)$, create the **MPSX** layout matrix for the MOLP:

$$\begin{aligned}
\min\ \{-3x_1 + x_2 &= z_1\}\\
\max\ \{4x_1 + x_2 + 5x_3 &= z_2\}\\
\max\ \{x_1 + 6x_2 + x_3 &= z_3\}\\
\min\ \{-3x_1 + x_2 - 2x_3 &= z_4\}\\
\text{s.t.}\quad x_1 + x_2 &\leqq 5\\
x_2 + x_3 &\leqq 5\\
x_1, x_2, x_3 &\geqq 0
\end{aligned}$$

15-5C Consider the MOLP and utility function of Section 15.6. Letting (a) $P = 5$ or (b) $P = 7$, solve the MOLP using either the augmented or lexicographic weighted Tchebycheff method.

15-6C Using a nonlinear programming code, solve:

$$\begin{aligned}
\max\ \{x_1x_2 + x_3 &= z\}\\
\text{s.t.}\quad x_1 + x_2 &\leqq 10\\
x_1 &\geqq 1\\
x_3 &\leqq 2\\
x_1, x_2, x_3 &\geqq 0
\end{aligned}$$

15-7C Using a nonlinear programming code, solve some of the problems in W. Hock and K. Schittkowski (eds.) (1981). "Test Examples for Nonlinear Programming Codes," *Lecture Notes in Economics and Mathematical Systems*, No. 187, Berlin: Springer-Verlag.

15-8C Solve the MOLP of this problem using either the augmented or lexicographic weighted Tchebycheff method. To simulate decision-maker judgment, let U be one of the following:

(a) $U = 100 - [(38 - z_1)^3 + 2(12 - z_2)^3 + (14 - z_3)^3 + (6 - z_4)^3]^{1/3}$

(b) $U = 1000 - [(40 - z_1)^4 + (15 - z_2)^4 + (15 - z_3)^4 + (10 - z_4)^4]^{1/4}$

(c) $U = (z_1 + 4)^{.2}(z_2 + 1)^{.3}(z_3 + 15)^{.4}(z_4 + 25)^{.1}$

(d) $U = (z_1 + 4)^{.5}(z_2 + 1)^{.1}(z_3 + 15)^{.1}(z_4 + 25)^{.3}$

What is the U value of the best efficient extreme point? What is the U value of the final solution found by the interactive procedure?

	x_1	x_2	x_3	x_4	x_5	x_6		
	6	5	4	−5	4	2	max	
	5			3	2	−2	max	
Objs	6	−3	−1	−5	1	6	max	
	−1	3	−3	1	1	−2	max	
	4	5		4	5	6	≤	9
	2	4	−1				≤	8
s.t.	4	3		5	6	3	≤	9
		1	−1	2	1		≤	8
		−1	−1		1		≤	10
			1				≤	6

15-9C Solve the MOLP of this problem using either the augmented or lexicographic weighted Tchebycheff method. To simulate decision-maker judgment, let:

$$U = 100 - \left[(10 - z_1)^4 + (10 - z_2)^4 + (60 - z_3)^4 + (35 - z_4)^4\right]^{1/4}$$

What is the U value of the best efficient extreme point? What is the U value of the final solution found by the interactive procedure?

	x_1	x_2	x_3	x_4	x_5	x_6		
	4	−1	−4	−3	−1	−3	max	
	−1	−4	4		5	5	max	
Objs	−4	−5	2	5		4	max	
				3		1	max	
		6		1			≤	10
	−1			−1			≤	6
	−1	−1	3			−1	≤	6
s.t.			3		5	4	≤	6
	3		1				≤	6
		4					≤	8

15-10C Solve the MOLP of this problem using the interactive weighted-sums/filtering method. To simulate decision-maker judgment, let U be one of the following:

(a) $U = 2z_1 + 5z_2 + z_3 + 6z_4 + 4z_5$
(b) $U = 2z_1 + 8z_2 + 8z_3 + 4z_4 + 6z_5$
(c) $U = 3z_1 + z_2 + 4z_3 + 2z_4 + 9z_5$
(d) $U = z_1 + 2z_2 + 8z_3 + 9z_4 + 2z_5$
(e) $U = 5z_1 + 7z_2 + 2z_3 + 2z_4 + z_5$

	x_1	x_2	x_3	x_4	x_5	x_6	x_7	x_8	x_9	x_{10}		
			3	1	3	−1	3	2	1	−1	max	
	−2	2	−2	−2		1	−1	2	−1	−2	max	
Objs	1		−2				−2		3		max	
		−2			1	2	−1		−2	−2	max	
	1		1		−1	−2	2		2		max	
	2				2			1			≤	6
			3	1						1	≤	6
				3					5		≤	9
						4	1			5	≤	6
					1						≤	10
	2				5						≤	6
		3	5		3					4	≤	5
s.t.		2	3	−1			3			3	≤	8
	5			3		5		−1			≤	5
										2	≤	5
		2				−1			3		≤	8
	2			3			1			4	≤	5
	5			3					2		≤	8
				2							≤	10

15-11C Solve the MOLP of this problem using the interactive weighted-sums/filtering method. To simulate decision-maker judgment, let U be one of the following:

(a) $U = 2z_1 + 9z_2 + 5z_3 + z_4 + 3z_5$

(b) $U = 5z_1 + 6z_2 + 4z_3 + 5z_4 + 6z_5$

(c) $U = z_1 + z_2 + 9z_3 + 5z_4 + 5z_5$

(d) $U = 6z_1 + 2z_2 + 3z_3 + 9z_4 + 4z_5$

(e) $U = 9z_1 + 6z_2 + z_3 + 7z_4 + z_5$

	x_1	x_2	x_3	x_4	x_5	x_6	x_7	x_8	x_9	x_{10}		
	3	3	−1		2	−1	−1			−1	max	
	−2	−1	2		−1				−2	−1	max	
Objs		2	1	1	−2		2	3		−1	max	
	3	3	−1				1	−1	1	3	max	
	−2				2	3	1	3	−1	−1	max	
									5		≤	10
	5				4					4	≤	10
		5		−1							≤	10
	5		3								≤	5
		5	3		3			5	4		≤	10
	3		4		2		−1				≤	6
	4										≤	9
s.t.		4			1	5		1			≤	7
	3						−1	4			≤	10
							2				≤	5
					5				5		≤	5
		−1		−1	5	4		1			≤	6
	2	5		4		3		−1			≤	6
		4		5				1			≤	8

15-12C Consider the MOLP:

$$\begin{aligned}
\max\{\ x_1 & & & = z_1\} \\
\max\{\ & x_2 & & = z_2\} \\
\max\{\ & & x_3 & = z_3\} \\
\text{s.t.}\quad x_1 & + x_2 & + x_3 & \leqq 8 \\
& x_1, x_2, x_3 & & \geqq 1
\end{aligned}$$

(a) What is the optimal point if $U = (z_1)^{2.1}(z_2)^{2.2}(z_3)^{2.3}$?

(b) What is the best U value obtained after three iterations of the augmented weighted Tchebycheff method?

15-13C Solve the MOLP of this problem using either the augmented or lexicographic weighted Tchebycheff procedure. To answer the questions posed by the algorithm, let U be one of the following:

(a) $U = 100 - [2(52 - z_1)^4 + (82 - z_2)^4 + 3(32 - z_3)^4]^{1/4}$

(b) $U = 100 - [4(52 - z_1)^4 + (82 - z_2)^4 + 2(32 - z_3)^4]^{1/4}$

(c) $U = (z_1/15)^3(z_2/15)^{2.5}(z_3/15)^{1.5}$

	x_1	x_2	x_3	x_4	x_5	x_6	x_7		
			1	4	3	3		max	
Objs	3	5	2	1			4	max	
		3	2					max	
	2		4		6	2		$\leqq$	46
				3	2	5	4	$\leqq$	33
	3		6		1			$\leqq$	44
s.t.	3	6		5		4	1	$\leqq$	47
		1	6	2				$\leqq$	26
			4			1	3	$\leqq$	47
	2		4		5			$\leqq$	45

CHAPTER 16

Applications

This chapter discusses three case studies and mentions two others to illustrate the applicability of the vector-maximum, filtering, and Tchebycheff techniques of multiple objective programming. The three case studies are as follows.

1. ***Sausage blending***. This problem is modeled as an MOLP. Because of the small size of the sausage blending MOLP, a vector-maximum/filtering solution procedure is applied. An $8 \times 16 \times 20$ computer example is used for purposes of illustration.
2. ***CPA Firm audit staff allocation***. This problem is modeled as an MOLP. A hybrid version of the interval criterion weights/vector-maximum and interactive weighted-sums/filtering procedures of Sections 13.4 and 13.5 is applied. This application demonstrates how the weighting vectors pertaining to the decision maker's final solution may be far different from what would have been guessed at the beginning of the problem. A $7 \times 32 \times 63$ computer example is used for purposes of illustration.
3. ***Managerial compensation planning***. This problem is modeled as a goal program except that the deviational variables are grouped to form multiple objectives. At each iteration, in a decision support systems context, the decision maker is allowed to adjust model coefficients. The lexicographic weighted Tchebycheff capability of the **MPSX**-based automated package of Chapter 15 is applied to permit convergence to a nonextreme final solution. A $5 \times 38 \times 33$ computer example is used for purposes of illustration.

16.1 SAUSAGE BLENDING

Large meat processing firms typically blend sausage meats (hot dogs, bologna, salami, etc.) with the aid of single objective cost minimization linear programming models. The decision variables are preblended cuts of meat (chuck, beef trim, beef head, beef cheek, pork trim, pork jowls, pork fat, etc.) along with spice and ice.

The constraints include (a) ingredient upper and lower bounds; (b) fat, protein, and moisture control constraints; (c) pork-to-beef balance constraints, and so forth. Because many of the model coefficients are subject to change from day to day, the LP models are run as frequently as several times a week.

Currently, the basic LP approach is to solve for the cost-minimizing solution. Then, postoptimal right-hand side parameterizations are conducted to explore tradeoff relationships among the measures (cost, color, fat, protein, moisture, proportion beef, proportion pork, etc.) by which the quality and profitability of the sausage product are adjudged. Using the output, the manager in charge of blending then determines the day's blend.

Because of the perishability of both the ingredient cuts of meat and finished products, the decision-making pace in sausage manufacturing is rapid. Thus, to mesh with the other duties involved in sausage manufacturing, any computer-assisted decision-aid approach such as the above must be able to perform its analysis and present its output within a reasonably short terminal session.

In this application, we study the *single formula* sausage blending case, in which a sausage manufacturer develops a small number of single formula models that, when used with proper RHS vectors, can blend, one after the other, all of the different sausage variations in the product line. The advantage of these models is that they are small, usually less than 40 constraints and 40 variables.

16.1.1 A Frankfurter Blending Problem

We now consider an adaptation of the canonical frankfurter formulation set out in Rust (1976, pp. 68–72). Whereas in the LP approach only cost is modeled with objective function status, the full list of criteria in the MOLP approach is

1. Cost/lb.
2. Color (on a scale of 0 to 10).
3. % Meat use.
4. % Fat.
5. % Protein.
6. % Moisture.
7. Beef to meat use %.
8. Pork to meat use %.

Let us assume that each of the cuts of meat

x_1	Chuck	Beef
x_2	Beef trim	Beef
x_3	Beef head	Beef
x_4	Beef cheek	Beef
x_5	Pork trim	Pork
x_6	Pork jowls	Pork
x_7	Pork fat	Pork
x_8	Mutton	

available for blending is stored in *preblended* form. That is, each cut of meat is ground; mixed with appropriate amounts of nitrate, salt, and possibly water; and chemically sampled for fat, protein, and moisture. Under normal storage conditions, it is anticipated that the preblends will be used within 72 hours.

From labeling requirements and quality control considerations, we have the following finished product (fp.) specifications.

1. Beef head use	$\leqq$ 10 lbs./cwt. fp.
2. Beef cheek use	$\leqq$ 9 lbs./cwt. fp.
3. Mutton use	$\leqq$ 9 lbs./cwt. fp.
4. Spice use	$=$ 9 lbs./cwt. fp.
5. Beef use	$\geqq$ 35 lbs./cwt. fp.
6. Pork use	$\geqq$ 30 lbs./cwt. fp.
7. Meat use	$\geqq$ 78 lbs./cwt. fp.
8. Fat content	$\leqq$ 28 lbs./cwt. fp.
9. Moisture content	$\leqq$ 4 $\times$ protein in fp. + 10 lbs./cwt. fp.
10. Beef to pork balance	55% beef to meat use target 45% pork to meat use target

16.1.2 Vector-Maximum / Filtering Solution Procedure

Recall that $\bar{\mathbf{x}} \in S$ is efficient if and only if there exists a

$$\boldsymbol{\lambda} \in \Lambda = \left\{ \boldsymbol{\lambda} \in R^k | \lambda_i > 0, \quad \sum_{i=1}^{k} \lambda_i = 1 \right\}$$

such that $\bar{\mathbf{x}}$ maximizes the weighted-sums LP

$$\max \{ \boldsymbol{\lambda}^T(\mathbf{Cx} + \boldsymbol{\alpha}) | \mathbf{x} \in S \}$$

The $\boldsymbol{\alpha}$-vector is included in the formulation because we will use objective function constant terms in this application.

Solving the weighted-sums LP for all $\boldsymbol{\lambda} \in \Lambda$ would generate the set of all extreme points efficient w.r.t. Λ. Let

$$\tilde{\Lambda} = \left\{ \boldsymbol{\lambda} \in R^k | \lambda_i \in \text{rel}\,[\ell_i, \mu_i], \quad \sum_{i=1}^{k} \lambda_i = 1 \right\}$$

Then, solving the weighted-sums LP for all $\boldsymbol{\lambda} \in \tilde{\Lambda}$ would generate the set of all extreme points efficient w.r.t. $\tilde{\Lambda}$. With reasonably wide $[\ell_i, \mu_i]$ criterion weight intervals, eight objectives, 40 constraints, and 40 variables, it would not be surprising for an MOLP to have up to 500 or 1000 extreme points efficient w.r.t. $\tilde{\Lambda}$. This is too many to present to the decision maker at one time. However, it is not too many to filter in order to present to the decision maker a few at a time.

In this case study, after assigning appropriate interval criterion weight bounds, we will use a vector-maximum algorithm to compute all extreme points efficient w.r.t. $\tilde{\Lambda}$. Then, we will use the forward and reverse filtering options to examine interactively the resulting extreme point solutions.

16.1.3 Frankfurter Blending Formulation

Of the eight frankfurter criteria in Section 16.1.1, cost and % fat are to be minimized. To be maximized are % meat use, % protein, and color. Beef to meat use % and pork to meat use % are *goal criteria* with targets of 55% and 45%, respectively. Moisture is a *free-floating* criterion in the sense that it is neither a maximization, minimization, nor goal criterion. That is, a given percentage of moisture is meaningless by itself. It can only be ascertained to be good or bad subjectively in conjunction with the other criterion values.

Using convenience (y_1 to y_6) and deviational ($d_7^+, d_7^-, d_8^+, d_8^-$) variables, the constraint coefficients are given in Table 16.1. By category, the constraints are as follows.

Batch Size Constraint. With B the number of pounds hundred weight, this constraint specifies the *batch size* (where x_9 is spice and x_{10} is ice).

Ingredient Upper Bound Constraints. From the product specifications, these constraints place upper bounds on beef head, beef cheek, and mutton use.

Ingredient Lower Bound Constraints. These constraints enforce the lower bound specifications for beef and pork use.

Spice. In this model, spice is a fixed ingredient at 9 lbs./cwt. fp.

Attribute Control Constraints. The $-100B$ coefficients are in the cost and color constraints so that the cost and color measures are expressed on a per pound basis. The B's and $-B$'s are in the meat use, fat, protein, and moisture constraints for these measures to be expressed in percentage terms. The other coefficients in the fat, protein, and moisture constraints come from the chemical sampling of the preblends. The last constraint in this category is

$$-4y_5 + y_6 \leqq 10$$

because of the specification that the moisture percentage be less than or equal to four times the protein percentage plus 10%.

Table 16.1
Constraint Coefficients

	x_1	x_2	x_3	x_4	x_5	x_6	x_7	x_8	x_9	x_{10}	y_1	y_2	y_3	y_4	y_5	y_6	d_7^+	d_7^-	d_8^+	d_8^-	RHS
Batch size	1	1	1	1	1	1	1	1	1	1											$=100B$
Beef head use			1																		$\leqq 10B$
Beef cheek use				1																	$\leqq 9B$
Mutton use								1													$\leqq 9B$
Beef use	1	1	1	1																	$\geqq 35B$
Pork use					1	1	1														$\geqq 30B$
Spice									1												$= 9B$
Cost	1.17	.93	.82	.97	.46	.36	.24	.96	2.52		$-100B$										$= 0$
Color	9.0	7.8	6.8	7.2	6.5	4.9	2.6	4.3	5.2			$-100B$									$= 0$
Meat use	1	1	1	1	1	1	1	1					$-B$								$= 78B$
Fat	.045	.237	.111	.201	.450	.689	.936	.130						B							$= 28B$
Protein	.202	.155	.179	.167	.110	.066	.014	.178							$-B$						$= 0$
Moisture	.740	.595	.697	.619	.427	.232	.037	.679		1						$-B$					$= 0$
Protein to moisture															-4	1					$\leqq 10$
Beef to meat use	1.2	1.2	1.2	1.2	1.2												$-B$	B			$= 55B$
Pork to meat use						1.2	1.2	1.2											$-B$	B	$= 45B$

Table 16.2
Criterion Coefficients

	x_1	x_2	x_3	x_4	x_5	x_6	x_7	x_8	x_9	x_{10}	y_1	y_2	y_3	y_4	y_5	y_6	d_7^+	d_7^-	d_8^+	d_8^-	α		(ℓ_i, μ_i)
z_1 Cost/lb											-1											max	(0,1)
z_2 Color												1										max	(0,1)
z_3 % Meat use													1								78	max	(0,1)
z_4 % Fat														1							-28	max	(0,1)
z_5 % Protein															1							max	(0,1)
z_6 % Moisture																1						—	(0,0)
z_7 Beef/meat use %																	1	-1			55	—	(0,0)
z_8 Pork/meat use %																			1	-1	45	—	(0,0)
z_9 —																	-1	-1				max	(0,1)
z_{10} —																			-1	-1		max	(0,1)

Pork to Beef Balance Constraints. From the product specifications, the beef to meat use and pork to meat use goal percentages are 55% and 45%, respectively. Using an *expected meat use* denominator value of 83% to avoid fractional constraints, we have the 1.2 ($= \frac{100}{83}$) constraint coefficients. The B's and $-B$'s in these constraints ensure that the beef to meat use and pork to meat use deviational variables (d_7^+, d_7^-, d_8^+, and d_8^-) measure percentages.

In Table 16.2, we have the 10 "objective functions" of the problem. Whereas the first eight are used for generating criterion values to be presented to the decision maker, only objectives 1 to 5, 9 and 10 are used for the vector-maximum generation of efficient extreme points. Note that objectives 9 and 10 minimize deviations from the beef to meat use and pork to meat use target percentages and that objectives 7 and 8 merely translate the deviations into the absolute percentages to be presented to the DM. For example, if d_7^- were 4, z_7 would display 51%.

Since objectives 7 and 8 have only been included for translation purposes and % moisture (objective 6) is free-floating, we set $\ell_i = \mu_i = 0$ for $i = 6, 7, 8$, so that objectives 6, 7, and 8 are ignored in the vector-maximum search for efficient extreme points.

16.1.4 Computer Results

To solve numerically the frankfurter blending model of Section 16.1.3 in accordance with the vector-maximum/filtering methodology of Section 16.1.2, the `ADBASE` and `FILTER` codes from Steuer (1983) are employed. With a batch size $B = 4$ (i.e., 400 lbs.), step by step we have the following.

1. With `ADBASE`, the frankfurter blending model is solved for all $\nu = 123$ efficient extreme points.
2. With sample size $P = 8$ and intended number of iterations $t = 5$, we have the r-reduction factor (see Section 11.7)

$$r = \sqrt[t-1]{P/\nu} = \sqrt[4]{\tfrac{8}{123}} = .505$$

3. Because the expected value of the denominator approach was used to linearize the pork to beef balance constraints, the nondominated criterion vectors generated are "corrected" before being further analyzed. Consider, for example, extreme point 116 for which we have

$$\mathbf{z}^{116} = \begin{bmatrix} -.957 \\ 6.710 \\ 84.833 \\ -17.393 \\ 13.930 \\ 58.574 \\ 55.000 \\ 36.000 \end{bmatrix} \qquad \mathbf{x}^{116} = \begin{bmatrix} 143.333 \\ .000 \\ 40.000 \\ .000 \\ 120.000 \\ .000 \\ .000 \\ 36.000 \\ 36.000 \\ 24.667 \end{bmatrix} \begin{matrix} \text{Chuck} \\ \text{Beef trim} \\ \text{Beef head} \\ \text{Beef cheek} \\ \text{Pork trim} \\ \text{Pork jowls} \\ \text{Pork fat} \\ \text{Mutton} \\ \text{Spice} \\ \text{Ice} \end{matrix}$$

TABLE 16.3
Criterion Vectors of the First Iteration

Criterion Vector	z_1	z_2	z_3	z_4	z_5	z_6	z_7	z_8
1	74.2	5.93	79.5	28.0	10.4	51.6	44.0	56.0
31	83.6	6.34	88.7	28.0	12.3	49.6	39.6	50.4
44	78.1	5.86	79.4	28.0	10.4	51.6	57.7	42.3
46	83.1	6.09	78.4	19.9	12.0	58.1	44.6	47.8
86	91.8	7.15	91.0	24.5	13.7	51.7	50.4	49.6
√ 95	88.9	6.92	91.0	28.0	12.8	49.1	58.8	41.2
100	87.6	6.46	78.0	17.2	12.6	60.2	58.8	41.2
119	102.4	7.27	91.0	17.7	15.2	57.0	57.1	33.0

With meat use being 339.333 lbs., beef use 183.333 lbs., and pork use 120.000 lbs., z_7 and z_8 are corrected to 54.02% and 35.36%, respectively. Stripping the negative signs from z_1 and z_4, the 123 corrected criterion vectors are stored in data set **MEATZ**. The decision variable values (x_1 to x_{10}) of the 123 efficient extreme points are stored in **MEATX**.

4. With **FILTER**, the 123 criterion vectors are forwardly filtered to obtain the first-iteration dispersed set of size eight in Table 16.3.
5. As the first-iteration selection, suppose criterion vector 95 is chosen as the most preferred in the sample of Table 16.3.
6. With **FILTER**, the criterion vectors in **MEATZ** are reverse filtered to obtain the 62 ($\cong .505 \times 123$) closest to criterion vector 95. The 62 criterion vectors are stored in **MEATZ1R**.
7. Forward filtering the 62 criterion vectors in **MEATZ1R** using criterion vector 95 as the seed point, we obtain the second-iteration dispersed set of size eight in Table 16.4.
8. As the second-iteration selection, suppose criterion vector 76 is chosen as the most preferred in the sample of Table 16.4.
9. We now reverse filter the criterion vectors in **MEATZ** to obtain the 31 ($\cong .505^2 \times 123$) closest to criterion vector 76. The 31 criterion vectors are stored in **MEATZ2R**.
10. Forward filtering the 31 criterion vectors in **MEATZ2R** using criterion vector 76 as the seed point, we obtain the third-iteration dispersed set of size eight in Table 16.5.
11. Suppose that criterion vector 76 is preferred to any of the vectors in the new sample of Table 16.5. Hence, criterion vector 76, which was the second-iteration selection, becomes the third-iteration selection also.
12. We now forward filter the first 16 ($\cong .505^3 \times 123$) criterion vectors in **MEATZ2R** to obtain the fourth-iteration dispersed set of size eight in Table 16.6.

TABLE 16.4
Criterion Vectors of the Second Iteration

Criterion Vector	z_1	z_2	z_3	z_4	z_5	z_6	z_7	z_8
95	*88.9*	*6.92*	*91.0*	*28.0*	*12.8*	*49.1*	*58.8*	*41.2*
24	87.3	6.99	91.0	28.0	12.8	49.0	50.4	49.6
31	83.6	6.34	88.7	28.0	12.3	49.6	39.5	50.4
38	82.8	6.51	83.9	28.0	11.3	50.6	54.6	45.4
√ 76	90.1	6.81	83.3	19.6	13.1	57.2	55.0	45.0
77	88.9	6.93	91.0	28.0	12.8	49.1	58.8	41.2
80	79.9	6.06	79.3	24.3	11.1	54.4	57.8	42.2
91	99.0	7.50	91.0	19.9	14.7	55.2	58.8	41.2
112	95.3	6.91	91.0	22.1	14.2	53.5	48.9	41.2

13. As the fourth-iteration selection, suppose criterion vector 96 is chosen as the most preferred in the sample of Table 16.6.

14. We now reverse filter the criterion vectors in **MEATZ** to obtain the eight ($= .505^4 \times 123$) closest to criterion vector 96 to obtain the final set of criterion vectors in Table 16.7 to be displayed to the decision maker.

15. Assuming that none of the new criterion vectors represent improvements, we conclude with criterion vector 96. Referencing the values of the

TABLE 16.5
Criterion Vectors of the Third Iteration

Criterion Vector	z_1	z_2	z_3	z_4	z_5	z_6	z_7	z_8
√ *76*	*90.1*	*6.81*	*83.3*	*19.6*	*13.1*	*57.2*	*55.0*	*45.0*
80	79.9	6.06	79.3	24.3	11.1	54.5	57.8	42.2
81	84.6	6.01	78.4	18.9	12.2	58.9	44.7	43.9
90	97.4	7.14	91.0	20.6	14.5	54.7	50.4	41.2
91	99.0	7.50	91.0	19.9	14.7	55.2	58.8	41.2
99	93.5	7.03	83.3	18.9	13.4	57.6	55.0	45.0
101	85.9	6.31	78.1	18.7	12.3	59.1	58.7	41.3
102	87.4	6.11	78.1	17.1	12.6	60.3	50.1	38.4
116	95.2	6.71	84.8	17.4	13.9	58.6	54.0	35.4

Table 16.6
Criterion Vectors of the Fourth Iteration

Criterion Vector	z_1	z_2	z_3	z_4	z_5	z_6	z_7	z_8
76	*90.1*	*6.81*	*83.3*	*19.6*	*13.1*	*57.2*	*55.0*	*45.0*
79	82.5	6.20	78.4	20.8	11.8	57.4	52.2	47.8
80	79.9	6.06	79.3	24.3	11.1	54.5	57.8	42.2
84	87.0	6.44	78.0	17.5	12.5	60.0	57.7	42.3
√ 96	88.2	6.64	83.3	21.0	12.8	56.1	55.0	45.0
99	93.5	7.03	83.3	18.9	13.4	57.6	55.0	45.0
103	87.3	6.55	78.0	18.7	12.3	59.0	51.9	48.1
105	88.2	6.30	78.0	16.7	12.7	60.6	55.5	38.5
116	95.2	6.71	84.8	17.4	13.9	58.6	54.0	35.4

Table 16.7
Criterion Vectors of the Fifth Iteration

Criterion Vector	z_1	z_2	z_3	z_4	z_5	z_6	z_7	z_8
√ *96*	*88.2*	*6.64*	*83.3*	*21.0*	*12.8*	*56.1*	*55.0*	*45.0*
48	84.1	6.35	78.2	19.4	12.1	58.5	52.1	47.9
50	79.6	6.12	79.5	24.7	11.0	54.2	57.7	42.3
61	81.8	6.32	83.3	26.2	11.6	52.2	55.0	45.0
76	90.1	6.81	83.3	19.6	13.1	57.2	55.0	45.0
79	82.5	6.20	78.4	20.8	11.8	57.4	52.2	47.8
80	79.9	6.06	79.3	24.3	11.1	54.5	57.8	42.2
83	85.6	6.44	78.0	19.0	12.1	58.8	55.0	48.1
99	93.5	7.03	83.3	18.9	13.4	57.6	55.0	45.0

decision variables for criterion vector 96 in **MEATX**, we have the blend:

x_1: Chuck	107.333 lbs.
x_3: Beef head	40.000 lbs.
x_4: Beef cheek	36.000 lbs.
x_5: Pork trim	150.000 lbs.
x_9: Spice	36.000 lbs.
x_{10}: Ice	30.667 lbs.
	400.000 lbs.

16. The manager in charge of blending may or may not wish to temper the above blend before going into production.

16.2 CPA FIRM AUDIT STAFF ALLOCATION

In this application, we study the audit staff planning problem in a CPA firm. This application is drawn from Balachandran and Steuer (1982) that, in turn, was motivated by Welling (1977).

In the audit staff planning problem, the decision maker will be referred to as the *partner in charge of personnel* whose objectives are several. They include such items as maximizing profit, accommodating projected bookings, avoiding unnecessary increases or decreases in the audit staff, minimizing excessive overtime, minimizing underutilization (i.e., higher skilled employees performing lower skilled audit tasks), and achieving professional development goals.

16.2.1 Description of Audit Staff Problem

The CPA firm audit staff will be assumed to consist of six levels. At the highest level is Partner *A* (auditor 1). In descending hierarchial order, we have Partner *B* (auditor 2), Manager (auditor 3), Staff Auditor *A* (auditor 4), Staff Auditor *B* (auditor 5), and Staff Auditor *C* (auditor 6).

In Fig. 16.1, we have the job/auditor allocation matrix where feasible assignments are indicated by *x*'s and notation for the parameters of the problem. We

		Type of Auditor 1	2	3	4	5	6	Billing rate/hr	Projected annual bookings
Type of job	1	x_{11}						b_1	w_1
	2	x_{21}	x_{22}					b_2	w_2
	3	x_{31}	x_{32}	x_{33}				b_3	w_3
	4		x_{42}	x_{43}	x_{44}			b_4	w_4
	5			x_{53}	x_{54}	x_{55}		b_5	w_5
	6				x_{64}	x_{65}	x_{66}	b_6	w_6
Annual base salary/empl/yr		c_1	c_2	c_3	c_4	c_5	c_6		
Base salary hrs/empl/yr		y_1	y_2	y_3	y_4	y_5	y_6		
Variable overhead/hr		v_1	v_2	v_3	v_4	v_5	v_6		
Overtime rate/hr		0	0	0	r_4	r_5	r_6		
Max reasonable overtime hrs/empl/yr		s_1	s_2	s_3	s_4	s_5	s_6		
Professional development hrs/empl/yr		p_1	p_2	p_3	p_4	p_5	p_6		
Staff augmentation cost/position		—	—	h_3	h_4	h_5	h_6		
Staff reduction cost/position		—	—	f_3	f_4	f_5	f_6		
Present employees		n_{11}	n_{12}	n_{13}	n_{14}	n_{15}	n_{16}		
Contemplated employees		n_{21}	n_{22}	n_{23}	n_{24}	n_{25}	n_{26}		
Organizational design ratios		—	—	g_3	g_4	g_5	g_6		

Figure 16.1. Job/auditor matrix and associated notation.

max	Profit
goal	Projected bookings
min	Staff augmentation
min	Staff reduction
min	Excessive overtime
min	Underutilization
goal	Professional development

s.t.	Projected bookings
	Staff augmentation and staff reduction constraints
	Professional development constraints
	Regular overtime and idle time constraints
	Excessive overtime constraints
	Organizational design ratio constraints

Figure 16.2. Audit staff MOLP.

assume that each employee is capable of working at his or her own level and two lower levels (except auditors 5 and 6).

In the job/auditor matrix, for example, x_{54} refers to the number of hours of job level 5 work to be allocated to auditor level 4 employees. In this case, there would be underutilization of auditor 4 capabilities. For this work, the firm would be compensating its employees at the base salary rate of c_4. However the client would be billed at the hourly rate of b_5. That is, we assume that any single auditor does not have the same billing rate on all audit activities.

16.2.2 Audit Staff Formulation

With multiple criteria, the audit staff problem is formulated as the MOLP with seven criteria and six categories of constraints in Fig. 16.2.

The constraints and objectives of the MOLP are as follows.

Projected Bookings Constraints

$$\sum_{\substack{\text{allowable} \\ j}} (x_{ij}) + d_{bi}^- = w_i \qquad (1 \leq i \leq 6)$$

The d_{bi}^- represent the hours by which planned allocations are lower than projected bookings. (Allowable j can be seen from Fig. 16.1. For example, for $i = 4$, allowable values for j are 2, 3, and 4.)

Staff Augmentation and Staff Reduction Constraints

$$n_{2j} - d_{nj}^+ + d_{nj}^- = n_{1j} \qquad (3 \leqq j \leqq 6)$$

The d_{nj}^+ represent the number of new people to be hired at each auditor level. The d_{nj}^- represent the amounts at each staff level by which the audit staff is to be reduced. The n_{1j} are the present numbers of employees at each auditor level. The model formulation assumes that the number of Partner *A* and Partner *B* auditors will remain constant.

Professional Development Constraints

$$x_{pj} + d_{pj}^{-} = p_j n_{2j} \qquad (1 \leqq j \leqq 6)$$

The quantities x_{pj} represent the number of professional development hours to be allocated to employees of auditor level j. The x_{pj} are nonbillable hours that take the form of orientation programs, seminars, conferences, and continuing education courses. The d_{pj}^{-} are underdeviation hours from goal at the different audit levels. The right side of the equation is the desired number of professional development hours for auditor level j. Note that the n_{2j}, which also appear in the staff augmentation and staff reduction constraints, are variables.

Regular Overtime and Idle Time Constraints

$$\sum_{\substack{\text{allowable}\\ i}} (x_{ij}) + x_{pj} - d_{oj}^{+} + d_{oj}^{-} = y_j n_{2j} \qquad (1 \leqq j \leqq 6)$$

In the above, d_{oj}^{+} is the overtime and d_{oj}^{-} is the idle time for auditors of level j. The right side gives the number of hours at regular time for auditor level j.

Excessive Overtime Constraints

$$\sum_{\substack{\text{allowable}\\ i}} (x_{ij}) + x_{pj} - d_{sj}^{+} \leqq (y_j + s_j) n_{2j} \qquad (1 \leqq j \leqq 6)$$

The d_{sj}^{+} represent the amount of overtime beyond the reasonable limits established by the s_j.

Organizational Design Ratio Constraints

$$n_{2j} \geqq \frac{1}{g_j} \sum_{i=4}^{6} n_{2i} \qquad (3 \leqq j \leqq 6)$$

The purpose of these constraints is to maintain a sound organizational structure. In accordance with the g_j, the number of auditors at each of the four auditor levels 3, 4, 5, and 6 must meet minimum proportions with regard to the total number of auditors at levels 4, 5, and 6. For example, consider $j = 3$ and let $g_3 = 10$. Then, the constraint says that there should be at least one Manager for every 10 employees at the levels of Staff Auditors *A*, *B*, and *C*. As another example, consider $j = 5$ and let $g_5 = 3$. Then, the constraint says that there should be at least one Staff Auditor *B* for every three employees at the levels of Staff Auditors *A*, *B*, and *C*.

z_1: Profit (in dollars). The profit function has six components. They are gross billings less annual base salary costs, overtime costs, staff augmentation and reduction costs, variable overhead, and fixed overhead. Gross billings are given by

$$\sum_{i=1}^{6} b_i \sum_{\substack{\text{allowable} \\ j}} x_{ij}$$

Annual base salary costs, overtime costs, staff augmentation and reduction costs, and variable overhead are respectively given by

$$\sum_{j=1}^{6} c_j n_{2j}$$

$$\sum_{j=4}^{6} r_j d_{oj}^{+}$$

$$\sum_{j=3}^{6} \left(h_j d_{nj}^{+} + f_j d_{nj}^{-}\right)$$

$$\sum_{j=1}^{6} v_j \left(\sum_{\substack{\text{allowable} \\ i}} (x_{ij}) + x_{pj} \right)$$

z_2: Projected Bookings (in hours below forecast)

$$\min \left\{ \sum_{i=1}^{6} d_{bi}^{-} = z_2 \right\}$$

z_3: Staff Augmentation (in employees)

$$\min \left\{ \sum_{j=3}^{6} d_{nj}^{+} = z_3 \right\}$$

z_4: Staff Reduction (in employees)

$$\min \left\{ \sum_{j=3}^{6} d_{nj}^{-} = z_4 \right\}$$

z_5: Excessive Overtime (in hours)

$$\min \left\{ \sum_{j=1}^{6} d_{sj}^{+} = z_5 \right\}$$

z_6: Underutilization (in hours)

$$\min \left\{ \sum_{j=1}^{4} \sum_{i=j+1}^{j+2} (x_{ij}) + x_{65} + \sum_{j=1}^{6} d_{oj}^{-} = z_6 \right\}$$

This objective minimizes the number of hours the different auditors would find themselves underutilized or idle.

z_7: ***Professional Development (in hours below goal)***

$$\min\left\{\sum_{j=1}^{6} d_{pj}^{-} = z_7\right\}$$

16.2.3 Interval Criterion Weights/Weighted-Sums/Filtering Solution Procedure

A hybrid version of the interval criterion weights/vector-maximum and interactive weighted-sums/filtering procedures of Sections 13.4 and 13.5 is used on this application. It is a hybrid version in the sense that we contract the criterion cone but do not use a vector-maximum capability on the last iteration. Instead, we keep sampling with weighted-sums LPs throughout.

To describe the hybrid procedure, we first rescale the objectives to form the $\mathbf{C}_1$ criterion matrix. Then, to obtain a dispersion of efficient extreme points over the efficient set, we solve the $(2k+1)$ weighted-sums LPs

$$\max\left\{\boldsymbol{\lambda}^{1,i}\mathbf{C}_1\mathbf{x} \mid \mathbf{x} \in S\right\} \qquad i = 1,\ldots,(2k+1)$$

where

$$\begin{aligned}
\boldsymbol{\lambda}^{1,1} &= (1,0,\ldots,0)\\
\boldsymbol{\lambda}^{1,2} &= (0,1,\ldots,0)\\
&\vdots\\
\boldsymbol{\lambda}^{1,7} &= (0,0,\ldots,1)\\
\boldsymbol{\lambda}^{1,8} &= \left(1/k^2, r,\ldots,r\right)\\
\boldsymbol{\lambda}^{1,9} &= \left(r, 1/k^2,\ldots,r\right)\\
&\vdots\\
\boldsymbol{\lambda}^{1,14} &= \left(r, r,\ldots,1/k^2\right)\\
\boldsymbol{\lambda}^{1,15} &= (1/k, 1/k,\ldots,1/k)
\end{aligned}$$

in which $r = (k+1)/k^2 = \frac{8}{49}$ because $k = 7$.

Letting P be the number of solutions to be presented to the decision maker at each iteration, we filter the criterion vectors of the extreme points resulting from the 15 optimizations to obtain the P most different. From the P efficient extreme point solutions, the decision maker selects his or her most preferred. Referencing Table 9.5 in conjunction with the $\boldsymbol{\lambda}$-vector that produced the DM's most preferred selection, we then construct the premultiplication matrix $\mathbf{T}_1$.

Now, to sample the second-iteration neighborhood in the efficient set, we solve the $(2k+1)$ weighted-sums LPs

$$\max\left\{\boldsymbol{\lambda}^{1,i}\mathbf{T}_1\mathbf{C}_1\mathbf{x} \mid \mathbf{x} \in S\right\} \qquad i = 1,\ldots,(2k+1)$$

Then, after filtering the results of the 15 optimizations, the DM makes his or her second-iteration selection. Knowing the $\boldsymbol{\lambda}$-vector corresponding to the second iteration, we construct the second premultiplication matrix $\mathbf{T}_2$.

To sample more closely about the DM's second-iteration selection, we now solve the $(2k + 1)$ weighted-sums LPs

$$\max\left\{\boldsymbol{\lambda}^{1,i}\mathbf{T}_2\mathbf{T}_1\mathbf{C}_1\mathbf{x} \mid \mathbf{x} \in S\right\} \qquad i = 1,\ldots,(2k+1)$$

We continue in this fashion until we complete the prechosen number of iterations t at which time we exit from the algorithm with a final solution.

16.2.4 Computer Solution

To illustrate the solution of the audit staff problem, consider a CPA firm that is characterized by the parameter values in Tables 16.8 and 16.9. This leads to an MOLP of size $7 \times 32 \times 63$. The 63 variables are as follows:

x_{ij}	15
d_{bi}^-	6
n_{2j}	4
d_{nj}^+, d_{nj}^-	8
x_{pj}	6
d_{sj}^+	6
d_{oj}^+, d_{oj}^-	12
d_{pj}^-	6

Of the 32 constraints, 10 are inequalities (excessive overtime and organizational design ratio) and 22 are equalities.

To solve the numerical example defined by Tables 16.8 and 16.9, the `ADBASE` and `FILTER` codes of Steuer (1983) were employed. Although any LP code could have been used for the weighted-sums optimizations, `ADBASE` was used (a) because of the relative ease with which sequences of weighted-sums problems can be solved via its interval criterion weights features and (b) because `ADBASE` can compute all optimal extreme points of the weighted-sums LPs (when alternative

TABLE 16.8
Job Parameters for CPA Firm Numerical Example

Type of Job i	b_i	w_i
1	\$40	3,500
2	35	5,800
3	30	25,000
4	25	45,000
5	20	30,000
6	15	85,000

TABLE 16.9
Auditor Parameters for CPA Firm Numerical Example[a, b]

Auditor j	c_j	y_j	v_j	r_j	s_j	p_j	h_j	f_j	n_{1j}	n_{2j}	g_j
1	$52,000	1,760	$18	$0	200	160	—	—	2	2	—
2	41,600	1,760	16	0	200	160	—	—	3	3	—
3	31,200	1,840	14	0	300	120	$20,000	$6,000	10	—	10
4	20,800	1,880	9	15	500	80	10,000	5,000	15	—	6
5	16,640	1,920	8	12	500	40	8,000	4,000	20	—	5
6	12,480	1,920	7	9	500	120	6,000	3,000	35	—	4

[a]Fixed firm overhead = $200,000/yr.
[b]The missing n_{2j} values are to be determined by the solution procedure.

optima exist). Using the interval criterion weights/weighted-sums/filtering capabilities, step by step we have the following.

1. Assume that negotiation between the DM and analyst results in the algorithm being calibrated with a sample size $P = 5$ and an intended number of iterations $t = 3$.
2. Rescale the first objective by 10^{-1} and the other six objectives by 10^3 to form criterion matrix $\mathbf{C}_1$. This causes the coefficients of largest absolute value in each of the objective functions to be of the same order of magnitude (between 1,000 and 2,000).
3. Using **ADBASE**, the 15 first-iteration weighted-sums LPs

$$\max\left\{\lambda^{1,i}\mathbf{C}_1\mathbf{x} \mid \mathbf{x} \in S\right\} \qquad i = 1, \ldots, (2k+1)$$

   are solved. Store the criterion vector results in **AUDIT1**.
4. Using **FILTER**, reduce the criterion vectors in **AUDIT1** to the $P = 5$ most different and present to the decision maker in the form of Table 16.10.

TABLE 16.10
Criterion Vector Solutions of the First Iteration[a]

	1–1	*1–2*	*1–3*	*1–4*	*1–9*
z_1: Profit	720,000	115,000	380,000	129,000	101,000
z_2: Projected bookings	−300	0	0	0	−400
z_3: Staff augmentation	0	9	0	7	9
z_4: Staff reduction	19	4	4	0	4
z_5: Excessive overtime	49,700	0	12,900	0	0
z_6: Underutilization	36,500	0	300	7,600	0
z_7: Professional development	−6,800	−400	−7,000	−400	0

[a]Negative values for z_2 and z_7 indicate underdeviations from goals.

5. From Table 16.10, assume that the DM selects solution 1-4 as the most preferred. We then identify

$$\boldsymbol{\lambda}^{1,4} = (0, 0, 0, 1, 0, 0, 0)$$

as the weighting vector that produced solution 1-4.

6. Contracting the original (first-iteration) criterion cone to $(1/k)$th its cross-sectional volume, we form criterion matrix $\mathbf{C}_2$ whose rows generate the second-iteration (enveloping reduced) criterion cone. $\mathbf{C}_2$ is given by

$$\mathbf{C}_2 = \mathbf{T}_1 \mathbf{C}_1$$

where

$$\mathbf{T}_1 = \begin{bmatrix} & & & p & & & 1-p \\ & & & p & & 1-p & \\ & & & p & 1-p & & \\ & & & 1 & & & \\ & & 1-p & p & & & \\ & 1-p & & p & & & \\ 1-p & & & p & & & \end{bmatrix}$$

according to Section 9.10.3.

7. With **ADBASE**, the 15 second-iteration weighted-sums LPs

$$\max\{\boldsymbol{\lambda}^{2,i}\mathbf{C}_1\mathbf{x} \mid \mathbf{x} \in S\} \equiv \max\{\boldsymbol{\lambda}^{1,i}\mathbf{T}_1\mathbf{C}_1\mathbf{x} \mid \mathbf{x} \in S\}$$

$$i = 1, \ldots, (2k+1)$$

TABLE 16.11

$\boldsymbol{\lambda}^{2,i}$ Vectors of the Second Iteration

	Components						
	1	*2*	*3*	*4*	*5*	*6*	*7*
$\boldsymbol{\lambda}^{2,1}$	.723			.277			
$\boldsymbol{\lambda}^{2,2}$		.723		.277			
$\boldsymbol{\lambda}^{2,3}$			.723	.277			
$\boldsymbol{\lambda}^{2,4}$				1.000			
$\boldsymbol{\lambda}^{2,5}$				.277	.723		
$\boldsymbol{\lambda}^{2,6}$				.277		.723	
$\boldsymbol{\lambda}^{2,7}$				.277			.723
$\boldsymbol{\lambda}^{2,8}$	.006	.045	.045	.769	.045	.045	.045
$\boldsymbol{\lambda}^{2,9}$	.045	.006	.045	.769	.045	.045	.045
$\boldsymbol{\lambda}^{2,10}$	.045	.045	.006	.769	.045	.045	.045
$\boldsymbol{\lambda}^{2,11}$	.045	.045	.045	.730	.045	.045	.045
$\boldsymbol{\lambda}^{2,12}$	.045	.045	.045	.769	.006	.045	.045
$\boldsymbol{\lambda}^{2,13}$	.045	.045	.045	.769	.045	.006	.045
$\boldsymbol{\lambda}^{2,14}$	.045	.045	.045	.769	.045	.045	.006
$\boldsymbol{\lambda}^{2,15}$	.004	.004	.004	.976	.004	.004	.004

TABLE 16.12
Criterion Vector Solutions of the Second Iteration

	2–1	*2–2*	*2–3*	*2–9*	*2–10*
z_1: Profit	652,000	129,000	321,000	101,000	113,000
z_2: Projected bookings	0	0	0	−400	0
z_3: Staff augmentation	0	7	0	9	9
z_4: Staff reduction	3	0	0	4	4
z_5: Excessive overtime	26,820	0	12,900	0	0
z_6: Utilization	25,220	7,600	7,600	0	100
z_7: Professional development	−7,840	−400	−2,800	0	−300

are solved with the criterion vector results stored in **AUDIT2**. The $\lambda^{2,i}$ are given in Table 16.11.

8. Using **FILTER**, reduce the criterion vectors in **AUDIT2** to the five most different and present to the DM in the form of Table 16.12.
9. From Table 16.12, assume that the DM selects solution 2-9 as the most preferred. We then identify

$$\lambda^{2,9} = (.045, .006, .045, .769, .045, .045, .045)$$

as the weighting vector that produced solution 2-9.

10. Contracting the second-iteration (enveloping reduced) criterion cone to $(1/k)$th its cross-sectional volume, we form criterion matrix $\mathbf{C}_3$ whose rows

TABLE 16.13
$\lambda^{3,i}$ Vectors of the Third Iteration

				Components			
	1	*2*	*3*	*4*	*5*	*6*	*7*
$\lambda^{3,1}$	.557		.033	.311	.033	.033	.033
$\lambda^{3,2}$	.033	.524	.033	.311	.033	.033	.033
$\lambda^{3,3}$	.033		.557	.311	.033	.033	.033
$\lambda^{3,4}$	.033		.033	.835	.033	.033	.033
$\lambda^{3,5}$	.033		.033	.311	.557	.033	.033
$\lambda^{3,6}$	.033		.033	.311	.033	.557	.033
$\lambda^{3,7}$	.033		.033	.311	.033	.033	.557
$\lambda^{3,8}$	.043	.087	.118	.398	.118	.118	.118
$\lambda^{3,9}$	.118	.012	.118	.398	.118	.118	.118
$\lambda^{3,10}$	.118	.087	.043	.398	.118	.118	.118
$\lambda^{3,11}$	.118	.087	.118	.323	.118	.118	.118
$\lambda^{3,12}$	.118	.087	.118	.398	.043	.118	.118
$\lambda^{3,13}$	.118	.087	.118	.398	.118	.043	.118
$\lambda^{3,14}$	.118	.087	.118	.398	.118	.118	.043
$\lambda^{3,15}$	.108	.075	.108	.385	.108	.108	.108

generate the third-iteration (enveloping reduced) criterion cone. $\mathbf{C}_3$ is given by

$$\mathbf{C}_3 = \mathbf{T}_2\mathbf{C}_2$$

where

$$\mathbf{T}_2 = \frac{p}{h}\begin{bmatrix} 1 & 0 & 1 & 1 & 1 & 1 & \frac{h}{p}-5 \\ 1 & 0 & 1 & 1 & 1 & \frac{h}{p}-5 & 1 \\ 1 & 0 & 1 & 1 & \frac{h}{p}-5 & 1 & 1 \\ 1 & 0 & 1 & \frac{h}{p}-5 & 1 & 1 & 1 \\ 1 & 0 & \frac{h}{p}-5 & 1 & 1 & 1 & 1 \\ 1 & \frac{h}{p}-6 & 1 & 1 & 1 & 1 & 1 \\ \frac{h}{p}-5 & 0 & 1 & 1 & 1 & 1 & 1 \end{bmatrix}$$

according to Section 9.10.3.

11. With `ADBASE`, the 15 third-iteration weighted-sums LPs

$$\max\left\{\boldsymbol{\lambda}^{3,i}\mathbf{C}_1\mathbf{x} \mid \mathbf{x} \in S\right\} \equiv \max\left\{\boldsymbol{\lambda}^{1,i}\mathbf{T}_2\mathbf{T}_1\mathbf{C}_1\mathbf{x} \mid \mathbf{x} \in S\right\}$$

$$i = 1, \dots, (2k+1)$$

are solved with the criterion vector results stored in `AUDIT3`. The $\boldsymbol{\lambda}^{3,i}$ are given in Table 16.13.

12. Using `FILTER`, reduce the criterion vectors in `AUDIT3` to the five most different and present to the DM in the form of Table 16.14.

13. From Table 16.14, assume that the DM selects solution 3-1 as the final solution. We then identify

$$\boldsymbol{\lambda}^{3,1} = (.557, .000, .033, .311, .033, .033, .033)$$

as the weighting vector that produced the final solution.

14. The audit staff allocation plan corresponding to solution 3-1 is given in Fig. 16.3.

TABLE 16.14
Criterion Vector Solutions of the Third Iteration

	3–1	*3–2*	*3–3*	*3–4*	*3–12*
z_1: Profit	160,000	115,000	170,000	101,000	108,000
z_2: Projected bookings	−9,700	0	−20,200	−400	0
z_3: Staff augmentation	5	9	0	9	9
z_4: Staff reduction	4	4	4	4	4
z_5: Excessive overtime	0	0	0	0	400
z_6: Underutilization	0	0	0	0	0
z_7: Professional development	0	−400	0	0	0

		Type of Auditor					
		1	2	3	4	5	6
Type of job	1	3,500					
	2		5,400				
	3			20,200			
	4				45,000		
	5					30,000	
	6						80,500
Contemplated employees		2	2	10	20	16	35
Overtime hours/employee		150	200	300	450	58	500

Figure 16.3. Audit staff plan for the final solution 3–1.

16.2.5 Unscaling the Final Weighting Vector

Suppose we wish to compute a weighting vector $\bar{\lambda} \in \Lambda$ which, when applied to the original "unscaled" criterion matrix **C**, would have caused the "unscaled" weighted-sums LP

$$\max \{\bar{\boldsymbol{\lambda}}^T \mathbf{C}\mathbf{x} \mid \mathbf{x} \in S\}$$

to generate the final solution 3-1 in "one shot."

To compute such a weighting vector, let us find the $\bar{\lambda}$ that causes $\bar{\boldsymbol{\lambda}}^T\mathbf{C}$ to point in the same direction as $\boldsymbol{\lambda}^{3,1}\mathbf{C}_1$. Thus,

$$\bar{\boldsymbol{\lambda}}^T\mathbf{C} \;\propto\; \boldsymbol{\lambda}^{3,1}\mathbf{C}_1 = \boldsymbol{\lambda}^{3,1}\bar{\mathbf{N}}\mathbf{C}$$

where $\propto$ means "is proportional to" and $\bar{\mathbf{N}}$ is a $k \times k$ diagonal matrix formed by the original scaling factors. Since the rows of **C** are linearly independent in the audit staff problem,

$$\bar{\boldsymbol{\lambda}} \;\propto\; \boldsymbol{\lambda}^{3,1}\bar{\mathbf{N}}$$

Normalizing the expression on the right so that the components sum to one, we have

$$\bar{\boldsymbol{\lambda}} = \frac{\boldsymbol{\lambda}^{3,1}\bar{\mathbf{N}}}{\|\boldsymbol{\lambda}^{3,1}\bar{\mathbf{N}}\|_1}$$

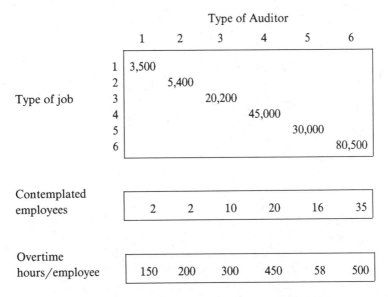

Numerically, for the audit staff problem, this yields

$$\bar{\lambda} = \begin{bmatrix} .0001 \\ .0000 \\ .0745 \\ .7019 \\ .0745 \\ .0745 \\ .0745 \end{bmatrix} \begin{matrix} \text{Profit} \\ \text{Projected bookings} \\ \text{Staff augmentation} \\ \text{Staff reduction} \\ \text{Excessive overtime} \\ \text{Underutilization} \\ \text{Professional development} \end{matrix}$$

This result is interesting. It serves to underscore the fact that specifying accurate weighting vectors can often be, and probably almost always is, exceedingly difficult to do. This is especially true when the number of objectives is more than three. Consider the values in $\bar{\lambda}$ above. It is highly unlikely that the partner in charge of personnel would have ever specified a similar weighting vector a priori. This is because it involves placing virtually no weight on what are probably the most important objectives: profit and projected bookings. Hence, we see the usefulness of methods that do not require the DM to specify weighting vectors.

16.3 MANAGERIAL COMPENSATION PLANNING

A problem in the field of wage and salary administration is how to deal with the conflicting forces that stress the design of wage and salary structures [Wallace and Fay (1983)]. The problem is especially vexing in the middle and top management areas. On one hand, there are internal (inside) forces. These include limited budgets, perceptions of equity, and considerations that argue for a salary structure consistent with each job's economic contribution to the firm. On the other hand, there are external (outside) forces. Along with inflation, these include short-run supply and demand conditions that typically put upward pressure on the salary structure.

What we have is essentially a curve-fitting problem. On the one hand, we have a profile of inside requirements, and on the other, a profile of outside requirements. In the interest of attracting and retaining qualified personnel, the challenge of the compensation planner is to find the salary structure that best meets the specifications of both profiles.

16.3.1 Current Practice

At present, compensation planners address conflicts among the forces affecting a salary structure in a two-phase fashion. In phase one, an assessment of each job's relative worth is performed by means of a process called *job evaluation*. The purpose of job evaluation is to establish relative rates of compensation based on each job's contribution to the organization.

In its most sophisticated form, job evaluation employs a point system to score each job according to its wage-making characteristics (educational requirements, degree of responsibility, span of authority, number of employees supervised, etc.). With the top job in the job structure judged to be 100%, the points are totaled to determine each job's inside relative worth by comparison. For instance, if the president's score is 150 points and individuals in one of the managerial echelons score 90 points, the inside relative worth of jobs in that level of management would be 60%.

Phase two involves bringing the job structure to "market" for pricing. This usually consists of conducting a wage survey to obtain salary information about comparable jobs in the external labor markets.

Most frequently, however, the inside relative worth percentages of phase one do not agree with the wage survey results of phase two. To resolve differences between the two phases, compensation planners have relied on a number of *remedial strategies*. To craft the most satisfactory salary structure, the remedial strategies are typically applied in an ad hoc, trail-and-error fashion in an iterative process of improvement. The intent is to edge the job evaluation judgments into greater alignment with outside market price realities. Among the most frequently used strategies are the following.

1. Reevaluate the job structure to determine if any jobs can be reclassified to bring them more into line with the external labor market.
2. Reexamine the job titles and job descriptions to see if any changes can be made to minimize friction with the forces stressing the salary structure.
3. "Red circle" positions that present unsurmountable problems. Recognize them as the result of unusual circumstances in order to prevent them from becoming a bad precedent, thereby jeopardizing the integrity of the rest of the salary structure.
4. Redesign troublesome jobs to allow them to be filled by lower level personnel.

In some firms, the task of designing salary structures is performed by their own in-house wage and salary specialists. However, for greater objectivity and enhanced credibility, outside wage and salary consultants (for example, from the large professional accounting firms) may be utilized. Either way, the design of salary structures is, in many respects, an art form.

16.3.2 Goal Programming / Tchebycheff Solution Procedure

A goal programming/Tchebycheff solution approach is applied to locate the salary structure that best minimizes the conflict among the different forces impinging upon the salary structure. In contrast to the conventional *Archimedian* and *preemptive* priority methods of goal programming, the method used on the managerial compensation planning problem is referred to as *multiple criterion function* goal programming (see Section 10.10). This is because the functions of the deviational variables that might otherwise have comprised the preemptive priority levels are written as separate objectives. Since the problem is linear, it is now expressed as an MOLP. The augmented weighted Tchebycheff option of the automated package of Chapter 15 is applied to solve the problem.

In this problem, because the compensation planner may wish to apply the remedial strategies of Section 16.3.1, it is possible for coefficient values in the A-matrix to change from iteration to iteration. As long as the changes are modest, they can be absorbed by the solution procedure. Thus, we can refine our feasible region at the same time we are narrowing our areas of search for a final solution. If the coefficient value changes are considered too abrupt, an algorithm restart is advised.

16.3.3 Managerial Compensation Formulation

Let

N ~ the number of job classes.
n_i ~ the number of employees in job class i.
p_i ~ the inside relative worth percentage for job class i.
m_i ~ the outside market price for employees in job class i.
s_i ~ the minimum dollar separation between job class i and job class $i + 1$.
f_i ~ the minimum salary (floor) for employees in job class i.

Values for N, the n_i and the p_i are determined by job evaluation. Values for the m_i are determined from a wage survey. Values for the s_i and f_i, on the other hand, are determined by subjective and other factors. For instance, for purposes of status, it may be necessary for a given job class to be paid at least a certain amount more than the job class just below it. Also, it may be necessary or desirable to place wage floors under particular job classes. For instance, it may be necessary to pay the lowest level managerial job class at least a certain amount because of what unionized employees are paid. With a total salary budget of B = \$800,000, let us assume that we have the initial salary structure parameter values of Table 16.15. We note that if during the solution process the compensation planner applies any of the remedial strategies, some of the values in Table 16.15 may change.

TABLE 16.15
Initial Salary Structure Parameter Values

Job Class i	n_i	p_i	m_i	s_i	f_i
1	1	1.00	80,000	12,000	—
2	2	.90	62,000	10,000	—
3	2	.65	52,000	6,000	—
4	5	.50	36,000	4,000	—
5	3	.35	38,000	8,000	—
6	10	.25	23,000	—	22,000

max	Reserve for individual salary adjustments
min	Average inside percentage deviation
min	Maximum inside percentage deviation
min	Average outside percentage deviation
min	Maximum outside percentage deviation

s.t.	Budgetary constraints
	Inside job evaluation constraints
	Job class separation constraints
	Inside minimax constraints
	Wage floor constraints
	External market price constraints
	Outside minimax constraints

Figure 16.4. Managerial compensation MOLP.

In concept, our managerial compensation objectives are to establish a salary structure that:

1. Attracts and retains sufficient numbers of qualified personnel.
2. Provides incentives for promotion from one job class to another.
3. Is perceived and accepted as equitable by all levels of management.

To operationalize these otherwise unquantifiable objectives, we use the five "surrogate" criteria in the context of the MOLP presented in Fig. 16.4.

The first, second, and sixth categories of constraints are "soft" in that they define the deviational variables out of which the five criterion functions are constructed. The constraints and criterion functions are specified as follows.

Budgetary Constraints. In this category, underachievement deviational variable d_1^- is used to take up the slack between the cost of the salary plan and the budget B. With an initial upper limit of 5% on d_1^-, we have the two constraints

$$\sum_{i=1}^{N} n_i x_i + d_1^- = B$$

$$d_1^- \leqq .05B$$

where x_i is the average salary to be paid to employees in the ith job class. Note that d_1^- measures the size of the discretionary reserve for individual salary adjustments and/or incentive payments.

Inside Job Evaluation Constraints. In this category, we have the following constraints

$$-p_i x_1 + x_i + \frac{p_i m_1}{100}(d_{2i}^- - d_{2i}^+) = 0 \qquad (1 \leqq i \leqq N)$$

in which we measure the percentages by which each of the jobs in the salary plan deviate from the inside relative worth structure. Since the p_i relate the different jobs to the top job in the salary structure, the $p_i m_1/100$ coefficients convert the d_{2i}^- and d_{2i}^+ values to percentages. Actually, it would be more appropriate to use the term $p_i x_1/100$, but this would cause quadratic problems so we use m_1 as an approximation of x_1.

Job Class Separation Constraints. These constraints ensure a minimum dollar separation between adjacent job classes. They are written

$$x_i - x_{i+1} \geqq s_i \qquad (1 \leqq i \leqq N-1)$$

Inside Minimax Constraints. To control the maximum percentage deviation from the inside relative worth structure, we have the following inside minimax constraints

$$d_{2i}^- \leqq \alpha_3 \qquad (1 \leqq i \leqq N)$$

$$d_{2i}^+ \leqq \alpha_3 \qquad (1 \leqq i \leqq N)$$

Wage Floor Constraints. These are lower bound constraints of the form

$$x_i \geqq f_i$$

for those job classes for which salary floors are specified.

External Market Price Constraints. Expressed as

$$x_i + \frac{m_i}{100}(d_{4i}^- - d_{4i}^+) = m_i \qquad (1 \leqq i \leqq N)$$

these constraints attempt to enforce compliance with outside market prices. Because of the $m_i/100$ coefficients, the deviational variable values are converted to percentages.

Outside Minimax Constraints. Assuming that overachievement deviations from market prices are not of concern, we have the following outside minimax constraints

$$d_{4i}^- \leqq \alpha_5 \qquad (1 \leqq i \leqq N)$$

Objectives. In terms of the deviational and minimax variables defined in the constraints, the five objectives of the MOLP are

$$\max \{ d_1^- = z_1 \}$$

$$\min \left\{ \frac{1}{N} \sum_{i=1}^{N} (d_{2i}^- + d_{2i}^+) = z_2 \right\}$$

$$\min \{ \alpha_3 = z_3 \}$$

$$\min \left\{ \frac{1}{N} \sum_{i=1}^{N} (d_{4i}^- + d_{4i}^+) = z_4 \right\}$$

$$\min \{ \alpha_5 = z_5 \}$$

With only one wage floor constraint, the model has $6N + 2$ constraints (including one upper bound and one lower bound) and $5N + 3$ variables.

16.3.4 Computer Results

To solve numerically the managerial compensation model of Section 16.3.3, the **AUTOPAKX** Tchebycheff/weighted-sums code of Steuer and Lambert (1984) and the **LAMBDA** and **FILTER** codes from Steuer (1983) were employed. With the number of job classes $N = 6$, step by step we have the following.

1. With the sample size $P = 6$, intended number of iterations $t = 5$, final iteration criterion weight interval width $w = .25$, let the $\bar{\Lambda}$-reduction factor $r = .7000$. Rescale first objective by 10^{-3}.
2. With **LAMBDA**, 250 ($50 \times k$) λ-vectors are generated from the intervals

$$[\ell^{(1)}, \mu^{(1)}] = \begin{bmatrix} .0000, & 1.0000 \\ .0000, & 1.0000 \\ .0000, & 1.0000 \\ .0000, & 1.0000 \\ .0000, & 1.0000 \end{bmatrix}$$

3. Using **FILTER**, reduce the 250 λ-vectors to 12 ($2 \times P$) dispersed representatives and store in **LWAGE1**.
4. Using **AUTOPAKX** with all $\varepsilon_i = 3$, the initial $\mathbf{z}^{**}$ ideal criterion vector is

$$\mathbf{z}^{**} = \begin{bmatrix} 43.00 \\ 2.73 \\ 6.30 \\ .81 \\ 5.78 \end{bmatrix}$$

5. In the same **AUTOPAKX** run, with $\rho = .001$, solve the 12 augmented weighted Tchebycheff programs associated with the λ-vectors in **LWAGE1**. Store the criterion vector results in **OWAGE1**.
6. Using **FILTER**, reduce the criterion vectors in **OWAGE1** to the six most different and present them, after unscaling the first objective, along with their decision variable values to the DM in the form of Table 16.16.
7. From Table 16.16, assume that the decision maker selects schedule 1–6 as the most preferred.
8. The weighting vector that has schedule 1–6 as the definition point of a weighted Tchebycheff contour is

$$\lambda^{(1)} = \begin{bmatrix} .0821 \\ .2298 \\ .2234 \\ .2949 \\ .1698 \end{bmatrix}$$

Table 16.16
Compensation Schedules of the First Iteration

	Schedules					
	1–1	*1–2*	*1–3*	*1–4*	*1–5*	*1–6*
z_1: Reserve	11,862	32,557	0	0	40,000	20,400
z_2: Ave inside	7.9	11.8	10.6	5.7	11.2	10.8
z_3: Max inside	12.1	30.6	22.2	15.5	14.8	14.6
z_4: Ave outside	5.3	10.1	6.9	7.1	9.8	7.1
z_5: Max outside	20.9	14.3	9.6	21.1	19.6	16.7
x_1: Job class 1	78,365	68,573	77,150	75,588	76,162	78,712
x_2: Job class 2	62,000	54,573	57,008	63,588	57,892	60,309
x_3: Job class 3	52,000	44,573	47,008	49,132	41,811	43,556
x_4: Job class 4	34,324	36,572	35,352	37,794	34,554	35,645
x_5: Job class 5	30,051	32,572	34,352	30,000	30,554	31,645
x_6: Job class 6	22,000	22,000	22,000	22,000	22,000	22,000

9. Generate 250 λ-vectors from the intervals

$$[\ell^{(2)}, \mu^{(2)}] = \begin{bmatrix} .0000, & .7000 \\ .0000, & .7000 \\ .0000, & .7000 \\ .0000, & .7000 \\ .0000, & .7000 \end{bmatrix}$$

10. Reduce the 250 λ-vectors to 12 dispersed representatives and store in `LWAGE2`.
11. Because of difficulties in meeting outside labor market prices for job class 5, the remedial strategies of Section 16.3.1 are applied to redesign job classes 4 and 5, resulting in the updated salary structure parameter values

Table 16.17
Updated Salary Structure Parameter Values for Second Iteration

Job Class i	n_i	p_i	m_i	s_i	f_i
1	1	1.00	80,000	12,000	—
2	2	.90	62,000	10,000	—
3	2	.65	52,000	6,000	—
4	6	.50	36,000	4,000	—
5	2	.32	32,000	8,000	—
6	10	.25	23,000	—	22,000

TABLE 16.18
Compensation Schedules of the Second Iteration

	Schedules					
	2–1	*2–2*	*2–3*	*2–4*	*2–5*	*2–6*
z_1: Reserve	0	36,152	0	11,831	29,733	35,055
z_2: Ave inside	9.9	13.2	7.9	11.2	10.3	9.9
z_3: Max inside	14.2	23.7	24.0	14.9	24.7	30.0
z_4: Ave outside	4.1	6.8	4.8	5.0	6.3	7.7
z_5: Max outside	6.7	13.5	6.8	12.6	7.5	12.8
x_1: Job class 1	82,355	80,000	74,521	81,822	73,963	69,719
x_2: Job class 2	63,864	54,962	62,521	62,904	58,076	57,719
x_3: Job class 3	48,518	44,962	48,438	45,430	48,076	45,317
x_4: Job class 4	35,480	34,000	37,260	34,947	34,000	34,859
x_5: Job class 5	30,000	30,000	30,000	30,000	30,000	30,000
x_6: Job class 6	22,000	22,000	22,000	22,000	22,000	22,000

of Table 16.17. The circled elements are those that are different from Table 16.15.

12. Using **AUTOPAKX**, recompute the $\mathbf{z}^{**}$ ideal criterion vector because the feasible region has probably changed. In the same **AUTOPAKX** run, solve the 12 augmented weighted Tchebycheff programs associated with the λ-vectors in **LWAGE2**. Store the criterion vector results in **OWAGE2**.

13. Reduce the criterion vectors in **OWAGE2** to the six most different and present to the decision maker in the form of Table 16.18.

14. From Table 16.18, assume that the decision maker selects schedule 2–4 as the most preferred.

15. The weighting vector that has schedule 2–4 as a definition point of a weighted Tchebycheff contour is

$$\lambda^{(2)} = \begin{bmatrix} .0492 \\ .1804 \\ .1783 \\ .3680 \\ .2241 \end{bmatrix}$$

16. Generate 250 λ-vectors from the intervals

$$[\ell^{(3)}, \mu^{(3)}] = \begin{bmatrix} .0000, & .4900 \\ .0000, & .4900 \\ .0000, & .4900 \\ .1230, & .6130 \\ .0000, & .4900 \end{bmatrix}$$

TABLE 16.19
Updated Salary Structure Parameter Values for Third Iteration

Job Class i	n_i	p_i	m_i	s_i	f_i
1	1	100	80,000	12,000	(76,000)
2	2	.90	62,000	10,000	—
3	2	.65	52,000	6,000	(48,000)
4	6	.50	36,000	4,000	—
5	2	.32	32,000	8,000	—
6	10	.25	23,000	—	22,000

17. Reduce the 250 λ-vectors to 12 dispersed representatives and store in `LWAGE3`.
18. Refining Table 16.17 with floors for job classes 1 and 3, we have the updated salary structure parameter values of Table 16.19. The circled elements are those that are different from Table 16.17.
19. Using `AUTOPAKX`, recompute the **z**** ideal criterion vector because the feasible region has probably changed. In the same `AUTOPAKX` run, solve the 12 augmented weighted Tchebycheff programs with the λ-vectors in `LWAGE3`. Store the criterion vector results in `OWAGE3`.
20. Reduce the criterion vectors in `OWAGE3` to the six most different and present to the decision maker in the form of Table 16.20.
21. Continuing in this fashion, the compensation planner will steer him or herself to a final salary structure after two more iterations.

TABLE 16.20
Compensation Schedules of the Third Iteration

	Schedules					
	3–1	*3–2*	*3–3*	*3–4*	*3–5*	*3–6*
z_1: Reserve	0	28,000	14,195	16,521	0	0
z_2: Ave inside	9.9	10.7	10.8	9.3	8.0	10.0
z_3: Max inside	14.2	22.2	15.8	22.2	22.2	14.1
z_4: Ave outside	4.2	5.9	4.2	4.5	4.0	4.3
z_5: Max outside	6.6	7.7	7.7	6.3	6.3	7.7
x_1: Job class 1	82,383	76,000	81,114	76,000	76,000	82,453
x_2: Job class 2	63,914	58,000	61,630	62,339	64,000	64,040
x_3: Job class 3	48,317	48,000	49,400	49,400	49,400	48,000
x_4: Job class 4	35,508	34,000	34,239	34,000	36,133	35,578
x_5: Job class 5	30,000	30,000	30,000	30,000	30,000	30,000
x_6: Job class 6	22,000	22,000	22,000	22,000	22,000	22,000

16.4 TWO ADDITIONAL APPLICATIONS

Two additional applications are reviewed. The first concerns a mixed 0–1 river basin water quality planning problem. The second concerns a Markov-based reservoir release policy problem, modeled as an MOLP.

16.4.1 River Basin Water Quality Planning

This application is drawn from Steuer and Wood (1986) which, in turn, came from the problem offered by Bishop et al. (1977). In this application, we study the river water quality planning problem of Fig. 16.5, in which the river is in violation of governmental standards at several surveillance points with regard to certain pollutants. The problem is modeled as a mixed 0–1 multiple objective program. Because of the 0–1 nature of the problem, the feasible region may contain unsupported efficient points. To compute such points, an augmented weighted Tchebycheff solution approach is applied.

As pictured in Fig. 16.5, the river is divided into five reaches, each having uniform hydraulic and nonpoint inflow characteristics. Along the river, there are four (point source) wastewater treatment plants and five surveillance points at which pollutant concentrations are measured. To meet the water quality requirements at the various surveillance points, it is necessary to determine the investments required to upgrade the municipal wastewater treatment plants along the river.

At each of the four wastewater treatment plants there are, potentially, nine levels of treatment. The pollutants in the river to be controlled are

1. Biochemical oxygen demand (BOD) that indicates the amount of oxygen required by bacteria to breakdown biodegradable wastes.

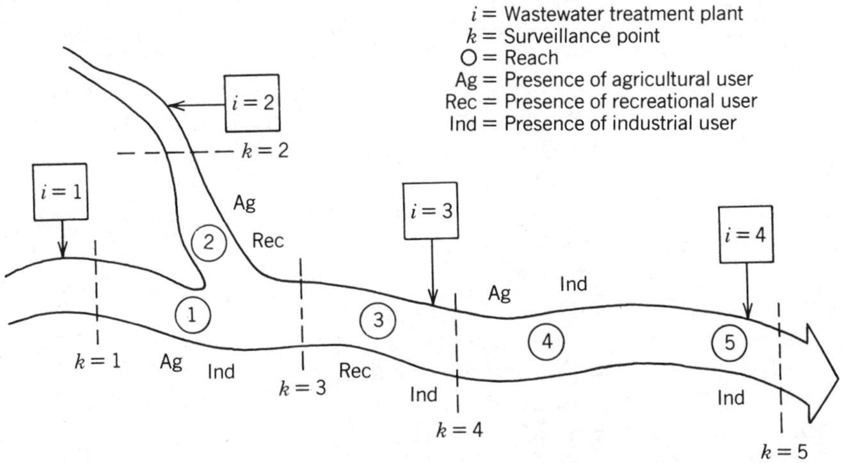

Figure 16.5. River basin system.

min	Average percentage deviation from agricultural water quality requirements
min	Maximum percentage deviation from agricultural water quality requirements
min	Average percentage deviation from recreational water quality requirements
min	Maximum percentage deviation from recreational water quality requirements
min	Average percentage deviation from industrial water quality requirements
min	Maximum percentage deviation from industrial water quality requirements
min	Capital investment
min	Annual operating expenses
s.t.	Treatment process 0–1 constraints
	Surveillance point stream standard constraints
	User water quality constraints
	Water quality minimax constraints

Figure 16.6. Water quality mixed 0–1 multiple objective program.

2. Ammonia (NH_4).
3. Phosphorus (P).
4. Dissolved oxygen deficit (DOD) that is an important measure of the ability of fish and other river life to survive.

The treatment options at each plant are characterized by their (upfront) capital requirements, annual operating expenses, and effluent concentrations.

At the five surveillance points, there are the government-imposed water quality standards for the four pollutants. Also, along the river there are three major beneficial users of water: agriculture, recreation, and industry. These users may have water quality requirements that are more strict than the governmental stream standards. The particular water quality requirements of the user depend on crops, types of recreation (e.g., water contact sports), and industrial processes.

The solution to the river basin planning problem is the configuration of wastewater plant treatment options that not only enables the river to meet the governmental stream standards at the different surveillance points, but best minimizes capital investment, annual operating expenses, and deviations from the user water quality requirements.

With 84 constraints and 89 variables (36 of which are 0–1), the problem is formulated as the 0–1 multiple objective program of Fig. 16.6.

To solve the water quality planning problem, a customized version of the `MIPZ1` mixed 0–1 code [McCarl, Barton, and Schrage (1973)] was used as the workhorse optimization software. The `MIPZ1` code has been customized in the sense that it has been enclosed in a master Fortran loop that automatically reloads `MIPZ1` with different λ-vectors. In this way, each iteration's repetitive optimizations can be solved in one job submission. Also used in this application were the `FILTER` and `LAMBDA` codes from Steuer (1983). Further details about

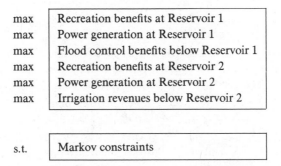

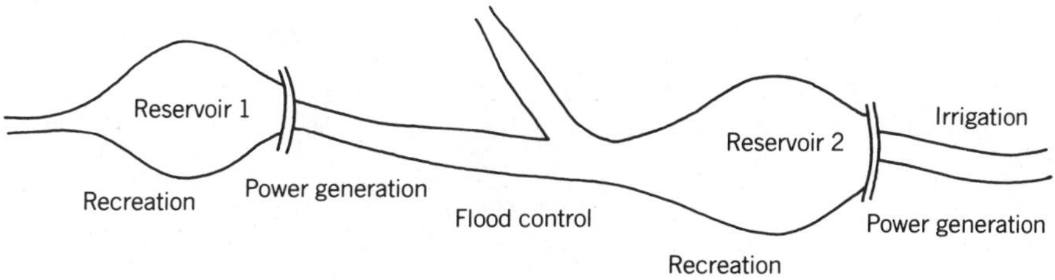

Figure 16.7. Two reservoirs in series.

the application can be obtained by consulting the Steuer and Wood (1986) source article.

16.4.2 A Markov Reservoir Release Policy Problem

This application is from Steuer and Wood (1986). The purpose of the problem is to determine the optimal reservoir release policy for the two reservoirs in series in Fig. 16.7. The problem is modeled as a multiple objective Markov decision process with three states for Reservoir 1 and four states for Reservoir 2. This defines 12 states for the system as a whole. Whereas the optimal policy of a single objective Markov decision process can be obtained by solving a single objective LP, the optimal policy of a multiple objective Markov decision process can be obtained by solving an MOLP.

With as many decisions for the system as states, the problem is formulated as the $6 \times 13 \times 144$ MOLP of Fig. 16.8. A description of how the Markov constraints are constructed is contained in Hillier and Lieberman (1980, Chapter 13).

The interactive weighted-sums method of Section 13.5 was used to solve the MOLP (because extreme points of the feasible region defined by the Markov constraints pertain to policy candidates). The **AUTOPAKX** code of Steuer and Lambert (1984) along with the **LAMBDA** and **FILTER** codes from Steuer (1983) were employed. Further details about the application can be obtained by consulting the Steuer and Wood (1986) source article.

max	Recreation benefits at Reservoir 1
max	Power generation at Reservoir 1
max	Flood control benefits below Reservoir 1
max	Recreation benefits at Reservoir 2
max	Power generation at Reservoir 2
max	Irrigation revenues below Reservoir 2

s.t.	Markov constraints

Figure 16.8. Reservoir release policy MOLP.

Readings

1. Ashton, D. H. and D. R. Atkins (1979). "Multicriteria Programming for Financial Planning," *Journal of the Operational Research Society*, Vol. 30, No. 3, pp. 259–270.
2. Balachandran, K. R. and R. E. Steuer (1982). "An Interactive Model for the CPA Firm Audit Staff Planning Problem with Multiple Objectives," *The Accounting Review*, Vol. LVII, No. 1, pp. 125–140.
3. Bishop, A. B., R. Narayanan, W. J. Grenney, and P. E. Pugner (1977). "Goal Programming Model for Water Quality Planning," *Journal of the Environmental Engineering Division*, ASCE, EE2, pp. 293–305.
4. Bitran, G. R. and K. D. Lawrence (1980). "Locating Service Offices: A Multicriteria Approach," *Omega*, Vol. 8, No. 2, pp. 201–206.
5. Bushenkov, V., F. Ereshko, J. Kindler, A. Lotov, and L. De Maré (1982). "Application of the Generalized Reachable Sets Method to Water Resources Problems in Southwestern Skane, Sweden," WP–82-120, International Institute for Applied Systems Analysis, Laxenburg, Austria.
6. Eatman, J. L. and C. W. Sealey (1979). "A Multiobjective Linear Programming Model for Commercial Bank Balance Sheet Management," *Journal of Bank Research*, Vol. 9, No. 4, pp. 227–236.
7. Grauer, M., A. Lewandowski, and L. Schrattenholzer (1982). "Use of the Reference Level Approach for the Generation of Efficient Energy Supply Strategies," In M. Grauer, A. Lewandowski, and A. P. Wierzbicki (eds.), *Multiobjective and Stochastic Optimization*, Laxenburg, Austria: International Institute for Applied Systems Analysis, pp. 425–444.
8. Grauer, M., A. Lewandowski, and A. P. Wierzbicki (1983). "Multiple-Objective Decision Analysis Applied to Chemical Engineering," *Angewandte Systemanalyse*, Vol. 4, No. 1, pp. 32–40.
9. Grauer, M. and E. Zalai (1982). "A Reference Point Approach of Nonlinear Macroeconomic Multiobjective Models," WP–82-134, International Institute for Applied Systems Analysis, Laxenburg, Austria.
10. Greis, N. P., E. F. Wood, and R. E. Steuer (1983). "Multicriteria Analysis of Water Allocation in a River Basin: The Tchebycheff Approach," *Water Resources Research*, Vol. 19, No. 4, pp. 865–875.
11. Hillier, F. S. and G. J. Lieberman (1980). *Introduction to Operations Research* (Chapter 13), Holden-Day, San Francisco.
12. Kok, M. and F. A. Lootsma (1985). "Pairwise-Comparison Methods in Multiple Objective Programming, with Applications in a Long-Term Energy-Planning Model," *European Journal of Operational Research*, Vol. 22, No. 1, pp. 44–55.
13. Lawrence, K. D. and G. R. Reeves (1982). "Combining Multiple Forecasts Given Multiple Objectives," *Journal of Forecasting*, Vol. 1, No. 3, pp. 271–279.
14. Leitch, R. A., R. E. Steuer, and J. T. Godfrey (1986). "An Interactive Multiple Objective Budget Model," College of Business Administration, University of Georgia, Athens, Georgia.

15. Lin, W. T. (1982). "An Accounting Control System Structured on Multiple Objective Planning Models," *Omega*, Vol. 8, No. 3, pp. 375–382.

16. McCarl, B., D. Barton, and L. Schrage (1973). "**MIPZ1**—Documentation on a Zero-One Mixed Integer Programming Code," Station Bulletin No. 24, Department of Agricultural Economics, Purdue University, West Lafayette, Indiana.

17. Rust, R. E. (1976). *Sausage and Meats Manufacturing*, AMI Center for Continuing Education, American Meat Institute, Washington, D.C.

18. Sakawa, M. and F. Seo (1982). "Interactive Multiobjective Decision Making in Environmental Systems Using Sequential Proxy Optimization Techniques (SPOT)," *Automatica*, Vol. 18, No. 2, pp. 155–165.

19. Silverman, J., R. E. Steuer, and A. W. Whisman (1985). "A Multi-Period, Multiple Criteria Optimization System for Manpower Planning," College of Business Administration, University of Georgia, Athens, Georgia.

20. Steuer, R. E. (1983). "Multiple Criterion Function Goal Programming Applied to Managerial Compensation Planning," *Computers and Operations Research*, Vol. 10, No. 4, pp. 299–309.

21. Steuer, R. E. (1983). "Operating Manual for the **ADBASE** Multiple Objective Linear Programming Package," College of Business Administration, University of Georgia, Athens, Georgia.

22. Steuer, R. E. (1984). "Sausage Blending Using Multiple Objective Linear Programming," *Management Science*, Vol. 30, No. 11, pp. 1376–1384.

23. Steuer, R. E. and E.-U. Choo (1983). "An Interactive Weighted Tchebycheff Procedure for Multiple Objective Programming," *Mathematical Programming*, Vol. 26, No. 1, pp. 326–344.

24. Steuer, R. E. and F. W. Harris (1980). "Intra-Set Point Generation and Filtering in Decision and Criterion Space," *Computers and Operations Research*, Vol. 7, No.'s 1–2, pp. 41–53.

25. Steuer, R. E. and T. D. Lambert (1984). "**AUTOPAKX**: An **MPSX**-Based Tchebycheff/Weighted-Sums Automated Package for Multiple Objective Linear Programming," College of Business Administration, University of Georgia, Athens, Georgia.

26. Steuer, R. E. and A. T. Schuler (1978). "An Interactive Multiple Objective Linear Programming Approach to a Problem in Forest Management," *Operations Research*, Vol. 25, No. 2, pp. 254–269.

27. Steuer, R. E. and A. T. Schuler (1981). "Interactive Multiple Objective Linear Programming Applied to Multiple Use Forestry Planning." In M. C. Vodak, W. A. Leuschner, and D. I. Navon (eds.), *Symposium on Forest Management Planning: Present Practice and Future Directions*, FWS-1-81, School of Forestry and Wildlife Resources, VPI, Blacksburg, Virginia, pp. 80–93.

28. Steuer, R. E. and E. F. Wood (1986). "On the 0–1 Implementation of the Tchebycheff Solution Approach: A Water Quality Illustration," *Large Scale Systems*, forthcoming.

29. Steuer, R. E. and E. F. Wood (1986). "A Multiple Objective Markov Reservoir Release Policy Model," College of Business Administration, University of Georgia, Athens, Georgia.

30. Steuer, R. E., L. R. Gardiner, and J. J. Bernardo (1986). "Multi-Period Production Planning Using Multicriteria Trajectory Optimization," College of Business Administration, University of Georgia, Athens, Georgia.

31. Tabucanon, M. T., P. Adulbhan, and R. A. Alivio (1983). "Application of Multiobjective Linear Programming in Deriving Tariff Adjustments," *Policy and Information*, Vol. 7, No. 2, pp. 139–163.

32. Wallace, M. J. and C. H. Fay (1983). *Compensation Theory and Practice*, Boston: Kent.

33. Wallenius, H., J. Wallenius, and P. Vartia (1978). "An Approach to Solving Multiple Criteria Macroeconomic Policy Problems and an Application," *Management Science*, Vol. 24, No. 10, pp. 1021–1030.

34. Welling, P. (1977). "A Goal Programming Model for Human Resource Accounting in a CPA Firm," *Accounting, Organizations and Society*, Vol. 2, No. 4, pp. 307–316.

CHAPTER 17

Future Directions

In this chapter, we discuss a number of topics from which further research and application developments are expected. In Sections 17.1 to 17.3, we discuss the computer/user interface, various screen displays, and trajectory optimization. In Section 17.4, we provide a bibliography of multiple objective applications in engineering management. Finally, in Section 17.5, we provide bibliographies on a range of other optimization topics that have, in various degrees of completeness, been undergoing a multiple objective treatment.

17.1 COMPUTER / USER INTERFACE

In many problems, the amount of multiple criteria information to be displayed at each iteration is large. For instance, in a problem with $k = 7$ criteria and a sample size of $P = 8$, eight criterion vectors involving 56 pieces of information would have to be communicated at each iteration. To facilitate the absorption of such large amounts of information, particular attention must be paid to the *computer/user interface*. Since this is where the user comes into contact with a method, the quality of the computer/user interface will be a major factor influencing the acceptability of the method. If a decision maker finds the interface rewarding and supportive of the way he or she does things, the DM is likely to embrace the solution method. If, on the other hand, a DM does not find the interface to be compatible with his or her way of thinking, the DM is likely to shy away from the solution methodology. We do not want to have a good method judged poorly because of neglect at the interface.

Consequently, we must be very careful about how iterative information is displayed. We want the information displayed attractively and professionally, in harmony with the decision maker's information processing abilities. We do not want the display of information to be intimidating or confusing. Rather, we want the display to be pleasant to the eye and stimulate the decision maker's interest and imagination. The idea is to avoid anything that might sidetrack the DM's mental energies and distract him or her from making the evaluations and judgments required by the solution procedure.

17.2 VARIOUS SCREEN DISPLAYS

Assume that we are working with an interactive procedure that displays each iteration's criterion vector information on a CRT. Some decision makers might like to view the information in a tabular format, some might like to view the information graphically, but most would probably like to view the information both ways if possible.

There is something to be said about viewing criterion vector information using *different* display formats. Aspects of the information that might be missed when viewing one type of display might become apparent when viewing another type. Probably, the best way to deal with this issue is to prepare the iterative information in several different formats. Then, by signaling to the computer, the user can view whichever types of displays are of interest. To illustrate, suppose that we are at a CRT where the first three *program function* (PF) keys are defined as follows:

PF1 Tabular display of criterion vector information.

PF2 Bar chart display of criterion vector information.

PF3 Value path display of criterion vector information.

	Criterion Vectors					
	1-1	1-2	1-3	1-4	1-5	1-6
Objective 1	40	80	95	30	40	20
Objective 2	55	50	25	60	20	85
Objective 3	85	30	45	70	60	40
Objective 4	10	75	70	95	30	60
Objective 5	30	65	80	15	90	10

PF1=Tabular PF2=Bar Chart PF3=Value Path

Figure 17.1. Tabular display of criterion vectors.

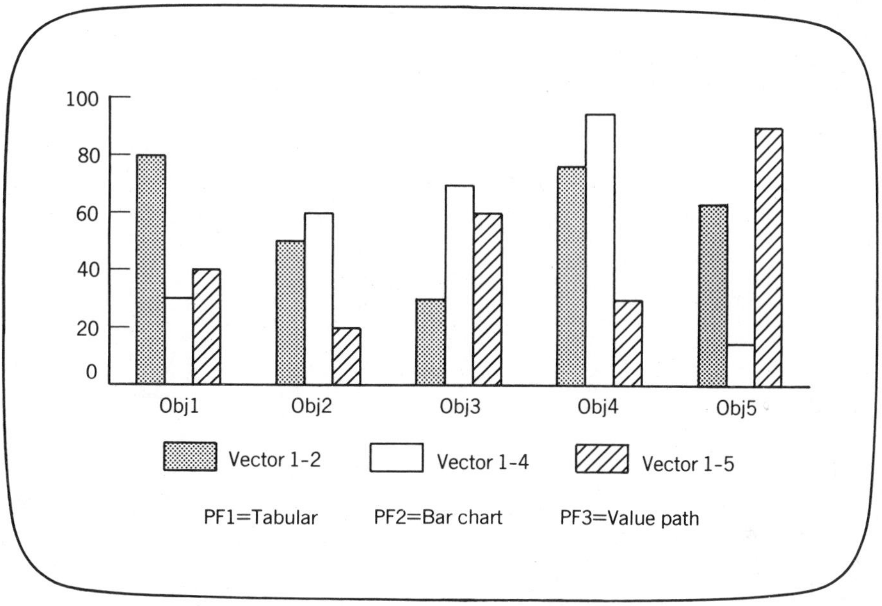

Figure 17.2. Bar chart display of criterion vectors.

Let us assume that the decision maker does the following. The DM presses the PF1 key and obtains the tabular display of Fig. 17.1. After reviewing the criterion vectors, the DM mentally rules out (at least for the time being) criterion vectors 1–1, 1–3, and 1–6. By using a cursor control device (such as the cursor positioning keys, a mouse, or a light pen), the DM makes a mark at the bottom of criterion vectors 1–2, 1–4, and 1–5 indicating that they are to be the subject of further analysis. Then, the DM presses the PF2 key to obtain the side-by-side bar charts of Fig. 17.2.

To view the three criterion vectors in a different graphical format, the DM presses the PF3 key to obtain the display of Fig. 17.3. In Fig. 17.3, the criterion vectors are displayed in *value path* format on a vertical bar background that portrays the ranges of the criterion values over the efficient set. These and other types of alphanumeric and graphical displays can be designed. Color graphics would be especially useful in the displays of Figs. 17.2 and 17.3 since a different color could be assigned to each different criterion vector.

Note that we have the PF key menu at the bottom of each of the displays. This is done so that the user will not have to memorize how the PF keys are defined. By using the PF keys, the user can flip back and forth among the displays. As long as the displays are attractive, the user should find this type of activity enjoyable and rewarding. The following are references about the power of color computer graphics and the display of criterion vector information.

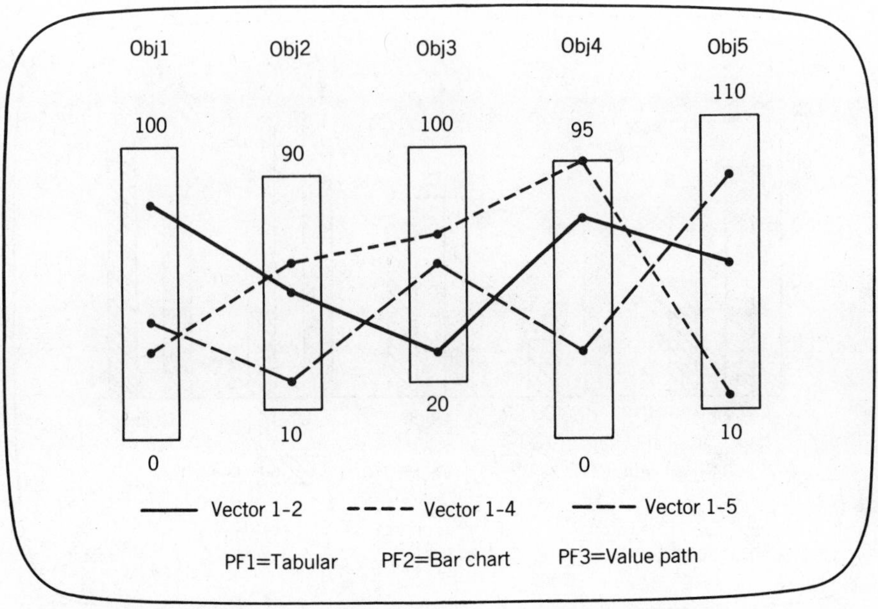

Figure 17.3. Value path display of criterion vectors.

1. IBM Document No. GC33–0100-5 (1984). "Graphical Data Display Manager (GDDM): General Information," IBM Corporation, Department 6R1H, Mechanicsburg, Pennsylvania.

2. Schilling, D. A., C. Revelle, and J. Cohon (1983). "An Approach to the Display and Analysis of Multiobjective Problems," *Socio-Economic Planning Sciences*, Vol. 17, No. 2, pp. 57–63.

3. Tufte, E. R. (1983). *The Visual Display of Quantitative Information*, Cheshire, Connecticut: Graphics Press.

17.3 TRAJECTORY OPTIMIZATION

Suppose we have a multitime period model in which we wish to monitor k criterion values in each of T time periods. Let us also assume that there is a goal level of achievement for each criterion in each time period. Then, for each objective, it is said that the goal levels form a *trajectory* over the T time periods. By the same token, for each solution, there is a trajectory of criterion values for each objective over the T time periods. In such problems, our purpose is to find the solution whose k *criterion value trajectories* most closely match the k *goal trajectories*. Hence, we have the term trajectory optimization.

How do we know whether one criterion value trajectory more closely matches a goal trajectory than another? It depends on the measure used. Unfortunately,

min	max percentage overdeviation from salary budget goal trajectory
min	max percentage deviation from strength-of-force goal trajectory
min	max percentage deviation from trajectory of promotion targets
min	max percentage underdeviation from trajectory of LOS targets

s.t. $\mathbf{x} \in S$

Figure 17.4. Trajectory optimization military manpower formulation.

there is no concensus about which measure to use. Candidate measures include average absolute deviation, maximum absolute deviation, average percentage deviation, and maximum percentage deviation.

With regard to the screen display of iterative information in a trajectory problem, consider a military manpower planning problem in which the following four objectives are to be monitored over 10 time periods.

1. Salary expenditures.
2. Strength of force (number of people in service).
3. Number of people promoted.
4. Average LOS (length of service, or in other words, the average number of years that a member of the force has been in the service)

Suppose that there is a goal trajectory for each objective and that the distance between a trajectory of criterion values and its goal trajectory is measured in terms of maximum percentage deviation. Forming the model of Fig. 17.4, let us apply a procedure that generates six candidate solutions to be presented to the decision maker at each iteration.

With 40 criterion values per candidate solution, 240 criterion values are generated each iteration. Since we cannot display this much information at once, we must find some way of *layering* the information. That is, we would want to have at our disposal different screens for viewing different levels of disaggregated detail. To illustrate, assume that we are on the third iteration. To obtain *aggregated* information corresponding to the objective functions in Fig. 17.4, we might have the display of Fig. 17.5. To obtain *disaggregated* trajectory information about—for example, solution 3–3—one would press program function key PF6 and obtain the four trajectory graphs displayed in Fig. 17.6.

Note how the trajectory format of Fig. 17.6 facilitates an "at-a-glance" comprehension of solution 3–3's 40 criterion values *viz a viz* the 40 goal values. Once again, a menu of PF keys is displayed to facilitate the flipping among screens that is expected at each iteration. Thus, by pressing PF12, we could zoom the strength-of-force trajectory graph to full-screen.

The trajectory optimization approach is not solely confined to multitime period models. For example, consider a river basin water quality problem in which we are concerned about the concentrations of k pollutants versus governmental

	Maximum Percentage Deviations					
	3-1	3-2	3-3	3-4	3-5	3-6
Salary	10	25	15	10	5	20
Strength of force	15	0	10	5	15	10
Promotions	12	8	4	0	5	10
Length of service	5	15	20	25	10	2

PF1=3-1 PF2=3-2 PF3=3-3 PF4=3-4 PF5=3-5 PF6=3-6

Figure 17.5. Display of aggregated information.

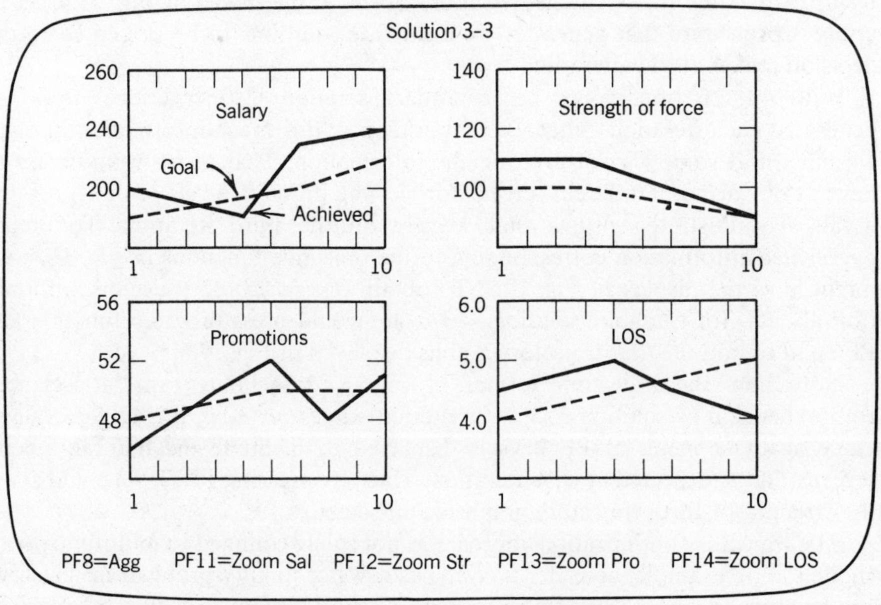

Figure 17.6. Trajectory graphs of solution 3–3.

standards at T surveillance points along the river. In this problem, the governmental standards and the criterion values of each solution have a trajectory representation over the multiple surveillance points. As another example, consider a sausage manufacturing model in which all of the products in the product line are blended simultaneously. Then, we would have trajectories of per lb. costs, fat percentages, protein percentages, and so on over the multiple products.

Most of the initial work on trajectory optimization was conducted at the International Institute for Applied Systems Analysis (Laxenburg, Austria) under the direction of Wierzbicki and Grauer. The following is a trajectory optimization bibliography.

1. Grauer, M. (1983). "A Dynamic Interactive Decision Analysis and Support System (DIDASS): User's Guide," WP–83-60, International Institute for Applied Systems Analysis, Laxenburg, Austria.
2. Grauer, M., A. Lewandowski, and L. Schrattenholzer (1982). "Use of the Reference Level Approach for the Generation of Efficient Energy Supply Strategies." In M. Grauer, A. Lewandowski, and A. P. Wierzbicki (eds.), *Multiobjective and Stochastic Optimization*, Laxenburg, Austria: International Institute for Applied Systems Analysis, pp. 425–444.
3. Kallio, M., A. Lewandowski, and W. Orchard-Hays (1980). "An Implementation of the Reference Point Approach for Multiobjective Optimization," WP–80-35, International Institute for Applied Systems Analysis, Laxenburg, Austria.
4. Kallio, M., A. Propoi, and R. Seppälä (1980). "A Model for the Forest Sector," WP–80-34, International Institute for Applied Systems Analysis, Laxenburg, Austria.
5. Kallio, M. and M. Soismaa (1982). "Some Improvements to the Reference Point Approach for Dynamic Multicriteria Linear Programming." In M. Grauer, A. Lewandowski, and A. P. Wierzbicki (eds.), *Multiobjective and Stochastic Optimization*, Laxenburg, Austria: International Institute for Applied Systems Analysis, pp. 411–423.
6. Silverman, J., R. E. Steuer, and A. W. Whisman (1985). "A Multi-Period, Multiple Criteria Optimization System for Manpower Planning," College of Business Administration, University of Georgia, Athens, Georgia.
7. Wierzbicki, A. P. (1982). "Multiobjective Trajectory Optimization and Model Semiregularization." In M. Grauer, A. Lewandowski, and A. P. Wierzbicki (eds.), *Multiobjective and Stochastic Optimization*, Laxenburg, Austria: International Institute for Applied Systems Analysis, pp. 3–38.

17.4 MULTIPLE OBJECTIVE APPLICATIONS IN ENGINEERING MANAGEMENT

An intriguing application area of multiple criteria optimization is in engineering management. Applications in engineering management are challenging and often involve nonlinearities. The following is a bibliography on multiple objective applications in engineering management.

1. Grauer, M., A. Lewandowski, and A. P. Wierzbicki (1983). "Multiple-Objective Decision Analysis Applied to Chemical Engineering," *Angewandte Systemanalyse*, Vol. 4, No. 1, pp. 32–40.
2. Grauer, M., P. Pawlak, and K. Hartmann (1979). "Polyoptimal Control of a Film-hardening Process," Messen–Steuern–Regeln, Vol. 22, No. 1, pp. 15–19.
3. Grauer, M., L. Pollmer, and B. Poltersdorf (1981). "Optimal Design and Operations of a Twin-Screw-Extruder by Multiple Criteria Analysis," Proceedings CHEM-CONTROL-81, pp. 2-21.
4. Ignizio, J. P. (1981). "Antenna Array Beam Pattern Synthesis via Goal Programming," *European Journal of Operational Research*, Vol. 8, No. 1, pp. 9–16.
5. McCammon, D. F. and W. Thompson (1983). "The Use of Goal Programming in the Design of Electro-Acoustic Transducers and Transducer Arrays," *Computers and Operations Research*, Vol. 10, No. 4, pp. 345–356.
6. Nakayama, H. and K. Furukawa (1985). "Satisficing Trade-Off Method with an Application to Multiobjective Structural Design," *Large Scale Systems*, Vol. 8, No. 1, pp. 47–59.
7. Sakawa, M. (1981). "Interactive Multiobjective Reliability Design of a Standby System by the Sequential Proxy Optimization Technique (SPOT)," *International Journal of Systems Science*, Vol. 12, No. 6, pp. 687–701.
8. Sophos, A., E. Rotstein, and G. Stephanopoulos (1980). "Multiobjective Analysis in Modeling the Petrochemical Industry," *Chemical Engineering Science*, Vol. 35, No. 12, pp. 2415–2426.
9. Sundaram, R. M. (1978). "An Application of Goal Programming Technique in Metal Cutting," *International Journal of Production Research*, Vol. 16, No. 5, pp. 375–382.
10. Tabak, D., A. A. Schy, D. P. Giesy, and K. G. Johnson (1979). "Application of Multiobjective Optimization in Aircraft Control Systems Design," *Automatica*, Vol. 15, No. 5, pp. 595–600.
11. Zanakis, S. H. and J. S. Smith (1980). "Chemical Production Planning via Goal Programming," *International Journal of Production Research*, Vol. 18, No. 6, pp. 687–697.

17.5 OTHER AREAS OF RESEARCH IN MULTIPLE CRITERIA OPTIMIZATION

In addition to the coverage in this book, there has been multiple objective research in a variety of other optimization areas. In most of these areas, much work remains to be done. For purposes of accessing the multiple objective literature in these areas, the seven bibliographies of Sections 17.5.1 to 17.5.7 are presented.

17.5.1 Bibliography on Bicriterion Mathematical Programming

1. Adulbhan, P. and M. T. Tabucanon (1977). "Bicriterion Linear Programming," *Computers and Operations Research*, Vol. 4, No. 2, pp. 147–153.
2. Aneja, Y. P. and K. P. K. Nair (1979). "Bicriteria Transportation Problem," *Management Science*, Vol. 25, No. 1, pp. 73–78.
3. Benson, H. P. (1979). "Vector Maximization with Two Objective Functions," *Journal of Optimization Theory and Applications*, Vol. 28, No. 2, pp. 253–258.
4. Cohon, J. L., R. L. Church, and D. P. Sheer (1979). "Generating Multiobjective Trade-Offs: An Algorithm for Bicriterion Problems," *Water Resources Research*, Vol. 15, No. 5, pp. 1001–1010.
5. Gearhart, W. B. (1979). "On the Characterization of Pareto-Optimal Solutions in Bicriteria Optimization," *Journal of Optimization Theory and Applications*, Vol. 27, No. 2, pp. 301–307.
6. Geoffrion, A. M. (1967). "Solving Bicriterion Mathematical Programs," *Operations Research*, Vol. 15, No. 1, pp. 39–54.
7. Kiziltan, G. and E. Yucaoglu (1982). "An Algorithm for Bicriterion Linear Programming," *European Journal of Operational Research*, Vol. 10, No. 4, pp. 406–411.
8. Pasternak, H. and U. Passy (1973). "Bicriterion Mathematical Programs with Boolean Variables." In J. L. Cochrane and M. Zeleny (eds.), *Multiple Criteria Decision Making*, Columbia, South Carolina: University of South Carolina Press, pp. 327–348.
9. Sadagan, S. and A. Ravindran (1982). "Interactive Solution of Bi-Criteria Mathematical Programs," *Naval Research Logistics Quarterly*, Vol. 29, No. 3, pp. 443–459.
10. Walker, J. (1978). "An Interactive Method as an Aid in Solving Bicriteria Mathematical Programming Problems," *Journal of the Operational Research Society*, Vol. 29, No. 9, pp. 915–922.

17.5.2 Bibliography on Duality in Multiple Objective Programming

1. Bitran, G. R. (1981). "Duality for Nonlinear Multiple-Criteria Optimization Problems," *Journal of Optimization Theory and Applications*, Vol. 35, No. 3, pp. 367–401.
2. Bitran, G. R. and T. L. Magnanti (1980). "Duality Based Characterizations of Efficient Facets," *Lecture Notes in Economics and Mathematical Systems*, No. 177, Berlin: Springer-Verlag, pp. 13–23.
3. Brumelle, S. (1981). "Duality for Multiple Objective Convex Programs," *Mathematics of Operations Research*, Vol. 6, No. 2, pp. 159–172.
4. Hannan, E. L. (1978). "Using Duality Theory for Identification of Primal Efficient Points and for Sensitivity Analysis in Multiple Objective Linear Programming," *Journal of the Operational Research Society*, Vol. 29, No. 7, pp. 643–649.
5. Hannan, E. L. (1979). "Using Duality Theory for Identification of Primal Efficient Points and for Sensitivity Analysis of MOLP: Reply," *Journal of the Operational Research Society*, Vol. 30, No. 3, pp. 287–288.
6. Ignizio, J. P. (1982). *Linear Programming in Single- & Multiple-Objective Systems* (Chapter 18), Englewood Cliffs, New Jersey: Prentice-Hall.

7. Isermann, H. (1976). "Existence and Duality in Multiple Objective Linear Programming," *Lecture Notes in Economics and Mathematical Systems*, No. 130, Berlin: Springer-Verlag, pp. 64–75.

8. Isermann, H. (1977). "The Relevance of Duality in Multiple Objective Linear Programming," *TIMS Studies in the Management Sciences*, Vol. 6, pp. 241–262.

9. Isermann, H. (1978). "Duality in Multiple Objective Linear Programming," *Lecture Notes in Economics and Mathematical Systems*, No. 155, Berlin: Springer-Verlag, pp. 274–285.

10. Isermann, H. (1978). "On Some Relations Between a Dual Pair of Multiple Objective Linear Programs," *Zeitschrift für Operations Research*, Vol. 22, No. 1, pp. 33–41.

11. Jahn, J. (1983). "Duality in Vector Optimization," *Mathematical Programming*, Vol. 25, No. 3, pp. 343–353.

12. Kornbluth, J. S. H. (1974). "Duality, Indifference and Sensitivity Analysis in Multiple Objective Linear Programming," *Operational Research Quarterly*, Vol. 25, No. 4, pp. 599–614.

13. Nakayama, H. (1985). "Duality Theory in Vector Optimization: An Overview," *Lecture Notes in Economics and Mathematical Systems*, No. 242, Berlin: Springer-Verlag, pp. 109–125.

14. Rödder, W. (1977). "A Generalized Saddlepoint Theory: Its Application to Duality Theory for Linear Vector Optimum Problems," *European Journal of Operational Research*, Vol. 1, No. 1, pp. 55–59.

15. Rosinger, E. E. (1977). "Duality and Alternative in Multiobjective Optimization," *Proceedings of the American Mathematical Society*, Vol. 64, No. 2, pp. 307–312.

16. Schönfeld, K. P. (1970). "Some Duality Theorems for the Non-Linear Vector Maximum Problem," *Unternehmensforschung*, Vol. 14, No. 1, pp. 51–63.

17. Tanino, T. and Y. Sawaragi (1977). "Duality Theory in Multiobjective Programming," *Journal of Optimization Theory and Applications*, Vol. 27, No. 4, pp. 509–529.

17.5.3 Bibliography on Multiple Objective Programming with Fuzzy Sets

1. Buckley, J. J. (1983). "Fuzzy Programming and the Pareto Optimal Set," *Fuzzy Sets and Systems*, Vol. 10, No. 1, pp. 57–63.

2. Carlsson, C. (1982). "Tackling an MCDM-Problem with the Help of Some Results from Fuzzy Set Theory," *European Journal of Operational Research*, Vol. 10, No. 3, pp. 270–281.

3. Hannan, E. L. (1979). "On the Efficiency of the Product Operator in Fuzzy Programming with Multiple Objectives," *Fuzzy Sets and Systems*, Vol. 2, No. 3, pp. 259–262.

4. Hannan, E. L. (1981). "Linear Programming with Multiple Fuzzy Goals," *Fuzzy Sets and Systems*, Vol. 6, No. 3, pp. 235–248.

5. Hannan, E. L. (1981). "On Fuzzy Goal Programming," *Decision Sciences*, Vol. 12, No. 3, pp. 522–531.

6. Hannan, E. L. (1981). "Some Further Comments on Fuzzy Priorities," *Decision Sciences*, Vol. 12, No. 3, pp. 539–541.

7. Ignizio, J. P. (1983). "Fuzzy Multicriteria Integer Programming via Fuzzy Generalized Networks," *Fuzzy Sets and Systems*, Vol. 10, No. 3, pp. 261–270.

8. Leberling, H. (1981). "On Finding Compromise Solutions in Multicriteria Problems Using the Fuzzy Min-Operator," *Fuzzy Sets and Systems*, Vol. 6, No. 2, pp. 105–118.

9. Saaty, T. L. (1978). "Exploring the Interface Between Hierarchies, Multiple Objectives and Fuzzy Sets," *Fuzzy Sets and Systems*, Vol. 1, No. 1, pp. 57–68.

10. Sakawa, M. (1983). "Interactive Computer Programs for Fuzzy Linear Programming with Multiple Objectives," *International Journal of Man-Machine Studies*, Vol. 18, No. 5, pp. 489–503.

11. Sakawa, M. and T. Yumine (1984). "Interactive Fuzzy Decisionmaking for Multiobjective Linear Fractional Programming Problems," *Large Scale Systems*, Vol. 5, No. 2, pp. 105–113.

12. Yager, R. R. (1977). "Multiple Objective Decision-Making Using Fuzzy Sets," *International Journal of Man-Machine Studies*, Vol. 9, No. 4, pp. 375–382.

13. Zimmermann, H.-J. (1978). "Fuzzy Programming and Linear Programming with Several Objective Functions," *Fuzzy Sets and Systems*, Vol. 1, No. 1, pp. 45–56.

17.5.4 Bibliography on Multiple Objectives in Game Theory

1. Bergstresser, K. and P. L. Yu (1977). "Domination Structures and Multicriteria Problems in N-Person Games," *Theory and Decision*, Vol. 8, No. 1, pp. 5–48.

2. Cook, W. D. (1976). "Zero-Sum Games with Multiple Goals," *Naval Research Logistics Quarterly*, Vol. 23, No. 4, pp. 615–622.

3. Freimer, M. and P. L. Yu (1974). "The Application of Compromise Solutions to Reporting Games." In A. Rapoport (ed.), *Game Theory as a Theory of Conflict Resolution*, Boston: D. Reidel Publishing, pp. 235–260.

4. Hannan, E. L. (1982). "Reformulating Zero-Sum Games with Multiple Goals," *Naval Research Logistics Quarterly*, Vol. 29, No. 1, pp. 113–118.

5. Yu, P. L. (1984). "Dissolution of Fuzziness for Better Decisions—Perspective and Techniques," *TIMS Studies in the Management Sciences*, Vol. 20, pp. 171–207.

6. Zeleny, M. (1976). "Games with Multiple Payoffs," *International Journal of Game Theory*, Vol. 4, No. 4, pp. 179–191.

17.5.5 Bibliography on Multiple Objective Integer Programming

1. Bitran, G. R. (1977). "Linear Multiple Objective Programs with 0–1 Variables," *Mathematical Programming*, Vol. 13, No. 2, pp. 121–139.

2. Bitran, G. R. (1979). "Theory and Algorithms for Linear Multiple Objective Programs with Zero-One Variables," *Mathematical Programming*, Vol. 17, No. 3, pp. 362–390.

3. Bitran, G. R. and K. D. Lawrence (1980). "Locating Service Offices; A Multicriteria Approach," *Omega*, Vol. 8, No. 2, pp. 201–206.

4. Bitran, G. R. and J. M. Rivera (1982). "A Combined Approach to Solve Binary Multicriteria Problems," *Naval Research Logistics Quarterly*, Vol. 29, No. 2, pp. 181–201.

5. Burkard, R. E., H. Keiding, J. Krarup, and P. M. Pruzan (1981). "Relationship Between Optimality and Efficiency in Multicriteria 0–1 Programming Problems," *Computers and Operations Research*, Vol. 8, No. 4, pp. 241–248.

6. Burkard, R. E., J. Krarup, and P. M. Pruzan (1982). "Efficiency and Optimality in Minisum, Minimax 0–1 Programming Problems," *Journal of the Operational Research Society*, Vol. 33, No. 2, pp. 137–151.

7. Deckro, R. F. and E. P. Winkofsky (1983). "Solving Zero-One Multiple Objective Programs Through Implicit Enumeration," *European Journal of Operational Research*, Vol. 12, No. 4, pp. 362–374.

8. Gabbani, D. and M. J. Magazine (1981). "An Interactive Heuristic Approach for Multi-Objective Integer Programming Problems," Working Paper No. 148, Department of Management Sciences, University of Waterloo, Waterloo, Ontario, Canada.

9. Karwan, M. H., S. Zionts, and B. Villarreal (1981). "An Improved Multicriteria Integer Programming Algorithm," Working Paper No. 530, School of Management, State University of New York at Buffalo, Buffalo.

10. Klein, D. and E. L. Hannan (1982). "An Algorithm for the Multiple Objective Integer Programming Problem," *European Journal of Operational Research*, Vol. 9, No. 4, pp. 378–385.

11. Kiziltan, G. and E. Yucaoglu (1983). "An Algorithm for Multiobjective Zero-One Linear Programming," *Management Science*, Vol. 29, No. 12, pp. 1444–1453.

12. Shapiro, J. F. (1976). "Multiple Criteria Public Investment Decision Making by Mixed Integer Programming," *Lecture Notes in Economics and Mathematical Systems*, No. 130, Berlin: Springer-Verlag, pp. 170–182.

13. Steuer, R. E. and E.-U. Choo (1983). "An Interactive Weighted Tchebycheff Procedure for Multiple Objective Programming," *Mathematical Programming*, Vol. 26, No. 1, pp. 326–344.

14. Villarreal, B. and M. H. Karwan (1981). "Parametric Multicriteria Integer Programming." In P. Hansen (ed.), *Studies on Graphs and Discrete Programming*, Amsterdam: North-Holland, pp. 371–379.

15. Villarreal, B. and M. H. Karwan (1981). "Multicriteria Integer Programming: A (Hybrid) Dynamic Programming Recursive Approach," *Mathematical Programming*, Vol. 21, No. 2, pp. 204–223.

16. Villarreal, B. and M. H. Karwan (1981). "An Interactive Dynamic Programming Approach to Multicriteria Discrete Programming," *Journal of Mathematical Analysis and Applications*, Vol. 81, No. 2, pp. 524–544.

17. Zionts, S. (1977). "Integer Linear Programming with Multiple Objectives," *Annals of Discrete Mathematics*, Vol. 1, pp. 551–562.

18. Zionts, S. (1979). "A Survey of Multiple Criteria Integer Programming Methods," *Annals of Discrete Mathematics*, Vol. 5, pp. 389–398.

17.5.6 Bibliography on Multiple Objectives in Networks, Markov Processes, Dynamic Programming, and Location

1. Bhatia, H. L., K. Swarup, and M. C. Puri (1976). "Time-Cost Trade-Off in a Transportation Problem," *Opsearch*, Vol. 13, No.'s 3–4, pp. 129–142.

2. Daellenbach, H. G. and C. A. DeKluyver (1979). "Note on Multiple Objective Dynamic Programming," *Journal of the Operational Research Society*, Vol. 31, No. 7, pp. 591–594.

3. Diaz, J. A. (1979). "Finding a Complete Description of All Efficient Solutions to a Multiobjective Transportation Problem," *Ekonomicko-Matematicky Obzor*, Vol. 15, No. 1, pp. 62–73.

4. Glickman, T. S. and P. Berger (1977). "Cost/Completion-Date Tradeoffs in the Transportation Problem," *Operations Research*, Vol. 25, No. 1, pp. 163–168.

5. Golabi, K. (1983). "A Markov Decision Modeling Approach to a Multi-Objective Maintenance Problem," *Lecture Notes in Economics and Mathematical Systems*, No. 209, Berlin: Springer-Verlag, pp. 115–125.

6. Hultz, J. W., D. Klingman, G. T. Ross, and R. M. Soland (1981). "An Interactive Computer System for Multicriteria Facility Location," *Journal of the Operational Research Society*, Vol. 8, No. 4, pp. 249–261.

7. Isermann, H. (1979). "The Enumeration of All Efficient Solutions for a Linear Multiple-Objective Transportation Problem," *Naval Research Logistics Quarterly*, Vol. 26, No. 1, pp. 123–139.

8. Kapur, K. C. (1970). "Mathematical Methods of Optimization for Multi-Objective Transportation Systems," *Socio-Economic Planning Sciences*, Vol. 4, No. 4, pp. 451–467.

9. Ross, G. T. and R. M. Soland (1980). "A Multicriteria Approach to Location of Public Facilities," *European Journal of Operational Research*, Vol. 4, No. 5, pp. 307–321.

10. Schilling, D. (1980). "Dynamic Location Modeling for Public-Sector Facilities: A Multicriteria Approach," *Decision Sciences*, Vol. 11, No. 4, pp. 714–724.

11. Schmee, J., E. L. Hannan, and M. P. Mirabile (1979). "An Examination of Patient Referral and Discharge Policies Using a Multiple Objective Semi-Markov Decision Process," *Journal of the Operational Research Society*, Vol. 30, No. 2, pp. 121–129.

12. Srinivasan, V. and G. L. Thompson (1976). "Algorithms for Minimizing Total Cost, Bottleneck Time and Bottleneck Shipment in Transportation Problems," *Naval Research Logistics Quarterly*, Vol. 23, No. 4, pp. 567–595.

13. Thomas, L. C. (1983). "Constrained Markov Decision Processes as Multi-Objective Problems." In S. French et al. (eds.), *Multi-Objective Decision Making*, London: Academic Press, pp. 77–94.

14. Viswanathan, B., V. V. Aggarwal, and K. P. K. Nair (1977). "Multiple Criteria Markov Decision Processes," *TIMS Studies in the Management Sciences*, Vol. 6, pp. 263–272.

15. Wendell, R. E., A. P. Hurter, and T. J. Lowe (1977). "Efficient Points in Location Problems," *AIIE Transactions*, Vol. 9, No. 3, pp. 238–246.

16. White, D. J. (1982). "The Set of Efficient Solutions for Multiple Objective Shortest Path Problems," *Computers and Operations Research*, Vol. 9, No. 2, pp. 101–108.

17.5.7 Bibliography on Multiple Objectives in Statistics

1. Charnes, A. and W. W. Cooper (1975). "Goal Programming and Constrained Regression—A Comment," *Omega*, Vol. 3, No. 4, pp. 403–409.

2. Freed, N. and F. Glover (1981). "Simple but Powerful Goal Programming Models for Discriminant Problems," *European Journal of Operational Research*, Vol. 7, No. 1, pp. 44–60.

3. Korhonen, P. (1984). "Subjective Principal Component Analysis," *Computational Statistics & Data Analysis*, Vol. 2, pp. 243–255.

4. Lawrence, K. D. and G. R. Reeves (1982). "Combining Multiple Forecasts Given Multiple Objectives," *Journal of Forecasting*, Vol. 1, No. 3, pp. 271–279.

5. Narula, S. C. and J. F. Wellington (1980). "Linear Regression Using Multiple Criteria," *Lecture Notes in Economics and Mathematical Systems*, Vol. 177, Berlin: Springer-Verlag, pp. 266–277.

6. Zeleny, M. (1982). *Multiple Criteria Decision Making* (Chapter 13), New York: McGraw-Hill.

Index

Achievement scalarizing functions, 399–406
Achilles, A., 7
ADBASE code, 254–266, 322–323, 330, 461, 467
  input format, 255–259
  numerical example, 261–263
  parameters, 254–255
  six-field format, 260–261, 264, 323, 330, 468
  versatility of, 259–260
  use in applications, 489, 498–500, 502
Ad hoc interactive procedures, 203, 365, 379
  defined, 362
Adjacent:
  efficient bases, 217–218
  extreme points, 33, 104–106
  optimal bases, 102
Admissibility, 149
Adulbhan, P., 518, 527
Aggarwal, V. V., 531
Alivio, R. A., 518
Alternative optima, 56, 61–64, 75–76, 99–114, 391, 433, 445–446, 498
Anderson, M. D., 407
Aneja, Y. P., 527
Applications:
  CPA firm staff allocation, 484, 493–504
  managerial compensation planning, 484, 504–512
  OCMM, 290–292
  reservoir release policy, 513, 515
  river basin water quality, 513–515, 525
  sausage blending, 484–492, 525
Archimedian goal programming, 285–292, 298–299
Armstrong, R. D., 303
Arrow, K. J., 159, 270
Arthur, J. L., 302, 303, 304
Ashton, D. J., 356, 516
Atkins, D. R., 354, 356, 451, 516
Aubin, J. P., 407
Augmented weighted Tchebycheff:
  metric, *see* Metrics
  procedure, *see* Tchebycheff interactive procedures
  program, *see* Programs
Automated package:
  AUTOPAKX program, 462, 509–512, 515
  implementation, 456–472, 475–477
  purpose of, 456
  structure of, 462–468
  warm starting, 460
Awerbuch, S., 356, 357

Balachandran, K. R., 493, 516
Barankin, E. W., 159, 270
Barton, D., 514, 517
Base:
  adjacent efficient, defined, 217
  adjacent optimal, 102–103
  coding, 103–104, 109–114
  decoding, 104
  defined, 29
  efficient, defined, 216
  initial efficient, 225
  optimal, 76
    tree of, 102–104
  w-efficient, defined, 223
Bazaraa, M. S., 91
Belensen, S. M., 408
Benayoun, R., 361, 362, 365, 408
Benson, H. P., 226, 232, 270, 275, 356, 451, 527
Benson's method, for finding an efficient point, 226, 232
Berger, P., 531
Bergstresser, K., 529
Bernardo, J. J., 518
Bertier, P., 409
Bhatia, H. L., 530
Bicriterion programming, bibliography in, 527
Big *M* method, 68, 85

Bishop, A. B., 304, 513, 516
Bitran, G. R., 270, 340, 356, 516, 527, 529
Blackwell, D., 159, 270
Blair, P. D., 7
Borwein, J., 451
Boundary point of a set, 31
Bowman, V. J., 451
Break point, in MOLFP, 349
Bres, E. S., 304
Broken edge, in MOLFP, 349
Brumelle, S., 527
Buckley, J. J., 528
Burkard, R. E., 530
Burns, D., 304
Bushenkov, V., 516

Calibrating interactive procedures, 325, 390, 394–396, 447, 462, 499
Campbell, H. G., 46
Carlsson, C., 7, 408, 528
Carr, C. R., 59
C-cone, *see* Criterion cone contraction
Change vector, 120, 123
Chankong, V., 7, 8, 208
Charnes, A., 7, 56, 91, 159, 271, 282, 286, 290, 304, 340, 356, 531
Chesser, D., 306
Childress, R. L., 46
Choo, E. -U., 354, 355, 356, 410, 419, 451, 478, 517, 530
Choo's MOLFP example, 354–355
Church, R. L., 527
Clayton, E. R., 306, 307
Closed set, defined, 31
Cochrane, J. L., 7, 9, 271, 273, 408, 409, 452, 527
Coded:
  basis, 103–104, 110–114
  extreme point, 104–105, 110–114
  unbounded edge direction, 105, 110–114
Cohon, J. L., 7, 8, 522, 527
Coincidence planes, in MOLFP, 349
Composite LP, 165. *See also* Weighted-sums program
Computational experience:
  D-cone generators, 249–250
  vector-maximum problems, 263–267
Computer graphics, 519–524
  color, 521–522
  in visual interactive procedure, 399, 403–404
Computer/user interface, 519
Cone, 38–42
  criterion, 170–181, 246–254
  dimensionality of, 39
  extreme rays of, 39
  of feasible criterion vector directions, 404–405
  generators of, 38–39
    essential, 38, 171–172
    minimal number, 38, 405
    nonessential, 38, 171, 249
  pointed, 39
  polar, 40–42
    nonnegative, 40–41
    semi-positive, 42, 349
    strictly positive, 151, 350
  polyhedral, 39
Connectedness:
  of efficient bases, 219
  of efficient set, 158–159, 220
Connected set, defined, 36–37
Constraints:
  active, defined, 64
  active λ-, 382
  goal, 286, 300
  redundant, defined, 64
  soft, 286, 507
Contini, B., 304
Contours:
  augmented weighted Tchebycheff metric, 423–431, 472–474
  in goal programming, 286–289
  L_p-metric, 44
  utility function, 146–147, 154–157, 170, 369, 372–374
  weighted L_p-metric, 45
  weighted Tchebycheff metric, 46
Convex combination operator, defined, 32
Convex dominated, *see* Unsupportedness
Convex hull, defined, 34
Convex set:
  characterization of, 36
  defined, 31
  image of, 34
Cook, W. D., 303, 304, 529
Cooper, W. W., 7, 56, 91, 159, 271, 282, 286, 290, 304, 340, 356, 531
Coordinatewise increasing utility function, 143–144, 394
Correlated objectives, defined, 198
Countably infinite set, 37
CPA firm staff allocation application, 1
Cramer's rule, 23
Crashing procedure, 107–109
  flowchart, 108
  missing variables, 107
  unneeded variables, 107

Criteria, examples of, 2-3, 337, 485, 494, 507, 514, 515, 523
Criterion cone, 127, 170-175, 180-181
  defined, 170
  dimensionality of, 171
  enveloping reduced, 257, 391-392
  extreme rays of, 171-172
  generators of, 170, 247, 391
  interval criterion weights, 247, 257
  in parametric programming, 127-129
  relative interior, 174-179
Criterion cone contraction, 246-254, 361, 389-393
  in ADBASE, 257
  C-cone, defined, 247
  in CPA firm application, 497-503
  critical weighting vectors, 247-248, 251, 253
  D-cone, defined, 247
  E-cone, defined, 250
  endpoint tables, 248-249, 251, 253
  enveloping reduced problem, 250, 392
  interval criterion weights problem, 246
  premultiplication **T**-matrices, 248-254, 391, 497-502
  subset **E**-cone, 252-254
Criterion matrix, 138, 165
Criterion space, defined, 145
Criterion values:
  lower bounds, insertion of, 476-477
  minimum values over efficient set, 267-270, 365, 476-477
  ranges over efficient set, 201, 475-476, 521-522
Criterion vectors, 138
  improperly nondominated, 420, 437-440, 444, 446
  nondominated, defined, 148
  optimal, 146
  stepwise, defined, 374
  supported-extreme nondominated, 432-435
  supported-nonextreme nondominated, 432-435
  unsupported nondominated, 420, 431-437, 513
Critical values, in parametric programming, 122, 125-127
Critical (weighting) vector, *see* Criterion cone contraction

Daellenbach, H. G., 531
Dantzig, G. B., 56, 91
Dauer, J. P., 304
Davis, W. J., 271
D-cone, *see* Criterion cone contraction
Decision space, defined, 145
Decision variables, defined, 55
Deckro, R. F., 305, 530
Decoding a basis, 104, 108
Definition point, of a Tchebycheff contour, 425-427, 469-471, 509, 511
Degeneracy, 64-65, 75
  degrees of, 65
De Kluyver, C. A., 531
De Mare', L., 516
Depraetere, P., 410
de Samblanckx, S., 410
Despontin, M., 8
Dessouky, M. I., 271
Determinants, 22-23, 29
Deviational variables, 143-145
  converting values to percentages, 300-301, 489, 508
  maximum absolute, 299-301, 523
  maximum percentage, 300-301, 508, 523
DeVoe, J. K., 304
Diagonal of a Tchebycheff metric, 425-431, 474
Diaz, J. A., 531
Dimensionality of:
  cone, 39-40
  criterion cone, 171-174
  set, 34-35
Dince, R. R., 305
Dinkelbach, W., 271, 356, 408
Direction finding program, *see* Programs
Divisibility factor, *see* Ideal criterion vector
Dominance:
  strong, defined, 147
  weak, defined, 147
Dominated criterion vector, 148
Domination sets, 150-154, 158, 175, 180-181, 349-251, 435-437
Dominey, J. M., 478
Dress, P. E., 305
Dual:
  program, 76
  simplex algorithm, 87-89, 90
Duality:
  lists, 76-77
  multiple criteria bibliography in, 527-528
  theory, 76-78
Duckstein, L., 8, 209
Duesing, E. C., 271
Duncan, C. P., 306
Dürr, W., 271
Dyer, J. S., 188, 305, 361, 367, 379, 408

Dynamic programming, multiple criteria
bibliography in, 530–531

Easton, A., 8, 188
Eatman, J. L., 516
Eckenrode, R. T., 188
Ecker, J. G., 226, 231, 232, 233, 235, 243, 249, 271, 274, 275, 276, 356, 451
Ecker-Kouada method, for finding an efficient point, 231–232
Ecker-Kouada (E-K) program, 231–232
Ecker's method, for nonbasic variable efficiency, 233, 235–236, 249
E-cone, *see* Criterion cone contraction
e-Constraint:
method, 202–206
program, 203–206
Edge:
of a polyhedral set, 33
unbounded, 32, 36
Efficiency:
defined, 149, 158, 213
goal, 296–297
graphical detection of:
in integer problems, 435–437
in MOLFP, 349–351
in nonlinear problems, 158
using criterion cones, 175–180
using domination sets, 150–155
strong, 347–355
weak, 133, 221, 223–224, 347–356
Efficient point, methods for finding an, 266
Benson's method, 232
Ecker-Kouada method, 231–232
lexicographic maximization, 228–231
weighted-sums, 227
Efficient set, 149
connectedness of, 158
disconnected of, 158–159, 220
minimum criterion values over, 267–270, 365, 476–477
Einhorn, H. J., 188
Elementary row operations, 24
Elster, K. -H., 7
Endpoint tables, *see* Criterion cone contraction
Engelberg, A., 188
Engineering management, multiple criteria bibliography in, 525–526
Enveloping reduced criterion cone, *see* Criterion cone
ϵ_i values, selecting, 420–422, 474
Ereshko, F., 516
Essential generators, of a cone, 38, 171–173
Evans, J. P., 159, 233, 249, 254, 271, 276
Evans-Steuer subproblem, 233, 254
Evans-Steuer test, for nonbasic variable efficiency, 233–234, 249, 254
Extreme point:
of Archimedian GP feasible region, 289–290
defined, 31
finding an initial efficient, 226–232
of greatest utility, 156–157
parametrically optimal, 121
Extreme ray, defined, 39–40

Facet:
defined, 181
efficient, 183, 187, 242–246, 383, 385
maximal, 182–183
maximally efficient, 182–183, 242–245
Family of sets, 13
Family of weighted-sums problems, 246, 254
Fandel, G., 8
Fay, C. H., 504, 518
Feasible region:
in criterion space, 145
in decision space, 145
Feinberg, A., 188, 361, 367, 379, 408
Ferguson, R. O., 304
Field, R. C., 305
Fielitz, B., 46
FILTER code, 322–324, 330–332, 396, 467, 469–470
use in applications, 489–490, 498–499, 501–502, 509, 514–515
Filtering, 311–326, 331–332, 383, 391–393, 396, 447–448, 476
forward, 497
closest point method, 318–320, 323–324
first point method, 314–318, 323–325
furthest point method, 320, 323
halving and doubling routine, 316–318
maximally dispersed subset, 320–321
seed points, 314, 316, 476
unattained set sizes, 321
interactive forward and reverse, 324–326, 487, 490
reduction factor, 325, 489
reverse, 321–325, 490–491
Filtering relationship, defined, 314
Final solution, defined, 4
Fortson, J. C., 305
Fractional objectives, 336–337
Frank-Wolfe algorithm, 367, 369–371, 379
Franz, L. S., 409

Freed, N., 305, 532
Freedman, H. T., 8
Freimer, M., 529
French, S., 8, 531
Functions, 15, 30
  achievement scalarizing, 399–406
  concave, 139–142
  nondecreasing, 139–141
  one-to-one, 15–16
  onto, 15–16
  pseudoconcave, 140–143
  quasiconcave, 140–143
  utility, *see* Utility function
Furukawa, K., 526
Fuzzy sets, multiple criteria bibliography in, 528–529

Gabbani, D., 477, 530
Gal, T., 8, 133, 188, 243, 271
Gale, D., 56, 91, 188
Game theory, multiple criteria bibliography in, 529
Gardiner, L. R., 518
Gass, S. I., 56, 91, 114, 133
Gear, A. E., 307
Gearhart, W. B., 527
Generators of a cone, *see* Cone
Geoffrion, A. M., 188, 271, 361, 367, 379, 408, 451, 527
Geoffrion-Dyer-Feinberg (GDF) procedure, 361, 367–379
  algorithm, 372–374
  flowchart, 375
  numerical example, 377–379
  pairwise comparison routine, 371–373
  sample output, 376–377
  stepwise criterion vectors, 374–375
Gershon, M., 188
Ghiassi, M., 271
Giesy, D. P., 526
Glickman, T. S., 531
Global optima, 155–156
Glover, F., 305, 532
Goal constraints, 286
Goal efficiency, 296–297
Goal programming, 283–284
  Archimedian, 285–292
  goal types, 283–284
  interactive, 298–299
  with Markov processes, 290–292
  multiple criterion function, 300–301, 505
  preemptive, 259, 285, 292–296
  rotating priorities, 298
  software, 301–303
  vector-maximum formulation, 297–298, 302–303
Godfrey, J. T., 477, 516
Goicoechea, A., 8, 209
Golabi, K., 531
Goldman, A. J., 46
Goodman, D. A., 305
Gradients, 57
  objective function:
    fractional, 343–345
    local, *see* Local
    in LP, 59
    parameterized, 120
    weighted-sums, 168–169
  radially dispersed, 391
  utility function, 168–170, 369–371
Graph of a function, defined, 17
Graphics, *see* Computer graphics
Grauer, M., 8, 408, 478, 516, 525, 526
Grawoig, D. E., 46
Green, G. I., 306
Greeney, W. J., 304, 516
Greenwood, A. G., 307
Greis, N. P., 516
Guideline relationships, 395–396, 447, 462

Hadley, G., 46, 56, 85, 91, 114, 133
Haimes, Y. Y., 7, 8, 208, 408
Half-space, 36
Hall, W. A., 8, 408
Halving and doubling routine:
  in forward filtering, 316–318
  in GDF procedure, 372
Hannan, E. L., 305, 356, 357, 527, 528, 529, 530, 531
Hansen, D. R., 8, 209
Hansen, P., 8, 530
Harris, F. W., 332, 451, 478, 517
Hartley, R., 8, 271
Hartmann, K., 526
Hastings, N. A. J., 332
Hebert, J. E., 305
Hegner, N. S., 243, 271. *See also* Shoemaker, N. E.
Hemming, T., 8, 408
Henn, R., 271
Hillier, F. S., 91, 515, 516
Ho, J. K., 408
Hobbs, B. J., 188
Hock, W., 480
Hogarth, R. M., 188
Howe, C. W., 59
Hultz, J. W., 531
Hurter, A. P., 531

Hwang, C. L., 8
Hyperplane:
 defined, 36
 supporting, 36, 181
 tangent, 169
Hypersphere, 31

IBM Corporation, 91, 133, 420, 451, 475, 477, 522
Ideal criterion vector:
 z** divisibility factor, 465–466
 z* in STEM, 363, 368
 z** in Tchebycheff method, 420–422, 459–460, 465, 474, 477
 in visual interactive procedure, 400, 404
Ignizio, J. P., 8, 282, 302, 305, 526, 527, 529
IIASA (International Institute for Applied Systems Analysis), 525
Ijiri, Y., 8, 282, 305
Improperly nondominated criterion vectors, 420, 437–440, 444, 446
Inconsistent:
 feasible region, 64, 74
 LP, 64
 system of linear equations, 30
Index set, 14–15
Indifference curves, *see* Contours
Information overload, 245, 476, 523
Initial:
 efficient basis, 225
 efficient extreme point, 226–232
Initially efficient edge, in MOLFP, 349
Integer programming, multiple criteria bibliography in, 529–530
Interactive procedures, 4, 361–407, 419–450, 456–477
 augmented weighted Tchebycheff, 419, 446–450, 456–460, 464, 506, 513
 e-constraint method, 203–204
 forward and reverse filtering, 324–326, 487
 Geoffrion-Dyer-Feinberg, 367, 369–379
 goal programming, 298–299
 interactive weighted-sums/filtering, 394–399, 484
 interval criterion weights/vector-maximum, 389–394, 484
 lexicographic weighted Tchebycheff, 419, 446–450, 456–461, 464, 472
 STEM, 361–368
 vector-maximum/filtering, 484, 486–487, 489–492
 visual interactive approach, 399–407
 Zionts-Wallenius method, 379–389
Interactive weighted-sums/filtering approach, 361, 394–399, 419, 447–448, 497
 algorithm, 396–397
 calibration, 394–396
 flowchart, 398
 implementation, 456–460
Interior point of a set, 31
Interval criterion weights, 246–257, 259–260, 266, 486–489
Interval criterion weights/vector-maximum approach, 361, 389–396, 399, 419, 477, 497
 algorithm, 390–392
 flowchart, 393
Intervals, 13–14
Inverse image, defined, 15
Inverse of a matrix:
 defined, 24
 Gauss-Jordan method, 24
 solving systems of linear equations, 25
Isermann, H., 233, 234, 235, 243, 249, 269, 271, 272, 306, 408, 451, 477, 528, 531
Isermann's method, for finding maximally efficient facets, 343–345
Isermann's test, for nonbasic variable efficiency, 233–235, 244, 249

Jääskeläinen, V., 306
Jahn, J., 272, 528
Jarvis, J. J., 91
Johnsen, E., 9
Johnson, K. G., 526

Kaliszewski, I., 451
Kallio, M., 408, 477, 525
Kapur, K. C., 408, 531
Karwan, K. R., 304
Karwan, M. H., 409, 530
Keeney, R. L., 5, 9, 188
Keiding, H., 530
Kendall, K. E., 306
Keown, A. J., 306, 307
Keuneman, D., 408
Killough, L. N., 306
Kim, C. I., 306
Kindler, J., 516
Kiziltan, G., 527, 530
Klein, D., 530
Klingman, D., 531
Knoll, A. L., 188
Knutson, D. L., 306
Kochetkov, Y., 7, 408
Kok, M., 409, 516
Köksalan, M. M., 409

Koopmans, T. C., 159
Korhonen, P., 306, 361, 399, 404, 405, 407, 409, 532
Kornbluth, J. S. H., 306, 355, 356, 528
Kouada, I. A., 226, 231, 232, 235, 243, 271, 274, 275
Krarup, J., 530
Krueger, R. J., 304
Kuhn, H. W., 46, 56, 159, 270
Künzi, H. P., 271

Laakso, J., 306, 361, 399, 404, 405, 409
LAMBDA code, 330-332, 396, 467-471
  use in applications, 508, 514-515
Lambert, T. D., 462, 478, 509, 515, 517
Laritchev, O., 362, 408
Lawrence, K. D., 9, 134, 303, 306, 516, 529, 532
Lawrence, S. M., 306
Learner, D. B., 304
Leberling, H., 188, 271, 529
Lee, D. N., 272
Lee, S. M., 9, 282, 302, 306, 307, 409
Leitch, R. A., 477, 516
Leitmann, G., 273
Leuschner, W. A., 410, 517
Lev, B., 357, 451
Level curves, 57-63
  of a linear fractional objective, 338-349
Lewandowski, A., 8, 477, 478, 516, 525, 526
Lexicographic maximization:
  for finding an efficient point, 226, 228-231, 254, 259, 269-270, 302
  in GP, 295-296, 302
Lexicographic goal programming, 259, 285, 292-296
Lexicographic simplex method, 295-296
Lexicographic weighted Tchebycheff:
  procedure, *see* Tchebycheff interactive procedures
  program, *see* Programs
Lieberman, G. L., 91, 515, 516
Lin, J. G., 209, 272
Lin, W. T., 307, 517
Linear algebra, review of, 18-30
Linear equations, systems of, 30
Linear fractional programming, single objective, 337-347
  rotation set, 338-340, 345-347
  updated objective function method, 343-345
  variable transformation method, 340-342
  when denominator vanishes, 345-347
Linearly independent set of vectors, 27-28
Linear programming, 55-91
  alternative optima, 56, 61-64, 75-76, 99-114
  artificial variables, 67-68
  basic variables, 69
  c_j-z_j row, 101-102
  classification of LPs, 99
  degeneracy, 64
  graphical approach, 59-65
  method of rectangles, 73-74, 101-102
  redundant constraints, 64
  simplex method, 66-76, 81-87
  two phase method, 85-87, 90
  warm starting, 89-91
Linear search, 369
Lipschutz, S., 46
L_p-metric, *see* Metric
L_p-norm, *see* Norms
Locally relevant weights, routine for inducing, 169, 371-373
Local objective function gradient:
  in Frank-Wolfe algorithm, 369-370
  in MOLFP, 343-345, 349-350
  of utility function, 168-170, 369-371
Local optima, 155-156
Location, multiple criteria bibliography in, 530-531
Lockett, A. G., 307
Lootsma, F. A., 409, 516
Lotov, A. V., 409, 516
Lowe, T. J., 531

McCammon, D. F., 526
McCarl, B., 514, 517
MacCrimmon, K. R., 9, 188
McKeown, P. G., 133, 134
Magazine, M. J., 477, 530
Magnanti, T. L., 270, 356, 427
Major, D. C., 9
Managerial compensation planning application, 484, 504-512
Mangasarian, O. L., 159, 180, 272
Mapping, defined, 15
March, R. G., 407
Marco Polo Caravans example, 59-60
Maret, M. W., 308
Marginal rate of substitution, 169
Markov processes:
  in goal programming, 290-292
  multiple criteria bibliography in, 530-531
  reservoir release policy application, 515
Marks, D. H., 8
Marquis, L. M., 306
Martos, B., 356

Master lists:
  of bases, 109–114
  of edge directions, 109–114
  of extreme points, 109–114
Masud, A. S. M., 8
Matrix, 18–25
  constraint, defined, 67
  criterion, defined, 138
  diagonal, 21, 215, 458
  in echelon form, 29
Matrix operations, 20–21
  addition and subtraction, 20
  multiplication, 20
  scalar multiplication, 20
Maximal facet, 182–183
Maximal index set, in Isermann's method, 235, 243–244
Maximally dispersed subsets, defined, 320
Maximally efficient facet, 182–183, 242–245
MCDM, *see* Multiple criteria decision making
Meeks, H. D., 91
Metrics:
  augmented weighted Tchebycheff, 422–431, 439
  L_p-, 44
  Tchebycheff (L_∞), 45
  weighted L_p, 45–46
  weighted Tchebycheff (L_∞), 46, 363–364, 419, 422–431
Middlemost:
  program, 383
  weighting vector, 380, 382
Military manpower planning trajectory example, 523
Miller, G., 332, 476, 478
Minch, R. A., 133
Minimum criterion values, *see* Efficient set
MINOS code, 456, 462, 475
MIPZ1 code, 514
Mirabile, M. P., 531
Miura, S., 272
MOLFP, *see* Multiple objective linear fractional programming
MOLP, *see* multiple objective linear programming
Monotonicity, 143–144
de Montgolfier, J., 362, 408
Moore, L. J., 307
Morgenstern, O., 159
Mori, N., 451
Morin, T. L., 451
Morris, R. L., 307
Morse, J. N., 9, 332
Moscarola, J., 8
Motzkin's Theorem of the Alternative, 214, 221
MPSX layout matrix, 458–460, 466
MPSX, 79–81, 122, 302, 420, 463
  input format, 79–81, 459–460, 464
MPSX-MIP, 420, 475
Muhlemann, A. P., 307
Muller, H., 410
Multiattribute decision analysis, 5
Multiple criteria decision making, 5
  multiattribute decision analysis, 5
  multiple criteria optimization, 4–5
Multiple criterion function GP, 300–301
Multiple objective integer programming, 433, 436
  in automated package, 456, 475
  bibliography in, 529–530
  with STEM, 363, 365
  with Tchebycheff procedures, 420, 440
Multiple objective linear fractional programming, 336–337, 348–356
  broken edges, 349
  Choo's example, 354–355
  coincidence planes, 349, 351–352
  graphical examples, 348–355
  initially efficient edges, 349, 353
  rotation set, 338–340, 345–347, 352
  vertices, 349
  w-efficient line segments, 354–355
Multiple objective linear programming:
  classification of MOLPs, 224
  defined, 138
  introductory discussions, 56, 120, 132–133, 173–174
  vector-maximum algorithms, 213–270
Multiple objective nonlinear programming:
  in automated package, 456, 475
  difficulties in, 158
  with GDF procedure, 379
  with STEM, 365
  with Tchebycheff procedures, 420, 440
  with visual interactive approach, 399
Multiple starting point property, 394, 399
Mulvey, J. M., 407
Murtagh, B. A., 456, 475, 478
Musselman, K., 209

Naccache, P. H., 272
Nair, K. P. K., 527, 531
Nakayama, H., 409, 526, 528
Narayanan, R., 304, 516
Narula, S. C., 532
Näslund, B., 407
Navon, D. I., 410, 517

Near optimality analysis, 206-208
Nehse, R., 7, 9
Networks, 399
 multiple criteria bibliography in, 530-531
Neumann, H. -W. 9
Niehaus, R. J., 7, 290, 304
Nijkamp, P., 9
Nonbasic variable efficiency, 233-240, 245, 249, 250
 defined, 217
 methods for determining, 233
  Ecker's method, 235-236, 249
  Evans-Steuer test, 233-234, 249, 254
  Isermann's test, 233-235, 249
  Zionts-Wallenius routine, 236-240, 249
Nondominated criterion vector:
 defined, 148
 improper, 420, 437-440, 444, 446
 supported-extreme, 432-435
 supported-nonextreme, 432-435
 unsupported, 420, 431-437, 513
Nondominated partial tradeoffs, 206
Nondominated set, defined, 148
Nonessential generators, of a cone, 38, 171, 249
Noninferiority, 149
Normalizing:
 an objective function, *see* Scaling objective functions
 a vector, 43, 200
Norms, L_p-, 43, 363
Notation, 5-7, 60, 151, 221, 349, 395, 462
Novaes, A. G., 340, 356
Null:
 set, 12, 31
 vector, defined, 22
Null objective function program, *see* Programs
Null vector condition, 41, 170-173, 177, 180
Nykowski, I., 357

Objective function values, *see* Criterion values
OCMM (Office of Civilian Manpower Management) models, 290-292
Olson, D. L., 307
Oppenheimer, K. R., 409
Optima:
 global, 155-156
 local, 155-156
Optimal:
 bases, 76
 criterion vector, defined, 146
 dual tableau, 77
 extreme point, 60, 76
 pivot, 102
 point (solution):
  in LP, defined, 56
  in MOLP, defined, 4, 146
 set, 60
  characterization of, 99, 114
 tableau, defined, 70, 77
 weighting vectors, 193-200
  difficulties in estimation, 169-170, 193-200, 484, 503-504
Optimality, defined, 4
Optimization:
 multiple criteria, 3-4, 5
 single objective, 1
Orchard-Hays, W., 56, 91, 114, 477, 525

PAGP code, 302
Paidy, S. R., 8
Pairwise comparison questions:
 in GDF procedure, 371-375
 in Z-W procedure, 380-381
Parametrically optimal set, 121
Parametric diagram, 132, 187, 385-386
Parametric programming:
 conventional, 121-123
 convex combination, 123-127
 criterion cone, 127-132
 MOLP approach, 132-133
 with visual interactive approach, 399, 401-404
Pareto, V., 159
Pareto optimal, 149
Partition, of a set, 15
Passy, U., 527
Pasternak, P., 527
Pawlak, P., 526
Payoff tables, 266-267, 363, 365-366
Peacock, J. B., 332
Perlis, J. H., 302, 305
Phase 0, *see* Criterion cone contraction
Phase I, 66, 85, 123, 220, 372
Phase II:
 in ADBASE, 254, 259
 in LP, 66, 85-87
 in MOLP, 330, 225
 in parametric LP, 123
Phase III:
 in ADBASE, 255
 bookkeeping system, 103, 114, 109, 220-221
 in LP, 99
  numerical example, 109-114
 in MOLP, 220, 225, 233
 in parametric LP, 123-127

Philip, J., 272
Piskor, W. G., 307
Pivot:
crashing, 108
efficient, defined, 217
optimal, 102
Pivot element, defined, 70
Pivoting procedure:
in LP, 70–74
in MOLP, 220–222
Polar cone:
nonnegative, 40–41
semi-positive, 42, 349
strictly positive, 151, 350
Pollmer, L., 526
Poltersdorf, B., 526
Preckel, P. V., 456, 475, 478
Predetermined convex combinations, 252, 327–328, 390, 497
Preemptive goal programming, 259, 285, 292–296
Premultiplication **T**-matrices, 248–254, 391, 497–502
Price, W. L., 307
Primal problem, 76
Priorities, rotating, 298
Priority factors, 292
Programs:
Archimedian goal, 286
achievement scalarizing, 400–401
achievement scalarizing parametric, 401–404
augumented weighted Tchebycheff, 424, 440–450, 457, 459, 474
convex combination LP, 391, 393
direction finding, 369–371, 374–375
dual, 76
Ecker-Kouada (E-K), 231
e-constraint, 203–206
lexicographic weighted Tchebycheff, 445–448, 450, 457–459, 469–471, 474
middlemost, 383
null objective function, 63, 110, 207, 237, 241, 303, 373
preemptive, 292
primal, 76
step-size, 369–371
utility function, 3, 146, 462
variable change, 341–342
weighted minimax, 364
weighted-sums, 165–167, 193, 227, 246, 260, 380, 391, 432, 458, 497
weighted Tchebycheff, 424, 440–441
Properly nondominated criterion vector, defined, 438
Propoi, A., 525
Pruzan, P. M., 530
Pseudoconcave function, *see* Functions
Pugner, P. E., 304, 516
Puri, M. C., 530

Quasiconcave function, *see* Functions

Radially dispersed:
diagonal directions, 447
gradients, 391
Raiffa, H., 5, 9, 188
Randolph, P. H., 91
Range equalization factors (weights), 201–202, 313–314
Ranges of criterion values, 201, 422–423, 475–476, 512, 521
Rank, of a matrix, 28–29
Rapoport, A., 529
Ravindran, A., 302, 304, 527
Real line, 13
Reduced cost matrix:
in MOLP, defined, 216
in parametric LP, 130–131
Reduced feasible region:
in e-constraint method, 202–206
in near optimality analysis, 206–208
in STEM, 362–365
Reduction factor:
in interactive filtering, 325, 489
weighting vector space, 395–396, 462, 509
Redundant constraint, defined, 64
Reeves, G. R., 306, 516, 532
Reference point, *see* Ideal criterion vector
Reinecke, W., 304
Relation, 17–18
Relative interior, of a cone, 174–179, 181
Relaxation quantities:
in GP, 294
in STEM, 364–365
Reservoir release policy application, 513–515
Revelle, C., 522
Rhode, R., 272
ρ values, selecting, 429–430, 446
Ricchiute, D. N., 306
Rietveld, P., 9
Rivera, J. M., 529
River basin water quality application, 513–515, 525
Robinson, J., 46
Rockafellar, R. T., 47
Rödder, W., 528
Rosinger, E. E., 528

Ross, G. T., 531
Rotation set, 338-340, 345-347, 352
Rotstein, E., 526
Roubens, M., 9
Roy, B., 9, 409
Ruefli, T., 307
Rust, R. E., 485, 517

Saaty, T. L., 9, 188, 529
Sadagan, S., 527
Sakawa, M., 357, 410, 451, 517, 526, 529
Salukvadze, M. E., 272
de Samblanckx, S., 410
Sample size, *see* Calibrating interactive procedures
Sartoris, W. L., 307
SAS Institute, 456, 475, 478
SAS/OR package, 456, 475
Saunders, G. J., 306
Saunders, M. A., 456, 475, 478
Sausage blending application, 484-492
Sawaragi, Y., 409, 528
Scaling objective functions:
 L_p normalization, 363, 379, 390, 396, 447
 range equalization weights, 201-202
 scale factors, 466
 10 raised to appropriate power, 200-201, 390, 396, 499
 unscaling, 466, 503, 509
Schaible, S., 357
Schilling, D. A., 522, 531
Schittkowski, K., 480
Schmee, J., 531
Schneider, D. M., 47
Schniederjans, M. J., 9, 282, 307
Schoemaker, P. J. H., 188
Schönfeld, K. P., 528
Schrange, L., 91, 514, 517
Schrattenholzer, L., 516, 525
Schubert, H., 271
Schuler, A. T., 410, 517
Schy, A. A., 526
Sealey, C. W., 516
Seed points:
 in forward filtering, 314-316, 476
 in reverse filtering, 321-322
s-efficiency, *see* Strong efficiency
Seiford, L., 272
Sensitivity analysis:
 in GP, 298
 on nondominated surface, 206
Seo, F., 517
Seppälä, R., 525
Set addition, defined, 34
Set discretization, 326-330, 383
Set operations:
 difference, 13
 intersection, 13
 union, 13
Sets:
 closed, 31
 connected, 36-37
 continuous, 37
 convex, 31
 countably infinite, 37
 dimensionality of, 34-35
 disconnected, 36-37
 discrete, 37
 disjoint, 13
 empty, 12
 infinite, 11
 null, 12
 open, 31
 polyhedral, 34
 unbounded, defined, 31
Set theory, 11-18, 31-37
Shapiro, J. F., 530
Sheer, D. P., 527
Shocker, A. D., 188
Shoemaker, N. E., 271, 451. *See also* Hegner, N. S.
Silverman, J., 478, 517, 525
Simplex method:
 lexicographic, 295-296
 single objective, 66-76, 81-87
Siu, J. K., 188
Slowinski, R., 9
Smith, J. S., 308, 526
Software, ADBASE, AUTOPAKX, FILTER, LAMBDA, MINOS, MIPZ1, MPSX, PAGP, SAS/OR, *see name of code*
Soismaa, M., 408, 409, 525
Soland, R. M., 148, 159, 410, 531
Sophos, A., 526
Souders, T. L., 306
Soyster, A. L., 357, 451
Spronk, J., 8, 9, 10
Spruill, M. L., 307
Srinivasan, V., 188, 531
Stamp-coin example, 154-155
Stancu-Minasian, I. M., 357
Standard equality constraint format, 67-68
Starr, M. K., 9
Statistics, multiple criteria bibliography in, 531-532
Steepest ascent algorithm, 369-370
STEM procedure, 361-368, 379, 389
 algorithm, 363-365

STEM procedure (*Continued*)
examples, 365, 367
flow chart, 366
sample output, 365, 368
Stephanopoulos, G., 526
Step-size program, *see* Programs
Steuer, R. E., 9, 134, 233, 234, 248, 249, 253, 254, 269, 271, 272, 276, 307, 332, 355, 356, 361, 377, 394, 410, 419, 451, 461, 462, 477, 478, 489, 493, 498, 509, 513, 514, 515, 516, 517, 525, 530
Stiemke's Theorem of the Alternative, 180
Strong efficiency, defined, 347
Subproblem for:
emanating edge efficiency, 218, 233
emanating edge *w*-efficiency, 223
extreme point efficiency, 225
extreme point *w*-efficiency, 226
unsupportedness, 434–435
Sum vector, defined, 22
Sundaram, R. M., 526
Supported criterion vectors, *see* Criterion vector
Supporting hyperplane, defined, 36
Swarup, K., 530

Tabak, D., 526
Tableau:
efficient, 233
optimal, 70, 77
Tabor, D., 46
Tabucanon, M. T., 518, 527
Tabular form (of presentation), 78
Talavage, J., 209
Tamura, K., 272
Tanino, T., 409, 528
Taylor, B. W., 306, 307
Tchebycheff interactive procedures:
algorithm, 446–449
augmented weighted, 419, 446–450, 456–460, 464, 506, 513
calibration, 394–396, 447, 462
computer example, 468–472
flowchart, 450
implementation, 456–477
lexicographic weighted, 446–450, 456, 461
managerial compensation application, 506, 509–512
Tchebycheff metric, *see* Metrics
Tchebycheff theory:
finite-discrete case, 440–443
nonlinear case, 444–446
polyhedral case, 443–444
Tell, B., 10
Tergny, J., 362, 408
Theorems of the Alternative:
Motzkin's, 214, 221
Stiemke's, 180
Tucker's, 214–216
Thiriez, H., 10
Thomas, L. C., 8, 531
Thompson, G. L., 531
Thompson, W., 526
Tigan, S. T., 357
Toof, D. I., 451
Törn, A. A., 7, 332, 451
Tradeoffs, 4, 168, 206, 364
Tradeoff vectors, 380–384, 387, 389
Trajectories of:
criterion values, 403–404, 522–523
goal values, 522–523
Trajectory optimization, 519, 522–525
bibliography, 525
display of output, 523–524
uses of, 525
Tree of optimal bases, 103–104, 114
Tucker, A. W., 46, 56, 159, 270
Tucker's Theorem of the Alternative, 214–216
Tufte, E. R., 522

Unattained set size, in forward filtering, 321
Unbounded line segment operator, defined, 32
Unbounded optimal z-value problem, defined, 61
Uniquely computable, 440, 444–445, 474
Unscaling, *see* Scaling objective functions
Unsupportedness, subproblem for detecting, 434–435
Unsupported nondominated criterion vectors, 420, 431–437, 513
Updated objective function method, in MOLFP, 343–345
Utility function:
approach, 3–4, 146–147
contours, *see* Contours
program, 3, 146, 462
Utility functions, 3–4, 139–148, 154–155, 362, 367, 379, 394, 433, 439
concave, 139
coordinatewise increasing, 143–144, 394
nonconcave, 154–155, 394
nondecreasing, 139–141
pseudoconcave, 140–143, 379, 399, 405
quasiconcave, 140–143
Utopian set, 283–285, 288–289

Value paths, 521–522
van de Panne, C., 115, 209
Van Horn, J. C., 9
Variables:
  artificial, 67–68
  basic, 69–70
  convenience, 457, 487
  decision, 55
  deviational, 143–144
  dual, 77
  minimax, 299, 425, 508
  missing, 107
  nonbasic, 69–70
  nonpositive, 78–79
  original, 55
  structural, 55
  unneeded, 107
  unrestricted, 76, 78–79
  z-, 457
Variable transformation method, in MOLFP, 340–342
Vector addition, defined, 26
Vector-maximum algorithms, 213–270
  computational experience, 263–267
  criterion cone contraction theory, 246–254
  minimum criterion value theory, 267–270, 365, 475–476
  Phase II theory, 225–232
  Phase III theory, 220–221
  w-efficiency theory, 221, 223–224, 226
Vector-maximum approach, for solving an MOLP, 245–246
Vector-maximum formulations:
  enveloping reduced, 250, 392
  of a GP, 297–298, 302–303
  interval criterion weights, 246
  of an MOLP, 213–214
Vector product, defined, 21
Vectors:
  convex combinations of, 26–27
  null, defined, 22
  sum, defined, 22
Vertex:
  of augmented GP feasible region, 289–290
  of a cone, 39
  in MOLFP, defined, 349
  of a Tchebycheff metric, 425–431, 448–449
Villarreal, B., 530
Vinso, J. D., 356
Visual interactive approach, 361, 399–407
  algorithm, 404–405
  flowchart, 406
  graphical display of output, 403
  projecting line segments, 401–403
Viswanathan, B., 531
Vodak, M. C., 410, 517
von Neumann, J., 159

Wacht, R. F., 307
Wagner, H. M., 91
Waid, C. C., 188
Walker, J., 527
Wallace, M. J., 504, 518
Wallace, W. A., 304, 356
Wallenius, H., 518
Wallenius, J., 9, 10, 233, 236, 237, 238, 240, 241, 249, 273, 276, 361, 379, 380, 382, 407, 409, 410, 452, 518
Warm Starting:
  in automated package, 460
  in LP when:
    constraint matrix changes, 90
    objective function changes, 89
    RHS changes, 90
Weak efficiency, 133, 221, 223–224, 347–356
Weakly efficient:
  line segments, 354–355
  set, disconnectedness of, 351–352
Weber, R., 272
w-efficiency, *see* Weak efficiency
Weglarz, J., 9
Weibull distribution, 329–330
Weighted L_p distance measure, defined, 312–313
Weighted-sums method, 165–187, 193–200, 206–208
Weighted-sums program, *see* Programs
Weighted Tchebycheff:
  interactive procedures, *see* Interactive procedures
  metric, *see* Metrics
  program, 424, 440–441
Weighting vectors:
  optimal, 193–200
  predetermined convex combinations, 252, 327–328, 390, 497
  randomly generated, 328–332, 394, 396
Weights:
  Archimedian, 282, 286–289, 292, 298–299
  estimating, 168–169
  in GDF procedure, 374
  interval criterion, 246–257, 259–260, 266
  locally relevant, 169, 373
  minimax weights in STEM, 364
  penalty, 286
  range equalization, 201–202, 313–314
  within priority level, 299–300

Weistroffer, H. R., 272
Welling, P., 307, 493, 518
Wellington, J. F., 532
Wendell, R. E., 134, 272, 531
Whisman, A. W., 478, 517, 525
White, D. J., 8, 10, 410, 531
Whitford, D. T., 307
Wierzbicki, A. P., 8, 400, 410, 452, 478, 516, 525, 526
Winkels, H. -M., 332, 479
Winkofsky, E. P., 305, 530
Wood, E. F., 394, 410, 451, 513, 514, 515, 516, 517

Yager, R. R., 529
Yoon, K., 8
Young, F. H., 47
Yu, P. L., 10, 134, 159, 243, 272, 273, 452, 529
Yucaoglu, E., 527, 530
Yumine, T., 357, 529

Zalai, E., 516
Zanakis, S. H., 308, 526
Zeleny, M., 7, 9, 10, 134, 243, 271, 273, 408, 409, 452, 527, 529, 532
Zimmermann, H. -J., 529
Zionts, S., 10, 91, 233, 236, 237, 238, 240, 241, 249, 273, 276, 357, 361, 379, 380, 382, 409, 410, 452, 530
Zionts-Wallenius (nondominance) routine, 236–242
- for nonbasic variable efficiency, 233, 236–240, 249
- other uses, 240–242
- in Zionts-Wallenius (MOLP) procedure, 380, 387–389

Zionts-Wallenius (MOLP) procedure, 361, 379–389
- algorithm, 380–385
- example, 385–387
- flowchart, 384

Zolkiewski, Z., 357